METAL-CONTAINING POLYMERIC SYSTEMS

METAL-CONTAINING POLYMERIC SYSTEMS

Edited by

John E. Sheats
Rider College
Lawrenceville, New Jersey

Charles E. Carraher, Jr.
Wright State University
Dayton, Ohio

and

Charles U. Pittman, Jr.
Mississippi State University
Mississippi State, Mississippi

PLENUM PRESS • NEW YORK AND LONDON

Library of Congress Cataloging in Publication Data

Main entry under title:

Metal-containing polymeric systems.

"Expanded version of papers presented at the Fourth Symposium on Organo-
metallic Polymers, held at the National Meeting of the American Chemical Socie-
ty, August 28–31, 1983, in Washington, D.C."—T.p. verso.
    Includes bibliographical references and index.
    1. Polymers and polymerization—Congresses. 2. Organometallic compounds
—Congresses. I. Sheats, John E. II. Carraher, Charles E. III. Pittman, Charles
U. IV. Symposium on Organometallic Polymers (4th: 1983: Washington, D.C.)
QD380.M48   1985   547′.05   84-26372
ISBN 0-306-41891-6

An expanded version of papers presented at the Fourth Symposium on Organometallic
Polymers, held at the National Meeting of the American Chemical Society, August 28–31,
1983, in Washington, D.C.

©1985 Plenum Press, New York
A Division of Plenum Publishing Corporation
233 Spring Street, New York, N.Y. 10013

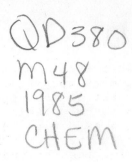

PREFACE

Research on metal-containing polymers began in the early 1960's when several workers found that vinyl ferrocene and other vinylic transition metal π -complexes would undergo polymerization under the same conditions as conventional organic monomers to form high polymers which incorporated a potentially reactive metal as an integral part of the polymer structures. Some of these materials could act as semi-conducters and possessed one or two dimensional conductivity. Thus applications in electronics could be visualized immediately. Other workers found that reactions used to make simple metal chelates could be used to prepare polymers if the ligands were designed properly. As interest in homogeneous catalysts developed in the late 60's and early 70's, several investigators began binding homogeneous catalysts onto polymers, where the advantage of homogeneous catalysis - known reaction mechanisms and the advantage of heterogeneous catalysis - simplicity and ease of recovery of catalysts could both be obtained. Indeed the polymer matrix itself often enhanced the selectivity of the catalyst.

The first symposium on Organometallic Polymers, held at the National Meeting of the American Chemical Society in September 1971 attracted a large number of scientists interested in this field, both established investigators and newcomers. Subsequent symposia in 1977, 1979 and 1983 have seen the field mature. Hundreds of papers and patents have been published. Applications of these materials as semiconductors and one-dimensional conductors, as radiation shields or as photo-resists, as catalysts, as controlled release agents for drugs and biocides and a wide variety of applications have been studied (see Chapter 1).

This book is an expanded version of the papers presented at the Fourth Symposium on Organometallic Polymers at the National Meeting of the American Chemical Society in Washington, D.C. in August 1983.

As in previous symposia there was a balance between
newcomers and established investigators. Of particular
interest is the appearance of a large number of papers
from abroad — Chile, Romania, Japan, France and South
Africa as new groups are established. We have been
happy to be a part of the development of this field and
are looking forward to seeing exciting new developments
in the next few years.

John E. Sheats
Charles E. Carraher, Jr.
Charles U. Pittman, Jr.

CONTENTS

CONTENTS

METAL-CONTAINING POLYMERS: AN INTRODUCTION

Charles E. Carraher, Jr.

Department of Chemistry
Wright State University
Dayton, Ohio 45435 USA

Charles U. Pittman, Jr.

University/Industry Chemical Research Center
Department of Chemistry
Mississippi State University
Mississippi State, Mississippi 39762 USA

INTRODUCTION

The classical polymer chemist concentrates on polymers containing only about eight elements (C,H,N,O,S,P,Cl,Br) while the
number of readily available metals is well over forty. Many of
these metals have several oxidation numbers and form several types
of compounds with organic moieties. Thus the number of new polymers
potentially available is large and often only limited by our own
intentions and imaginations. This chapter is intended to present
a general framework which allows the reader to place the remainder
of the book into perspective. It provides an overview of organometallic polymers. We hope it helps the reader appreciate the
unique opportunities which metal-containing and organometallic
polymer science offers in supplying materials for the coming decades,
materials in a multitude of areas, performing numerous tasks in
diverse areas.

This book will not cover the topic of classic inorganic polymers
such as polymeric sulphur, selenium and tellurium; borate, silica
and silicate glasses; boron, aluminum, titanium, zirconium, tin,
silicon and cerium phosphates. Please consider references 1-3 for
review of these topics. Instead we focus on metal-containing polymers with organic moieties and on polymers that behave as metals

1

with regards to electrical conductivity. Recent reviews of organometallic and metal-containing polymers are found in references 4-13.

The potential importance of metal-containing organic polymers
is seen by considering organosiloxanes. Organosiloxanes are characterized by combinations of chemical, mechanical and electrical properties which taken together are not common to any other commercially
available class of polymers. They possess a relatively high thermal
and oxidation stability, low power loss, high dielectric strength,
and unique rheological properties and are relatively inert to most
ionic reagents. While a wide range of siloxanes are available and
much research continues to be done, this book and chapter will largely
focus on other current, emerging polymer types.

The topic of metal-containing polymers can be divided by many
means. Here the topic will be divided according to the type of reaction employed to incorporate the metal and/or metal-containing moiety
into the polymer chain - namely addition, condensation and coordination polymerization. Emphasis is given to unifying factors.

CONDENSATION ORGANOMETALLIC POLYMERS

Condensation reactions exhibit several characteristics such as
explusion of a smaller molecule (often H_2O or HX) on reaction leading
to a repeat unit containing fewer atoms than the sum of the two reactants. Most reactions can be considered in terms of polar (Lewis
acid-base; nucleophilic-electrophilic) mechanisms. The reaction site
can be at the metal atom, adjacent to the metal or somewhat removed
from the metal atom. The difference between coordination and condensation reactions is often in the "eye of the beholder". Reactions
directly involving a metal atom already chemically bonded to an
organic moiety (such as $\underline{1}$) will be considered condensation polymerizations, whereas reactions involving metal ions and metal ion oxides
(such as $\underline{2}$) in simple chelations or electrostatic attractions will be
covered under the topic coordination metal polymerizations.

$$\text{HAR'AH} + \text{XMX} \rightarrow \begin{matrix} R \\ | \\ \left(AR'AM\right) \\ | \\ R \end{matrix} + \text{HX} \qquad\qquad \underline{1}$$

$$\text{HAR'AH} + \text{MO}_2^{+2} \rightarrow \begin{matrix} O \\ \| \\ \left(ARAM\right) \\ \| \\ O \end{matrix} \qquad\qquad \underline{2}$$

Research in this area was catalyzed by the observation that many organometallic halides possess a high degree of covalent character within their composite structure and that they can behave as organic acids in many reactions such as hydrolysis (c.f. reactions 3, 4).

$$RCOCl + H_2O \rightarrow RCOOH + HCl \qquad\qquad 3$$

$$-\overset{|}{\underset{|}{M}}Cl + H_2O \rightarrow -\overset{|}{\underset{|}{M}}-OH + HCl \qquad\qquad 4$$

Thus many organometallic polycondensations can be considered as extensions of organic polyesterifications (5,6) polyaminations, etc.

$$ClOCRCOCl + HOR'OH \rightarrow \{\overset{O\ O}{\overset{\|\ \|}{CRCOR'O}}\} + HCl \qquad 5$$

$$R_2MCl_2 + HOR'OH \rightarrow \{\underset{\underset{R}{|}}{\overset{\overset{R}{|}}{M}}-ORO\} + HCl \qquad 6$$

Further, some of the polycondensations are based on analogy to reactions known to occur with small monomeric systems. Thus Cp_2ZrCl_2 is known to react with benzene thiol and 1,2-dithiolbenzene in the presence of triethylamine giving monomeric products 7 and 8 (13,14). Conceptionally, it is a simple extension to employ dithiols where internal bridging is not possible. While the analogous polymerization reaction does not produce high molecular weight, zirconium polythiols, such polymers were formed using an interfacial reaction system (16). Thus the existance in literature of prior information concerning the formation of the Zr-S-R moiety provided chemical and physical property data useful in the characterization of 9.

Table 1.   Lewis Bases Employed in the Synthesis of
Condensation Organometallic Polymers

$$\underset{\substack{R}}{\overset{HON}{\diagdown}}\underset{\substack{R}}{\overset{NOH}{\diagup}}CRC$$

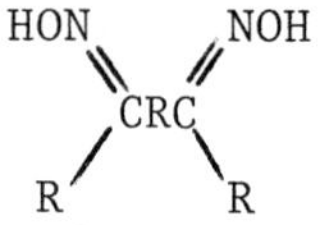

dioximes

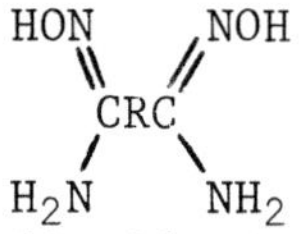

$$\underset{H_2N}{\overset{HON}{\diagdown}}\underset{NH_2}{\overset{NOH}{\diagup}}CRC$$

diamidoximes

$$\overset{S}{\underset{\parallel}{H_2N-C-NH_2}}$$

urea

thiourea

$$\overset{X\ X}{\underset{\parallel\ \parallel}{H_2NCRCNH_2}}$$

diamide
dithioamide

$$\overset{X\ X}{\underset{\parallel\ \parallel}{H_2NNHCRCNHNH_2}}$$

dihydrazide
dithiohydrazide

$$^-O_2CRCO_2^-$$

acid salts

HOROH

diol

HSRSH

dithiol

H_2NRNH_2

diamine

H_2NRNH_2

(primary) amines

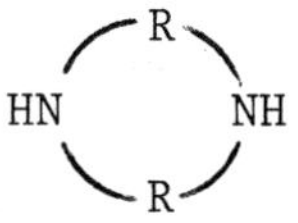

$$\overset{R}{\underset{R}{\overset{\diagup\diagdown}{HN\qquad NH}}}$$

(secondary) amines

HORNH$_2$

amine alcohol and other
appropriate mixed Lewis
bases

The organometallic moiety has been introduced into polymer
chains through reactions involving the metal atom (6,9); c.f. Tables
1 and 2 for lists of Lewis acids and bases already successfully
polymerized) or through condensations where the reaction site is
removed from the metal atom (as 10 where M can be Fe, Ru, Rh$^+$X$^-$,
Co$^+$X$^-$).

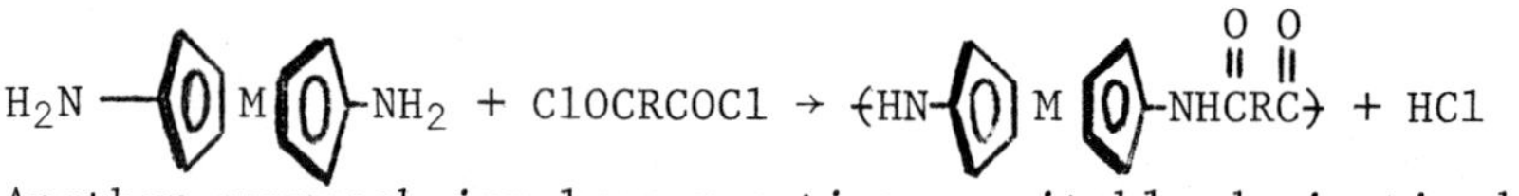

$$H_2N-\hexagon M \hexagon-NH_2 + ClOCRCOCl \rightarrow \{HN-\hexagon M \hexagon-\overset{O\ O}{\underset{\parallel\ \parallel}{NHCRC}}\} + HCl \quad \underline{10}$$

Another approach involves reacting a suitably derivatized metal-
containing compound with suitable performed natural (dextran, cellulose, wool) and synthetic (polyvinyl alcohol, polyethyleneimine)

Table 2.   Lewis acids employed in the Synthesis of
Organometallic Polymers

R_2TiX_2	R_2SiX_2	R_3As	R_3BiX_2
R_2ZrX_2	R_2GeX_2	R_3AsX_2	R_2MnX_2
R_2HfX_2	R_2SnX_2	R_3Sb	
	R_2PbX_2	R_3SbX_2	

polymers that contain appropriate reactive sites (such as 11,12).

$$\{CH_2-CH\} + R_3SnX \rightarrow \{CH_2-CH\} + HX \qquad\qquad 11$$
$$\quad\;\; | \qquad\qquad\qquad\qquad\qquad | $$
$$\quad\;\; OH \qquad\qquad\qquad\qquad\quad O$$
$$\qquad\qquad\qquad\qquad\qquad\qquad\quad |$$
$$\qquad\qquad\qquad\qquad\qquad\qquad SnR_3$$

$$\{CH_2-CH\} + \quad \text{(M cyclopentadienyl)}-OH \rightarrow \{CH_2-CH\} \qquad 12$$
$$\quad | \qquad\qquad\qquad\qquad\qquad\qquad\qquad\qquad\qquad |$$
$$\quad COCl \qquad\qquad\qquad\qquad\qquad\qquad\qquad\quad C=O$$
$$\qquad\qquad\qquad\qquad\qquad\qquad\qquad\qquad\qquad\quad |$$
$$\qquad\qquad\qquad\qquad\qquad\qquad\qquad\qquad\qquad\quad O$$
$$\qquad\qquad\qquad\qquad\qquad\qquad\qquad\qquad\qquad\quad M$$

Effort has been exerted to develop new (novel) reaction systems and to modify existing reaction systems. Since many of the reactants and products are especially thermally sensitive, low temperature (solution and interfacial condensation systems) have typically been employed (for example in the synthesis of 13 and 14). There exists a number of possible solution systems, the sole requirement being that the reaction system is homogeneous. Such solution systems may include surfactants, supposed catalysts, several different solvents, salts, bases and other additives.

When stable aqueous organometallic reactants are formed, water based solution systems are attractive. Thus Cp_2TiCl_2 dissolves (actually undergoes hydrolysis) in water forming various aquated species containing the yellow colored Cp_2Ti moiety. If a polycondensation reaction is carried out within several minutes after the Cp_2TiCl_2 is dissolved, the titanium–containing moiety reacts as though it were simply Cp_2Ti^{+2} forming polyesters (13), polythioethers and polyamines from salts of diacids, dithiols and diamines, respectively (4,9,17,20,21).

$$\qquad\qquad\qquad\qquad\qquad\qquad Cp \quad O\;\;O$$
$$\qquad\qquad\qquad\qquad\qquad\qquad\;| \qquad \| \;\|$$
$$Cp_2TiCl_2 + {}^-O_2CRCO_2^- \;\rightarrow\; \{Ti-OCRCO\} \qquad\qquad 13$$
$$\qquad\qquad\qquad\qquad\qquad\qquad\;|$$
$$\qquad\qquad\qquad\qquad\qquad\qquad Cp$$

For reactants which are not water soluble or which hydrolyze forming "inert products", nonaqueous solution systems have been developed. For some water insoluble monomers, aqueous-containing systems have been developed. For example, R_2PbX_2 compounds are insoluble in most organic solvents and in water. However, polyesters were formed by initially dissolving the salt of the diacid in a minimum amount of water followed by addition of one to ten fold (by volume) of DMSO. The lead reactants were dissolved in DMSO. The two solutions, when brought together, result in the formation of lead polyesters ($\underline{14}$;$\underline{22}$).

$$R_2PbX_2 + {}^-O_2CR'CO_2^- \rightarrow \text{(Pb-OCRCO)} \qquad \underline{14}$$

Two general aqueous interfacial systems have been successfully employed, each differing only in the location of the two reactants. In classical interfacial systems, the Lewis base (ie. a diamine) is located in the aqueous phase, while the organic phase contains the Lewis acid such as Cp_2TiCl_2. In the reverse, or inverse, interfacial system the Lewis acid (Cp_2TiCl_2) is dissolved in the aqueous phase and the Lewis base dissolved in the organic phase. The usual inter-facial system with its many reaction variables (as stirring rate, stirring time, order of addition of reactants, nature and amount of reactants, nature and amount of added base, pH, reactant mole ratio, reactant concentration, nature and amount of other additives as phase transfer agents, etc.) has thus far proven to be the most effective of the various reaction systems employed (4,9,17,19).

Other systems have been developed. When one of the reactants is a liquid it can be used alone as its own phase. Thus dibutyltin dibromide is a liquid which is immiscible with water. One phase is simply the organostannane and the second aqueous phase contains the Lewis base and any other additive. An analogous nonaqueous system consists of a liquid monomer such as ethylene glycol, along with a suitable liquid base such as triethylamine, layered over a typical organic (immiscible) phase containing the Lewis acid. These two systems require that one of the reactants be in the liquid state at or near room temperature.

Other nonaqueous systems that can accomodate solids consist of a nonpolar liquid such as hexane, octane, nonane or carbon tetra-chloride containing the Lewis acid and the second phase composed of a solvent (check immiscibility) such as nitrobenzene, acetonitrile, DMSO, 2,5-hexanedione or tetrahydrothiophene-1,1-dioxide containing the Lewis base (17).

Each new set of reactants represents potential new opportunities for innovation. The general construction of polymerization systems remains a major area in need of research in the further synthesis of potentially useful metal and nonmetal-containing polymers, including many natural polymers.

We presently utilize a limited amount of our potential reserves. We have limited our sojourn to collect, gather our feedstocks to within about two miles of the earth's surface and even here we

severely limit our search to emphasize largely carbon based resources
(oils, gases, coal). We have a need to reuse products developed from
these feedstocks such as reuse of polymer-based materials. One
requirement for large scale reclaiming will be ready identification
of select materials. Organometallic polydyes (23-25) offer one solution to the problem of content identification since such polydyes
can be "permanently" incorporated into polymeric materials as plastics, paper-based products, coatings, rubbers and fibers. Further,
due to the variety of metals, organometallic sites and dyes, a large
array of combinations, chromaphoric sites are readily available.
Further, the presence of the metal allows for additional detection
techniques to be employed. The organometallic polydyes is the name
associated with products of metal-containing compounds and dyes.
Below, 15, is one such dye, based on mercurochrome and dibutyltin
dichloride, which is both colorful and antifungal. Such polydyes
may also be useful in photosensitive media for photocopying, as
permanent coloring agents in cloth, paper, coatings and plastics;
in biomedical applications as specialized stains and toxins; as
laser materials and as radiation enhancers for curing applications.

15

   The condensation organometallic polymers also have been employed
as biologically active agents (as described by Hickley, chapter 10),
the biological activity derived from the polymer chain in total,
oligomeric portions of the chain or through control release of mono-
meric-like fragments (26-30). The desired drug may be either or both
comonomers. In the case of 16 derived from reacting Cp_2MCl_2 with
the dioxime of vitamin K, the drug may be only the vitamin K moiety
where M is Zr or both moieties where M is Ti since Cp_2Ti-containing
compounds are known to be antitumoral agents. Product 17 is employed
to deliver the dibutyltin moiety where as in product 18 the pyrimi-
dine moiety was included to entice, encourage host acceptance and
subsequent demise through release of the arsenic moiety (30). Thus,
there exists a wide latitude within which to tailor-make products
with desired biological activities.

   Many organometallic condensation polymers are semiconducting,

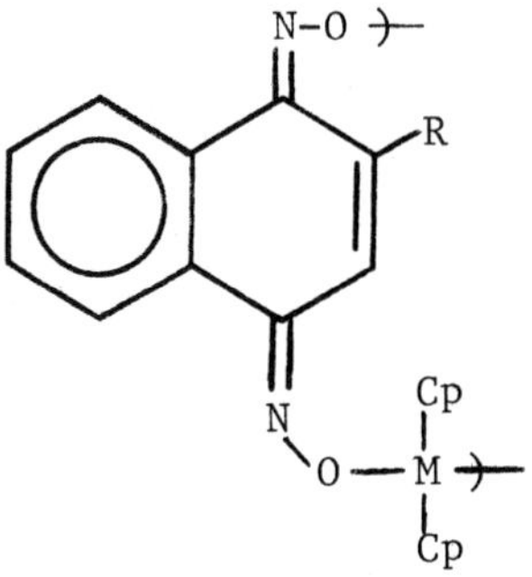

16

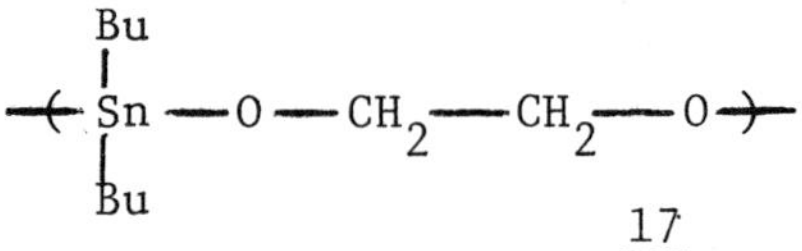

17

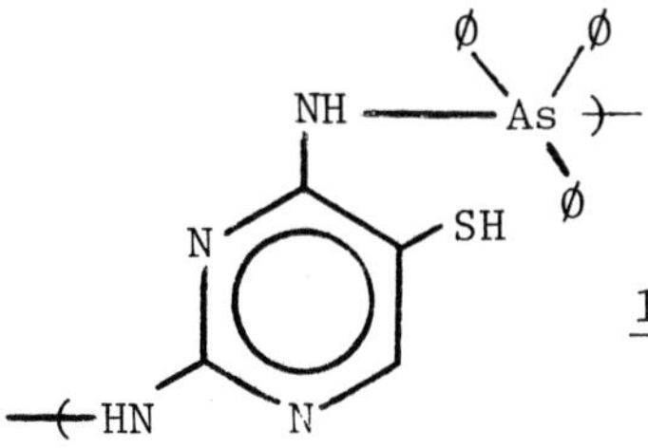

18

exhibiting bulk restivities typically within the range of 10^2 to 10^{12} ohm-cm for both AC and DC measurements (for instance 31). A number of polumers containing the Cp_2Ti moiety exhibit a phenomena called "anomolous fiber formation", reminiscence of "metal whiskers" (for instance 32). Others have good thermal properties with some retaining 80% weight to 1200°C, although the structure of the organic portions has been destroyed at this temperature. Most show poor low temperature stabilities (with less initial weight losses occurring between 50 to 250°C) but many retain significant organic portions to the 600 to 800°C range. The topic of organometallic polymer thermal stabilities is reviewed in reference 33. Still other products have exhibited good adhesive and fiber forming properties.

The area of polymer modification through condensation reactions has also received attention and aspects of this topic are covered in chapter

The work of Currell and Parsonage is particularly exciting since it involves the generation of usuable oligomeric and polymeric materials from renewable and/or quite abundant, almost untapped feedstocks. The technique involves the application of trimethyl-silylation to mineral silicates and is aptly reviewed in reference 34. Products suitable as greases, cement and masonry waterproofing agents have already been realized.

Monomers containing transition metals have been polymerized by polycondensation processes where the metal atom, while in the backbone, was not the site of the condensation reaction. For example, melt phase polytransesterifications of 1,1´-bis(carbethoxy)cobalt-icinium hexafluorophosphate, _19_, and diols such as 1,10-decanediol and 1,4-bis(hydroxymethy)-benzene gave fairly low molecular weight ionic polyesters ($\underline{20}$,35). Attempts to prepare analogous polyamides from alkyl diammonium salts were not successful since decomposition occurred. However, polyamides, $\underline{21}$, were successfully prepared by

Neuse (36) who reacted aromatic diamines with 1,1′-dicarboxycobalti-cenium chloride in molten antimony trichloride at 150-175°. The cobalticenium unit in the polymer occurs as an $SbCl_4^-$ salt which can be isolated as the PF_6^- salt upon treatment with NH_4PF_6.

Ferrocene diepoxide deivatives $\underline{22}$(37) and $\underline{23}$(38) have been made in good yields and they have been incorporated into epoxy resins.

Many ferrocene-containing polyesters, $\underline{24}$, polyamides, $\underline{25}$, and polyurethanes have been reported where classic polycondensation methods have been employed (39-42).

Polyferrocenylphosphine oxides, $\underline{26}$, and sulfides, $\underline{27}$, result upon the reaction of phenyldichlorophosphonic acid or phenylphos-phonothioic dichloride with ferrocene in the presence of $ZnCl_2$(43). However, ring cleavage occurs and cycloalkyl-bridged ferrocene polymer units also are formed.

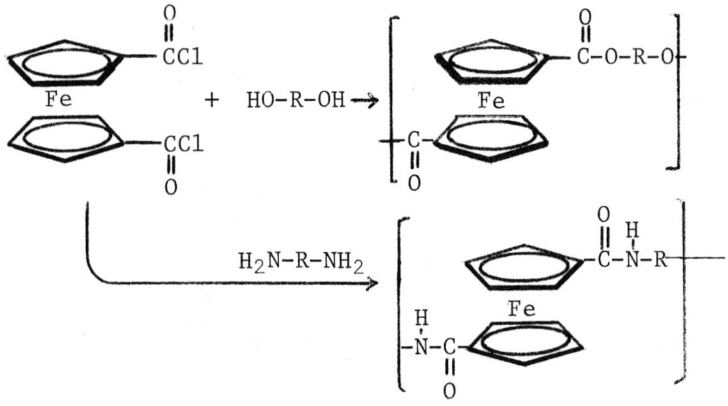

24

25

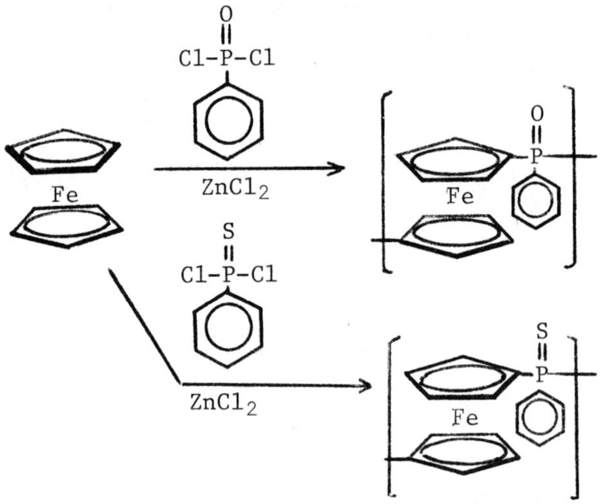

26

27

Marvel synthesized a boron-containing polymer (benzborimidazoline), 28, from ferrocene-1,1´-diboronic acid and 3,3´-diaminobenzidine as part of an effort to produce polymers of superior thermal stability (44).

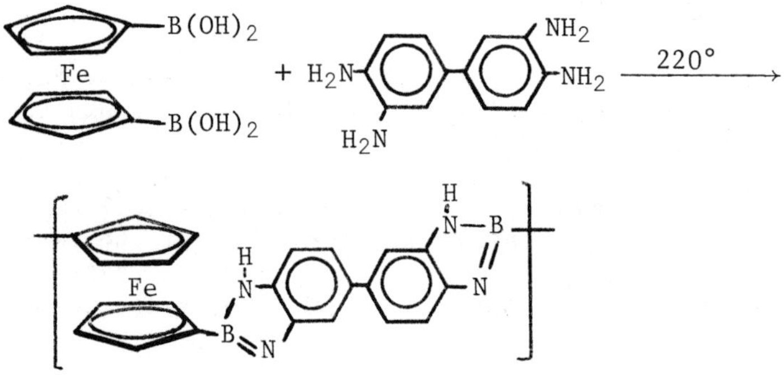

28

As a final class of organometallic polycondensation polymers
(where the metal atom is in the backbone not involved in the condensation reaction) we will mention silicon-containing fenocene polymers.
A very early patent by Wilkus and Berger (45) reported that trimethyl-
silylalkyl derivatives of ferrocene, 29, gave high yields of polysiloxanes, 30, in H_2SO_4 or $CH_2Cl_2/AlCl_3$. Greber and Hallensleben (46)
produced a variety of ferrocene/silicon-containing polyamides, polyesters, polyurethanes and polysiloxanes starting with the monomer
series 31a-g. These polymers, by virtue of the presence of silicon,
were liquids or elastomers.

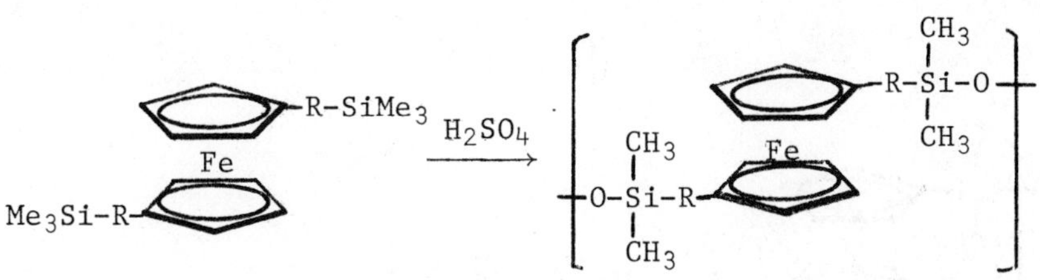

29   30

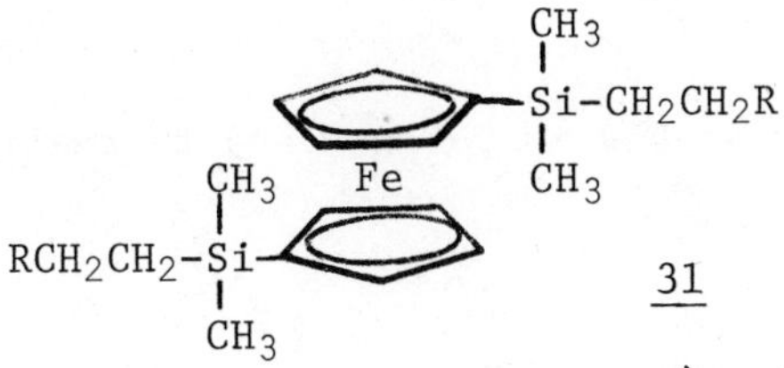

31

a)= $-CO_2CH_3$ e) $-CH_2NH_2$
b)= $-COOH$
c) $-CH_2-O-SiMe_3$
d) $-CH_2OH$ f) $-CH_2(SiO)_3Si-Cl$

g) $-CH_2-O-CH_2-CH-CH_2$

    Among the best characterized ferroncene-containing siloxane
polymers were prepared via a bis(dimethylamino)silane-disilanol polycondensation (47). The monomer, 1,1'-bis(dimethylaminodimethyl-
silyl)ferrocene, 32, was polymerized with three aryl disilanols:
dihydroxydiphenylsilane, 33, 1,4-bis(hydroxydimethylsilyl)benzene,
34, and 4,4´-bis(hydroxydimethylsilyl)biphenyl, 35. Melt polymerizations at 1 torr and 100°C gave the highest molecular weights
(~50,000). These oxysilane polymers had structures such as 37.

$$\underset{32}{(CH_3)_3N-\overset{\overset{\displaystyle CH_3}{|}}{\underset{\underset{\displaystyle CH_3}{|}}{Si}}-\text{(ferrocene)}-\overset{\overset{\displaystyle CH_3}{|}}{\underset{\underset{\displaystyle CH_3}{|}}{Si}}-N(CH_3)_3}$$

$$\underset{33}{HO-\overset{\overset{\displaystyle Ph}{|}}{\underset{\underset{\displaystyle Ph}{|}}{Si}}-OH}$$

$$\underset{34}{HO-\overset{\overset{\displaystyle CH_3}{|}}{\underset{\underset{\displaystyle CH_3}{|}}{Si}}-\text{(C}_6\text{H}_4\text{)}-\overset{\overset{\displaystyle CH_3}{|}}{\underset{\underset{\displaystyle CH_3}{|}}{Si}}-OH}$$

$$\underset{35}{HO-\overset{\overset{\displaystyle CH_3}{|}}{\underset{\underset{\displaystyle CH_3}{|}}{Si}}-\text{(C}_6\text{H}_4\text{-C}_6\text{H}_4\text{)}-\overset{\overset{\displaystyle CH_3}{|}}{\underset{\underset{\displaystyle CH_3}{|}}{Si}}-OH}$$

$$\underset{37}{\left[\overset{\overset{\displaystyle CH_3}{|}}{\underset{\underset{\displaystyle CH_3}{|}}{Si}}-\text{(ferrocene)}-\overset{\overset{\displaystyle CH_3}{|}}{\underset{\underset{\displaystyle CH_3}{|}}{Si}}-O-\text{(C}_6\text{H}_4\text{-C}_6\text{H}_4\text{)}\right]}$$

They were hydrolytically stable in THF/H_2O and thermally stable to over 400°C (47).

COORDINATION POLYMERS

Coordination polymers have served mankind since before recorded history. The tanning of leather and generation of select colored pigments depend on the coordination of metal ions. Many biological agents are polymers with metals coordinated; for example, hemoglobin. Many of these coordination polymers have unknown and/or irregular structures.

Coordination polymers can be prepared by a number of routes, the three most common being:

a) Preformed metal coordination complexes polymerized through functional groups where the actual polymer forming step may be a condensation or addition reaction (38).

$$H_2NR-C\underset{\substack{N\\H}}{\overset{N-O}{\diagdown}}M\underset{\substack{N\\H}}{\overset{O-N}{\diagdown}}C-RNH_2 + ClOCROOCl \rightarrow \left(NHRC\underset{\substack{N\\H}}{\overset{N-O}{\diagdown}}M\underset{\substack{N\\H}}{\overset{ON}{\diagdown}}C\overset{O\;\;O}{\overset{||\;\;||}{RNHCRC}}\right)$$

$$\underset{38}{}$$

b) Coordination of a metal ion by preformed polymers containing chelating groups (39), and

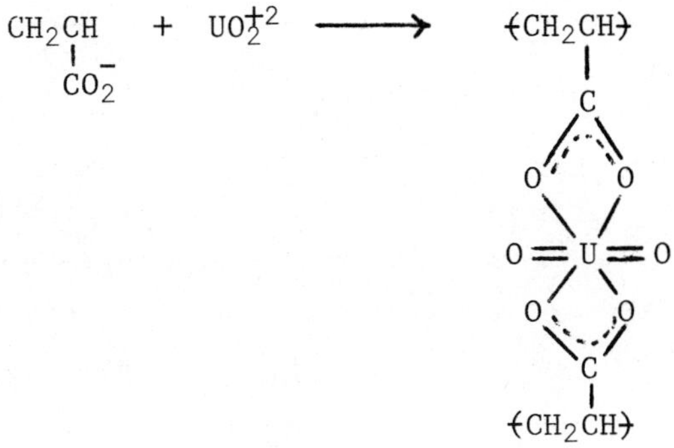

$$CH_2CH \quad + \quad UO_2^{+2} \quad \longrightarrow \quad \text{-}(CH_2CH\text{)-}$$

39

c) Polymer formation through reaction of metal donor atom co-ordination.

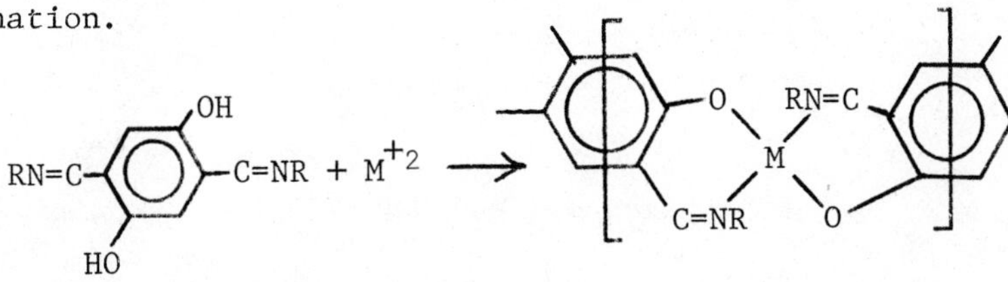

40

Table 3 summarizes a number of the chelating functional groups typically employed to complex metal ions, oxides, etc. Polymers formed through routes a and c can contain definable "exact" repeat units while structures of polymers formed through route b (and sometimes c) will vary with reaction conditions including nature of reactants, temperatures, amount and nature of additives and monomer ratio and concentration. Ionomers are industrial examples of coordination polymers formed through route b.

Tannebaum, Goldberg and Flenniken (chpt. 18) describe the dissolution or mixing of the metal carbonyls into polymers followed by decomposition of the metal carbonyl in solid polymer matrices. This leads to products with variable microstructures. Roman et al. (chpt 8), Neckers (chpt 22) and Bergbreiter (chpt 23) employ route b giving structures of varying regularity. In Figure 1, chapter 22, Neckers outlines the approach often employed for the synthesis of metal-containing polymeric catalysts. Khor and Taylor (chpt 21) describe a doping technique widely employed in attempts to generate conductive polymeric bulk solids and coatings (films). This also leads to structures with variable microstructures with the structures varying both on the surface and according to the depth within the polymeric matrix.

Table 3. Partial Listing of Functional Groups Typically Employed to Complex Metal Ions, Oxides, Etc.

General Structure	Name	General Structure	Name
R–OH	Alcohol	R–SH	Thiol
$\overset{\displaystyle O}{\overset{\|}{R-C-R}}$	Ketone	$\overset{\displaystyle S}{\overset{\|}{R-C-R}}$	Thiocarbonyl
R–O–R	Ether	R–S–R	Thioether
$R-CO_2H$	Acid	$R-CS_2H$	Thiocarbamate
$\overset{\displaystyle O}{\overset{\|}{RNHCR}}$	Amide	$\overset{\displaystyle S}{\overset{\|}{RNHCR}}$	Thioamide
$\overset{\displaystyle O}{\overset{\|}{RNHCNHR}}$	Urea, Semicarbazone	$\overset{\displaystyle S}{\overset{\|}{RNHCNHR}}$	Thiourea, Semithiocarbazone
$\overset{\displaystyle O\ \ OH}{\overset{\|\ \ \ \|}{RC-C-R}}$	α–Hydroxyketo	$\overset{\displaystyle S}{\overset{\|}{ROCSR}}$	Xanthane
RNH_2, R_2NH, R_3N, $\diagdown\!\!N\diagup$	Amines	RCN	Nitriles
$\overset{\displaystyle \|}{\underset{\|}{-P=N-}}$	Phosphazenes	$R_2C=NOH$	Oxime
PO_4^{-2}	Phosphates	$\underset{NH_2}{\overset{\|}{RC=NOH}}$	Amidoxime

Brittain (chpt 25) describes products synthesized via route c. The products may be either "exact" or variable structures depending on the particular reactants employed. The products described by Gressier, Levesque, Patin and Varret (chpt 17) are precomplexed and then polymerized employing the addition process, illustrating route a. Archer, et al. (chpt 20) describe materials formed through polymerizations involving the metal atom as a site of reaction and are thus of form c. These materials typically have known repeating units. Carraher, et al. (chpt 9) describes the synthesis of materials employing both routes b and c. The products produced through chelation by polyethyleneimine exhibit varying microstructures while those produced through chelations involving monomeric nitrogen-containing products have regular, reoccurring structures.

Bailar listed a number of principles which can be considered in designing coordination polymers (48). Briefly there are: a) little flexibility is imparted by the metal ion or its immediate environment. Thus, flexibility must arise from the organic moiety; (b) metal ions only stabilize those ligands in the immediate vicinity; thus the chelates should be robust and close to the metal atom; (c) thermal, oxidative and hydrolyticstability are not directly related. Polymers must be designed specifically for the properties desired; (d) metal-ligand bonds have enough ionic character to permit them to rearrange more readily than typical "organic bonds"; (e) flexibility increases as the covalent nature of the metal-ligand bond increases; (f) polymer structure (such as square planar, octahedral; planar, linear, network) is dictated by the coordination number and sterochemistry of the metal ion or oxide; and solvents employed should not form a strong complex with the metal ion or chelating agent or they will be incorporated into the polymer structure and/or prevent desired reactions from occurring.

The drive for the synthesis and characterization of coordination polymers was catalyzed by work supported and conducted by the USA Air Force in a search for materials which exhibited high thermal stabilities. Attempts to prepare stable, tractable coordination polymers which would simulate the exceptional thermal and/or chemical stability of model monomeric coordination compounds such as cooper ethylenediaminobis acetylacetonate (II) or copper phthalocyanine (I) have been disappointing at best. Typically only short chains were formed and the thermal stability associated with the monomeric chelate did not seem to stand up in the corresponding coordination polymers. Another major problem was processability. Coordination polymers prepared in the early work were insoluble and high molecular weight species were not obtained since precipitation from solution occurred at an early stage.

Despite the early difficulties, interest in coordination polymers continued. Carraher utilized formation of coordination polymers as a model for uranylion reterival for both environmental and industrial purposes. Select salts of diacids, digxines, etc. are capable of removing the uranyl ion to 10^{-4} to 10^{-8} molar levels (49-51). (The uranyl ion is the most common naturally occurring water soluble form of uranium.) This has been extended to removal utilizing salts of polyacrylic acids (41), polyacrylic acid itself and a wide variety of carboxylic acid sulfonate and sulfate containing resins etc(50). Almost all of these compounds containing the complexed uranyl moiety have greatly reduced toxicities to a wide range of bacteria and fungi compared to UNHH itself (25, 26).

$$UO_2^{+2} \cdot 2H_2O + {}^-O_2CRCO_2^- \rightarrow$$

41

The use of resins to remove, concentrate and detect metal ions,

including the uranyl ion, has been practiced for some time (52).
Such reactions are critical analytical, chromatographic polymeric
reactions. Also there are a number of plants known to chelate
various ions (for instance see 53-55).

Block and coworkers (for instance 56) made numerous single-,
double- and triple-bridged phosphinate polymers through route C
containing Al, Be Co, Cr, Ni, Ti, and Zn metal atoms. The metal
phosphinates are typically prepared from metal salts and dialkyl or
diarylphosphonic acids employing melt or solution systems. The
products can give films and fibers. They can be employed as additives to thicken and improve high pressure physical properties of
silicones.

Allcock and Carraher (57-60) have synthesized platinum coordination polymers as antineoplastic agents. Rosenberg in 1964 discovered that cells failed to divide and eventually that the cause for
inhibition of cell division was <u>cis</u>-dichlorodiamine platinum II,
<u>cis</u>-DDP (61). <u>Cis</u>-DDP is now licensed under the tradename
Platinol and is employed extensively in conjunction with other
drugs in the treatment of a wide variety of cancers. Chapters
11 and 12 describe this work in greater detail.

Extensive work has been done with bis-beta-diketones (for
instance 62, 63). Polymers are formed through the bulk polymerization of the diketones with metal acetylacetonates or metal
acetates.

Vallee and coworkers (for instance 64) studied metal binding
proteins and their affect on enzymatic and membrane interactions.

As previously noted, the area receiving the greatest recent
activity concerns the "dopeing" of materials with metal salts and
oxides, organometallic compounds, metal carbonyls, etc. which may
permit electrical, photo, or pizoelectrical conductive materials.
While much science is involved, art still plays an important role in
this area where small changes markedly affect the conductive
properties and where the structures are typically quite varied and
dependent on reaction conditions.

In some cases, preformed polymers have been prepared with
efficient coordination sites for metals. Subsequently, these
polymers are used to coordinate specific metal ions. An example
of this type of polymer is the poly(thiosemicarbazide), $\underline{42}$,
prepared by Donaruma (65). This class of polymer was extremely
efficient in complexing Cu^{+2} ions from waste effluents from brass
mills. Cu^{+2} ions were specifically complexed, as illustrated by
structure $\underline{43}$, even in the presence of high concentrations of other
metal ions. Cu^{+2} was tenaciously held by the polymer even on
elution with mineral acids but elution with 1,4-benzoquinone

releases cooper in purities of 96-99% in field trials.

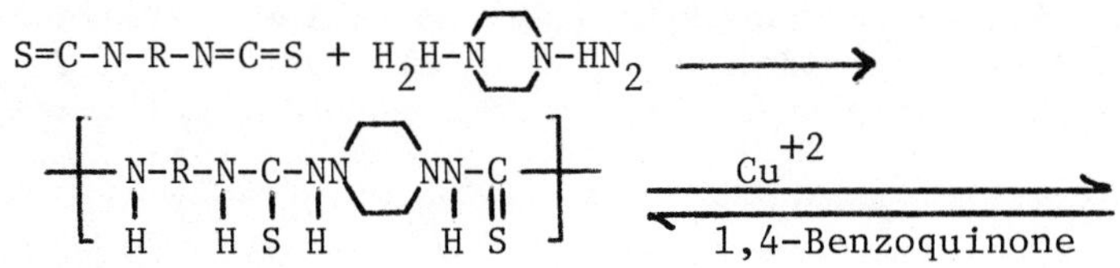

42

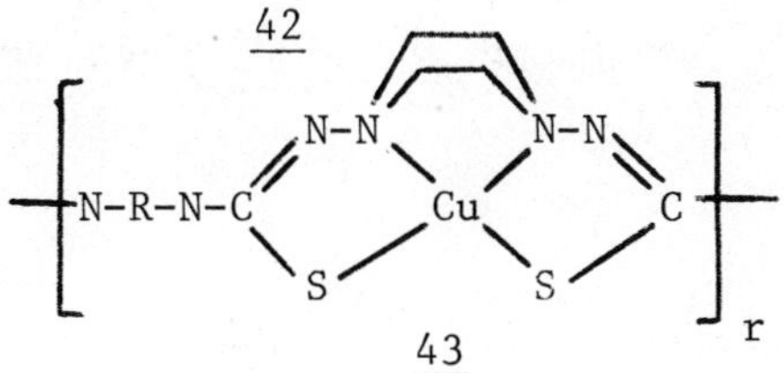

43

It is interesting to note that the copper-containing polymer, 43, can be used as a reagent for the oxidation of aldehydes to carboxylic acids and the coupling of nitroso compounds to azoxy derivatives.

ADDITION ORGANOMETALLIC POLYMERS

A large number of vinyl organometallic monomers have been prepared, homopolymerized and copolymerized with other classic vinyl monomers. The synthesis and polymerization of many organometallic vinyl monomers has been reviewed (66-69). To provide a view of the variety of vinylorganometallic structures which have been prepared and polymerized examine Table 4. The vinylcyclopentadienyl monomers have typically been propared by acetylation of the cyclopentadiene ring followed by reduction of the keto function and dehydration as illustrated below for (η^5-vinylcyclopentadienyl)tricarbonylmanganese, 45.

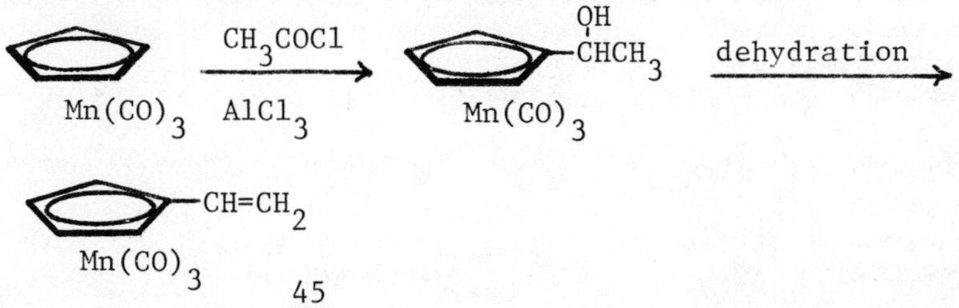

45

This general method has been used to make vinylferrocene, 44, (70,71), (η^5-vinylcyclopentadienyl) dicarbonylnitrosylchromium 46 (72-73), vinylruthenocene, 49 (74) and 3-vinylbisfulvalenediiron 51, (75,76). However, many cyclopentadienyl metal derivatives are not stable to Freidel Crafts reaction conditions and other synthetic methods have been sought.

The preparation of (η^5-vinylcyclopentadienyl)tricarbonyl-
methyltungsten, 48, was effected (77, 78) using the novel reagent,
sodium formylcylopentadienide, which when refluxed with
hexacarbonyltungsten displaced three carbon monoxide molecules

Table 4.   Vinyl Organometallic Monomers Which Have Been
           Prepared.

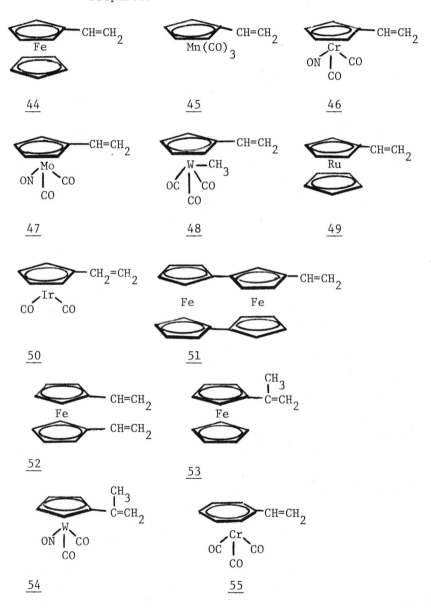

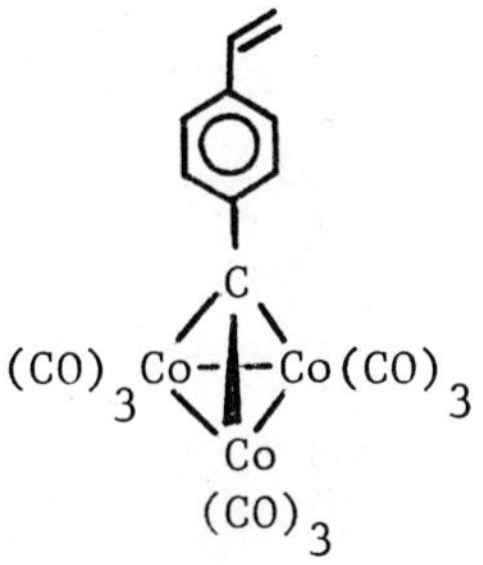

56

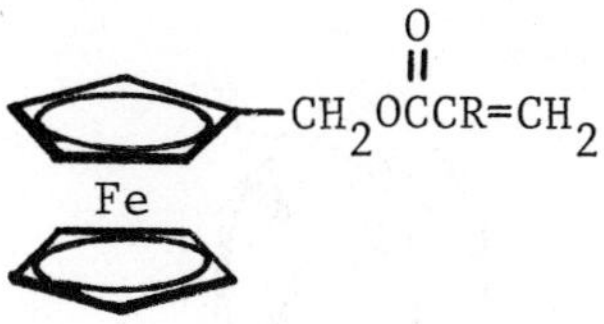

57 R=H
58 R=CH$_3$

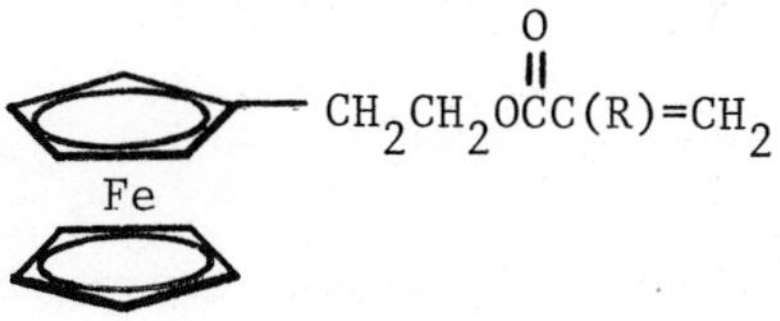

59 R=H
60 R=CH$_3$

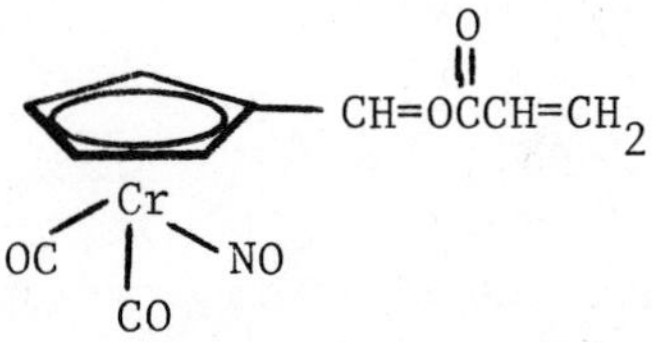

61

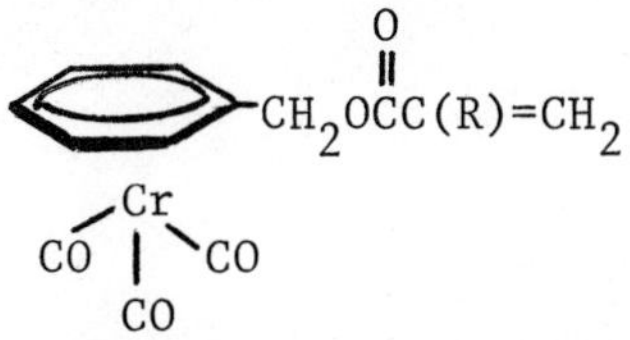

62 R=H
63 R=CH$_3$

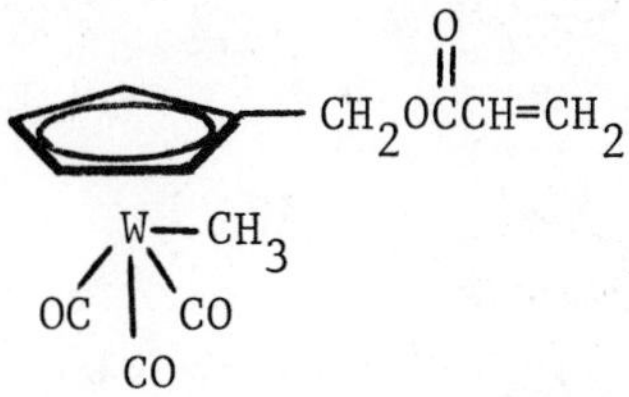

64

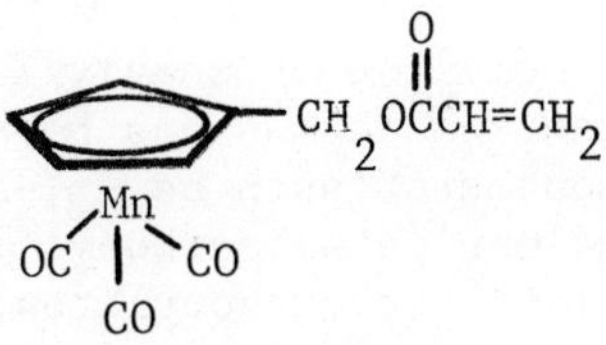

65

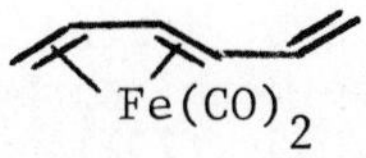

66

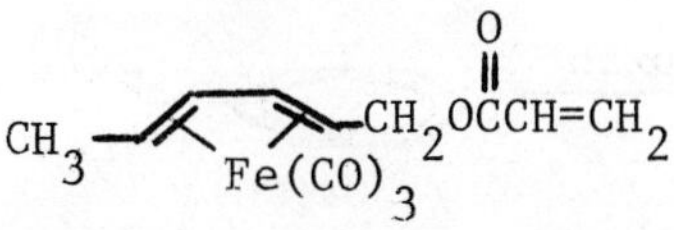

67

(continued)

Table 4 (continued)

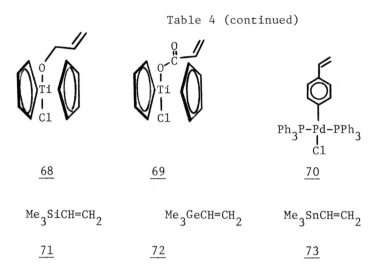

| 68 | 69 | 70 |

Me₃SiCH=CH₂   Me₃GeCH=CH₂   Me₃SnCH=CH₂

| 71 | 72 | 73 |

to give the (η^5-formylcyclopentadienyl)-tricarbonyltungsten anion, 49. Methylation at tungsten and Wittig synthesis resulted in good yields of the vinyl monomer, 48.

The displacement of either carbon monoxide or weakly ligated molecules from metals by π-bonding organic moieties has been used to make both η^6-arene metal carbonyl monomers (such as (η^6-styrl) tricarbonylchromium, 55, (79)) and η^4-diene metal carbonyl monomers (such as (η^4-2,4-hexadien-1-ylmethacrylate) tricarbonyliron, 67 (80)).

$$CH_2=CHCOCl \longrightarrow CH_3 \text{---} \underset{Fe(CO)_3}{\text{---}} CH_2O\overset{O}{\overset{\|}{C}}CCH=CH_2 \qquad \underline{67}$$

These methods have been extended to make acrylic organometallic monomers. For example, benzyl alcohol or 2-phenylethanol, when refluxed in DME with $Cr(CO)_6$, give the corresponding η^6-arene tricarbonylchromium complexes (ie $\underline{74}$) which can be esterified with methacrylyl chloride to give acrylate, $\underline{75}$ (81).

$$\bigcirc\text{---}CH_2CH_2OH + Cr(CO)_6 \xrightarrow[DME]{\Delta} \bigcirc\text{---}CH_2CH_2OH \quad \underset{Cr(CO)_3}{} \qquad \underline{74}$$

$$\xrightarrow[Bz,Pyr.]{CH_2=CHCOCl} \bigcirc\text{---}CH_2CH_2O\overset{O}{\overset{\|}{C}}C=CH_2 \quad \underset{Cr(CO)_3}{} \quad \underset{CH_3}{} \qquad \underline{75}$$

Acrylates or methacrylates of ferrocene (such as $\underline{57}$ and $\underline{58}$) were prepared from N,N-dimethylaminoferrocene by methylation, displacement to give the ferrocenylmethanol and esterification (82–84). Similar routes have been used to make other organometallic acrylates and their homo- and copolymerization have been amply described (85–92).

$$\underset{Fe}{\bigcirc}\text{---}CH_2N(CH_3)_2 \xrightarrow{CH_3I} \xrightarrow{NaOH} \underset{Fe}{\bigcirc}\text{---}CH_2OH$$

$$\xrightarrow{CH_2=CHCOCl} \underset{Fe}{\bigcirc}\text{---}CH_2O\overset{O}{\overset{\|}{C}}CCH\text{---}CH_2$$

$$\underset{}{\text{P}}\text{---}\bigcirc + Cr(CO)_6 \xrightarrow{DME} \text{P}\text{---}\bigcirc \quad \underset{Cr(CO)_3}{}$$

Recently, the number of organometallic monomers available for polymerization has been expanded by the use of lithium vinylcyclopentadienide, 76, and lithium isopropenylcyclopentadienide (93,94). This route, which leads to a variety of Ti, Mo, Cu W, Co, Rh and Ir-containing monomers, is described by Rausch, Pittman et al. (Chapter 2).

REACTIVITY OF VINYL ORGANOMETALLIC MONOMERS

The effect that organometallic functions exert in vinyl polymerizations is beginning to become clear in some instances. One might expect a transition metal with its various readily available oxidation states and large steric bulk to exert unusual electronic and steric effects during polymerization. First, we will consider the vinylcyclopentadienyl monomers and use vinyl-ferrocene, 44, as an introductory example. Its homopolymerization has been initiated by radical (70, 71, 95–98) cationic (99) coordination (100) and Ziegler-Natta (99) initiators. Unlike the classic organic monomer, styrene, vinylferrocene undergoes oxidation at iron when peroxide initiators are employed. Thus, azo initiators (such as AIBN) are commonly used. Here we see one difference between an organic and an organometallic monomer. The stability of the ferricinium ion makes ferrocene readily oxidizable by peroxides whereas styrene, for example, would undergo polymerization.

Unlike most vinyl monomers, the molecular weight of polyvinyl-ferrocene does not increase with a decrease in initiator concentration (98). This is the result of vinylferrocene's anomalously high chain-transfer constant (C_m=8 x 10^{-3} versus 6 x 10^{-5} for styrene at 60°C) (19). Finally, the rate law for vinylferrocene homopolymerization is first order in initiator in benzene (101). Thus, intramolecular termination occurred. Mössbauer studies support a mechanism involving electron transfer from iron to the growing chain radical to give a zwitterion which terminates and further results in a high spin Fe(III) complex (102).

$$Vp = 5.64 \times 10^{-4}[\text{vinylferrocene}]^{1.12}[\text{AIBN}]^{1.1} \text{ in benzene} \quad (101)$$
$$Vp = 5.99 \times 10^{-5}[\text{vinylferrocene}]^{0.97}[\text{AIBN}]^{.42} \text{ in dioxane} \quad (102)$$

where $Vp = \text{mol } L^{-1} s^{-1}$ and $k - \text{mol}^{-1} s^{-1}$

Unusual homopolymerization kinetic behavior was also observed for the radical-initiated polymerizations of (η^5-vinylcyclopentadienyl)tricarbonylmanganese. In benzene, benzonitrile and acetone the rate was half order in AIBN and three halves order in monomer (103). Specifically at 60°C the rate expressions were.

$$Vp = 1.303 \times 10^{-4}[M]^{1.45}[\text{AIBN}]^{0.48} \text{ in benzene}$$
$$Vp = 1.980 \times 10^{-4}[M]^{1.58}[\text{AIBN}]^{0.47} \text{ in benzonitrile}$$
$$Vp = 1.500 \times 10^{-4}[M]^{1.54}[\text{AIBN}]^{0.47} \text{ in acetone}$$

A rate equation of the form $Vp = k[M]^{1.5}[I]^{0.5}$ is derived if the initiator efficiency is low and therefore the initiator is pproportional to [M] (ie. $f - f'[M]$). This rate equation requires that the degree of polymerization follow the expression:

$$DP = Vp/Vi = [k_p/(2f'k_t k_d)^{0.5}][[M]/[I]].^{0.5}$$

Molecular weight measurements (103) confirmed that the degree of polymerization was proportional to $([M]/[I])^{0.5}$.

Radical initiated homopolymerization kinetic studies (104) of (η^5-vinylcyclopentadienyl)methyltricarbonyltungsten were carried out in benzene where the rate expression was found to be:
$Vp = 1.13 \times 10^{-2}[M]^{0.8}[I]^{2.3}$ The homopolymerizations were sluggish and several reinitiations were required to obtain good conversions. Chain transfer and chain termination by hydrogen abstraction from (a) the tungsten-bound methyl group (b) the cyclopentadienyl ring or (c) the backbone methine groups were ruled out as causes of the sluggishness (78, 104).

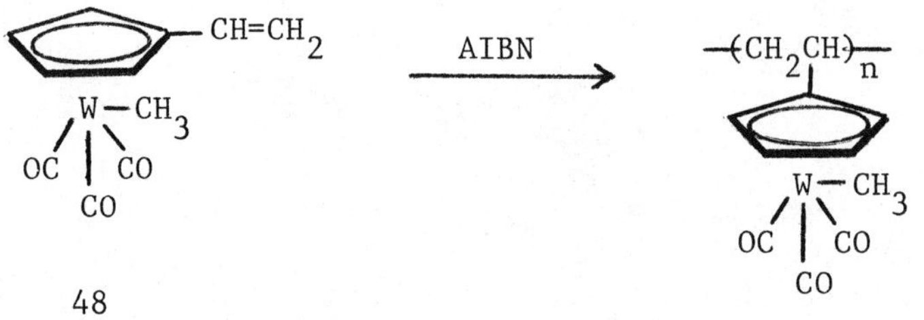

48

Titanium allyl and methacrylate monomers **68** and **69** (Table 4) gave only very low molecular weight materials under benzoyl peroxide initiation and in styrene copolymerizations only small amounts of **68** and **69** were incorporated (105). This is in accord with a low reactivity and a high chain transfer activity for these monomers (106). Even less reactive is (η^4-hexatrienyl)tricarbonyl-iron, **66** (80). It does not undergo either radical-initiated homopolymerization or copolymerization. Indeed, it inhibited the polymerization of both styrene and methyl acrylate. Presumably, the radical, **77**, resulting from chain addition to the vinyl group of **66**, is stable and does not undergo propagation.

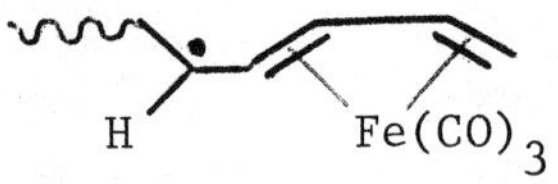

77

The organometallic acrylates and methacrylates undergo ready radical initated homopolymerization and copolymerizations. Unlike the unusual kinetic behavior of the η^5-vinylcyclopentadienyl monomers, the acrylic systems appear to behave naturally. Thus, homopolymerizations of ferrocenylethyl acrylate, **59**, and ferro-cenylethyl methacrylate, **60**, were found to be first order in monomer and half order in initiator (85).

Mixed oxidation state polymers have been formed from (ferrocenylmethyl methacrylate) and poly(vinylferrocene), **78**, by treatment with electron acceptors such as dichlorodicyanoquinone or iodine (82, 83, 88, 107, 108). Mössbauer spectroscopy has been quite useful in analyzing the fraction of ferrocene moieties which have been oxidized to ferricenium groups (83, 88).

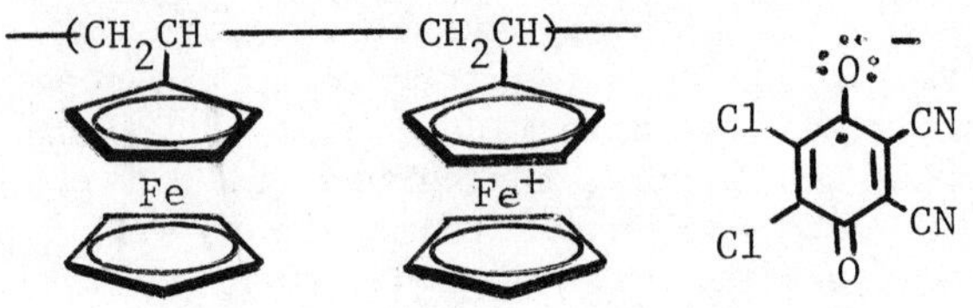

The anionic initiation of vinylferrocene, 44, (η^5-vinylcyclo-pentadienyl)tricarbonylmagnagese, 45, (η^5-vinylcyclopentadienyl) dicarbonylnitrosylchromium, 46, and (η^5-vinylcyclopentadienyl)tricar-bonylmethyltungsten, 47, all resist anionic initiation (109). For example n-BuLi/hexane, n-BuLi/toluene-hexane, n-BuLi/THF-hexane, EtMgBr/THF, sodium naphthaline/THF and LiAlH$_4$/THF all failed to initiate 45 at temperatures from -78 to 20°C (109). Similarly, η^4-(2,4-hexadien-1-ylmethacrylate)tricarbonyliron was not anioni-cally initiated (109). The vinylcyclopentadienyl monomers resist anionic initiation because the vinyl group is exceptionally electron rich in every case (68). Copolymerization of 45 with methyl-acrylate initiated by n-BuLi or sodium naphthalide gave copolymers with very low molar incorporation of 45.

Ferrocenylmethyl methacrylate, 60, and acrylate, 59, are initiated using a variety of anionic systems (109-111). The effect of the ferrocenylmethyl methacrylate/LiAlH$_4$ mole ratio on molecular weight was regular. By varying this ratio from 17 to 300 the values of $\bar{M}_n$ and $\bar{M}_w$ increased from 3,000 and 5,400 to 277,000 and 724,000, respectively (111). Using vacuum line techniques, a solution of ferrocenylmethyl methacrylate which had been polymerized using LiAlH$_4$ initiation was shown to exist as 'living' polymers (110). This led to the use of LiAlH$_4$-tetramethylethylenediamine initiation to make block copolymers of ferrocenylmethyl methacrylate, 58, with methyl methacrylate and acrylonitrile (110)(see 79 and 80). The polyferrocenylmethyl methacrylyl anion was unable to initiate styrene (110). Therefore, 58 was added to THF solution of living polystyrene, originally prepared by sodium naphthalide initiation to give 60-styrene block copolymers, 81 (110).

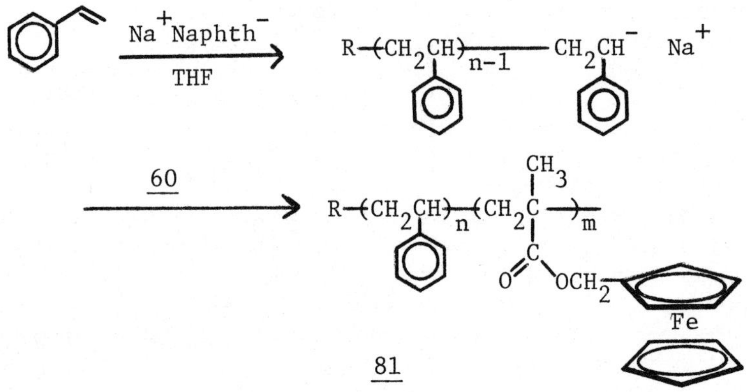

81

    The electron rich cyclopentadienyl ring in vinyl monomers 44-54 (Table 4) is able to stabilize adjacent positive charge. This has long been noted in ferrocene systems where the exceptional stability of α-ferrocenyl carbenium ions is well known. Therefore, one might expect the vinyl group in these monomers to be quite electron rich and behave, as such, both in radical and cationic initiated homo- and copolymerizations. This is illustrated by the ready cationic polymerization of 1,1'-divinyl-ferrocene, 52. Molecular weights up to 35,000 have been obtained using $BF_3(Et_2O)$ initiation (112). One of the interesting points about this polymerization is the formation of cyclolinear structures, 82, due to intramolecular electrophilic attack of the α-ferrocenyl-carbenium ion on the adjacent vinyl group (112, 113). Cationic initation of some isopropenyl monomers is discussed by Rausch and Pittman (Chpt. 2)

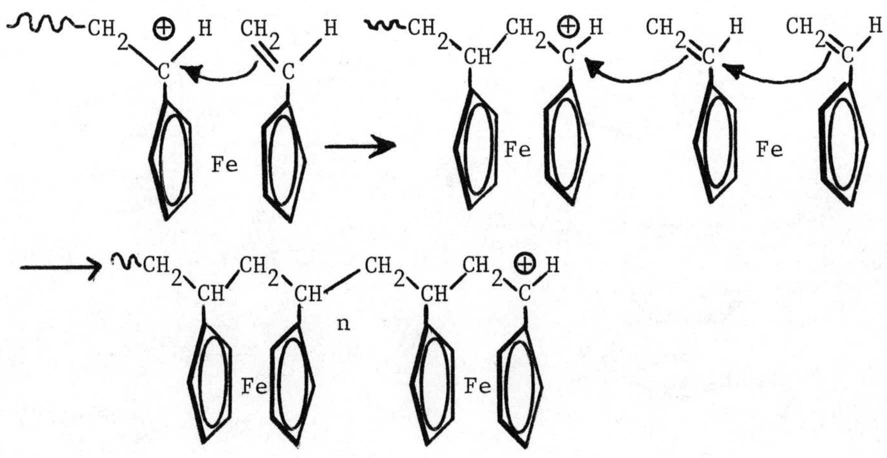

82

The great electron richness of vinylferrocene as a monomer
was illustrated in its copolymerizations with maleic anhydride
where 1:1 alternating copolymers were obtained over a wide range
of M_1/M_2 feed ratios and $r_1 \cdot r_2 = 0.003$ (96). Subsequently, a
large number of detailed copolymerization studies were carried out
between vinylferrocene and classic organic monomers such as styrene
(113), N-vinyl-2-pyrrolidone (97), methyl acrylate (113, methyl
methacrylate (113), N-vinylcarbazole (114) and acrylonitrile (113).
The relative reactivity ratios (r_1 and r_2) were obtained and from
them the values of the Alfrey Price Q and e parameters were
obtained. The value of e is a semiemperical measure of the electron
richness of the vinyl group. The best value of e for vinylferrocene
is about -2.1 which when compared to the e values of maleic
anhydride (+2.25), p-nitrostyrene (+0.39) styrene (-0.80), p-N,N-
dimethylaminostyrene (-1.37) and 1,1'-dianisylethylene (-1.96)
illustrates the exceptional electron richness of vinylferrocene's
vinyl group.

Remarkably, the presence of different metals or the presence
of electron withdrawing carbonyl groups on metal atoms attached to
the η^5-vinylcyclopentadienyl group does not markedly diminish the
electron richness of the vinyl group as measured by the Alfrey
Price e value. This is illustrated below. Also illustrated

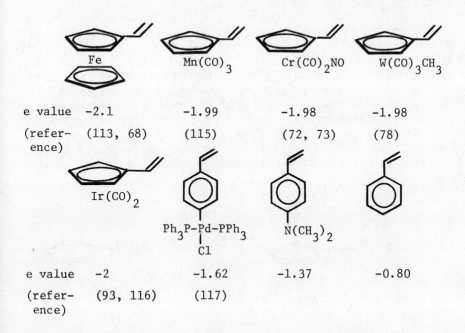

	Fe	Mn(CO)$_3$	Cr(CO)$_2$NO	W(CO)$_3$CH$_3$

e value   -2.1   -1.99   -1.98   -1.98

(refer-   (113, 68)   (115)   (72, 73)   (78)
ence)

         Ir(CO)$_2$ Ph$_3$P-Pd-PPh$_3$ N(CH$_3$)$_2$
                        Cl

e value   -2   -1.62   -1.37   -0.80

(refer-   (93, 116)   (117)
ence)

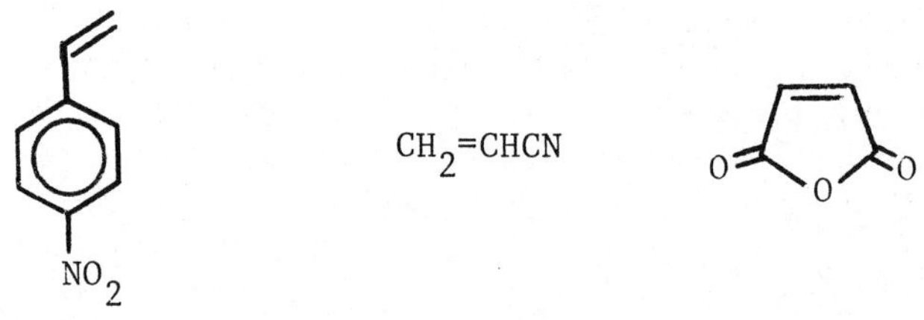

e value	+0.39	+1.21	+2.25

CH$_2$=CHCN

is the powerful electron donating effect of the p-Pd(PPh$_3$)$_2$Cl group in monomer 70. Notice that it is a significantly stronger donor than the p-(CH$_3$)$_2$N- group. All indications are that the p-CCo$_3$(CO)$_9$ function in monomer 56 (Table 4) is also strongly electron donating (118). Further studies of vinyl organometallic monomers are covered by Rausch and Pittman (Chpt. 2).

A large number of copolymerization studies have been conducted on organometallic monomers and a substantial number of reactivity ratios are now available for various monomer pairs in radical-initiated copolymerizations. An extensive discussion of these is beyond the scope of this chapter. However, a number of these have been collected in Table 5. It is instructive to examine a few of these in order to place the reactivity of certain organometallic monomers into perspective.

First, the r_1 values of ferrocenylmethyl acrylate, 57, are smaller than those of 2-ferrocenylethyl acrylate, 59, in copolymerizations with corresponding monomers. Similarly, the r_1 values of ferrocenylmethyl methacrylate, 58, are smaller than the corresponding values of 2-ferrocenylethyl methacrylate, 60, for corresponding comonomers. This is a manifestation of the steric effect of the large bulky ferrocene moiety. When it is removed further from the reaction center the monomer will become more reactive. In general the steric effects of the organometallic groups in monomers 57-63 and 67 reduces their reactivity relative to the replacement of this group with a hydrogen.

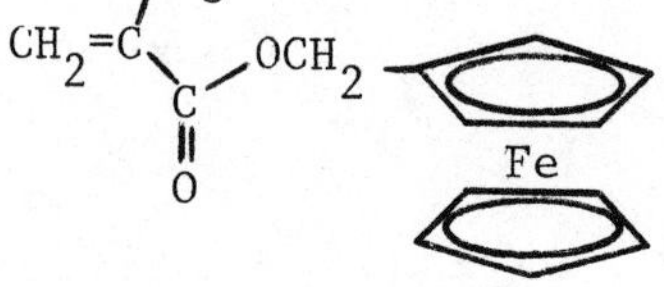

58

Less reactive

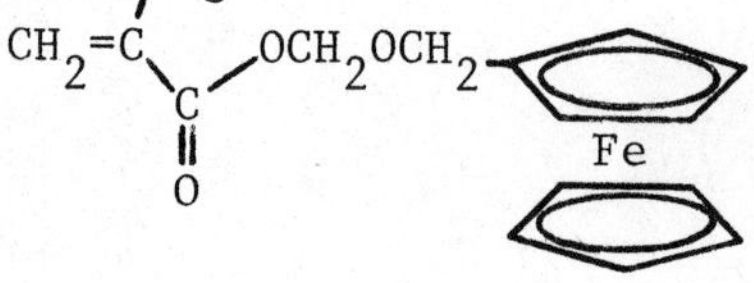

60

More reactive

Table 5. Collected Reactivity Ratios For Selected
          Copolymerizations of Vinyl Organometallic
          Monomers

Organometallic Monomer M_1	Comonomer M_2	Reactivity Ratios r_1	r_2	Reference
Vinylferrocene, 44	Styrene	0.08	2.50	113
"	Methyl methacrylate	0.52	1.22	113
"	Methyl acrylate	0.82	0.62	113
"	Acrylonitrile	0.15	0.16	113
"	N-Vinylpyrolidone	0.66	0.42	97
"	N-Vinylcarbazole	0.47	0.20	114
(η^5-Vinylcyclopentadienyl)tricarbonylmanganese, 45	Styrene	0.1	2.50	115
	vinyl acetate	2.35	0.06	115
	methyl acrylate	0.19	0.47	115
"	Acrylonitrile	0.19	0.22	115
"	vinylferrocene	0.44	0.49	115
"	N-Vinylpyrrolidone	0.14	0.09	97
(η^5-Vinylcyclopentadienyl)dicarbonylnitrosylchromium 46	styrene	0.30	0.82	72,73
	N-Vinylpyrrolidone	5.3	0.08	72,73
(η^5-Vinylcyclopentadienyl)tricarbonylmethyltungsten, 48	Styrene	0.16	1.55	78
(η^5-Vinylcyclopentadienyl)dicarbonyliridium, 50	Styrene	0.28	0.76	93,116
(η^6-Stryryl)tricarbonylchromium, 55	Styrene	0.0	1.35	79
	Methyl acrylate	0.0	0.70	79

(continued)

Table V (cont.)

Ferrocenylmethyl acrylate, 57	Styrene	0.02	2.3	84
"	Methyl acrylate	0.14	4.46	84
"	Methyl methacrylate	0.08	2.9	84
"	Vinyl acetate	1.44	0.46	84
"	Maleic anhydride	0.61	0.11	91
Ferrocenylmethyl methacrylate, 58	Styrene	0.03	3.7	84
"	Methyl acrylate	0.08	0.82	84
"	Methyl Methacrylate	0.12	3.27	84
"	Vinyl acetate	1.52	0.20	84
"	N-Vinylpyrro- lidone	3.71	0.05	91
"	Acrylonitrile	0.30	0.11	91
"	Maleic Anhy- dride	0.28	0.10	91
2-Ferrocenylethyl Acrylate 59	Styrene	0.41	1.06	86
"	Vinyl Acetate	3.4	0.07	86
"	Methyl Acrylate	0.76	0.69	86
2-Ferrocenylethyl Methacrylate, 60	Styrene	0.08	0.58	86
"	Vinyl Acetate	8.79	0.06	86
"	Methyl Methacrylate	0.20	0.65	86
(η^6-Benzylacrylate)- tricarbonylchromium 62	Styrene	0.10	0.34	87
	Methyl Acrylate	0.56	0.63	87
(η^6-2-Phenylethyl Acrylate)tricarbonyl- chromium	Styrene	0.1	0.5	92
	Methyl Acrylate	0.3	1.0	92
"	Acrylonitrile	0.6	0.2	92

Table V (cont.)

(η^6-2,4-Hexadien-1-yl Acrylate)tricarbonyl- iron, 67	Styrene	0.26	1.81	80
	Vinyl Acetate	2.0	0.05	80
	Methyl			
"	Acrylate	0.30	0.74	80
	Acrylonitrile	0.34	0.74	80
(η^6-2-Phenylethyl Methacrylate)- tricarbonylchromium 75	Styrene	0.04	1.35	81
	Methyl Methacry-			
"	late	0.09	1.19	81
	Acrylonitrile	0.07	0.79	81

DERIVITIZING PREFORMED POLYERMS WITH ORGANOMETALLIC FUNCTIONS

Preformed vinyl polymers and resins have often been further derivatized with organometallic functions. A very large body of work exists on the functionalization of polymers with organo-metallic catalytic groups to make polymer-bound catalysts. This topic has been extensively reviewed (119-121). Therefore, only a few recent examples will be given here. Polystyrene-divinylbenzene resins, modified with triphenylphosphine groups, were used to prepare heterogenized nickel catalysts for use in ethylene oligomerization (122). Polymers 83 and 84 catalyzed the linear oligomerization of olefins in 99 + % linearity and in 93 to 99% selectivity to alpha-alkenes at rates > 3 moles C_2H_4 per mole of Ni per sec (112).

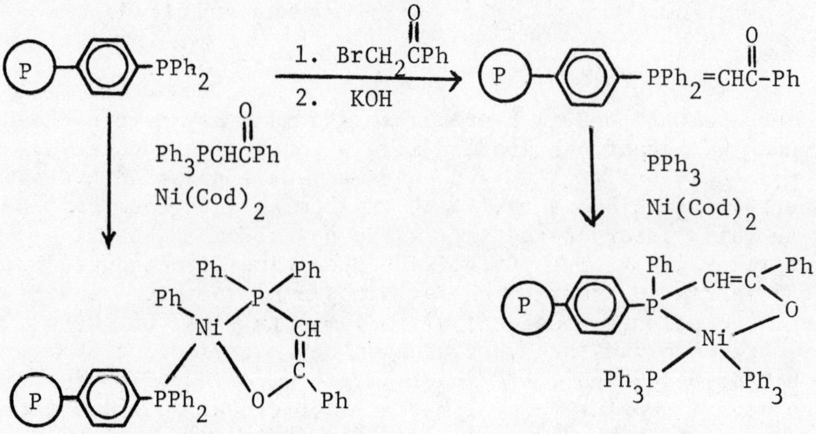

83

84

Polystyrene resins have also been reacted with asymmetric chelating ligands such as (-)DIOP. For example, the asymmetric catalytic hydroformylation of styrene has been carried out using a polystyrene resin-bound (-)DIOP, 85, to which PtCl$_2$ and SnCl$_2$ have been attached (123, 124). The Pt atoms are chelated by the asymmetric (-)DIOP ligand, 86, and therefore exist in a chiral environment. Using the dibenzophosphole analog of DIOP bound to a polymer together with PtCl$_2$ and SnCl$_2$ gave 87. The highest enantiomeric excesses ever achieved in hydroformylation (up to ~80% ee have been obtained) resulted when the homogeneous analog of 87 was used as a catalyst.

Dibenzophosphole System

Since polystyrene is hydrophobic and many asymmetric rhodium-catalyzed hydrogenations proceed more effectively in polar or hydroxylic solvents, Stille and coworkers have prepared hydrophilic resins, 88, from hydroxyethyl methacrylate and ethylene dimethacrylate which incorporated asymmetric pyrrolidine phosphine ligands (125, 126). The pyrrolidine phosphine ligands were built into the matrix in order to chelate rhodium in a chiral environment. These polymer-bound chiral rhodium catalysts, 89, were employed to catalyze the hydrogenation of α-acrylamidocarboxylic acids at 800 psi g and 20°. Enantiomeric excesses of 83 to 91% were achieved. The use of polymeric bipyridines, to which metals are coordinated, as hydrogenation catalysts is discussed by Neckers (Chpt. 22).

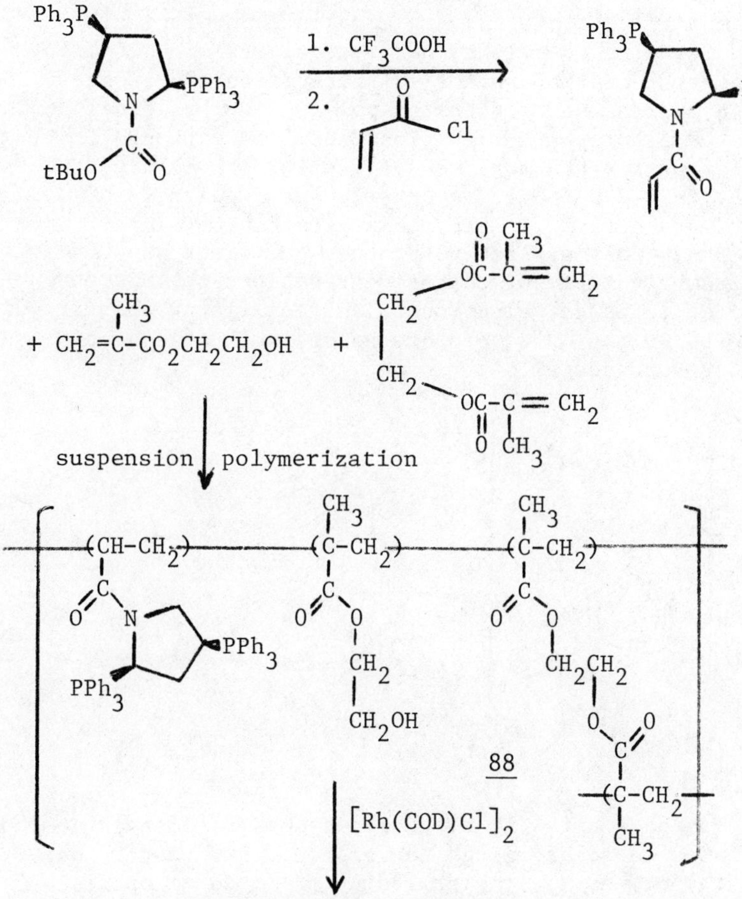

$$\text{(pyrrolidine with } Ph_3P \text{ and } PPh_3, \text{ N-C(=O)-OtBu)} \xrightarrow[\text{2.}\ CH_2=CH-C(=O)-Cl]{\text{1.}\ CF_3COOH} \text{(pyrrolidine with } Ph_3P, PPh_3, \text{ N-C(=O)-CH=CH}_2\text{)}$$

$$+\ CH_2{=}\underset{CH_3}{C}{-}CO_2CH_2CH_2OH\ +\ \begin{array}{c} CH_2{-}O\!-\!\!\underset{O}{\overset{\ }{C}}{-}\underset{CH_3}{C}{=}CH_2 \\ | \\ CH_2{-}O\!-\!\!\underset{O}{\overset{\ }{C}}{-}\underset{CH_3}{C}{=}CH_2 \end{array}$$

suspension ↓ polymerization

$$\left[\text{—(CH-CH}_2)\text{—(C-CH}_2)\text{—(C-CH}_2)\text{—} \right]$$

88

[Rh(COD)Cl]$_2$

↓

Chiral Polymer-Bound Rhodium Catalysts

89

Polymer-supported rhodium (HRh(CO)(PPh$_3$)$_3$) complexes have been used in <u>gas phase</u> hydroformylations of propylene (127). Thus, H$_2$, CO and CH$_2$=CHCH$_3$ were pumped through fixed catalyst beds where the reaction took place at the gas/solid interface. Polystyrene XAD-2 resins were functionalized at the surface with -PPh$_2$ groups which, in turn, ligated the Rh complexes. No deactivation was observed over a 500 m. period of use.

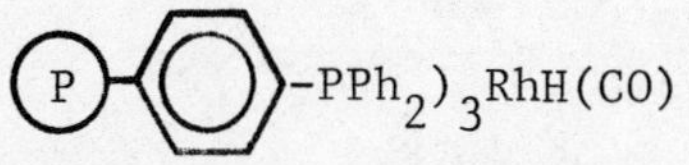

(P)-(O)-PPh$_2$)$_3$RhH(CO)   90

...u use of modified polymers, with covalently or ionically bound organometallic functions, has recently become a focal point of interest in the area of polymer-coated electrodes (128). Thus polyvinylferrocene, protonated poly(4-vinylpyridine) $Fe(CN)_6^{-3}$, the sulfonated fluoropolymer, Nafion, modified with cobalt(II) tetraphenylporphyrin and more recently with both cobalt(II)tetra-phenylporphyrin and $Ru(NH_3)_6^{3+}$ counterions and polymers of tris(4-vinyl-4'-methyl-2,2'-bipyridine)-ruthenium(II), 91 have been used. Such polymers act to protect the electrodes from oxygen or from a variety of corrosive reactions which destroy the electrode, to effect electrocatalysis or electrochemistry in surrounding solutions or to promote hole/electron separation at semiconductor surfaces (128).

91

Kaneko and Yamada (Chpt. 15) discuss polymers for use at semicon-ductor surfaces in solar energy conversion. Rare earth ions chelated into polymer systems are being examined as optical pumps for lasers by Okamoto (Chpt. 24). The entrapment of col-loidal metal catalysts and semiconductors in microemulsions, and polymer vehicles and the use of such species in photo-catalyzed water splitting is discussed by Fendler (Chpt. 19).

One class of polymers contains metals which are directly bonded to the polymers via metal-carbon sigma bonds. These are quite rare. Usually, transition metal-carbon bonds are subject to β-metal hydride elimination. However, this is not possible when β-hydrogens are absent. Pittman and Felis (129,130) showed that the treatment of both linear and crosslinked chloromethylated polystyrenes with $Mn(CO)_5^-$ $(n^5-C_5H_5)Mo(CO)_3^-$ or $(n^5-C_5H_5)W(CO)_3^-$ resulted in displacement of chloride and formation of a stable metal-carbon bond within the polymers (ie 92 and 93).

$$\text{P}-\bigcirc-CH_2Cl + Mn(CO)_5^- \longrightarrow \text{P}-\bigcirc-CH_2Mn(CO)_5$$

$$\underline{92}$$

$$\text{P}-\bigcirc-CH_2Cl +$$

$$\xrightarrow{M=W \text{ or } Mo}$$

$$\text{P}-\bigcirc-CH_2-M(CO)_3Cp$$

$$Cp = \eta^5-C_5H_5$$

$$\underline{93}$$

Thermal decomposition of crosslinked resins containing these structures was carried out to see if the metal decomposition products could be dipersed within the polymer matrix. The decomposition of $\underline{92}$ released $Mn_2(CO)_{10}$ within the resin while $\underline{93}$ decomposed to produce the metal-metal dimer structure, $\underline{94}$. The decomposition of metal carbonyls dissolved in polymers is discussed by Goldberg (Chpt. 18).

$$\underline{93} \xrightarrow{160°} \text{P}-\bigcirc-CH_2 \cdots$$

$$\underline{94}$$

POLYSILANES

A new class of organometallic polymers are polysilanes. These have just recently been prepared (131-134). Polysilanes are polymers which contain an all silicon backbone. For 60 years chemists dreamed of the existance of silicon backbone polymers in analogy to carbon but, despite the success in preparing polysiloxanes, poly-silanes remained elusive. Polysilane homopolymers were finally prepared by the reaction of dichlorosilanes with sodium but the materials were insoluble, highly crystalline and intractable. Permethylpolysilane, $\underline{95}$, fits this description.

$$\begin{array}{c} CH_3 \\ | \\ Cl-Si-Cl \\ | \\ CH_3 \end{array} + 2Na \longrightarrow \begin{array}{c} CH_3 \\ | \\ -(Si)_n \\ | \\ CH_3 \end{array} \qquad \underline{95}$$

West and coworkers have now resolved this problem by introducing large organic substituent groups and by copolymerizing
two monomers to mix these types of substituents. This approach decreases crystallinity and increases solubility. New polysilanes can
be prepared in high yields which are tractable, soluble in a
variety of solvents, easily shaped, readily purified, easily
molded and castable into films. Structures 96 - 99 fit this
category.

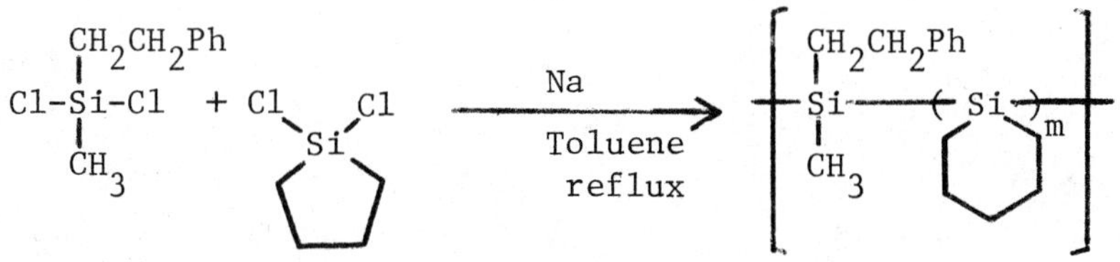

96

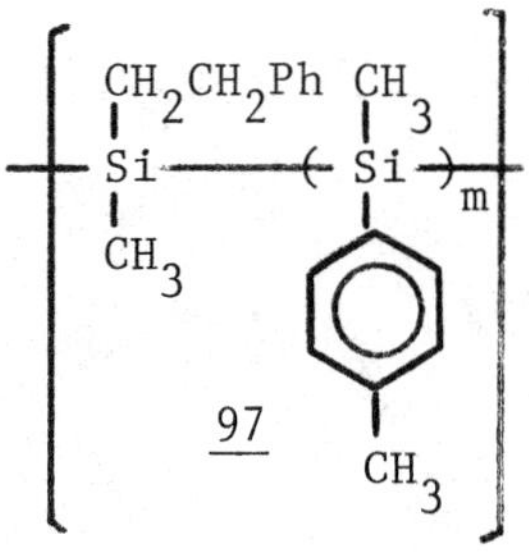

97 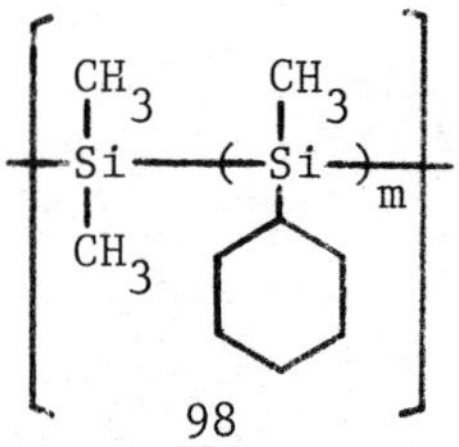   98 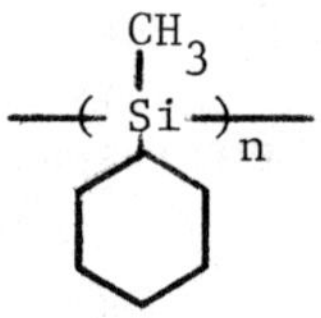   99

The properties of polysilanes are now under study. With the
recent availability of defined high molecular weight samples
(for example, values of $\bar{M}_w$ to 400,000 were reported in reference
134), photochemical studies were carried out. Exposure to 350 or
300 nm leads to rapid photolytic degradation as shown by incre-
mental decreases in molecular weight with exposure. Copolymers
with phenyl side chains directly attached to Si have strong
absorptions near 330 nm due to a conjugatuve interaction
between the phenyl groups and the silicon backbone which acts
as a $\sigma-\sigma^*$ or $\sigma-\pi$ chromophore. In fact, questions about a possible
conjugation along the silicon backbone still exist long after
being postulated (135-136). The cyclohexylmethylsilane homopolymer,
99, exhibits an absorption band at 326 nm (133) suggesting interaction among the chain silicon atoms which does not occur in
carbon chains.

In conclusion, polymer chemistry now encompassess a variety
of metal-containing systems. No longer should we consider polymer
chemistry the domain of just C, H, N, O, S, Si (as polysiloxanes),
P, Cl and Br. We hope that the rest of this book emphasizes this
important fact.

REFERENCES

1. N. H. Ray, "Inorganic Polymers," Academic Press, N.Y., 1978.
2. F. G. R. Gimblett, "Inorganic Polymers," Blackwell, Oxford,
    1963.
3. D. V. Hunter, "Inorganic Polymers," John Wiley, N.Y., 1963.
4. C. Carraher, J. Chemical Education, 58 (11), 92 (1981).
5. C. Pittman, "Organometallic Reactions" (E. Becker and
    M. Tsutsui, Eds.), Plenum Press, N.Y., 1977, vol. 6.
6. J. Sheats
7. C. Pittman, and G. Marlin, J. Polymer Sci. Chem. Ed., 11,
    2753 (1973).
8. C. Carraher, J. Sheats and C. Pittman, Eds. " Advances in
    Organometallic and Inorganic Polymer Science," Dekker, N.Y.,
    1982.
9. C. Carraher, J. Sheats and C. Pittman, Eds. "Organometallic
    Polymers," Academic Press, N.Y., 1978.
10. K. A. Andrinnov, "Metallorganic Polymers," Wiley, N.Y., 1965.
11. H. R. Allcock, "Phosphorus-Nitrogen Compounds," Academic Press,
    N.Y., 1972.
12. E. W. Neuse and H. Rosenberg, Metallocene Polymers, Dekker,
    N.Y., 1970.
13. F. G. A. Stone and W. A. G. Graham, "Inorganic Polymers,"
    Academic Press, N.Y., 1962.
14. H. Kopf, J. Organometal. Chem., 14, 353 (1968).
15. R. King and C. Eggers, Inorg. Chem., 7, 340 (1968).
16. C. Carraher and R. Nordin, J. Applied Polymer Sci. 18, 53
    (1974).
17. C. Carraher, "Interfacial Synthesis, Vol. II Technology and
    Applications" (F. Millichow, C. Carraher, Eds.),
    Dekker, N.Y., 1978.
18. P. W. Morgan, "Condensation Polymers: By Interfacial and
    Solution Methods," Wiley, N.Y., 1965.
19. C. Carraher and J. Preston (Eds.), "Interfacial Synthesis,
    Vol. III. Recent Advances," Dekker, N.Y., 1981.
20. C. Carraher and S. T. Bajah, Polymer (Br.), 14, 42 (1973).
21. C. Carraher, Makromoleculare Chemie, 166, 31 (1973).
22. C. Carraher and C. Deremo-Reece, Angew. Makromolekulare Chemie,
    65, 95 (1972).
23. C. Carraher, R. A. Schwarz, J. A. Schroeder and M. Schwarz,
    J. Macromol. Sci. Chem., A15 (5), 773 (1981).
24. C. Carraher, R. A. Schwarz, J. A. Schroeder, M. Schwarz, and
    H. M. Moloy, Organic Coat. Plast. Chem., 43, 798 (1980).
25. C. Carraher, R. S. Venkatachalam, T. O. Tiernan and
    M. L. Taylor, Applied Polymer Sci. Proc., 47, 119 (1982).
26. C. Carraher, and C. Gebelein, Eds. "Biological Activities of
    Polymers," ACS, Washington, DC 1983.
27. C. Gebelein and C. Carraher, Eds. "Biologically Active
    Polymer Systems," Plenum Press, N.Y., 1984.

28. C. Carraher, J. Schroeder, W. Venable, C. McNeely, D. Giren, W. Woelk and M. Feddersen, "Additives for Plastics, Vol. 2" (R. Seymour, Ed.) Chpt. 8, Academic Press, N.Y., 1978.

29. C. Carraher, J. Schroeder, C. McNeely, J. Workman and D. Giren, "Modification of Polymers" (Eds. C. Carraher and M. Tsuda), ACS, Washington, D.C., 1980.

30. C. Carraher, W. Moon and T. Longworthy, Polymer P., 17, 1 (1976).

31. C. Carraher, R. Linville, T. Manck, H. Blaxall, J. Taylor and L. Torre, "Conductive Polymers" (R. Seymour, Ed.) Plenum Pubs., N.Y., 1981; Chpt. 8.

32. C. Carraher, Chem. Tech., 141 (1972).

33. C. Carraher, J. Macromol. Sci.-Chem., A17 (8), 1293 (1982).

34. B. Currell and Parsonage, "Advances in Organometallic and Inorganic Polymer Science," (C. Carraher, J. Sheats and C. Pittman, Eds.), Dekker, N.Y., 1982, Chpt. 4.

35. C. U. Pittman, Jr., O. E. Ayers, B. Suryanarayanan, S. P. McManus and J. E. Sheats, Die Makromolekular Chemie, 175, 1427 (1974). Also see C. U. Pittman, Jr., O. E. Ayers, S. P. McManus, J. E. Sheats and C. E. Whitten, Macromolecules, 4, 360 (1971).

36. E. W. Neuse, in "Organometallic Polymers," C.E. Carraher, Jr., J. E. Sheats and C. U. Pittman, Jr. Eds., Academic Press, New York, 1978 p. 95.

37. S. L. Sosin, V. P. Alekseeva, M. D. Litvinova, V. V. Korshak and A. F. Zhigach, Vysokomol. Soedin. XVII B (9), 703 (1976).

38. H. Watanabe, J. Motoyama, K. Hata, Bull. Chem. Soc. Japan., 39, 784 (1966).

39. C. U. Pittman, Jr. J. Polym. Sci. A-1., 6, 1687 (1968).

40. H. Valot, Compt. Rend. C, 265, No. 5, 403 (1966).

41. M. Okawara, Y. Takemoto, H. Kitaoka, E. Haruki and E. Imoto, Kogyo Kagaku Zasshi, 65, 685 (1962).

42. P. Petrovitch and H. Valot, Compt. Rend., 263, 214 (1966).

43. C. U. Pittman, Jr. J. Polym. Sci. A-1,5 2927 (1967).

44. J. E. Mulvaney, J. J. Bloomfield and C. S. Marvel, J. Polym. Sci., 62 59 (1962).

45. E. V. Wilkus and A. Berger, French Pat. 1,396,274, 16 April 1965.

46. Von G. Greber and M. L. Hallensleben, Makromolekulare Chemie, 92, 137 (1966).

47. W. J. Patterson, S. P. McManus and C. U. Pittman, Jr., J. Polym. Sci. Polym. Chem. Ed., 12, 837 (1974).

48. J. C. Bailor, "Preparative Inorganic Reactions, Vol 1" (Ed. W. Jolly), Interscience Pubs., N.Y., 1964.

49. C. Carraher and J. Schroeder, Polymer Letters, 13, 215 (1975).

50. C. Carraher S. Tsuji, W. A. Feld and J. E. DiNunzio, "Modification of Polymers (C. Carraher and J. Moore, Eds.), Plenum Press, N.Y., 1983; chapts 15, 16.

51. C. Carraher and J. Schroeder, Polymer P., 19, 619 (1978).

52. J. Korkish, "Modern Methods for the Separation of Rarer Metal Ions", Pergamon Press, N.Y., 1969.
53. A. Knight, W. Crooke and R. Inkson, Nature, 192, 142 (1961).
54. R. Clymo, Ann. Bot. Land, N.S., 27 309 (1963).
55. J. Craigie and W. Maass, Ann. of Botany, N.S., 30, 153 (1966).
56. B. P. Block, Inorg. Macromol. Chem., 1 (2), 155 (1970).
57. H. Allcock, W. Cook and D. Mack, Inorg. Chem., 4, 2584 (1972)
58. H. Allcock, "Organometallic Polymers" (Eds. C. Carraher, J. Sheats and C. Pittman), Chapt. 28, Academic Press, N.Y., 1978.
59. C. Carraher, D. Giron, W. Scott and J. Schroeder, J. Macromol. Sci. Chem., A 15 (4), 625 (1981).
60. C. Carraher, J. Fortman, D. Giron, C. Adamu-John, "Biologically Active Polymer Systems" (C. Gebelein and C. Carraher, Eds.), Plenum, N.Y., 1984.
61. B. Rosenberg, L. Van Camp and T. Krigas, Nature 205, 698 (1965).
62. R. Charles, J. Polymer Sci., A-1, 267 (1963) and J. Phys. Chem., 64, 1747 (1960).
63. V. Korshak and S. Vinogradora, Dokl. Akad. Nauk. SSSR, 138, 1353 (1961).
64. B. L. Vallee and J. R. Riordan, "Proceedings of the International Symposium on Proteins" (C. H. Li, Ed.), Academic Press, N.Y., 1978.
65. L. Donaruma, Polymer Preprints, 22 (1), March 1981.
66. C. U. Pittman, Jr. J. Paint Technol., 43 (561), 29 October (1971) and Chemtech, 1, 416 (1971).
67. C. U. Pittman, Jr., P. Grube and R. M. Hanes, J. Paint Technol., 46 (597), 35 October (1974).
68. C. U. Pittman, Jr. in "Organometallic Polymers", C. E. Carraher, Jr., J. E. Sheats and C. U. Pittman, Jr. Eds., Academic Press 1978 pp 1-11.
69. C. U. Pittman, Jr. in "Organometallic Reactions", E. Becker and M. Tsutsui Eds., Vol 6, p. 1-62, Marcel Dekker, 1977.
70. F. S. Arimoto and A. C. Haven, Jr., J. Am. Chem. Soc., 77 6295 (1955).
71. J. C. Lai, T. Rounsefell and C. U. Pittman, Jr. J. Polym. Sci. A-1, 9, 651 (1971).
72. E. A. Mintz, M. D. Rausch, B. H. Edwards, J. E. Sheats, T. D. Rounsefell and C. U. Pittman, Jr., "J. Organometal. Chem., 137, 199 (1977).
73. C. U. Pittman, Jr., T. D. Rounsefell, E. A. Lewis, J. E. Sheats, B. H. Edwards, M. D. Rausch and E. A. Mintz, Macromolecules, 11, 560 (1978).
74. J. E. Sheats and T. C. Willis, Org. Coatings and Plastics Pre-prints, 41 (2), 33 (1979) and J. Polym. Sci. Polym. Chem. Ed. 22, 1077 (1984).
75. C. U. Pittman, Jr., B. Surynarayanan and Y. Sasaki, in "Inorganic Compounds with Unusual Properties." Adv. in Chem. Ser. No. 150, R. B. King Ed., American Chemical Society 1976 pp. 46-55.

76. C. U. Pittman, Jr. and B. Surynarayanan, J. Am. Chem. Soc., 96, 7916 (1974).
77. D. W. Macomber, M. D. Rausch, T. V. Jayaraman, R. D. Priester and C. U. Pittman, Jr., J. Organometal Chem., 205, 353 (1981).
78. C. U. Pittman, Jr., T. V. Jayaraman, R. D. Priester, Jr., S. Spencer, M. D. Rausch and D. Macomber, Macromolecules, 14, 237 (1981).
79. C. U. Pittman, Jr., P. L. Grube, O. E. Ayers, S. P. McManus, M. D. Rausch and G. A. Moser, J. Polym. Sci. A-1, 10, 379 (1972).
80. C. U. Pittman, Jr., O. E. Ayers and S. P. McManus, J. Macromol. Sci. Chem., A7 (8), 1563 (1973).
81. C. U. Pittman, Jr., O. E. Ayers and S. P. McManus, Macromolecules, 7, 737 (1974).
82. C. U. Pittman, Jr., J. C. Lai and D. P. Vanderpool, Macromolecules, 3, 105 (1970).
83. C. U. Pittman, Jr., J. C. Lai, D. P. Vanderpool, M. Good and R. Prado, Macromolecules, 3, 746 (1970).
84. J. C. Lai, T. D. Rounsefell and C. U. Pittman, Jr., Macromolecules, 4, 155 (1971).
85. C. U. Pittman, Jr., R. L. Voges and W. R. Jones, Macromolecules, 4, 291 (1971).
86. C. U. Pittman, Jr., R. L. Voges and W. R. Jones, Macromolecules, 4, 298 (1971).
87. C. U. Pittman, Jr., R. L. Voges and J. Elder, Macromolecules, 4, 302 (1971).
88. C. U. Pittman, Jr., J. C. Lai, D. P. Vanderpool, Mary Good and R. Prados in "Polymer Characterization: Interdisciplinary Approaches", C. D. Craver, Editor, Plenum Press, 1971, pp. 97-124.
89. C. U. Pittman, Jr. and J. C. Lai, Macromol. Synth., 4, W. J. Bailey, Editor, John Wiley, 1972, p. 161.
90. C. U. Pittman, Jr. and R. L. Voges, Macromol. Synth., 4., W. J. Bailey, Editor, John Wiley, 1972, p. 175.
91. O. E. Ayers, S. P. McManus and C. U. Pittman, Jr., J. Polym. Sci. Polym. Chem. Ed., 11, 1201 (1973).
92. C. U. Pittman, Jr. and G. V. Marlin, J. Polym. Sci. Polym. Chem. Ed., 11, 2753 (1973).
93. D. W. Macomber, W. P. Hart, M. D. Rausch, R. D. Priester and C. U. Pittman, Jr., J. Am. Chem. Soc., 104, 884 (1982).
94. C. U. Pittman, Jr., T. V. Jayaraman, R. D. Priester, Jr., M. D. Rausch, D. W. Macomber, and W. P. Hart, Polymer Preprints, 23 (2), 73 (1982).
95. Y. H, Chen, M. Fernandez-Refojo and H. G. Cassidy, J. Polym. Sci., 40, 433 (1959).
96. C. U. Pittman, Jr., R. L. Voges and J. Elder, Polym. Lett., 9, 191 (1971).
97. C. U. Pittman, Jr. and P. L. Grube, J. Polym. Sci. A-1, 9, 3175 (1971).

98. Y. Sasaki, L. L. Walker, E. L. Hurst and C. U. Pittman, Jr.,
     J. Polym. Sci. Polym. Chem. Ed., 11, 1213 (1973).
99. C. Aso, T. Kunitake and T. Nakashima, Makromol. Chem., 124,
     232 (1969).
100. C. U. Pittman, Jr., Polym. Lett., 6, 19 (1968).
101. M. H. George and G. F. Hayes, J. Polym. Sci. Chem. Ed., 13,
     1049 (1975).
102. M. H. George and G. F. Hayes, J. Polym. Sci. Chem. Ed., 14,
     475 (1976).
103. C. U. Pittman, Jr., C. C. Lin and T. D. Rounsefell, Macro-
     molecules, 11, 1022 (1978).
104. C. U. Pittman, Jr., R. D. Priester, Jr. and T. V. Jayaraman
     J. Polym. Sci. Polym. Chem. Ed., 19, 3351 (1981).
105. V. V. Korshak, A. M. Sladkov, L. K. Luneva, and A. S.
     Girshovich, Vysokomol. Soedin., 5, 1284 (1963).
106. R. Ralea, et al, Rev. Roumaine Chim., 12, 523 (1967).
107. D. O. Cowan, J. Park, C. U. Pittman, Jr., Y. Sasaki, T. K.
     Mukherjee, and N. A. Diamond, J. Am. Chem. Soc., 94, 5110
     (1972).
108. C. U. Pittman, Jr., Y. Sasaki and T. K. Mukherjee, Chem. Lett.
     383 (1975).
109. C. U. Pittman, Jr. and C. C. Lin, J. Polym. Sci. Polym. Chem.
     Ed., 17, 271 (1979).
110. C. U. Pittman, Jr. and A. Hirao, J. Polym. Sci. Polym. Chem.
     Ed., 16, 1197 (1978).
111. C. U. Pittman, Jr. and A. Hirao, J. Polym. Sci. Polym. Chem.
     Ed., 15, 1677 (1977).
112. S. L. Sosin, L. V. Jashi, B. A. Antipova and V. V. Korshak,
     Vysokomol. Soedin., XII (9), 699 (1970).
113. J. C. Lai, T. D. Rounsefell and C. U. Pittman, Jr., J. Polymer
     Sci. A-1, 9, 651 (1971).
114. C. U. Pittman, Jr. and P. L. Grube, J. Appl. Polym. Sci., 18,
     2269 (1974).
115. C. U. Pittman, Jr., G. V. Marlin and T. D. Rounsefell, Macro-
     molecules, 6, 1 (1973).
116. M. D. Rausch, D. W. Macomber, F. G. Fang, C. U. Pittman, Jr.
     and T. V. Jayaraman in "New Monomers and Polymers", B. M.
     Culbertson and C. U. Pittman, Jr., Eds. Plenum Press, 1984.
117. N. Funita and K. Sonogashira, J. Polym. Sci. A-1, 12, 2845 (1974).
118. Unpublished work of C. U. Pittman, Jr., D. Seyferth and S. Massad.
119. C. U. Pittman, Jr. in "Comprehensive Organometallic Chemistry",
     G. Wilkinson, F. G. A. Stone and E. W. Abel, Editors, Perga-
     mon Press, Chapter 55, 553 (1982).
120. C. U. Pittman, Jr. in "Polymer Supported Reactions in Organic
     Synthesis", P. Hodge and D. C. Sherrington, Editors, John
     Wiley and Sons, 1980, Chapt. 5, p. 249.
121. Y. Chauvin, D. Commereuc and F. Dawans, Prog. Polym. Sci., 5,
     95 (1977).
122. M. Peuckert and W. Keim, J. Mol. Catal., 22, 289 (1984).

123. C. U. Pittman, Jr., Y. Kawabata and L. I. Flowers, J. Chem. Soc. Chem. Commun., 473 (1982).
124. G. Consiglio, P. Pino, L. I. Flowers and C. U. Pittman, Jr., J. Chem. Soc. Chem. Commun., 612 (1983).
125. G. L. Baker, S. J. Fritschel and J. K. Stille, ACS Polymer Preprints, 22, 155 (1981).
126. G. L. Baker, S. J. Fritschel and J. K. Stille, J. Org. Chem., 46, 2954 (1981).
127. P. DeMunik and V. R. Scholten, J. Mol. Catal., 11, 331 (1981).
128. L. R. Faulkner, Chem. Eng. News, 28, Feb. 27, 1984.
129. C. U. Pittman, Jr. and R. F. Felis, J. Organometal. Chem., 72, 389 (1974).
130. C. U. Pittman and R. F. Felis, J. Organometal Chem., 72, 399 (1974).
131. J. P. Wesson and T. C. Williams, J. Polym. Sci. Polym. Chem. Ed., 19, 65 (1981).
132. R. West, L. D. David, P. I. Djurovich, H. Yu and R. Sinclair, Ceram. Bull., 62 (8), 899 (1983).
133. Xing-Hua Zhang and R. West, J. Polym. Sci. Polym. Chem. Ed., 22, 159 (1984).
134. Xing-Hua Zhang and R. West, J. Polymer Sci. Polym. Chem. Ed., 22, 225 (1984).
135. K. M. MacKay and R. Watt, Organometal. Chem. Rev., 4, 137 (1969).
136. E. Hengge, Fortschr. Chem. Forsch., 51, 1 (1974).

THE SYNTHESIS AND POLYMERIZATION OF VINYL ORGANOMETALLIC MONOMERS

M. D. Rausch,[*] D. W. Macomber, K. Gonsalves, F. G.
Fang, and Z.-R. Lin

Department of Chemistry
University of Massachusetts
Amherst, Massachusetts 01003

C. U. Pittman, Jr.[*]

Department of Chemistry
Mississippi State University
Mississippi State, Mississippi 39762

INTRODUCTION

There is currently a considerable research effort in both
academic and industrial laboratories concerning the synthesis and
properties of metal-containing polymers. These new materials are
not only of intrinsic interest, but they offer potential utility
as catalysts, semiconductors, lithographic photoresists, electrode
coating materials, anti-fouling agents, and for a variety of other
applications.[1,2] In principle, there are two basic routes to the
formation of metal-containing polymers. One approach has been to
derivatize preformed organic polymers with organometallic functions.[3] A second approach has involved the synthesis of organometallic compounds which contain polymerizable functional groups,
followed by homopolymerization of these monomers or copolymerization with conventional organic monomers. We have developed this
second route to metal-containing polymers in a joint program of
research over the past several years, and report here a summary of
our recent results.

MONOMER SYNTHESES AND RADICAL-INITIATED POLYMERIZATIONS

The first organometallic monomer which contained a transition metal was vinylferrocene (1). This monomer was prepared in 1955,[4] and its polymerization behavior has been extensively studied under radical,[4-7] cationic[8] and Ziegler-Natta conditions[9] (1 is inert to anionic initiation). Several other vinyl-type organometallic monomers, including minylcymantrene (2)[10-12] and vinylcynichrodene (3) [13-14] have more recently been prepared. They have been homopolymerized and copolymerized with a variety of electron-donating and -attracting organic comonomers (i.e., styrene, methyl methacrylate, acrylonitrile and N-vinylpyrrolidine).[11-14]

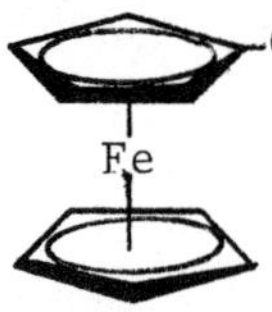

      1 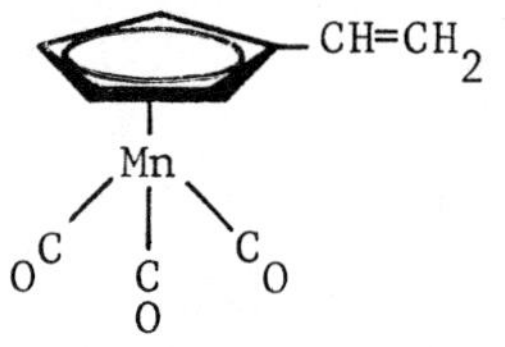   2 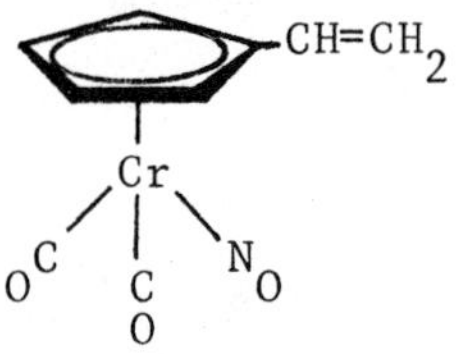   3

Monomers 1-3 are easily synthesized in good yield, since the parent compounds [viz., $(\eta^5-C_5H_5)_2Fe$, $(\eta^5-C_5H_5)Mn(CO)_3$, and $(\eta^5-C_5H_5)Cr(CO)_2(NO)$] are metallo-aromatic, i.e., they readily undergo electrophilic aromatic substitution reactions such as Friedel-Crafts acylation. Moreover, the acylated products also undergo standard organic transformations including reduction and dehydration to readily form the desired vinyl monomers. This synthetic route is illustrated below for monomer 3:[13]

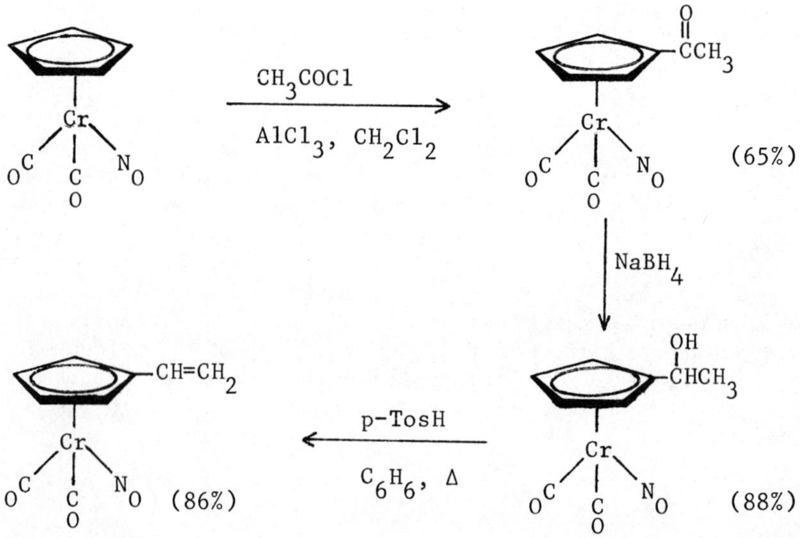

In connection with polymerization studies on monomers 1-3, it has been desirable to ascertain their Alfrey-Price "Q,e" values, as determined from relative reactivity ratios in radical-initiated copolymerizations with styrene. These values together with those for two organic monomers, styrene and 1,1-di(p-anisyl)ethylene, are tabulated below.

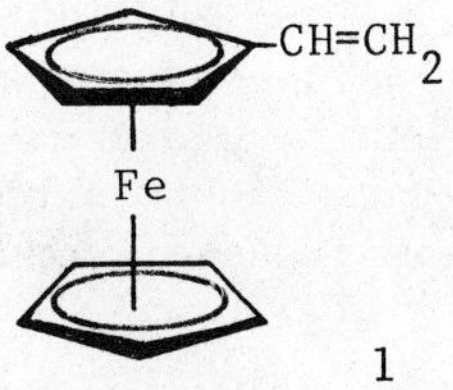

$e = -2.1$

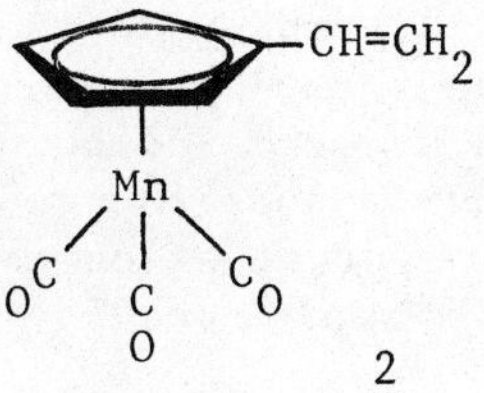

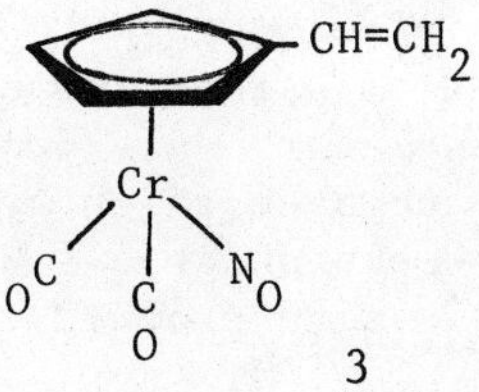

$Q = 2.3$

$e = -1.99$

$Q = 2.9$

$e = -1.98$

$Q = 3.1$

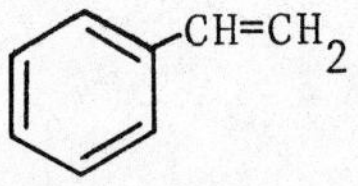

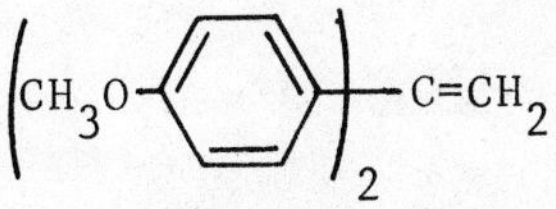

$e = -0.80$

$Q = 1.00$

$e = -1.96$

$Q = 1.46$

Since e is a measure of the electron richness of a given monomer in radical-initiated copolymerization reactions, it is clear that monomers 1-3 have exceptionally electron-rich vinyl groups, equal to or exceeding that in the highly electron-rich 1,1-di(p-anisyl)-ethylene. The Q values for 1-3 as determined in these studies likewise indicate a high degree of resonance interaction of the vinyl groups in 1-3 with the attached organometallic substituents.

The above and related studies on vinyl monomers 1-3 demonstrated that they possess unusual properties and behavior in polymerization reactions. However, in order to further explore this area of organometallic polymers, new synthetic routes to additional vinyl-containing organometallic monomers had to be developed, since only a very few η^5-cyclopentadienyl-metal compounds other than those mentioned above undergo electrophilic aromatic substitution.[15] Consequently, the development of alternate routes to vinyl-substituted η^5-cyclopentadienyl-metal compounds became a major goal of our research.

Several years ago, we found that cyclopentadienylsodium reagents containing aldehyde, ketone and ester substituents could readily be obtained from reactions of cyclopentadienylsodium with ethyl formate, ethyl acetate, and dimethyl carbonate respectively.[16] These organosodium reagents could then be utilized in the synthesis of cyclopentadienyl-transition metal compounds which likewise contained the functional substituent. One such compound was $(\eta^5\text{-}C_5H_4CHO)W(CO)_3CH_3$ (4), which could be obtained in 82% yield from a reaction between formylcyclopentadienylsodium and tungsten hexacarbonyl, followed by methylation of the resulting organotungsten anion by methyl iodide.[17] We subsequently found that 4 would undergo a Wittig-type reaction under phase transfer conditions to produce the vinyl analogue (5) in 80% yield.[17]

Extensive homopolymerization and copolymerization studies of 5 were carried out.[17,18] Homopolymerizations of 5 were sluggish under radical initiation, however, good yields of copolymers were obtained with acrylonitrile, methyl methacrylate, N-vinyl-2-pyrrolidone and styrene. From the latter copolymerizations, the e value for 5 was determined to be -1.98, whereas the Q value was 1.66. It is again quite clear that the vinyl group in the organotungsten monomer 5 is very electron-rich. Detailed chemical and spectral investigations of the homopolymerization process also showed that internal hydrogen atom abstraction from the $W\text{-}CH_3$ group followed by termination or further polymerization by $L_nW\text{-}CH_2\cdot$ addition to another vinyl group did not occur.

Although synthesis of the organotungsten vinyl monomer 5 via the formyl intermediate 4 proved to be successful, attempts to extend this sequence of reactions to vinyl monomers containing other transition metals were less successful, due primarily to variable, low yields during the Wittig reaction or related transformations.[19] We therefore set out to develop a more direct, general synthetic route to η^5-vinyl- as well as η^5-isopropenyl-cyclopentadienyl-metal monomers. We subsequently found that a reaction between 6-methylfulvene and lithium diisopropylamide in tetrahydrofuran (THF) solution at $25°C$ afforded the new organolithium reagent vinylcyclopentadienyllithium (6) in 80-90% yield.[20] In a similar manner, isopropenylcyclopentadienyllithium (7) could be obtained in high yield from a reaction between 6,6-dimethyl-fulvene and lithium diisopropylamide under analogous conditions.[20,21]

Both 6 and 7 are air-sensitive solids, as is cyclopentadienyl-lithium itself, and both have been characterized by their ^{1}H NMR spectra as recorded in THF-d_8 solution. Thus, the ^{1}H NMR spectrum of 6 exhibits two apparent triplets due to the non-equivalent cyclopentadienyl protons, as well as a typical ABX pattern for the terminal vinylic protons.

Both 6 and 7 reacted with a variety of organotransition metal compounds to afford the corresponding η^5-vinyl- and η^5-isopropenyl-metal monomers. For example, reactions between 6 and either molybdenum hexacarbonyl or tris(dimethylformamide)tricarbonyl-tungsten in refluxing THF gave η^5-vinylcyclopentadienyl monomers of these metals.[22] Nitrosylation of the intermediate metal carbonyl anions with N-methyl-N-nitroso-p-toluenesulfonamide afforded vinyl monomers 8 and 9, respectively. In a similar manner starting with 7, the isopropenyl analogues (10) and (11) could be obtained.

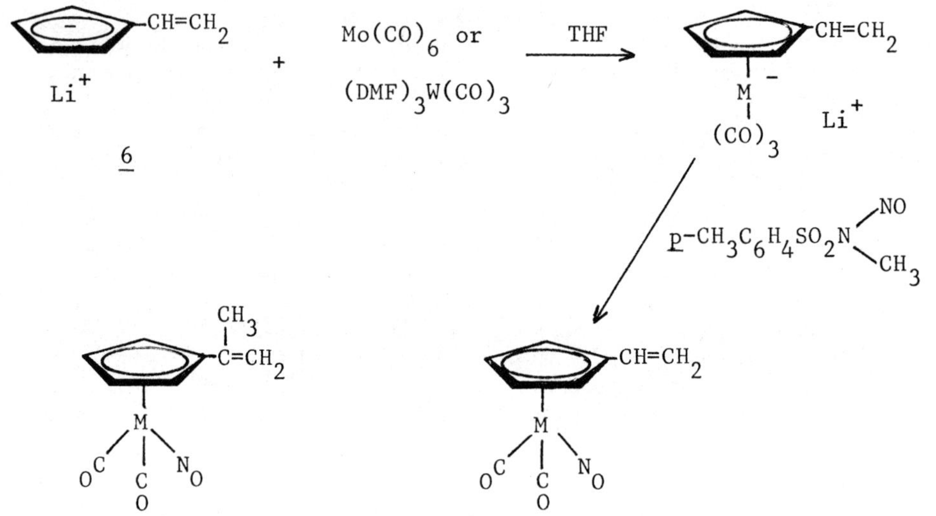

-CH=CH$_2$

Li$^+$

6

+

Mo(CO)$_6$ or

(DMF)$_3$W(CO)$_3$

$\xrightarrow{\text{THF}}$

-CH=CH$_2$

M$^-$

(CO)$_3$ Li$^+$

p-CH$_3$C$_6$H$_4$SO$_2$N$\diagdown$NO / CH$_3$

CH$_3$
|
-C=CH$_2$

M

OC   C   NO
     O

10   M = Mo (93%)

11   M = W (34%)

-CH=CH$_2$

M

OC   C   NO
     O

8   M = Mo (85%)

9   M = W (36%)

The new vinyl monomers containing molybdenum (**8**) and tungsten (**9**) are of special interest, since they cannot be obtained by electrophilic substitution routes involving the parent systems, as can the chromium-containing vinyl monomer **3**. Moreover, the availability of **3**, **8** and **9**, each containing a different group 6B transition metal, allows a comparison of the vinyl reactivity of these monomers as a function of the metal.

The recent use of polystryrene-bound (η^5-cyclopentadienyl)-dicarbonylcobalt as a catalyst[23-25] prompted us to synthesize monomers containing this metal. A reaction of **6** with an equimolar mixture of Co$_2$(CO)$_8$ and I$_2$ in THF at 25°C gave the organocobalt vinyl monomer (**12**) in low yield. A corresponding isopropenyl monomer containing cobalt could likewise be obtained starting with **7**.

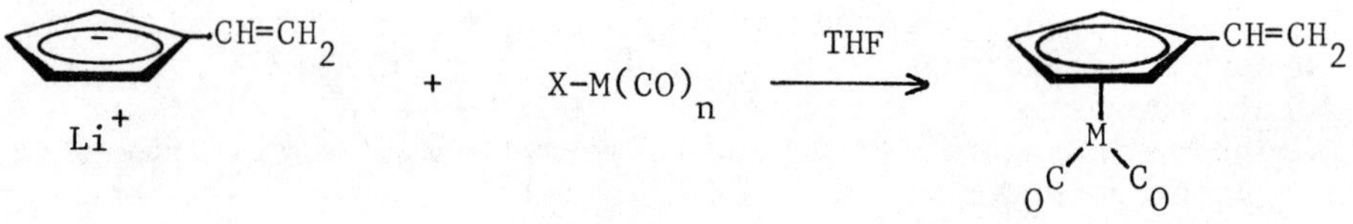

-CH=CH$_2$

Li$^+$

+

X-M(CO)$_n$

$\xrightarrow{\text{THF}}$

-CH=CH$_2$

M

OC   CO

Co$_2$(CO)$_8$ + I$_2$ **12**   M = Co (17%)

[Rh(CO)$_2$Cl]$_2$ **13**   M = Rh (47%)

Ir(CO)$_3$Cl   **14**   M = Ir (91%)

Organolithium reagent 6 is also a convenient precursor for the analogous rhodium (13) and (14) monomers. Thus, the reaction of 6 with [Rh(CO)$_2$Cl]$_2$ in THF at 25°C or with Ir(CO)$_3$Cl in refluxing hexane afforded the respective vinyl monomers 13 and 14.

Using reagents 6 and 7, we have likewise been able to synthesize the first organometallic monomers of copper. For example, reactions of 6 or 7 with [ClCu(PEt$_3$)]$_4$ gave monomers (15) and (16), respectively.

| 6 | R = H | | 15 | R = H (90%) |
| 7 | R = CH$_3$ | | 16 | R = CH$_3$ (66%) |

A reaction between 6 and (η^5-cyclopentadienyl)titanium trichloride has also provided the first vinyl monomer of titanium, (η^5-vinyl-cyclopentadienyl)(η^5-cyclopentadienyl)dichlorotitanium (17) in low yield.

6   17 (16%)

It is therefore evident that a wide range of cyclopentadienyl-metal compounds which contain potentially polymerizable vinyl substituents can be prepared by this new direct route involving organolithium reagents 6 and 7.

Because the vinyl monomers derived from intermediates 6 and 7 via the direct routes are relatively new, and have not yet been produced in large amounts, we have only recently begun to investigate their behavior under polymerization conditions. However, both homopolymerizations as well as copolymerizations of the

molybdenum-containing monomer 10 and the iridium-containing monomer
14 have been carried out under radical initiation conditions in
benzene (50-75°C).[20] Monomer 10 readily copolymerized with styrene
under these conditions. Copolymers (18) containing from 15 to 70
mol percent of 10 have been isolated with molecular weights of
ca. 3×10^4. Bulk polymerizations led to higher molecular weights.

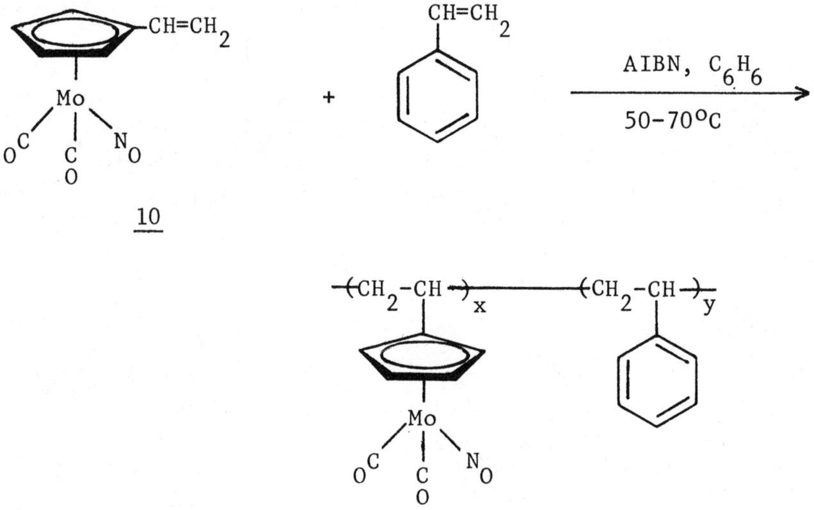

10

18

The Alfrey-Price value of e for 10, as determined from relative
reactivity ratios derived from styrene copolymerizations, was
-1.97, demonstrating that 10 is an exceptionally electron-rich
vinyl monomer like its chromium analogue 3 (e = -1.98). Moreover,
the fact that the e values for 10 and 3 are virtually identical
suggests that the identity of the metal did not effect the polymerization behavior when Cr was changed to Mo. Interestingly,
the Q values for both of these group 6B metal vinyl monomers are
3.1, consistent with substantial electron delocalization in both
α-cyclopentadienyl-metal radicals during the radical-initiated
polymerization process.

Radical-initiated homopolymerizations of 14 in benzene were
successful, and produced polymers (19) which contain over 50
weight percent iridium. These polymers after reduction by NaBHEt$_3$
were shown to be useful as catalysts for the hydrogenation of
1,5-cyclooctadiene at 100°C. Copolymerizations of 14 and styrene
have also been studied (AIBN, C$_6$H$_6$). The e value for iridium
monomer 14, calculated from this data, was -2.08, whereas Q = 4.1.
As in the case of all the other η^5-vinylcyclopentadienyl monomers
studied thus far, 14 has a very electron-rich vinyl group.

CATIONIC-INITIATED POLYMERIZATIONS

In contrast to the radical-initiated polymerization studies of organometallic monomers described above, very few instances of cationic polymerization have been reported in the literature. Kunitake et al.[8] first reported the cationic polymerization of vinylferrocene 1, and Korshak and Sosin[26] polymerized 1,1'-diisopropenylferrocene via cationic means. Since these early reports, this area of organometallic polymer chemistry has been relatively dormant.

As mentioned previously, we have been able to produce a wide range of η^5-cyclopentadienyl-metal monomers which contain isopropenyl units.[15,16,19,22,27] These new organometallic monomers would appear to be potentially attractive candidates for polymers produced under cationic initiation conditions. As a prelude to conducting further research in this uncharted area, however, we decided to initially study the cationic polymerization of isopropenylferrocene (20).

Although monomer 20 has been reported previously by several groups of investigators,[2,28-30] we found it necessary to develop a facile, dependable, high-yield synthesis for this compound. This route has involved the Friedel-Crafts acetylation of ferrocene to form acetylferrocene (21), followed by reaction of 21 with methylmagnesium iodide in THF and subsequent hydrolysis to form 2-ferrocenyl-2-propanol (22). Using a procedure developed earlier by Rausch and Siegel[31] for the dehydration of 1-ferrocenylethanol,

the carbinol 22 was dehydrated by pyrolysis-sublimation in a vacuum
sublimer at ca. 120°C. Monomer 20 was produced in good yield and
in very high purity by this technique. Elemental analysis and
spectroscopic data (^{1}H NMR, ^{13}C NMR, IR) were used to fully characterize the monomer.

     Homopolymerizations of 20 have been investigated using four
different cationic initiators, $BF_3 \cdot OEt_2$, $SnCl_4$, $AlCl_3$ and
Ph_3C^+ $SbCl_6^-$, employing methods similar to those used by Kunitake
et al.[8] Polymerization temperatures ranged from -78°C to ambient
(20°C), utilizing flame-dried Schlenk tubes which had been purged
with argon. The charge in each experiment was 2-4 g of monomer
in 20-40 ml of dry, deoxygenated solvent (methylene chloride, THF
or toluene). Initiator used was 2-5% by weight of monomer.

     At low temperatures (-78°C to -30°C), no polymerization
occurred. However, when polymerization with $BF_3 \cdot OEt_2$ or with $SnCl_4$
was initiated at 0°C for 1 h and the system was then allowed to
attain ambient temperature over a period of 24 h, low molecular
weight polymers, insoluble in methanol, were produced. With $AlCl_3$
and Ph_3C^+ $SbCl_6^-$ as initiators, unsatisfactory results were
obtained. The number average molecular weight ($\overline{M}_n$) of these homopolymers ranged from 1100 to 3900 as determined by vapor phase
osmometry using toluene as the solvent.

The resulting materials have been characterized by ^{1}H NMR and ^{13}C NMR spectrometry at ambient temperature and at 90°C, as well as by IR spectroscopy. The homopolymers of 20 were determined to have a structure (23) in which normal polymerization through the isopropenyl units has occurred. Fairly conclusive evidence for structure 23 was obtained from IR and ^{13}C NMR data. The IR spectra of the products exhibited two sharp absorptions near 1000 and 1100 cm^{-1}, indicative of the presence of unsubstituted cyclopentadienyl rings.[32] The ^{13}C NMR spectra showed similarities to spectra for predominantly syndiotactic polymethyl methacrylate,[33,34] suggesting a similar structure for 23. This feature was elicited for the triad of resonances observed for the methyl carbons in 23 at δ30.08, 30.29 and 30.64 ppm. Assignments for the quaternary carbon were not discernible. The spectroscopic data thus indicates that at ambient temperatures, the mode of propagation is via isopropenyl addition:

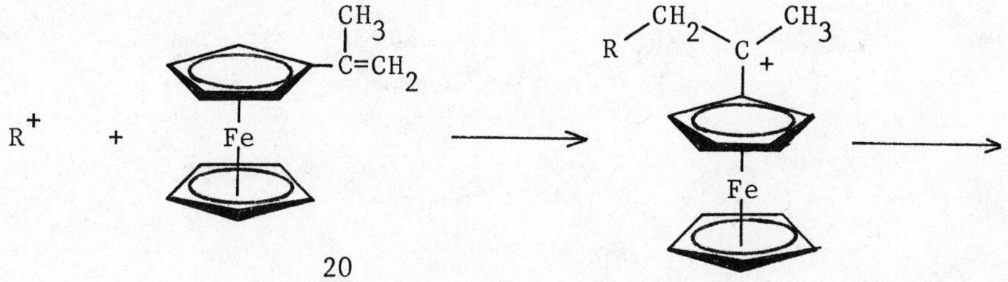

20

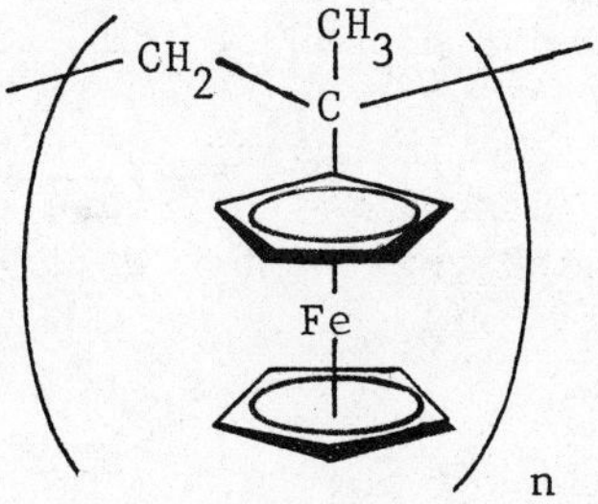

23

Initial studies concerning possible copolymerization reactions of 20 under cationic conditions have been undertaken. In attempted copolymerizations of 20 and α-methylstyrene (isopropenyl-benzene) at 0°C and under varying molar ratios of these monomers, traces of polymer formation were evident only at a 70/30 ratio. Since α-methylstyrene has a very low ceiling temperature, we decided to employ styrene as a comonomer under conditions as utilized with α-methylstyrene but conducted at ambient temperature. In a typical experiment, 20 and styrene in a 70/30 molar ratio in

methylene chloride were mixed together at ambient temperature and
polymerization was initiated by $BF_3 \cdot OEt_2$. After a reaction period
of 20 h, analysis of the resulting copolymer (24) ($\bar{M}_n$ = 2900,
Yield = 0.30%) via 250 MHz ^{1}H NMR indicated a composition of 27%
20 and 73% styrene units. This composition was also confirmed by
a total elemental analysis of the copolymer.

Both homo- and copolymerization reactions of 20 are therefore
very sluggish under the normal conditions employed for cationic
polymerization of other vinyl monomers. In fact, it appears that
carbenium ions derived from 20 act as inhibitors toward cationic
polymerization in many instances, and these results are most likely
due to the well-documented very high stabilities of α-ferrocenyl
carbenium ions.[35]

CONCLUSIONS

Several new and general synthetic routes to vinyl organometallic monomers have been developed in our laboratories. Many
of these monomers undergo radical-initiated homo- and copolymerizations to produce a variety of new metal-containing polymers.
The polymerization behavior of these monomers has now been defined
on a semi-quantitative basis. Isopropenylferrocene has been both
homo- and copolymerized using cationic initiation methods,
although these reactions are very sluggish under normal polymerization conditions.

ACKNOWLEDGEMENTS

Acknowledgement is made to the Donors of the Petroleum Research Fund, administered by the American Chemical Society for a grant to M.D.R., to the Army Research Office for a grant to C.U.P., and to a National Science Foundation grant to the Materials Research Laboratory, University of Massachusetts. All of these agencies have played important roles in developing the research programs described in this article.

REFERENCES

1. C. E. Carraher, Jr., J. E. Sheats and C. U. Pittman, Jr., "Organometallic Polymers," Academic Press, New York, 1978.

2. C. U. Pittman, Jr., Organomet. React. and Synth., 6, 1 (1977).

3. C. U. Pittman, Jr., "Polymer Supported Reactions in Organic Synthesis" (P. Hodge and D. C. Sherrington, Eds.), Wiley, New York, 1980.

4. F. S. Arimoto and A. C. Haven, Jr., J. Am. Chem. Soc., 77, 6295 (1955).

5. Y. Sasaki, L. L. Walker, E. L. Hurst and C. U. Pittman, Jr., J. Polym. Sci. Chem. Ed., 11, 1213 (1973).

6. M. H. George and G. F. Hayes, J. Polym. Sci. Chem. Ed., 13, 1049 (1975).

7. M. H. George and G. F. Hayes, J. Polym. Sci. Chem. Ed., 14, 475 (1976).

8. C. Aso, T. Kunitake and T. Nakashima, Macromol. Chem., 124, 232 (1969).

9. C. R. Simionescu, Macromol. Chem., 163, 59 (1973).

10. A. N. Nesmeyanov, K. N. Anisimov, N. E. Kolobova and I. B. Zlotina, Dokl. Akad. Nauk SSSR, 154, 391 (1964).

11. C. U. Pittman, Jr., and T. D. Rounsefell, Macromolecules, 9, 937 (1976).

12. C. U. Pittman, Jr., and T. D. Rounsefell, Macromolecules, 11, 1022 (1978).

13. E. A. Mintz, M. D. Rausch, B. H. Edwards, J. E. Sheats, T. D. Rounsefell and C. U. Pittman, Jr., J. Organomet. Chem., 137, 199 (1977).

14. C. U. Pittman, Jr., T. D. Rounsefell, E. A. Lewis, J. E.
     Sheats, B. H. Edwards, M. D. Rausch and E. A. Mintz.
     Macromolecules, 11, 560 (1978).

15. D. W. Macomber, W. P. Hart and M. D. Rausch, Adv. Organometal.
     Chem., 21, 1 (1982).

16. W. P. Hart, D. W. Macomber and M. D. Rausch, J. Am. Chem. Soc.,
     102, 1196 (1980).

17. D. W. Macomber, M. D. Rausch, T. V. Jayaraman, R. D. Priester,
     Jr., and C. U. Pittman, Jr., J. Organomet. Chem., 205,
     353 (1981).

18. C. U. Pittman, Jr., T. V. Jayaraman, R. D. Priester, Jr.,
     S. Spencer, M. D. Rausch and D. W. Macomber, Macromolecules,
     14, 237 (1981).

19. D. W. Macomber and M. D. Rausch, J. Organomet. Chem., in
     press.

20. D. W. Macomber, W. P. Hart, M. D. Rausch, R. D. Priester, Jr.,
     and C. U. Pittman, Jr., J. Am. Chem. Soc., 104, 884 (1982).

21. An earlier synthesis of 7 from 6,6-dimethylfulvene and
     potassium t-butoxide has been reported: J. Hine and D. B.
     Knight, J. Org. Chem., 35, 3946 (1970).

22. D. W. Macomber and M. D. Rausch, J. Organomet. Chem., 250,
     311 (1983).

23. G. Bubitosa, M. Boldt and H. H. Brinzinger, J. Am. Chem. Soc.,
     99, 5174 (1977).

24. B. H, Chang, R. H. Grubbs and C. H. Brubaker, J. Organomet.
     Chem., 172, 81 (1979).

25. P. Perkins and K. P. C. Vollhardt, J. Am. Chem. Soc., 101,
     3985 (1979).

26. S. L. Sosin and V. V. Korshak, Dokl. Akad. Nauk SSSR, 179,
     1124 (1968).

27. D. W. Macomber and M. D. Rausch, Organometallics, in
     press.

28. K. L. Rinehart, Jr., P. A. Kittle and A. F. Ellis,
     J. Am. Chem. Soc., 82, 2083 (1960).

29. A. F. Ellis, Dissertation Abstr., 24, 510 (1963).

30. W. P. Fitzgerald, Jr., Dissertation Abstr., 24, 2687 (1964).

31. M. D. Rausch and A. Siegel, J. Organomet. Chem., 11, 317
     (1968).

32. M. Rosenblum and R. B. Woodward, J. Am. Chem. Soc., 80,
     5443 (1958).

33. L. F. Johnson, F. Heatley and F. A. Bovey, Macromolecules,
     3, 175 (1970).

34. J. C. Randall, "Polymer Sequence Determination. Carbon-13
     NMR Method," Academic Press, New York, 1977.

35. W. E. Watts, "Comprehensive Organometallic Chemistry" (Eds.
     G. Wilkinson, F. G. A. Stone and E. Abel), Pergamon Press,
     New York, 1982, Vol. 8, p. 1052.

MODELS FOR HIGHLY PHENYLATED TRANSITION METAL-CONTAINING POLYMERS:

DERIVATIVES OF THE PENTAPHENYLCYCLOPENTADIENYL LIGAND

D. W. Slocum*, S. Duraj and M. Matusz

Department of Chemistry and Biochemistry
Southern Illinois at Carbondale
Carbondale, IL 62901

J. L. Cmarik, K. M. Simpson and D. A. Owen*

Department of Chemistry
Murray State University
Murray, KY 42071

INTRODUCTION

Numerous η^5-cyclopentadienyl derivatives of most of the transition metals are known.[1] Over the past decade attention has been directed to transition metal complexes containing peralkylated η^5-cyclopentadienyl groups, particularly those involving the pentamethylcyclopentadienyl ligand.[2-6] Such complexes usually possess, when compared to the relevant cyclopentadienyl analogs, properties reflecting their increased bulk and steric character, e.g., enhanced oxidative, reductive and thermal stability, coupled with reduced volatility. These peralkylated complexes have proved to be more easily isolable and their chemical reactivity and catalytic activity more amenable to study.

Significant enhancement of the relatively modest gains in stability and reactivity of these peralkylated complexes might be anticipated by use of perarylated cyclopentadienyl systems, with their much higher molecular weights and reduced volatility and solubility. Surprisingly, little of the chemistry of such systems has been reported. Examples of complexes containing the parent ligand, i.e., the pentaphenylcyclopentadienyl group, are few. Those described in the literature include red-brown $Ph_5C_5Fe(CO)_2Br$, available from $Fe(CO)_5$ and Ph_5C_5Br;[7] $(Ph_5C_5)_2Mo$, a low-yield product of the reaction of $Mo(CO)_6$ and diphenylacetylene;[8] (μ-PhC≡CPh)-$(Ph_5C_5Pd)_2$, from $Pd(O_2CCH_3)_2$ and diphenylacetylene;[9] and $(Ph_5C_5)_2Ni$ from $Ni(O)$ and the Ph_5C_5 radical.[10] Recently, a series of Ph_5C_5Pd (η^3-allyl) complexes have been described[11] as well as an initial disclosure of a synthesis of $(C_5Ph_5)_2Sn$.[12] Each of these last two disclosures is noteworthy in that they involved syntheses beginning with the pentaphenylcyclopentadienide anion.[13]

Contributions of the pentaphenylcyclopentadienyl moiety to the properties of appropriate complexes have been investigated. A cyclic voltammetry study has led to the conclusion that the elec-tron-withdrawing ability of the phenyl rings in such complexes renders low oxidation states more accessible by virtue of increased potentials of the redox couples.[14] Both Geiger[14] and Zuckermann[12] have observed that this ligand imparts greater kinetic stability to these complexes. In many respects these pronouncements may be ap-plicable to another recently "perderivatized" cyclopentadienyl ligand, namely, the pentacarboxymethylcyclopentadienyl group.[15]

Our discovery of a three step, high-yield economical route to pentaphenylcyclopentadiene from inexpensive, commercially available starting materials coupled with a new set of solvents and condi-tions for handling its lithium derivative[13] has afforded for the first time the opportunity to explore in depth the extraordinary chemistry of this ligand and its transition metal derivatives. This ligand is in fact easier to prepare than C_5Me_5H. Appropriate to this discussion the pentaphenylcyclopentadienyl group offers the potential of several modes of complexation and derivatization of transition metals thereby affording the opportunity of cross-linking an existing complex to form polymers. In addition syn-thesis of substituted organic derivatives, i.e., vinyl, alkynyl, etc. of the phenyl or cyclopentadienyl rings in such complexes suggests that highly phenylated transition metal cyclopentadienyl polymers can be formed by use of essentially conventional poly-merization techniques.

Prior systems studied by our consortium include vinyl and alkynyl metallocenes,[16] which have served a monomers and co-monomers for pendant group metallocene polymers[17] and tetra-cyclones and tetraaryl borates which have served as model com-pounds for, respectively, highly phenylated diene and arene transi-tion metal polymers.[18] The present ligand serves as a model for perphenylated η^5-cyclopentadienyl-complexed transition metal poly-mers. Polymers from such complexes will avoid one of the major drawbacks of more conventionally ligated polymers, vis., the gradual leaching of metal from the polymer complex.

I. Ligand Preparation

Our synthesis of pentaphenylcyclopentadiene and its corresponding bromide is shown in Scheme I. This synthesis is suitable for scale up to >100 g quantities. This relatively acidic diene reacts readily with n-butyllithium in a variety of solvents (tetrahydrofuran, glyme, diglyme, di-n-butyl ether, toluene and xylene) to form the colorless lithium pentaphenylcyclopentadienide[13] ion which is stable under a nitrogen atmosphere and is moderately air- and water-stable.

II. Half-Sandwich Complexes

The chemistry of yellow 1,2,3,4,5-pentaphenyl-1,3-cyclopentadiene and the colorless lithio salt was investigated by the formation of the half-sandwich complexes of several transition metals. Reaction of the lithio derivative with group VI metal hexacarbonyls gave the corresponding, yellow to orange salts (I a-c). These could be isolated as their quaternary ammonium salts (equation 1) or converted directly with iodomethane at reflux into the yellow σ-methyl derivatives (IIa and IIb) (equation 2):

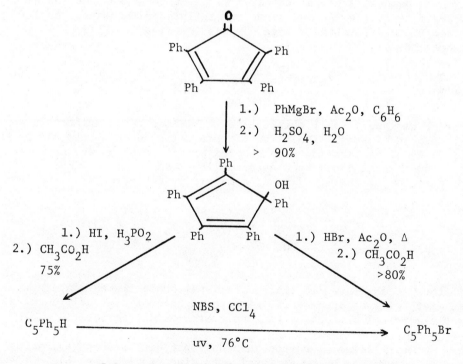

Scheme I: Synthesis of C_5Ph_5H and C_5Ph_5Br

$$M(CO)_6 + LiC_5Ph_5 \xrightarrow[16h \ (-3 \ CO)]{Glyme, \ 85°C} [C_5Ph_5M(CO)_3]^-Li^+$$

Ia, M = Mo

Ib, M = W

Ic, M = Cr

$$\xrightarrow[\substack{CH_2Cl_2 \\ (-LiBr)}]{(CH_3CH_2)_4N^+Br^-} [C_5Ph_5M(CO)_3]^-[(CH_3CH_2)_4N]^+ \quad (1)$$

M = Cr, Mo, W

$$[C_5Ph_5M(CO)_3]^-Li^+ + CH_3I \xrightarrow[LiI \ (-LiBr)]{Glyme, \ \Delta, \ 2h} C_5Ph_5M(CO)_3CH_3 \quad (2)$$

IIa, M = Mo   63%

IIb, M = W   82%

An analogous iron complex, $(C_5Ph_5)Fe(CO)_2CH_3$ (III) was prepared in similar fashion from iron pentacarbonyl in 67% yield.

To demonstrate that the aryl rings in these systems can also be complexed, the neutral diene was refluxed with hexacarbonyl chromium to afford the orange η^6-phenyl derivative, III (equation 3):

$$C_5Ph_5H + Cr(CO)_6 \xrightarrow[\substack{160°C, \ 5h \\ (-3CO)}]{diglyme} C_5Ph_4H(\eta^6\text{-}C_6H_5)Cr(CO)_3 \quad (3)$$

III

The ^{1}H-nmr spectrum of this complex exhibited two distinct aromatic proton signals. Isolation of this compound suggested the possibility of preparation of a complex simultaneous bonded in both η^5- and η^6- fashion. Such a possibility was realized with the preparation of the mixed metal complex, IV (equation 4):

$$\begin{array}{c} C_5H_5FeC_5Ph_5 \\ + \\ (OC)_3Mo(NCCH_3)_3 \end{array} \xrightarrow[\substack{85°, \ 9h \\ (-3 \ CH_3CN)}]{Glyme} C_5H_5FeC_5Ph_4(\eta^6\text{-}C_6H_5)Mo(CO)_3 \quad (4)$$

It thus seems plausible that multi-metal cross-linked complex polymers will be preparable, with the myriad of physical, chemical and electrical properties such polymers would offer.

Group VIII metal carbonyls also react with pentaphenylcyclopentadiene or its bromo derivative to give stable, isolable derivatives. Reaction of bromopentaphenylcyclopentadiene with a neutral transition metal precursor afforded a stable, non-volatile pale yellow ruthenium derivative, VIa, (equation 5):

$$3 \ C_5Ph_5Br + Ru_3(CO)_{12} \xrightarrow[(-6 \ CO)]{\Delta} 3 \ C_5Ph_5Ru(CO)_2Br \qquad (5)$$

$$VIa, \ 66\%$$

In analogous fashion, the corresponding Fe derivative, $C_5Ph_5Fe(CO)_2Br$ (VIb) was prepared. Also prepared in this manner from anionic cobalt carbonyl precursors was moderately stable, orange, cobalt derivative, VII (equation 6):

$$3 \ C_5Ph_5Br + Co(CO)_4^- \xrightarrow[(-Br^-, \ -2CO)]{THF, \ 0-20°C} C_5Ph_5Co(CO)_2 \qquad (6)$$

$$VII, \ 35\%$$

A combination of analytical, infrared, ^{1}H and ^{13}C nmr spectra strongly support the formulations and structures of compounds IIa, IIb, III-VII; in addition, mass spectroscopy at elevated temperatures and low impact electron energies not only pinpointed all molecular ions but also provided much information about the fragmentation of these species at elevated temperatures. In every case, save that of compound V which did not volatilize, sequential loss of CO groups with formation of the stable $C_5Ph_5M^+$ ion was observed. TABLE I provides the ^{13}C spectra of complexes Ia-c, IIa,b, III and VIa,b. The ring carbon shifts recorded here are similar to those observed for "free pentaphenylcyclopentadienyl" in its alkali metal salts.[13]

III. Sandwich Complexes

The preparation of several stable, nonvolatile open-faced η^5- and η^6-pentaphenylcyclopentadienyl complexes encouraged the attempt to synthesize the even more stable bis-(pentaphenylcyclopentadienyl) transition metal complexes which would be more easily incorporated into polymer systems or be more easily adsorbed on suitable (e.g. graphite) polymer supports. However, attempts at preparation of mixed (sym pentaphenyl) and symmetrical decaphenyl-metallocenes by standard procedures in ethereal solvents failed to give the desired $C_5H_5M(C_5Ph_5)$ or $(C_5Ph_5)_2M$ derivatives. In fact, the "high steric volume" and "lower base strength" postulated for the pentaphenylcyclopentadienyl anion had been invoked by other workers who have apparently obtained similar results.[7,13]

The use of a modified experimental procedure employing a much less basic solvent system allowed preparation of several noteworthy examples of such complexes. Reaction of titanium (or zirconium) tetrachloride with LiC_5Ph_5 afforded

TABLE I

^{13}C NMR Spectra of Pentaphenylcyclopentadienyl
Transition Metal Carbonyl Complexes[a]

Complex	$-\underline{C}O$	$-\underline{C}_6H_5$	$M\underline{C}_5Ph_5$	$-\underline{C}H_3$
Ic [b]	243.87	136.18, 132.4 126.3, 125.1	103.77	--
Ia [c]	236.6	137.3, 134.1 127.0, 125.6	111.79	--
Ib [b]	226.45	134.9, 132.9 126.7, 125.6	108.66	--
IIa	227.25	132.2, 131.8 127.6, 127.3	113.0	-5.26
IIb	217.5	132.4, 131.3 127.8, 127.6	111.7	-17.3
III	217.6	132.0, 131.8 127.6, 127.3	101.9	-7.7
Va	196.6	132.3, 129.6 128.4, 127.8	106.4	--
Vb	215	132.6, 130.0 129.8, 128.5	100.8	--

[a] Chemical Shifts in ppm with TMS as internal standard.
Solvent is $CDCl_3$ unless otherwise specified.

[b] in DMSO $-d_6$

[c] in C_6D_6, THF mixture

the corresponding deep red-violet titanium (VIIIa) and pale red-
orange zirconium (VIIIb) derivatives (equation 7).

$$MCl_4 + 2\ C_5Ph_5Li \xrightarrow[\text{6h (-2 LiCl)}]{\text{xylenes, 50-135°}} (C_5Ph_5)_2\ M\ Cl_2 \quad (7)$$

VIIIa, M = Ti   27%

VIIIb, M = Zr   40%

Isolation of these complexes and certain of the ones described below casts doubt on the touted "steric control" of their formation.
More likely it is the favorable basicity of the weakly basic pentaphenylcyclopentadienide ion in nonionic, inert solvents that
allows for their production in such solvents and not in the more
basic ethers.

These rationalizations prompted reaction of cyclopentadienyliron dicarbonyl iodide, pentamethylcyclopentadienyliron dicarbonyl
iodide and the aforementioned pentaphenylcyclopentadienyliron dicarbonyl bromide' with lithio pentaphenylcyclopentadienide ion
under similar conditions with the resulting formation, for the
iron compounds, of sym-pentaphenylferrocenes IXa and Xa and pale
yellow decaphenylferrocene, XIa, respectively (equations 8, 9 and
10).

$$C_5H_5M(CO)_2I + LiC_5Ph_5 \xrightarrow[\substack{18\text{-}24h \\ (-LiI, -2CO)}]{\text{xylenes, } 135°} C_5H_5MC_5Ph_5 \qquad (8)$$

$$\underset{\sim\sim\sim}{IXa}, \text{ M = Fe, } 55\%$$

$$\underset{\sim\sim\sim}{IXb}, \text{ M = Ru, } 18\%$$

$$C_5Me_5M(CO)_2I + LiC_5Ph_5 \xrightarrow[\substack{18\text{-}24 \text{ h} \\ (-LiI, -2CO)}]{\text{xylenes, } 135°} C_5Me_5MC_5Ph_5 \qquad (9)$$

$$\underset{\sim\sim\sim}{Xa}, \text{ M = Fe}$$

$$\underset{\sim\sim}{Xb}, \text{ M = Ru}$$

$$C_5Ph_5M(CO)_2Br + LiC_5Ph_5 \xrightarrow[\substack{18\text{-}24 \text{ h} \\ (-LiI, -2CO)}]{\text{xylenes, } 135°} (C_5Ph_5)_2M \qquad (10)$$

$$X \quad \underset{\sim\sim\sim}{XIa}, \text{ M = Fe, } 40\%$$

$$\underset{\sim\sim\sim}{XIb}, \text{ M = Ru, } 69\%$$

These compounds all possess to an extraordinary degree the anticipated thermal and oxidative stability associated with metallocenes
in general. They are unchanged in air at 315°C and volatilize
under mass spectrometric conditions only at 250-300°C. Ir, nmr
and mass spectra and elemental analytical data confirm their formulations (TABLE II). The ^{1}H nmr spectrum of each in CHCl3 is
surprisingly simple, with singlet and near singlet resonances
being recorded for all protons in these complexes. One might have
anticipated more structure to these spectra if a hindrance to rotation about the ring metal bond were to be manifested. The slight
upfield shift of the resonances of all phenyl hydrogens is consistent with some shielding from the neighboring.π-clouds of adjacent
phenyl rings, all of which are apparently equivalently staggered
on the ^{1}H nmr time scale. Low temperature studies of these complexes are not possible due to their very low solubilities.

Conventional aromatic ring type ferrocene chemistry can be
carried out on the unsubstituted ring in such complexes. For
example, three derivatives of sym-pentaphenylferrocene have been
prepared as illustrated in equation 11:

TABLE II

Physical and Spectral Data of
Metallocene Complexes of Iron and Ruthenium

Complex	$-\underline{C}_6H_5$	$\underline{C}_5Ph_5$	$\underline{C}_5(CH_3)_5$	$(\underline{C}H_3)_5$	Mass Spectra m/e
IX a	136.0, 132.3 127.0, 126.0	75.1	87.6	--	566
X a	135.0, 133.0 126.4, 125.8	80.7	86.2	9.8	636
XI a	--	--	--	--	946
IX b	135.6, 132.5 126.9, 126.0	78.4	93.8	--	612
X b	134.8, 132.4 127.0, 125.7	85.5	91.4	9.7	682
XI b	--	--	--	--	992

$$C_5Ph_5FeC_5H_5 \xrightarrow[\text{BF}_3, \ 100°, \ 1 \ \text{hr}]{(CH_3CO)_2O} C_5Ph_5FeC_5H_4COCH_3$$

$$\xrightarrow[\text{H}_2\text{O, THF, 15 min}]{\text{NaBH}_4, \ \text{NaHCO}_3} C_5Ph_5FeC_5H_4CHOHCH_3 \xrightarrow[\text{alumina}]{\Delta} \qquad (11)$$

$$C_5Ph_5FeC_5H_4CH{=}CH_2$$

These observations offer the distinct possibility of the preparation of polymers and copolymers from the vinyl monomer and any other such monomers.

CONCLUSIONS

Complexes of several types of the now routinely available pentaphenylcyclopentadienyl ligand have been prepared for the first time from the ligand itself. Complexes of Cr, Mo, W, Co, Fe, Ru, Ti and Zr have been isolated and characterized. One example of an η^6-ligated species as well as an η^5-, η^6- mixed-metal ligated species have also been prepared. Although these complexes possess extraordinary molecular weights, they have proven to be sufficiently air- and water-stable and volatile and soluble enough for analysis by conventional spectral techniques. As anticipated, most of these complexes possess high thermal and oxidative stability with $C_5Ph_5Co(CO)_2$ (VII) proving a notable exception. All of these results are encouraging in our attempts to evaluate the conditions and limitations of production of highly aromatic, stable macromolecules containing arylated ligands.

Acknowledgements:

Many of the initial experiments involving preparation of pentaphenylcyclopentadiene were performed by Mr. Steve Johnson. Thanks are due Susan Caputo of the University of Pittsburgh for the typing of this manuscript. DWS is grateful for the support accorded him by the University of Pittsburgh and Carnegie-Mellon University during the tenure of his joint lectureship.

Addendum:

Description of the preparation of pentaphenylcyclopentadiene has been submitted to Organic Syntheses.

REFERENCES

1. "Comprehensive Organometallic Chemistry", G. Wilkinson,
    F. G. A. Stone and E. W. Abel, editors, Pergamon Press,
    New York, N.Y. (1982).
2. P. T. Wolczanski and J. E. Bercaw, Acc. Chem. Res., 13,
    121 (1980).
3. T. J. Marks, Science, 217, 989 (1982).
4. T. J. Marks, J. M. Manoiguez, P. J. Fagan, W. V. Day,
    C. S. Day and S. H. Vollmer, ACS Sym. Series, No. 3,
    131 (1980).
5. P. M. Maitlis, Acc. Chem. Res., 11, 301 (1978).
6. A. L. Wayda and W. J. Evans, Inorg. Chem., 19, 2190 (1980).
7. S. McVey and P. L. Pauson, J. Chem. Soc., 4312 (1965).
8. W. Hubel and R. Mérényi, J. Organometal. Chem., 2, 213 (1964).
9. T. R. Jack, C. J. May and J. Powell, J. Am. Chem. Soc., 99,
    4707 (1977).
10. A. Schott, H. Schott, G. Wilke, J. Brandt, H. Hoberg and
    E. G. Hoffmann, Liebigs Ann. Chem., 508 (1973).
11. J. Powell and N. I. Dowling, Organometallics, 2, 1742 (1983).
12. M. J. Heeg, C. Janiak and J. J. Zuckerman, personal
    communication.
13. R. Zhang, M. Tsutsui and D. E. Bergbreiter, J. Organometal.
    Chem., 229, 109 (1982).
14. K. Broadley, G. A. Lane, N. G. Connelly and W. E. Geiger,
    J. Am. Chem. Soc., 105, 2486 (1983).
15. M. I. Bruce, J. R. Rodgers and J. K. Walton, Chem. Commun.,
    1253 (1981).
16. D. W. Slocum, M. D. Rausch and A. Siegel in Organometallic
    Polymers, C. U. Pittman, C. E. Carraher and J. Sheats, eds.,
    Academic Press, New York, N.Y. (1978), pp. 39-51.
17. cf. for example C. U. Pittman in Organometallic Reactions
    and Synthesis, vol. 6, E. I. Becker and M. Tsutsui, eds.,
    Plenum Press, New York, N.Y. (1977), p. 1.
18. D. W. Slocum, B. Conway, M. Hodgman, K. Kuchel, M. Moronski,
    R. Noble, K. Webber, S. Duraj, A. Siegel and D. A. Owen,
    J. Macromol. Sci.-Chem., A16 (1), 357 (1981).

FERROCENE POLYMERS

Cr.Simionescu, Tatiana Lixandru, Lucia Tataru,
I.Mazilu, M.Vata and D. Scutaru

Department of Organic and Macromolecular
Chemistry, Polytechnic Institute, Jassy
Romania

ABSTRACT

The study of some homo- and heterochain polymers
with ferrocenyl residues either in the backbone or as a
substituent, synthesized by polymerization, polycondensation or reactions on polymers such as cellulose,
poly(vinyl alcohol) and polyvinyl-pyrrolidone is
reported.
The following monomers were used: ethynylferrocene, 1-chloro-1' -ethynylferrocene, p-ferrocenylsty-
rene, p-ferrocenylacetophenone, 1,1' -diacetylferro-
cene, 1,1' - bis [-(2-furyl)acryloyl]ferrocene.
The 4,4' -diaminodiphenylether, 4,4' -diamino-
diphenylthioether, 4,4' -diamino-2,2' -dinitrodi-
phenyldisulplhide were used as diamino components.
Polymers with linear or tridimensional structure
showing semiconducting properties and a rather good
thermal stability were obtained. Some of them were
found to catalyse the polymerization of chloroformylated vinylic esters.

INTRODUCTION

The homo- and heterochain polymers containing
ferrocene units in the main chain or as a substituent
exhibit certain particular electric and magnetic
properties. They were used as accelerators in redox

systems, as catalysts in polymerization and dehydration
reactions as well as thermo- and photo-sensitizers etc.
    In the last years a particular attention was paid
to the biologically active ferrocene derivatives since
due to their low toxicity and easy metabolization they
can be used as an iron source for the body.
    In the present paper a part of our results concern-
ing the synthesis and characterization of some ferrocene polymers obtained by polymerization, polycondensation and reactions on polymers are reported.

RESULTS AND DISCUSSION

    The polymers containing ferrocene in the backbone
were obtained by polycondensation reactions while those
with ferrocene residues as substituent by both polycondensation and polymerization reactions as well as by
reactions on polymers.

Polymers with Ferrocenyl Residues in the Backbone

    Polyferrocenylazomethines Few papers are reported
concerning the polyazines (1-4) and polyazomethines
(2,5-7) obtained by polycondensation of dicarbonyl-
ferrocene compounds with various amines.
    Starting from the fact that the resonance effects
in ferrocene are not transmitted through the central
iron atom and the conjugation extension is limited to
the length of the system

The polycondensation of diacetylferrocene (DiAcFc) and
1,1' -bis [β-(2'furyl)acryloyl] ferrocene (DiAcFcF) was
carried out with aromatic diamines of benzidine type,
where Y is an oxygen atom, sulphur, a disulphide bridge
or is absent(8). The influence of Y group nature on
the conversion and polymer properties was followed.
The reaction conditions and some characteristics of the
obtained macromolecular compounds are listed in Table
1.
    Polyferrocenyleniminoimides were obtained by
condensing diacetylferrocene and 1,1' -bis[
-(2-furyl)acryloyl] ferrocene with biuret (9). The
reaction conditions are given in Table 2.

Macromolecular Compounds with Ferrocenyl Substituent

Polymers obtained by polymerization   Few papers, reported only during the last decade, concern the polymerization and co-polymerization of ferrocenylacetylene. Thus Rosenblum (10) and Schlogl (11, 12) tried the thermal and catalytic polymerization obtaining a cyclic trimer only.

We obtained polyferrocenylacetylene by polymerizing ferrocenylacetylene and its derivatives in the presence of benzoyl and lauroyl peroxides, triiopropylboron and complex catalysts of $P(C_6H_5)_3$ $_2$ NiX_2 type (13,14).

Table 1   Polycondensation of DiAcFc and DiAcFcF
                         with Aromatic Diamines

Type of Polymer		Reaction Conditions			Conversion %	
		Temperature ($^\circ$C)	Time (h)	Soluble	Insoluble	Total
A_1		120	20	17.40	15.20	32.60
A_1	(b)	150	21	--	52.60	52.60
A_2	(b)	150	21	--	88.00	88.00
B_1		120	10	6.38	50.21	56.59
B_1	(c)	120	10	14.80	4.60	19.40
B_2		120	10	14.50	18.10	32.60
C_1		120	10	14.50	9.16	23.66
D_1		120	10	13.70	24.50	38.20
D_1	(c)	120	10	4.08	41.60	45.68
D_1	(d)	120	10	2.60	23.70	26.30
D_1		120	20	--	41.60	41.60
D_1		120	35	--	68.80	68.80
D_1	(b)	150	23	--	90.16	90.16
D_2		120	10	22.21	48.80	71.01
D_2		120	20	--	69.10	69.10
D_2		120	35	--	85.30	85.30
D_2	(b)	200	20	--	51.00	51.00

(a) catalyst $ZnCl_2$, dicabonylic/diaminic component mole ratio 1/1; A_1 = DiAcFc + benzidene, A_2 = DiAcFcF + Benzidine; B_1 =DiAcFc + 4,4' -diaminodiphenylether, B_2 = DiAcFcF = 4,4' - diaminodiphenylether; C_1 = DiAcFc + 4,4' -diaminodipenylsulphide; D_1 = DiAcFc + 4,4' -diamino-2,2' -dinitrodiphenyldisulphide; (b) reaction in melt; (c) p-toluenesulphonic acid as catalyst; (d) dicarbonylic component/diaminic component mole ratio 1/2

Concomitantly Korshak (15, 16) and Nakashima (17)
obtained polymers along with the cyclic trimer.
Pittman accomplished the polymerization in the presence
of AIBN and RhCl$_2$ (18). Copolymers of ferrocenylacetylene with phenylacetylene (19), isobutene (20), and
isoprene (21) have also been obtained.

Finally, the p-ferrocenylphenylacetylene was
polymerized (22) under the same conditions as
ferrocenylacetylene.

The thermal polymerization of α-chloro-β-formyl-p
-ferrocenylstyrene (ClFFcSt) leads to infusible and
insoluble products, copolymers containing p-ferro-
cenylphenylacetylenic residues.

Comparative data for radical polymerization of
ethynylferrocene, 1-chloro-1' -ethynylferrocene
(Cl-EFc) and of α-chloro-β-formyl-p-ferrocenylstyrene
are listed in Table 3.

Table 2   Polycondensation of DiAcFc and DiAcFcF with
             Biuret

Dicarbonylic component	Temperature (°C)	Solvent	Conversion (%)		
			Soluble	Insoluble	Total
DiAcFc	140	CH$_3$COOH	–	8.1	8.1
DiAcFcF	140	CH$_3$COOH	–	77.9	77.9
DiAcFc	120	Dioxan	–	24.3	24.3
DiAcFcF	120	Dioxan	52.4	26.0	78.4
DiAcFc	120	DMSO	–	78.8	78.8
DiAcFc	90	DMSO	39.4	2.7	42.1
DiAcFcF	90	DMSO	70.3	8.3	78.6

Dicarbonylic component/biuret mole ratio 1/1; reaction
duration 20 hours

Table 3   Radical Polymerization of Ethynylferrocene (EFc), 1-Chloro-1' -ethynylferrocene and α-Cnloro-β-formyl-p-ferrocenylstyrene

Monomer	Initiator		Temp.	Conversion
	Type	Conc.(%)	(°C)	(%)
EFc	PB	1.00	140	14.83 (a)
EFc	PB	1.00	190	91.67 (b)
EFc	PL	1.00	170	40.20 (b)
EFc	PB	1.00	230	94.15 (b)
EFc	PL	1.00	200	76.76
EFc	$B(i-C_3H_7)_3$	5.32	190	25.87 (a)
EFc	$B(i-C_3H_7)_3$	2.72	230	29.46 (a)
Cl-EFc	PB	1.00	140	11.46 (a)
Cl-EFc	PB	1.00	190	41.50 (b)
Cl-FFcSt	-	-	180	94.70
Cl-FFcSt	-	-	200	95.00
Cl-FfcSt	-	-	230	95.20

(a) benzene soluble polymer; (b) benzene insoluble polymer; PB = benzoyl peroxide, PL = lauroyl peroxide

Polymers obtained by polycondensation Though by this type of reactions polymers of a regular structure are obtained, homo- and heteropolycondensation of acetyl and diacetylferrocene are less investigated. Using various ferrocene monomers, polyenes (17,18,24-27), polycyclotrimers (11,12), polyazines (1, 2) and poly-Schiff bases were obtained (2).

We synthesized polyferrocenylphenylacetylenes (28) by polycondensing the p-ferrocenylacetophenone in the presence of metallic halides. The study of catalytic reactivity in the polycondensation reactions indicated the following succession: $HgCl_2 > SnCl_2 > AlCl_3 > FeCl_3 > CuCl_2 > MgCl_2 > CoCl_2 > NiCl_2 > ZnCl_2 > CdCl_2$

The metallic salts increase the reactivity of carbonyl compounds, labilizing the hydrogen in α position by the formation of a coordination complex (29,30).

Ferrocene polymers obtained by reactions on polymers Sensoni and Sigmund (31) obtained poly-p-ferrocenyl-styrene by reactions on polystyrene. We introduced the ferrocenyl residue in cellulose and poly(vinyl alcohol) molecules by esterification with ferrocenecarboxylic acid in the presence of triphenylphosphite and pyridine (32).

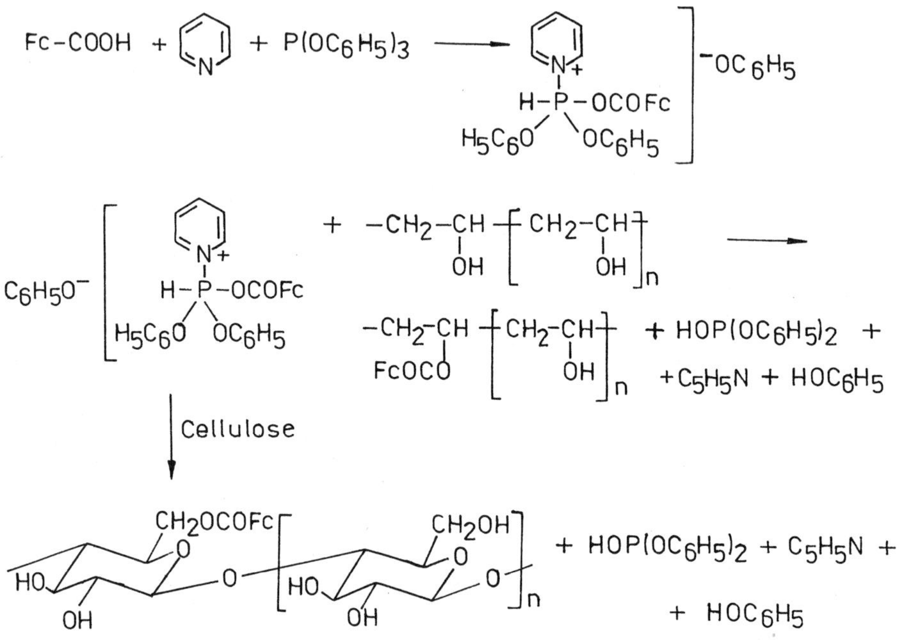

Figure 1   Reaction scheme for Esterification of
            Poly(vinyl alcohol) and Cellulose with
            Ferrocenecarboxylic Acid

The samples taken at various reaction durations
indicate the reaction time does not influence the
frequency of ferrocenyl residue occurrence in the chain.
The "n" values for PVA and cellulose are 64 and 40,
respectively.
The modified poly-N-vinylpyrrolidone was obtained by
the condensation with ferrocene carboxaldehyde in
pyridine, using piperidine as catalyst at 100°C under
inert atmosphere.

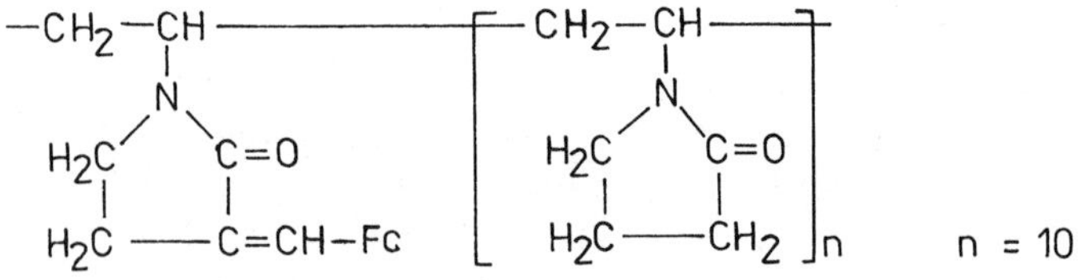

Figure 2   Poly-N-vinylpyrrolidone Modified with
            Ferrocenecarboxaldehyde

Molecular Structure of Polymers

The molecular structure of polymers has been established by means of ir spectral measurements and elemental analysis. The soluble polyferrocenyl-acetylones have linear structures (I,II), while the insoluble ones have tridimensional structure (III, IV) (Fig. 3).

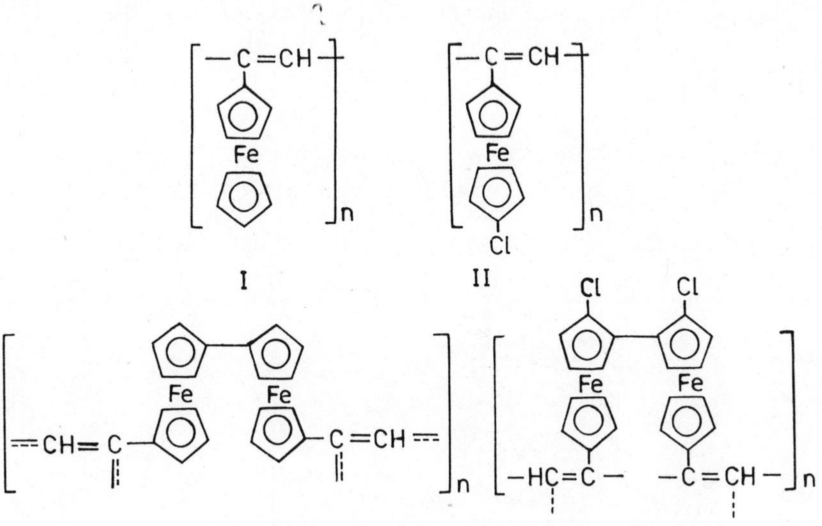

Figure 3 The structure of polymers obtained by polymerization of ferrocenylacetylene

Poly-p-ferrocenylacetylenes obtained by polymerization of α-chloro-β-formyl-p-ferrocenylstyrene have tridimensional structures (V and VI, Fig. 4).

The structure VI was confirmed by the fact that the reaction is accompanied by denydrohalogenation or dehaloformylation of monomer or polymer.

Poly-p-ferrocenylphenylacetylenes obtained by polycondensation are tridimensional polymers containing polyene units (a) and segments of the type obtained by the ferrocene reaction with ketones in the presence of Lewis acids (33), (b,c, Fig. 5).

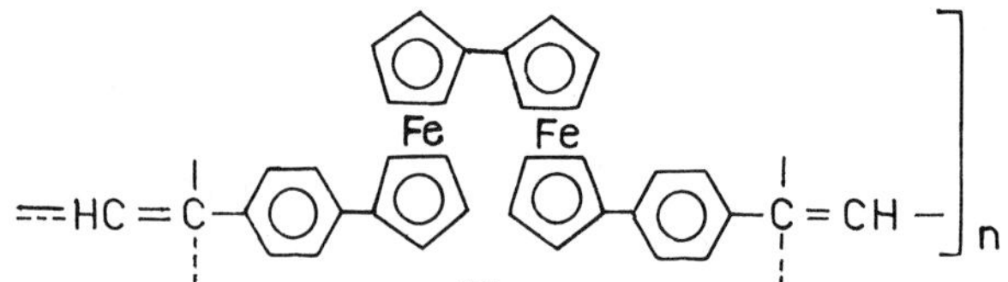

Figure 4   The Structure of Polymers Obtained
                by Polymerization of Cl-FFcSt

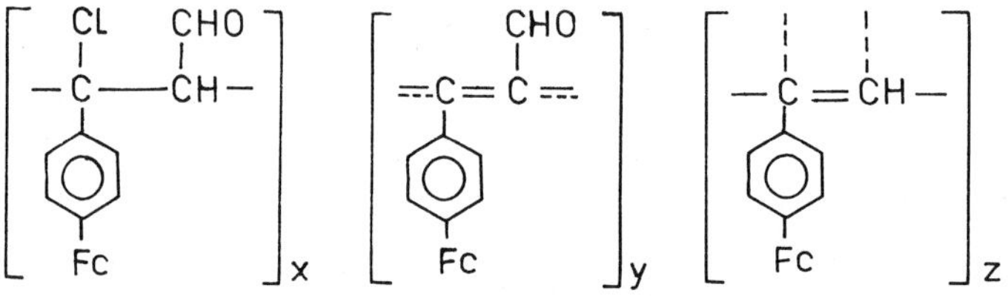

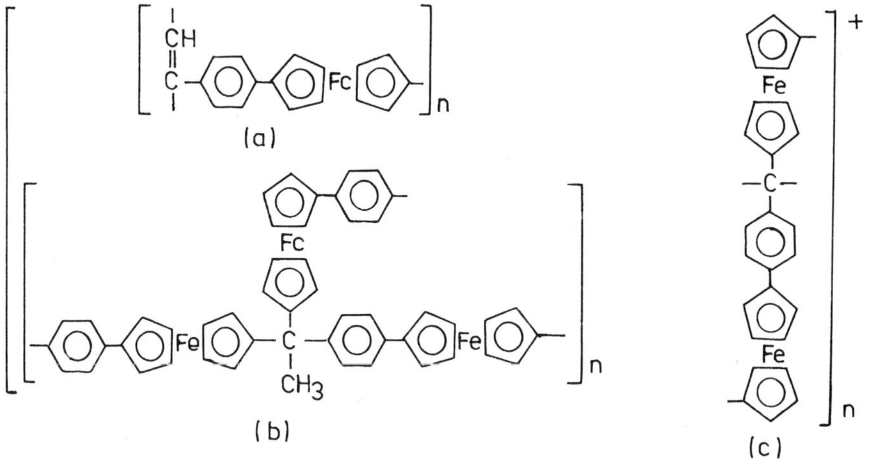

Figure 5   The Poly-p-ferrocenylphenylacetylene
                by Polycondensation of Cl-FFcSt

        The polyferrocenyliminoimides and polyferrocenylena-
zomethines have complex structures due to certain concurrent reactions. Thus DiAcFc reacts with itself
leading to polyferrocenylenevinylenes or triferrocenyl-
benzenes (1590 - 1600 cm^{-1} in the ir spectrum). The
insoluble products exhibit a tridimensional structure
containing units of "a", "b", "c", and "d" type (fig.
6).

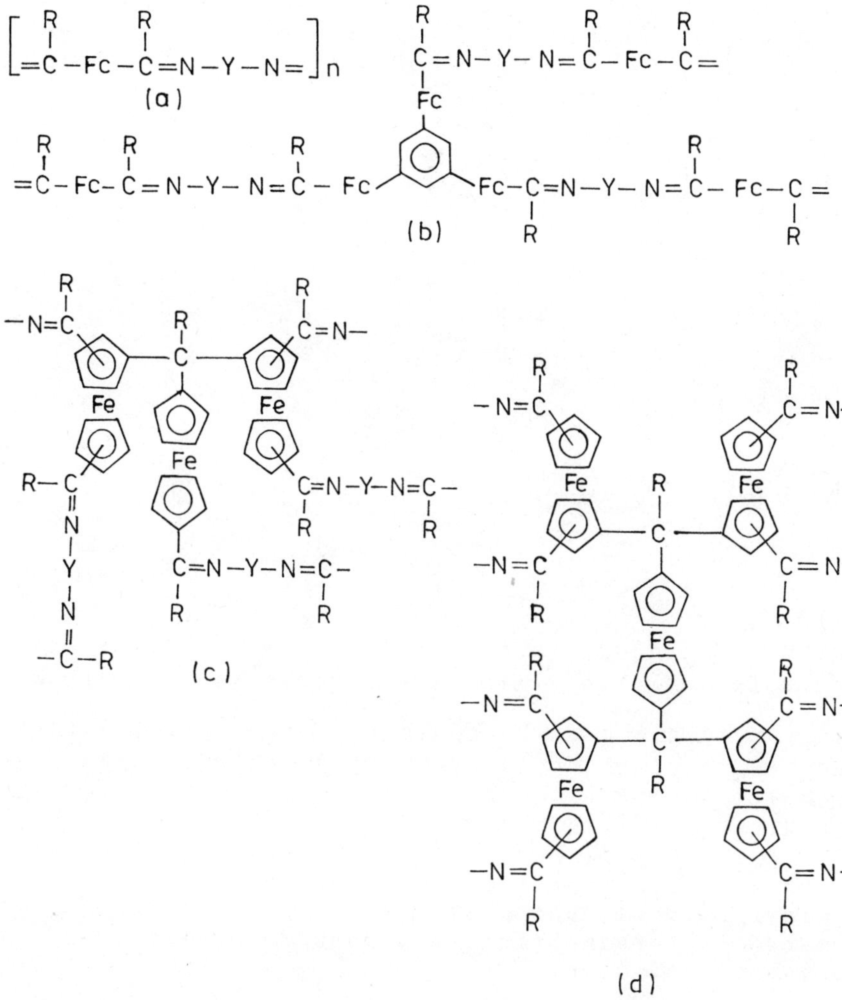

Figure 6   The Structure of Polyferrocenyleniminoimides and Polyferrocenylenazomethines

In order to avoid the complex structure favoured by Lewis acids used as catalysts in polycondensation as well as to investigate a new method for obtaining polyferrocenylenazomethines, N,N' -disulphynyl-bis-(4-amino-2-nitrophenyl)disulphide was prepared initially according to Westphal and Henklein's new method (34) and submitted then to polycondensation with diacetylferrocene and 1,1' -bis[-(2-furyl)-acryloyl]ferrocene. According to the reaction

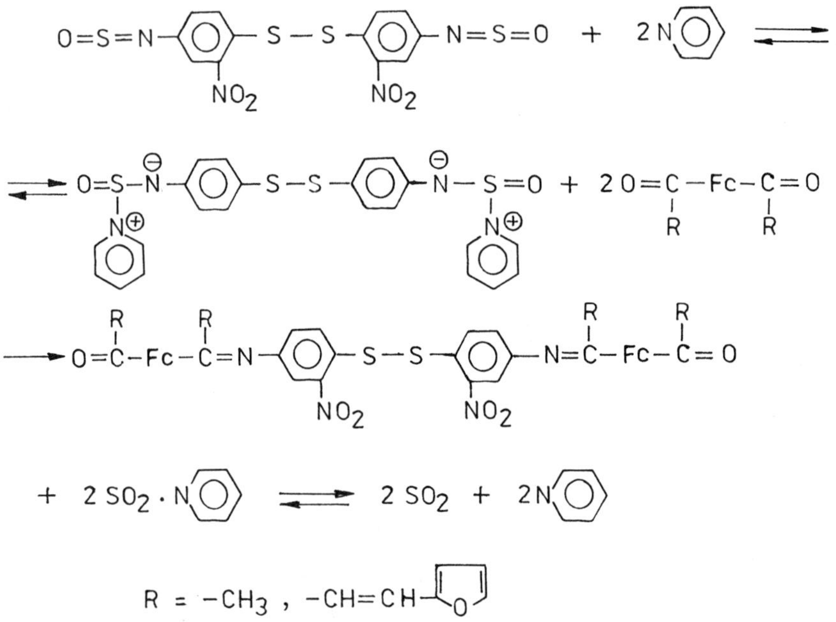

$$O=S=N-\langle\bigcirc\rangle-S-S-\langle\bigcirc\rangle-N=S=O \ + \ 2N\langle\bigcirc\rangle \ \rightleftharpoons$$

$$\text{(NO}_2\text{)} \quad \text{(NO}_2\text{)}$$

$$\rightleftharpoons \ O=S-\overset{\ominus}{N}-\langle\bigcirc\rangle-S-S-\langle\bigcirc\rangle-\overset{\ominus}{N}-S=O \ + \ 2\,O=\overset{}{C}-Fc-\overset{}{C}=O$$

$$\overset{\oplus}{N} \qquad \overset{\oplus}{N} \qquad R \qquad R$$

$$\longrightarrow \ O=\overset{R}{\underset{|}{C}}-Fc-\overset{R}{\underset{|}{C}}=N-\langle\bigcirc\rangle-S-S-\langle\bigcirc\rangle-N=\overset{R}{\underset{|}{C}}-Fc-\overset{R}{\underset{|}{C}}=O$$

$$\text{NO}_2 \qquad\qquad \text{NO}_2$$

$$+ \ 2\,SO_2\cdot N\langle\bigcirc\rangle \ \rightleftharpoons \ 2\,SO_2 \ + \ 2N\langle\bigcirc\rangle$$

$$R = -CH_3 \ , \ -CH=CH-\langle\text{O}\rangle$$

Figure 7   Polycondensation of DiAcFc and DiAcFcF with
N,N'-disulphynyl-bis(4-amino-2-nitrophenyl) disulphide

mechanisms advanced by Hornhold (35) the pyridine forms
an aduct with disulphynil derivatives which reacts with
dicarbonyl component leading to a poly-Schiff base with
linear structure (Fig. 7).
     The frequency of occurence of ferrocene residue in
the macromolecular chain in modified cellulose,
poly(vinyl alcohol) and polyvinylpyrrolidone was
determined by gravimetrical and photocolorimetrical
determination of iron.

Polymer Properties

     The conjugated system and the ferrocenyl residue
give to polymers certain properties such as thermal
stability, electrical conductivity and magnetic
properties.
     Thermal stability Though ferrocene is very stable
towards action of temperature, the ferrocenyl residue
introduced in polymer does not modify its thermal
stability. Most of polymers are stable up to $300^\circ C$.
The action of temperature on the polyferrocenyleniminoimides and polyferrocenylenazomethines was
studied in comparison with that of dicarbonyl and
diamine components (36).

The polyferrocenylenazomethines are more stable than the polyferrocenyleniminoimides, the degradation temperature having values between 200 - 280°C for polyferrocenylenazomethines and 184 - 198°C for polyferrocenyleniminoimides.

The substantial weight loss at 300 - 400°C (in air) of polyferrocenylenazomethines is similar to that found by Neuse and co-workers (6) for polyferrocenylenazomethines obtained by the polycondensation of 1,1' -diformylferrocene with p-phenylendiamine.

The thermogravimetrical analysis of cellulose, PVA and PVP modified with ferrocene indicate the thermal decomposition to and at 800°C; above this temperature the amount of residue remains constant.

Electrophysical properties All polymers synthesized by polymerization and polycondensation exhibit paramagnetic properties and electrical conductivity.

If in case of polyferrocenylacetylenes obtained by polycondensation and heterochain polymers, the electrical conductivity can be due to the catalyst traces included in polymer and/or to the mixed valence states of Fe (Fe^{II} - Fe^{III}) (37), in case of polyferrocenylacetylenes obtained by polymerization it results only from the conjugated system.

The introduction of oxygen or sulphur atoms as ether (-O-), thioether(-S-) or disulphide (-S-S-) bonds does not modify significantly the thermal stability and electrophysical properties of polymers since the oxygen and sulphur atoms contribute with their unshared electrons to the resonance hybrids (pseudoconjugated polymers) (38).

The values of electrical conductivity ($10^{-12}-10^{-10}$ ohm^{-1} cm^{-1}) as well as those of the concentration in paramagnetic particles (10^{17} - 10^{22} spin/g)place these polymers among the organic semiconductors.

Catalytic activity The reversible conversion of ferrocene to ferricinium ion makes possible the use of ferrocene polymers as catalysts in reactions with electronic transfer between substrates and catalyst.

The catalytic action of the same polymers was also investigated in the polymerization of chloroformylated vinyl compounds of the

$$X - C_6H_4 - \underset{Cl}{C} = CH - CH = 0 \qquad X = Cl, Br, CH_3, C_2H_5$$

type, the catalysts proving to be efficient in case of the less reactive monomers. (39,40).

REFERENCES

1. Yu.V.Korshak, T.A.Proniuk and B.E.Davydov,
    Neftekhimya, 5, 677 (1963).
2. R.G.Gomper, P.T Funke and A.A Volpe, J. Polym.Sci.,A
    1, 9, 2137 (1971).
3. B.E. Davydov, A.Z.Zagharyan, G.P.Karpocheva,
    B.A.Krentsel, G.A.Lapitskii and G.V. Khutareva,
    Dokl.Akad.Nauk SSSR, 160, 650 (1965).
4. K.Nata, I.Motoyama and T.Masuko, Jap. Pat.9679
    (1967); C.A.68, 99582 (1968).
5. K.Sonogashira and N.Hagihara, Kogyokagaku Zasshi,
    66, 1090 (1963).
6. E.W. Neuse, H.Rosenberg and R.R.Carlen,
    Macromolecules, 1,424 (1968).
7. B.Hetnorski and Z.Grabowski, Bull. Akad.Sci.,
    Ser.Sci.Chem.,17, 391 (1969).
8. L.Taratu, M.Vata, I.Mazilu, T.Lixandru and
    Cr.Simionescu, Ang. Makromol Chem.,101,19 (1981).
9. L. Taratu,I.Mazilu, M.Vata, T.Lixandru and
    Cr.Simioinescu, J.Organometal.Chem.,214, 107 (1981).
10. M.Rosenblum, N.Brown and B.King, Tetrahedron
    Letters, 4421 (1967).
11. K.Schogl and H.Soukup, Tetrahedron Letters, 1181
    (1967).
12. K.Schogl and H.Soukup, M., 99, 927 (1968).
13. Cr.Simionescu, T.Lixandru, I.Mazilu and L.Tataru,
    Makromol.Chem., 147, 69 (1971).
14. Cr.Simionescu, T.Lixandru, I.Negulescu, I.Mazuli and
    L.Tataru, Makromol.Chem., 163, 59 (1973).
15. G.A.Yurlova, B.Yu.Ciumakov, M.T.Ejova,V.L.Dzashi,
    S.L.Sosin and V.V.Korshak, Vysokomal.Soedin.,13,
    2761 (1971).
16. V.V.Korshak, V.L.Dzashi and S.L.Sosin, Nuova Chim.,
    49, 31 (1973); C.A. 79,5644(1973).
17. T.Nakashima, T.Kunitake and C.Aso, Makromol.Chem.,
    157, 73 (1972).
18. Ch.U.Pittman,Jr., Y.Sasaki and P.L.Grube,
    J.Macromol.Sci.Chem., A 8, 5, 923 (1974).
19. B.F.Sokolov, T.P.Vishnyakova and A.S.Kislenko,
    Vysokomol.Soedin Ser A, 15, 2709 (1973).
20. Ya.M.Paushkin, B.F.Sokolov, T.P.Vishnyakova and
    O.G.Glazkova, Dokl.Akad.Nauk SSSR, 206,664 (1972).
21. T.P.Vishnyakova, L.I.Tolstykh, G.M.Ignatieva,
    B.F.Sokolov and Ya.M.Paushkin, Dokl.Akad.Nauk SSSR,
    208,853 (1973).
22. Cr.Simionescu, T.Lixandru, I.Mazilu and L.Tataru,
    J.Organometal.Chem.,113, 23 (1976).

23. Cr.Simionescu, T.Lixandru, L.Tataru, I.Mazilu and
    I.Cocirla, J.Organometal.Chem., 102,219 (1975).
24. Ya.M.Paushkin, T.P.Vishnyakova, I.I.Patalah,
    T.A.Sokolinskaya, and F.F.Machus, Dokl.Akad.Nauk
    SSSR, 149, 856 (1963).
25. Ya.M.Paushkin, L.S.Polak, T.P.Vishnyakova,
    I.I.Patalah, F.F.Machus and T.A.Sokolinskaya,
    Vysokomol.Soedin., 6, 545 (1964).
26. Ya.M.Paushkin, L.S.Polak, T.P.Vishnyakova,
    I.I.Patalah, F.F.Machus and T.A.Sokolinskaya,
    J.Polym.Sci., C 4, 1481 (1964).
27. Ya.M.Paushkin, T.P.Vishnyakova, S.A.Nisova,
    A.F.Lunin, O.I.Omarov, I.I.Markov, F.F.Machus,
    A.I.Golubeva, L.S.Polak, I.I.Patalah, V.A.Stychenko
    and T.A.Sokolinskaya, J.Polym.Sci., A 1, 5, 1203
    (1967).
28. Cr.Simionescu, T.Lixandru, L.Tataru and I.Mazilu,
    J.Organometal.Chem., 73, 375 (1974).
29. Cr.Simionescu and S. Vasiliu, Plaste und Kautchuk,
    18, 260 (1971).
30. Cr.Simionescu and S. Vasiliu, Acta
    Chim.Acad.Sci.Hung., 74, 461 (1972).
31. B.Sansoni and O.Sigmund, Angaw.Chem., 73, 299
    (1961).
32. Cr.Simionescu, T.Lixandru, L.Tataru, I.Mazilu,
    M.Vata and D.Scutaru, "The Fourth International
    Symposium on Organometallic and Inorganic Polymer
    Science for the National American Chemical
    Societies", Washington, August 28, 1983.
33. E.W.Neuse and H.Rosenberg, J.Macromol.Sci., 4, 1
    (1970).
34. G.Westphal and P.Henklein, Z.Chem.,9, 1, 26 (1969).
35. H.Horhold, Z.Chem., 2, 41 (1972).
36. L.Tataru, M.Daringa, M.Vata, T.Lixandru and
    Cr.Simionescu, J.Therm.Anal., 22, 259 (1981).
37. C.U.Pittman,Jr.and Y.Sasaki, Chem.Letters, 383
    (1975).
38. G.F.D'Alelio, J.V.Crivello, R.K.Schooning and
    T.F.Huemmer, J.Macromol.Sci., Chem., A 1, 7, 1161
    (1967).
39. Cr.Simionescu and I.Mazilu, Rev.Roum.Chim., 23, 753
    (1977).
40. Cr.Simionescu, I.Mazilu, L.Tataru and T.Lixandru,
    Rev.Roum.Chim., 25, 2, 277 (1980).

USE OF ORGANOMETALLIC POLYMERS FOR PRE-HEAT SHIELDS

FOR TARGETS IN INERTIAL-CONFINEMENT NUCLEAR FUSION

John E. Sheats, Fred Hessel, Louis Tsarouhas,
Kenneth G. Podejko, Thomas Porter, L.B. Kool [1a],#
and R.L. Nolan, Jr. [1b],#

Chemistry Department   # KMS Fusion, Inc.
Rider College   P.O. Box 1567
Lawrenceville, NJ   Ann Arbor, MI

INTRODUCTION

Nuclear fusion, the energy process operating in the sun, offers promise of production of almost unlimited energy without the toxic and radioactive wastes associated with nuclear fission. Harnessing nuclear fusion, however, has proven to be a challenging task that may not be completed for another thirty years. Because of the strong repulsive forces to be overcome in order for nuclei to fuse, the process will take place only at temperatures above 50,000,000 degrees. No known materials can contain matter at this temperature. Thus the fusion reaction must be confined without its touching the walls of its container. Two methods of confining this super-heated material are being studied, Magnetic Confinement and Inertial Confinement.

In the Magnetic Confinement process, currently being studied at the Forrestal Campus of Princeton University, a high energy beam of ionized particles or plasma, as it is commonly known, is accelerated to velocities approaching the speed of light within a toroidal magnetic field in such a manner that the particles never touch the sides of the container. When the particles reach the necessary velocity, the beam can be compressed and nuclear fusion will take place.[2] The apparatus for producing the toroidal magnetic field is called a TOKAMAK, a term borrowed from the Russian scientists who first suggested this approach.

The Inertial Confinement process[3] is being investigated
at KMS Fusion, Inc. in Ann Arbor, Michigan, at the
University of Rochester, at the National Laboratories at Los
Alamos, New Mexico, and Livermore, California, and elsewhere. In this process, commonly called Laser Fusion, a
small hollow sphere containing deuterium and tritium at high
pressure is placed at the focal point of a high-intensity
laser beam. The energy - eight trillion watts per square
centimeter (100 billion light bulbs on your fingernail)
causes the spherical shell to vaporize outward and a shock
wave to proceed inward at one-tenth the speed of light, com-
pressing the hydrogen to 1/1000th of its original volume and
raising temperatures up to one-hundredmillion degrees (Fig.
1). Nuclear fusion takes place, producing helium and a neutron which can later be captured by a lithium shield,[4]
releasing usable energy.

$$^2_1H + ^3_1H --> ^4_2He + ^1_0n$$

deuterium + tritium = helium + neutron

$$^6_3Li + ^1_0n --> ^7Li + energy$$

Theory of Target Design

In Inertial Confinement Fusion, two major problems are
encountered, obtaining sufficient energy to heat the fuel
hot enough to trigger fusion and delivering that energy
efficiently to the fuel.[5] Smaller targets require less
energy but also allow less time for the energy to be
absorbed. The first targets were simple hollow spheres of
silica and plastic.[6-11] These were sufficient for low
energy studies where compression of fifty-fold or less was
obtained but reflection of the energy from the surface and
too-quick destruction of the shell prevented greater compression at higher energies. A double shell design was next
employed with an outer layer composed of material of low
density and low atomic number which would absorb the energy
of the laser beam and ablate away, generating a compressive
force, and an inner layer which would contain the fuel.
More advanced designs utilized x-rays emitted from a metal
of high atomic number for the final compression of the
fuel.[6] Still more advanced designs contained multiple

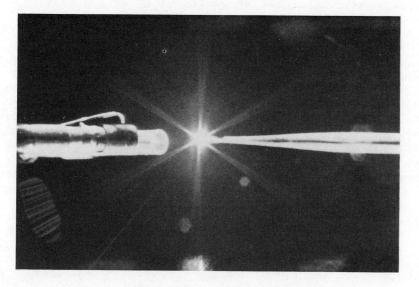

Fig. 1 View of Inertial Confinement
Nuclear Fusion in Progress

layers of materials with different responses toward the developing shock wave of high energy plasma.[6-11] Between the layers are cushions of low-density foam [7, 12, 24] which maintain the proper spacing. The spacing of the layers is critical since it controls the timing of the different events in the implosion-fusion sequence. Too long a time interval would lower the final temperature obtained and too short a time interval would not allow the energy of the laser beam to be fully utilized. Indeed fusion and expansion might begin to take place while the laser beam was still delivering energy. This failure in timing is analogous to the knock in an automobile engine, which is caused by a premature explosion of the gasoline-air mixture while compression is still taking place. A great reduction in efficiency results.

A second type of inefficiency called pre-heat[13] occurs when high energy (supra thermal) electrons and other types of radiation are released when the outer shells begin to disintegrate and head toward the center ahead of the main shock front, heating the fuel before compression is complete. Again a great loss in efficiency results. In order to prevent pre-heat an additional shell is added containing a material of low average atomic number $\bar{z}$ with 1-4 atom % of atoms of atomic number 50-85 uniformly distributed on the molecular level.[6] Clusters of the high -z atoms would cause inhomogeniety in the plasma as the ablation process progresses and would also leave gaps which would allow the electrons to penetrate to the center.

A diagram of one of the more recent target designs is shown in Fig. 2.

Fabrication of Targets for Inertial-Confinement Nuclear Fusion.

In order for the fusion process to take place in the manner described, the targets must be fabricated with extremely high precision. Each shell must be perfectly spherical, with variations in thickness no greater than 1%. The inner and outer surfaces must be concentric within 1% and no surface roughness greater than 100 A$^\circ$ must be present.[8] In order to prepare the tiny (less than 1 mm) multishell targets to such tolerances entire new technology for preparing polymer films has had to be developed.[9, 14] A complete discussion of the technology of target fabrication is beyond the scope of this paper. Good summaries are available in references [9-10]

Preparation of Materials for Pre-Heat Shields

The primary interest of the authors of this paper has been in developing suitable high -z polymer films and techniques for either preparing free-standing hemispheres or coating target spheres with a uniform layer of the materials. Several approaches to this problem will be reviewed.

The first approach, undertaken by R. Liepins and co-workers at the Los Alamos Scientific Laboratory has been the glow-discharge polymerization of p-xylene or cyclooctatetraene, COT, or the vapor-phase pyrolysis polymerization of p-xylylene in the presence of an organometallic monomer.[6]

In the glow discharge polymerization process a vapor of p-xylene or COT at 0.2 - 0.4 torr mixed with 0.02-0.10 torr of an inert carrier gas is ionized to a plasma by microwave radiation. The plasma is passed between electrodes with a potential difference of 200V. A constant low power density of 0.01 - 0.10 w/cm2 is maintained. Collision of the charged particles with neutral p-xylylene produces ionization and polymerization in a uniform layer on the electrodes and on microspheres levitated within the cavity. If deposition proceeds too rapidly, nodular growths appear in the coating as a result of crystallinity in the polymer. Crystallinity can be reduced by admitting controlled amounts of air or oxygen to terminate chain growth and by conducting the coating process with surface temperature of -10 degrees C or below. By this procedure coatings 10-100 μm thick with surface roughness as low as 30nm have been obtained.

If organometallic monomers are introduced along with the hydrocarbon, polymer films incorporating the metal can be produced (Table I). In addition to the metals listed in Table I, Au, Pd, Pt, Os, W and Ag have been incorporated into polymer films. The metal content of the films has been determined by neutron-activation analysis which can detect small amounts of the metal atom and is insensitive to changes in oxidation state, chemical form or location. By varying the pressure of the organometallic monomer as deposition proceeds, a gradient in metal content can be produced. Usually the coatings are formed with higher metal content outermost. This gradient causes a less abrupt change in density and Z between shells.

Vapor phase pyrolysis of p-xylylene occurs at 600°C to produce the reactive monomer, which is introduced at pressures of 0.05 -0.10 Tor along with the organometallic monomer. The advantage of this process is that thick uniform

TABLE I

Polymers obtained by Copolymerizing Organometallic Monomers
with Hydrocarbons

Monomer	Hydrocarbon	Process	wt% metal	at % metal
$Ta(OC_2H_5)_5$	COT	GD	45	2.4
$(CH_3)_2$ Hg	COT	GD	87	11.1
$(C_2H_5)_2$ Hg	COT	GD		
Tl (HCO_2)	PX	VPP	1	0.03
$(CH_3)_4$ Pb	COT	GD	78	5.9
$(C_2H_5)_4$ Pb	COT	GD		
$(CH_3)_3$ Bi	COT	GD	82	7.7
$(C_6H_5)_3$ Bi	PX	VPP	16	0.6

GD = Glow Discharge   VPP = Vapor Phase Pyrolysis

PX = p-Xylylene   COT = Cyclooctatetraene

layers can be deposited rapidly. Surface finish, however,
is not so good as in glow-discharge polymerization. The
finish can be improved greatly if substituents such as
ethyl, isopropyl, n-butyl or chloro are added to the
p-xylylene.[15]

Incorporation of the high Z element directly into the
monomer has been studied by investigators at KMS
Fusion.[16-17] Iodo-p-xylene alone or mixed with p-xylene can
be glow-discharge polymerized to form films with up to 5.2
atom % (55.6 wt %) of iodine. The films absorbed oxygen when
exposed to the air. Composition of the film is not always
uniform. Again a gradient is also possible if the relative
amounts of iodo-p-xylene and p-xylene are varied as
deposition proceeds.

A second approach, investigated at KMS Fusion and at
Rider College by the authors of this paper involves synthesis
of organometallic vinylic monomers which can be polymerized
in solution or in the vapor phase.[17-18] Extensive research
(19-21) on vinylic monomers and polymers containing first
row transition elements has been conducted, but little
previous work has been done with metals of high atomic
number. In this study the monomers in Table II were
prepared. Since

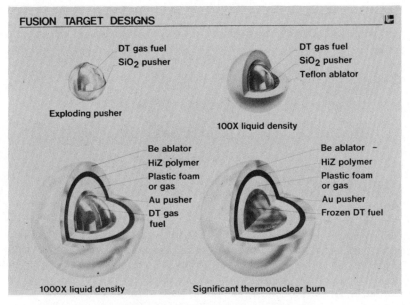

Fig. 2   Design for Targets for Use in
         Inertial Confinement Nuclear Fusion

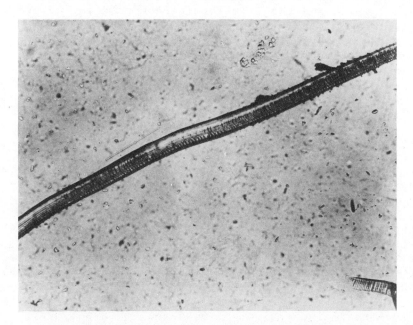

Fig. 3   Tool Mark in Tungsten Polymer 100X

Table II

Vinyl Organometallic Monomers to be Used for Preparing Pre-heat Shields

$(C_6H_5)_3Pb\ C_6H_4CH=CH_2$ $C_5H_5\ W(CO)_3CH_2C_6H_4CH=CH_2$

$C_5H_5\ Ru\ C_5H_4CH=CH_2$ $(CO)_3CH_3WC_5H_4CH=CH_2$

$C_5H_5\ Os\ C_5H_4CH=CH_2$ $(CO)_2NO\ WC_5H_4CH=CH_2$

$(CO)_3\ ReC_5H_4CH=CH_2$ $(CO)_3\ W\ C_6H_5CH_2O_2C\ CH=CH_2$

the procedure for synthesis and the properties of the monomers and polymers have been reported previously[17-18], only the most important points will be summarized here. The monomers undergo free radical polymerization either neat or in benzene at 60-80° with AIBN as the initiator. Organic peroxide initiators can not be used because they oxidize the monomers. The two most extensively studied monomers are triphenyl-p-styryllead, TSL (Table III) and vinyl-ruthenocene, VR[22] (Table IV). The homopolymers have rela-tively low molecular weight and high Tg. Films can be cast from solution but they are too brittle to serve as coatings or to be fabricated into hemispheres. The glass transition can be lowered and the film-forming characteristics greatly improved by copolymerization with conventional monomers such as styrene, methyl acrylate, acrylonitrile or lauryl or octadecyl methacrylate. The long hydrocarbon chains of the latter two act as internal plasticizers.

A simple test was developed to measure the machinabili-ty of the polymer.[17] A thin film was cast on a glass micro-scope slide and a stylus dragged across it. If sharp, clearly defined scratches were obtained, with no flaking at the edges and no subsequent closure of the scratches by film-creep, (Fig. 3), the polymer was suitable for further study.

The best films were obtained by copolymerizing a 3:1 weight ratio of TSL to octadecylmethacrylate. The resulting polymer contained approximately 25% by weight of Pb (0.9 atom %) and had Tg below room temperature. Films were cast on a mandrill and hemispheres machined by a diamond point stylus[14] (Fig. 4).

Hemishells were also produced from this polymer by vacuum molding. This was accomplished by placing a free-

TABLE III

Polymers and Copolymers of Triphenyl-p-Styryllead, TSL

Comonomer	Wt ratio TSL: comonomer	Yield	Wt % TSL	Atom % Pb	T_g (DSC) °C	T_s, °C
none[a]		67%	100	2.0	98 140(TBA)	--
none[b]		--	100	2.0	>200	
isopropylstyrene[a]	25:1	--	---	---	141	160-165
acrylonitrile[a]	3.5:1	--	70	1.3	159	165
octadecyl-methacrylate[a]	3:1	45	65	0.9	<25	46
lauryl-methacrylate and divinylbenzene[a]	5:1:0.15	--	75	1.2	--	90
n-butyl methacrylate[a]	5:1	--	---	---	114	105

a   benzene is solvent   b   no solvent

TABLE IV

Polymers and Copolymers of Vinylruthenocene, VR

Comonomer	Wt. Ratio VR: Comonomer	Yield %	M_n	M_w	Wt. % VR	Atom % Ru	n dlg^{-1}	T_g °C
none		95	5,860	18,670	100	4.0		>250
"		93	19,660	118,660	100	4.0	0.15	>250
Methyl acrylate	1:1	76	26,210	139,670	45.2	1.4	0.103	91
	2:1	85	8,650	114,000	60.1	2.0	0.095	111
Styrene	1:1	45	6,520	26,830	34.0	1.1	0.096	99-117
	2:1	57	5,980	29,870	54.5	1.9	0.129	105-131
Vinylpyrrolidinone	2:1	--	3,130	9.650	55.9	1.8	---	203-217

Fig. 4   S E M Photomicrograph of Machined
         Hemishell

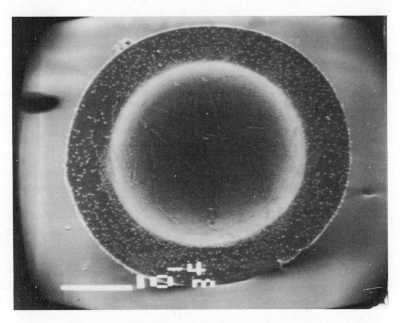

Fig. 5   Inside View of Hemishell
         Produced by Vacuum Molding

standing polymer film on a silicon hemispherical cavity mold.[14] The mold was then evacuated to remove any trapped air. Next, the mold was heated above the softening point of the polymer and atmospheric pressure was admitted, causing the film to be pushed into the mold. Finally, the inter-connecting matrix was removed by grinding with calcium carbonate, which is softer than silicon, and the hemishells were removed from the mold with ultrasonic agitation and/or immersion in liquid nitrogen. Fig. 5 is a photomicrograph of a hemishell made by this method.

Since the organometallic monomers in Table II have appreciable vapor pressures, the glow discharge polymerization or vapor-phase pyrolysis techniques discussed previously may also be applicable. An attempt to copolymerize ferrocene with p-xylene was only partially successful.[6] The ferrocene deposited along with the polymer film but was not bound to it and therefore could be leached out. The vinyl groups should cause the monomers to be bound into the polymer film.

Polymer films containing colloidal clusters of metal atoms or metal oxide can be prepared by co-depositing a volatile transition metal carbonyl with the polymer film and subsequently decomposing the metal carbonyl thermally or photochemically.[23-25] Particles of iron $5.0 - 100$ nm in diameter homogeneously dispersed within the polymer film were obtained. Exposure to oxygen leads to rapid oxidation of the metal particles. Although dispersion of the metal is not as homogeneous as the methods described previously, it may still prove satisfactory. Only metals from groups IVA-VIII A which form stable carbonyls may be employed.

A different approach, employing metal ions to crosslink commercially available sulfonated polystyrene in aqueous solution has been studied at the University of Rochester.[23] Solutions containing 30% of polystyrene sulfonic acid with molecular weight of 120,000 were neutralized with aqueous CsOH or Ba (OH)$_2$ Films could be cast containing up to 5 atom % Cs or 2.5 atom % Ba. The average atomic number $\bar{z}$ varies from 4.8 to 7.5. Ion-containing polystyrene films tend to form ion aggregates when the metal content exceeds 3 mole % but at the high degree of substitution of these polymers aggregation is less likely, since there is not sufficient unsubstituted styrene to create hydrophobic domains. Films cast from these solutions provided satisfactory coatings for glass microspheres.

TABLE V

Preparation of Methyl Methacrylate- Methacrylic Acid Polymer
Films Containing the Lanthanides

	Metal Compound, g	Polymer, g	% Metal by weight	Atom %
Sm_2O_3	0.15	1.00	10.1	0.51
	0.30	1.00	16.5	0.93
	0.60	1.00	24.3	1.54
	1.00	1.00	29.9[b]	2.10
	1.50	1.00	33.8[b]	2.57
	2.0	1.00	46.2[b]	2.89
Eu_2O_3	0.15	1.00	10.4	0.48
Tb_2O_3	0.20	1.00	7.9	0.35
$UO_2(C_2H_3O_2)_2$[a]	0.30	0.50	21.0	0.35

a   Ethanol was used on the solvent
b   Phase separation is observed

A similar approach was taken by the authors at Rider [18] College to incorporate the lanthanides into polymer films. Solutions containing 2% of a 3:1 copolymer of methyl acrylate and methacrylic acid and up to 3% of a rare earth oxide or acetate in glacial acetic acid were evaporated to produce clear, glossy films containing up to 3 atom % of the rare earth elements. Similar procedures have been employed for the lanthanides by Okamoto et al.[26-7] at Polytechnic Institute of New York and for Tl by workers at IBM.[28] The polymer films produced are summarized in Table V. The weight % and atom % of the metal in the polymer film are estimated from the stoichiometry employed. A sample of the polymer containing 0.15g of Sm_2O_3 per 1.00 g of polymer was submitted to Galbraith Laboratories. The results C 47.39; H 6.93, Sm 9.55; O (by difference) 36.13 correspond to 0.48 atom % Sm in good agreement with the results in Table V, 10.1 wt % and 0.51 atom % Sm. The best films were obtained with Sm. At higher metal concentrations phase separation takes place. At the low concentrations Okamoto employed no visual evidence of phase separation was observed but luminescence studies indicated ion aggregation in some of the polymer films.[26-7] The advantage of using the lanthanides is that Z can be varied from 57-71 without major changes of experimental conditions or mixtures of the elements can be employed to cover the entire range. Since only preliminary investigations were conducted, no attempts to coat spheres or prepare hemispheres were made.

As this paper is being written no one of these methods of preparing pre-heat shields has emerged as clearly superior. None of the high-Z polymer films has been tested in an actual target shot to determine its effectiveness as a pre-heat shield. Thus much more research is needed before a preferred technique will emerge.

ACKNOWLEDGEMENTS

Support for research at KMS Fusion and at Rider College described in this paper and in our previous publication[17] was provided by grants from the Department of Energy DOE 1981 -DE-AC08 - 78 DP 40030 and DOE 1982 DE-AC08-DP 40152.

REFERENCES

1a. Present address: Chemistry Dept. University of Massachusetts, Amherst, MA 01002

1b. Present address: Radian Co. P.O. Box 9948 8501 Mo Pac
     Blvd., Austin, TX 78766

2.   H.P. Furth, Sci. American, 241, 50-61 (1979).

3.   R.J. Nuckolls, L. Wood, A. Thiessen and G. Zimmerman,
     Nature 239, 139 (1972). J.H. Nuckolls, Physics Today,
     Sept 1982, 25-31.

4.   W.R. Meier and W. B. Thomson, Proc. Top. Meet. Technol
     Controlled Nucl. Fusion Vol 3, 297-307 (1978). Chem.
     Abs. 92, 118069 (1980).

5.   S. Bardwell and U. Parpart, Fusion, Oct.Nov. 1981 pp
     22-32.

6.   R. Liepins, M. Campbell, J.S. Clements, J. Hammond and
     R.J. Fries, J. Vac. Sci Technol., 18 (3), 1218-26
     (1981).

7.   R.C. Kirkpatrick, C.C. Cremer, L.C. Madsen, H.H. Rogers
     and R.S. Cooper, Nucl. Fusion, 15, 333 (1975).

8.   R. Liepins, M. Campbell and R.J. Fries, Prog. Polym.
     Sci., 6, 164-86 (1980).

9.   H.W. Deckman and G.M. Halpern, Laser Interact. Relat.
     Plasma Phenom., 5, 651-85 (1981).

10. R.J. Fries, AICHE Symposium Series, Vol. 75, no. 191,
     1979 pp 208-18.

11. C.D. Hendricks, J.K. Crane, E.J. Hsieh and S.F. Meyer,
     Thin Solid Films, 83, 61-72 (1981).

12. A.T. Young, D.K. Moreno and R.G. Marsters, J. Vac. Sci.
     Technol., 20 (4), 1094-7 (1982).

13. R.E. Kidder, Nucl. Fusion, 21, 145 (1981).

14. H.W. Deckman, J. Vac. Sci. Technol. 18 (3),
     1171-4(1981).

15. D.G. Peiffer, T.J. Corley, G. M. Halpern. and B.A.
     Brinker, Polymer, 22, 450-60 (1981).

16. R.L. Crawley, L.B. Kool and R.L. Nolen, J. Vac. Sci.
     Technol., 18 (3), 1255-7 (1981).

17. J.E. Sheats, F. Hessel, L. Tsarouhas, K.G. Podejko,
     T. Porter, L.B. Kool and R.L. Nolen, Jr., New and
     Unusual Monomers and Polymers, Plenum Press, New York,
     1983.

18. J.E. Sheats, F. Hessel, L. Tsarouhas, K.G. Podejko,
     T. Porter, L.B. Kool and R.L. Nolen, Jr., Polymeric
     Mat. Sci.and Eng., 49(2), 363-7 (1983).

19. C.U. Pittman, Jr. in E. Becker and M. Tsutsui, eds.,
     Organometallic Reactions, Vol 6, Plenum Press, New
     York, 1977 pp 1-62.

20. J.E. Sheats, J. Macromol Sci. Chem., A15(6) 1173-99
     (1981).

21. C.E. Carraher, Jr., J.E. Sheats and C.U. Pittman, Jr.,
     Advances in Organometallic Polymer Science, J. Macromol
     Sci. Special Symposium Issue A16, 1981, hardback,
     Marcel Dekker, New York, 1982.

22. J.E. Sheats and T.C. Willis, Org. Coatings and Plastics
     Chem., 41 (2), 33-7(1979) Full Paper J. Polymer Sci.,
     Polymer Chem Ed., 22, 1077 (1984).

23. H. Kim, J. Mason and J.R. Miller, J. Vac. Sci. Technol.
     A(2) 890-93 (1983).

24. M.A. de Paoli in Advances in Inorganic and
     Organometallic Polymer Science, Marcel Dekker, New York
     1982 pp. 251-60.

25. R. Tannenbaum, E.P. Goldberg and C.L. Flenniken in J.E.
     Sheats, C.E. Carraher, Jr. and C.U. Pittman, Jr.,
     Metal-Containing Polymeric Species, Plenum Press, New
     York, 1984, in press.

26. E. Banks, Y. Okamoto and Y. Ueba, J. Appl. Polymer Sci,
     25, 359 and 2007 (1980).

27. Y. Okamoto, Y Ueba, I. Nagata and E. Banks,
     Macromolecules, 14, 807 (1981).

28. I. Haller, R. Federo, M. Hatzakis and E. Spiller, J.
     Electrochem. Soc., 126, 154 (1979).

29. K. Okada, T. Mochizuki, S. Sakabe, H. Shiraga, T. Yabe
     and C. Yamanaka, Appl. Phys. Lett. 42, 231-3 (1983).

HYDROPHILIC MACROMOLECULAR FERRICENIUM SALTS – PART I: PREPARATIVE AND SPECTROSCOPIC FEATURES OF SELECTED MONONUCLEAR FERRICENIUM COMPOUND

Eberhard W. Neuse

Department of Chemistry
University of the Witwatersrand
Johannesburg, 2000, Republic of South Africa

INTRODUCTION

The chemical and spectroscopic properties of ferricenium (Fc^+) salts have been under investigation ever since the first communication from Woodward's laboratory.[1] In more recent years, ferricenium research has received a special impetus with the emergence of theoretical and experimental interest in mixed-valence and electron transfer phenomena. Even more significantly, certain ferricenium salts have just been found[2] to exhibit cancerostatic properties against Ehrlich ascites murine tumor, and this observation opens up an entirely new and extremely challenging vista in metallicenium research. The ferricenium system represents a cation radical, and it is this radical cationic character that may be implicated in the biological activity of the ferricenium complex (and, possibly, of related metallocenes possessing unpaired d electrons). Although the literature abounds with analytical, spectroscopic and other physico-chemical data for ferricenium (and, somewhat less so, for bi- and oligoferricenium) salts, a closer screening of the reported details reveals countless inconsistencies with respect to both composition and properties of the compounds studied.

For biomedical applications a precise knowledge of preparative details and compositional features of the ferricenium salts is a vital prerequisite, as is an assessment of the degree of product purely attainable in the various known syntheses, which in turn calls for the availability of unambiguous spectroscopic means of characterization. In an effort to provide such knowledge and information as part of a major biomedical investigation of non-polymeric and polymeric ferricenium compounds in this laboratory, we decided to repeat and modify as required some of the more

99

significant published ferrocene oxidation work, identify the obtained
ferricenium salts analytically, and reexamine their IR and Raman
spectroscopic features.   This effort is covered in the present
Part I.   A future publication (Part II) will deal with the attachment of the ferricenium cation to biocompatible macromolecules
with the ultimate aim of developing polymer-anchored ferricenium
salts as antitumor agents.

RESULTS   AND   DISCUSSION

A total of 12 ferricenium salts were investigated (Tables 1-3).
The compounds were initially prepared in repetitive experiments
exactly duplicating the literature prescriptions so as to establish
reproducibility of the reported results.   In a number of instances,
particularly whenever the findings failed to confirm the literature
data, preparative conditions were modified as required.   Product
yields generally proved to be far from quantitative, and further
losses had to be accepted in subsequent purifiction steps owing
to the lability of most ferricenium salts (proneness to cation
reduction by dioxygen and other reactants[3-8]) in the dissolved
state.   It is obvious, however, that other authors have experienced
similar problems, as yield data are conspicuously wanting in the
literature.   An individual treatment of salts 1-12 follows.

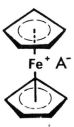

$$\text{Fe}^+ \text{A}^-$$

1	A^-	=	PF_6^-
2	A^-	=	BF_4^-
3	A^-	=	$C_6H_2N_3O_7^-$ (picrate)
4	A^-	=	I_3^-
5	A^-	=	$CCl_3COO^- \cdot CCl_3COOH$
6	A^-	=	$CCl_3COO^- \cdot 2\ CCl_3COOH$
7	A^-	=	$SbCl_4^-$
8	A^-	=	$BiCl_4^-$
9	A^-	=	$FeCl_4^-$
10	A^-	=	$FeBr_4^-$
11	A^-	=	$\left[Cr(NH_3)_2(SCN)_4\right]^-$
12	A^-	=	$C_8HCl_2N_2O_2^-$ (DDQH$^-$)

Table 1. Some properties of Ferricenium Salts 1 – 12[a]

Compound	Formula	Composition	Molecular Mass	Color	Melting Point /°C		Anal./% (Found / Calcd.)				Preparative Refs.[c]
						C	H	Fe	hal	N	
1	$C_{10}H_{10}F_6FeP$	$Fc^+ PF_6^-$	331.0	Purplish black	> 300[d]	36.35 / 36.29	3.10 / 3.05	16.52 / 16.87	33.98 / 34.44		6, 10, 12, 16, 17
2	$C_{10}H_{10}BF_4Fe$	$Fc^+ BF_4^-$	272.8	Purplish black	> 300	44.15 / 44.02	3.61 / 3.69	20.80 / 20.47	26.94 / 27.85		10, 33, 34
3	$C_{16}H_{12}FeN_3O_7$	$Fc^+ C_6H_2N_3O_7^-$	414.1	Black	181 – 182 dec.[e]	46.22 / 46.40	2.81 / 2.92	13.30 / 13.49		10.11 / 10.15	10, 24, 36, 38
4	$C_{10}H_{10}FeI_3$	$Fc^+ I_3^-$	566.7	Black	186 – 190 dec.[f]	21.09 / 21.19	1.79 / 1.78	9.66 / 9.85	67.29 / 67.17		10, 11, 41, 43
5	$C_{14}H_{11}Cl_6FeO_4$	$Fc^+ CCl_3COO^- \cdot CCl_3COOH$	511.8	Bluish black	127 – 128 dec.[g]	32.81 / 32.85	2.19 / 2.17	10.78 / 10.91	41.43 / 41.56		52, 53
6	$C_{16}H_{12}Cl_9FeO_6$	$Fc^+ CCl_3COO^- \cdot 2CCl_3COOH$	675.2	Bluish black	143 – 144 dec.[g,h]	28.57 / 28.46	1.72 / 1.79	8.32 / 8.27	47.30 / 47.26		21, 34, 51, 53
7	$C_{10}H_{10}Cl_4FeSb$	$Fc^+ SbCl_4^-$	449.6	Bluish black	154 – 156[i]	26.78 / 26.71	2.26 / 2.24	12.86 / 12.42	31.21[j] / 31.54		56
8	$C_{10}H_{10}BiCl_4Fe$	$Fc^+ BiCl_4^-$	536.8	Bluish black[k]	210 – 212	22.34 / 22.37	1.80 / 1.88	10.11 / 10.40	25.77[j] / 26.42		57b
9	$C_{10}H_{10}Cl_4Fe_2$	$Fc^+ FeCl_4^-$	383.7	Bluish black	> 300[l]	31.33 / 31.30	2.78 / 2.63	28.91 / 29.11	36.79 / 36.96		3, 62 – 64
10	$C_{10}H_{10}Br_4Fe_2$	$Fc^+ FeBr_4^-$	561.5	Brownish black	240 – 242 dec.[m]	21.51 / 21.39	1.78 / 1.80	19.08 / 19.89	56.01 / 56.92		69, 70
11	$C_{14}H_{16}CrFeN_6S_4$	$Fc^+ [Cr(NH_3)_2(SCN)_4]^-$	504.1	Reddish gray-blue	> 300	32.04 / 33.35	3.13 / 3.20	11.89 / 11.08		15.95 / 16.67	24
12	$C_{18}H_{11}Cl_2FeN_2O_2$	$Fc^+ C_8HCl_2N_2O_2$	414.0	Black	179 – 180[n]	52.29 / 52.21	2.50 / 2.68	13.82 / 13.49	16.97 / 17.12	6.42 / 6.77	73

[a] This work (pertinent literature data cited as footnotes). Fc^+ = ferricenium. Data are reported for recrystallized salts with the exception of 6, 8, 10 and 11 evaluated as crude products (see text). [b] Observed only up to 300°C. [c] Selected references only. [d] 160°C dec. (ref. 18). [e] Explosive degradation. 166–167°C dec. (ref. 34); 170–180°C dec. (ref.38). [f] Degrades sharply within 1–2°C. 169–171°C dec. (ref.41); 140°C dec. (ref. 5); 192°C dec. (ref. 43). [g] Mild decomposition; 129–130°C (ref. 21); 126–128°C dec. (refs. 34, 38). [h] Infusible (ref. 3). [i] Metallic luster. [j] Additionally found %O: 0.41 (7), 0.38 (8). [k] Metallic luster. [l] > 300°C (ref. 56). [m] 244°C dec. (ref. 70). [n] 178°C (ref. 73).

Table 2. IR absorption bands of the ferricenium cation[a]

Compound	Frequency/cm^{-1} [b]												Selected References
	ν_8, ν_{17}	ν_{20}	ν_{10}	ν_{25}, ν_{31}	ν_{18}	(ν_{27}, ν_{33})	ν_{19}	ν_9	(ν_2, ν_{14})	ν_{21}	(ν_{16})	
1	3120 m	1420 m	1114 vw	1060 w	1012 wm	-[c]	-[c]	-[c]	-[c]	505 vw	421 wm[d]	18
2	3115 m	1421 m	-[c]	-[c]	-[e]	890 vw	855 s	820 w, sh	804 vw	501 vw	422 w	27, 32
3	3105 wm[f]	1420 m	1112 vw	1057 m	1012 m	-[g]	861, 850, 842 s	820 vw[d]	801[h]	498 vw[h]	419 vw, br	24
4	3105 wm	1415 m	1115 wm	1054 vw	1011 wm	-[g]	854 s	-[g]	808 vw	502 vw, br	422 vw[h]	46
5	3110 m	1421 m	-	1060 w, br	1012 m	-[c]	-[c]	-[c]	-[c]	-[g]	423 m[d]	53
6	3113 m	1418 m	1112 vw	1063 vw, br	1011 wm	-[g]	-[c]	-[c]	-[c]	486 m[i]	427 wm[d]	53
7	3090 wm[j]	1414 m	1110 vw	1060 w	1010 wm[k]	-[g]	854, 845 s	-[g]	801 vw	505 vw, br	-[g]	
8	3100 wm	1413 m	1118, 1110 w	1061 w	1010 wm[k]	885 vw	855, 843 s	820 vw, sh	801 vw	501 vw	-[g]	
9	3102 wm	1415 m	1112 vw	1060 vw	1010 wm	883 w	851 s	-[g]	802 vw	503 vw	425 vw[e]	23, 64
10	3100 wm	1414 m	1112 vw	1055 vw	1009 wm	880 vw	855 s	-[g]	802 vw	503 vw	420 vw[h]	
11	3102 w	1416 m	1115 vw[h]	1060 vw	1010 w	885 vw	855 s	-[g]	805 vw	-[c]	-[g]	24
12	3120 wm	1416 m	1115 vw	1060 w	1016, 1001 w	880 w	849 s	-[g]	-[g]	-[c]	-[g]	73

[a] This work, KBr matrix (Nujol mull for 2). Very similar band positions in Nujol spectra. [b] IR active modes (IR inactive or Raman active modes in parentheses); numbering per ref. 26c. [c] Buried in anion absorption. [d] Major contribution by anion contribution. [e] Partly merged with anion absorption. [f] Shoulder at 3080 cm^{-1}. [g] Not observed. [h] Not always detectable. [i] Additional peak at 472 cm^{-1} (m; ν_{11}). [j] Shoulder at 3105 cm^{-1}. [k] Multiplet.

Table 3. Raman shifts of the ferricenium cation[a]

Compound	Frequency/cm^{-1}[b]							Selected References
	ν_3	ν_{25}	ν_2	ν_{28}	(ν_{21})	ν_{16}	ν_4	
1	1560 w, br[c]	1070 vw, br	805 wm, br	610 w	510 w[d]	—[e]	308 vs	12
2	1530 w, br	1096, 1080 w[f]	820 w, br	610 vw, br[c]	510 vw[c]	420 vw[d,g]	310 vs	29b, 35
3	—[e]	1075 vw, br	—[j]	—[e]	508 vw	420 vw[g]	308 s	35
4	—[e]	1070 vw, br	810 w, br	620 vw[d]	—[e]	408 vw[d]	310 s	35
5	1550 w, br[c]	1070 w, br[d]	810 w	—[e]	520 vw	—[e]	302 s	
6	1550 w, br	1060 vw, br[d]	808 w, br	610 vw[d]	505 vw[d]	415 w[d]	304 s	12
7	1545 w, br	1070 vw, br	810 w, br	600 vw[d]	510 vw[d]	412 vw[d,g]	305 vs	
8	1540 w, br	1080, 1060 wm	805 vw	600 vw, br	—[e]	—[e]	302 s	
9	1560 w, br[c]	1063 w	805 w, br	608 w	—[e]	410 vw		35
10	1535 vw	1065 w, br	820 vw, br	605 vw, br	504 vw[h]	—[e,i]	310 vs	
11	1525 wm, br[c]	1070 vw, br[c]	802 w, br	—[e]	506 vw	—[e]	308 vs	
12	—[j]	1090, 1060 vw	805 w, br	608 vw	504 vw	406 vw[d,g]	306 s	

[a] In rotating capillaries; excitation frequency λ$_0$ = 514.5 nm. [b] IR active mode in parentheses; numbering per ref. 26c.
[c] Observable only with λ$_0$ = 488.0 nm. [d] Not always observable. [e] Not observed. [f] 1075 w, br with λ$_0$ = 488.0 nm.
[g] In addition, 480 – 485 vw (possibly vw gaining R allowedness). [h] Observable only with λ$_0$ = 647.1 nm.
[i] 470 vw, observable with λ$_0$ = 647.1 nm. [j] Buried in anion signal.

Ferricenium hexafluorophosphate   ($\underset{\sim}{1}$)

The preparation of $Fc^+PF_6^-$ has commonly been brought about by oxidation of ferrocene in concentrated sulfuric acid, followed by precipitation of the salt from the diluted acidic solution in the presence of hexafluorophosphate anion.[6,9-13]   The generation of elemental sulfur attests to the active oxidant role of the medium in this process.   However, reaction with dissolved dioxygen doubtlessly contributes to the overall oxidation process, and this autoxidation has been suggested to proceed <u>via</u> the (metal-prononated) ferrocenonium ion,[14] possibly[15] involving the stoichiometry given in eq. 1 (Fc = ferrocene).   Other

$$4 \ Fc \ + \ 4H^+ \ + \ O_2 \longrightarrow \ 4 \ Fc^+ \ + \ 2H_2O \qquad\qquad (1)$$

reported preparations include the treatment of ferricenium tetra-fluoroborate ($\underset{\sim}{2}$) with PF_6 anion in aqueous solution[10] and the oxidation of ferrocene by nitric acid[16] or silver hexafluorophosphate.[17]

In our hands, the sulfuric acid oxidation proved efficient and reproducible, giving a crude product salt that was readily purified by recrystallization.   The nitric acid oxidation method[16] was found to be less satisfactory, being accompanied by side reactions. Similarly unsatisfactory proved to be the reaction proceeding <u>via</u> the tetrafluoroborate salt,[10] as conversion remained incomplete and the product crystallizing from solution tended to include inorganic PF_6 salt, requiring additional purification steps.   [17] Discouraging results were also obtained by oxidation with $AgPF_6$; we found the crude product in all experiments to contain metallic silver difficult to remove because of its tendency to disperse in colloidal form in the solvents used and thus accompany the ferricenium salt through several recrystallization cycles.

Properly purified $\underset{\sim}{1}$, as obtained in this study, does not melt or decompose below 300°C.   This behavior contrasts with literature reports[13,18] giving a decomposition point of 160°C.   The salt is stable in the solid state;   it shows reasonable resistance to degradation when dissolved in aqueous acidic media, yet tends to decompose slowly in neutral organic solvents.   Such behavior is common to most ferricenium salts and has been abundantly observed elsewhere.[19-22]   The compound, in agreement with the literature,[18] gives the characteristic[23,24] ferricenium pattern in the IR spectrum with prominent bands at 3120 (ν_8, ν_{17}; CH stretch), 1420 (ν_{20}; antisym. CC stretch), and 1012 cm^{-1} (ν_{18}; in-plane CH bending) (Table 2).   The typical out-of-plane CH bending absorption near 850 cm^{-1} (ν_{19}),[23,24] of appreciable diagnostic value with other ferricenium compounds,[25] is not observable here because of overlap with the powerful anion absorption in the 800-900 cm^{-1} range.[26]

In the low-frequency region the spectrum displays a band of weak to moderate intensity at 421 cm^{-1}. Absorption at this position was previously observed[24] in $\underset{\sim}{1}$ and ascribed[24,27,28] to the antisymmetric metal-ring stretching mode ν_{11} of the ferricenium complex. However, in view of the much lower strength observed by us in the spectra of other ferricenium salts, we assign most of this peak's intensity to an anion vibration; ammonium hexafluoro-phosphate absorbs weakly at 410 cm^{-1}. Moreover, the small component expected to be contributed by the ferricenium complex is likely to have its origin in the symmetric ring tilt mode ν_{16}, which may have gained some IR allowedness in the cation because of mild anion-induced distortion from D_{5h} symmetry. The previous assignment[24,27,28] to ν_{11} is questionable in light of the large frequency difference between this peak and ν_{11} in ferrocene (478 cm^{-1}), as was pointed out before by Duggan and Hendrickson.[12]

The room-temperature Laser-Raman spectrum, in the region of higher frequencies, shows a broad feature of unknown origin[29] in the vicinity of 1555 cm^{-1}, a shift of low to moderate intensity at 1118 cm^{-1} associated with symmetric ring breathing (ν_3), and a very weak and broad signal centered at 1070 cm^{-1}, which is assignable to an out-of-plane CH bending mode (ν_{25}). Another out-of-plane CH bending vibration (ν_2) gives rise to a broad band at 805 cm^{-1} (w-m), and a feature near 610 cm^{-1} (vw) can reasonably be assigned to the ring distortion mode ν_{28}. The corresponding Raman shifts of the neutral ferrocene complex are typically[30] observed at 1106 (ν_3), 1062 (ν_{25}), 815 (ν_2), and 600 cm^{-1} (ν_{28}). The low frequency region of the spectrum is dominated by the intense resonance-enhanced shift at 308 cm^{-1}; the peak, previously reported,[10,12] is due to the symmetric, skeletal stretching mode ν_4. The corresponding tilting mode ν_{16} should manifest itself as a maximum in the neighborhood of 400 cm^{-1}; We have not been able, however, to detect this band clearly and reproducibly in the more than ten scans examined.

The Raman shifts cited in the foregoing for the hexafluoro-phosphate $\underset{\sim}{1}$ have similarly been found to characterize the spectra of most of the other ferricenium salts of this study (Table 3) and will not receive major attention in the subsequent text.

Ferricenium Tetrafluoroborate ($\underset{\sim}{2}$)

Although the tetrafluoroborate, $Fc^+BF_4^-$, represents one of the more frequently quoted ferricenium salts, not much detailed prepar-ative information is available. In the older literature, only two synthetic procedures, both involving the oxidation of ferrocene with an excess of p-benzoquinone in an acidic environment as earlier indicated by Wilkinson and collaborators,[1] were outlined in some detail.[10,31] Whereas both Stölzle and Nesmeyanov et al.[31] used boron trifluoride etherate as the acidic reactant, Hendrickson et al.[10]

employed aqueous tetrafluoroboric acid.   In our hands the method of the last-named authors[10] proved somewhat superior, notably with ethereal tetrafluoroboric acid used in place of the aqueous acid, giving a product anion structure more closely approaching the BF_4 composition than obtained by the method of Stölzle and Nesmeyanov et al.[31]   A procedure suggested more recently,[32] viz. oxidation of ferrocene by copper(II) tetrafluoroborate, was found to be inferior as the product invariably contained copper(I) tetrafluoroborate and required meticulous fractionating crystallization for decontamination.   Similarly discouraging proved to be an attempt at oxidation by silver tetrafluoroborate, which, as in the hexafluorophosphate case, entailed high contents of difficultly removable metallic silver in the product salt.   The most recent literature contains a paper by Schumann[33] describing the preparation of 2 by ferrocene oxidation with nitromethane in the presence of tetrafluoroboric acid etherate.   In our hands, this method turned out to be the most efficient and reproducible one, giving consistently high crude product yields.   In this and all other procedures examined, it proved necessary to add repeated recrystallization steps to obtain a satisfactory elemental composition in accord with 2.   The oxidation of ferrocene by air in ether/tetrafluoroboric acid[34] was not reinvestigated by us.

Pure 2, as obtained in our laboratory by ferrocene oxidation with either nitromethane or benzoquinone, is infusible up to 300°C and stable in the absence of moisture.   The compound's IR spectrum (Table 2), substantially in accord with reported data,[27,32] shows the known ferricenium bands with the exception of the peaks due to ν_{10}, ν_{18}, and $\nu_{25,31}$ (merged with broad anion absorption in 1000–1100-cm^{-1} region).   The spectrum must be recorded on a Nujol mull; pelletizing in a KBr matrix results in partial reduction of the ferricenium cation even when performed under an argon blanket.   The room-temperature Raman data are summarized in Table 3. The shifts due to ν_4 and ν_{16} were previously observed under low-temperature conditions, albeit at slightly different positions, by Gächter et al.[29,35]   The electronic Raman transition manifested as a strong band at 213 cm^{-1} in the low-temperature spectrum of these authors is not observable under our room-temperature conditions.

Ferricenium Picrate (3)

The picrate 3, $Fc^+C_6H_2N_3O_7^-$, can be prepared readily by oxidation of ferrocene with sulfuric acid, trichloroacetic acid/dioxygen, or p-benaoquinone and subsequent treatment with picric acid.[1,10,24,34,36-38]   Most of the products of the cited investigations were described as ferricenium picrate (3), possessing a mononuclear anion structure(confirmed in one study by an X-ray structure analysis[37]).   Decomposition points of 166–167°C[34] and 170–180°C[38] were reported for the purified salt.   One of the papers[34] lists an additional salt, mp 119–120°C, for which analytical data suggest a dipicrate composition, (ferricenium picrate-picric acid solvate) $Fc^+C_6H_2N_3O_7^- \cdot C_6H_3N_3O_7$.

Reinvestigation of the published procedures in our laboratory confirmed the formation of both mono- and dipicrates under most of the preparative conditions specified, and even the existence of a tripicrate was evidenced in some of the experiments.   Thus, reactions proceeding in benzene medium, with ferrocene and picric acid present in equimolar quantities,[36] gave almost exclusively the monopicrate 3, whereas a one-molar excess of the acid over the metallocene in the same medium[10] led to the formation of 3 as the predominant product in addition to a small proportion of the dipicrate.   In aqueous acidic phase, with approximately equimolar quantities of acid and metallocene employed,[24] the primary product was not 3 as claimed,[24] but rather represented a mixture predominantly composed of the dipicrate, and a tri-picrate by-product was occasionally isolated as well.   In all those instances, however, the di- and tripicrates converted readily to the monopicrate upon repeated, careful recrystalli-zation from water, the excess picric acid remaining in the mother-liquors.

Pure ferricenium picrate (3) melts at 181-182°C (dec.) and shows good stability in the solid state.   The IR spectrum, although dominated by a multitude of high-intensity peaks of the trinitrophenoxide system, shows most of the ferricenium bands without encumbrance (Table 2).   Of particular interest is the out-of-plane bending absorption near 850 cm^{-1}.   While emerging as a broad apparent singlet in the KBr spectrum of crude material, it exhibits appreciable site-splitting in the spectrum of the recrystallized salt, appearing now as a triplet feature.   Reduced splitting is observed in the Nujol spectrum. An additional very weak band in this region, emerging at 820 cm^{-1}, is assignable to the out-of-plane bending mode ν_9, and there is probably a contribution to this peak by an anion mode (picric acid absorbs weakly at 830-820 cm^{-1}).   However, we found less pure material, especially first-generation recrystallizates, to show this band in increased intensity (together with enhanced ν_{10} at 1114 cm^{-1} and ν_{11}, ν_{21} at 500-480 cm^{-1}), suggesting con-tamination by ferrocene in these samples.   By and large the spectrum resembles the one described by Pavlík and Klikorka,[24] although these authors reported a high intensity for the 1116-cm^{-1} peak (ferrocene impurity, vide supra) and listed medium-strong to strong bands at 1598, 1462 (doubtlessly Nujol artifacts), 1195, 1127, and 405 cm^{-1} not observed by us in either KBr or (properly compensated) Nujol matrix.

The Raman spectrum (Table 3) offers no unusual features. The sharp peak due to ν_4, previously observed by Gächter et al.,[35]

emerges as the prominent signal.   Again, the strong electronic
Raman band reported by these authors to occur at 213 cm^{-1} has not
been detected by us.

Ferricenium Triiodide   (4)

Ferricenium   polyiodides, $Fc^+I_x^-$ (x = 2-20), have been the
subject of numerous investigations.[4,5,10,11,39-46]   In nearly
all of these, elementary iodine served as both the oxidant and
the anion source.   The oxidation of ferrocene by iodine is smoothly
achieved in a variety of organic solvents, such as hydro- and
halocarbons, ethanol, or acetone.   It proceeds both under
thermal-equilibrium conditions[40]   and with photoassistance,[40,45]
the ultimate result in either case being the complexation of iodide
anion with variable molar equivalents of iodine as exemplified
in eq. 2 for the case of the triodide 4.   The last-named salt,
being the prototype of the class, has been characterized by an
X-ray crystal structure analysis,[42] which indicates the I_3 anion
to be linear and the ferricenium cation to be rotationally
disordered.

$$Fc \; + \; {\textstyle\frac{3}{2}} I_2 \; \rightleftharpoons \; Fc^+ I_3^- \qquad\qquad (2)$$

Contrasting with the syntheses of salts 1 - 3, which are reasonably
straightforward and have not provoked major controversy in the
literature, the ferrocene oxidation with iodine has been accomplished in several laboratories[4,10,11,39-43] with variable, and
sometimes contradictory, results with respect to composition and
properties of the ferricenium iodides obtained.   A comprehensive
study of the ferrocene/iodine reaction recently completed in our
laboratory has shown the generation of the prototype 4 to be
considerably less straightforward than heretofore assumed.[46]   In
fact, as it turns out,[46]   the formation of polyiodides essentially
conforming to the composition of 4 is strictly conditional upon
the use of molar I_2/Fc ratios not exceeding the value of 1.5
prescribed by eq. 2, irrespective of the solvent medium employed.
While confirming the observations of one group of authors (working
with benzene solvent),[11] these results contradict other literature
reports that indicate 4 to be formed at I_2/Fc ratios as high
as 2.5 (in cyclohexane)[10]   and even 3.0 (in 1,2-dichloroethane).[43]
Also at variance with published data, we have been unable[46]
to substantiate the existence of a diiodide, $Fc^+I_2^-$,[47] or,
at the other extreme, of an icosaiodide, Fc^+I_{20},[39]   even though
the preparative conditions chosen in each case (large excess
of, respectively, ferrocene and iodine) should have been conducive
to the formation of these species.   On the other hand, in agreement
with previous observations, all higher-order polyiodides have been
found to convert ultimately to the more stable triiodide by
repeated recrystallization from acetone.

Ferricenium triiodide ($\underset{\sim}{4}$), as typically prepared from iodine and ferrocene ($I_2/Fc = 1.5$) in ethanol, melts with decomposition at 186-190°C. Contrasting with the higher-order ferricenium poly-iodides,[46] it possesses excellent stability in the solid state and good stability in solution. The IR and Raman spectra of $\underset{\sim}{4}$ display the characteristic ferricenium bands (Tables 2,3), the comparatively high intensity of the 1112-cm^{-1} IR band (three- to four-fold enhancement relative to other ferricenium salts) being a notable feature. The Raman shift at 310 cm^{-1} (ν_4) was previously reported;[35] however, it was assigned[35] to an electronic Raman transition.[48]

Ferricenium Trichloroacetates $\underset{\sim}{5}$ and $\underset{\sim}{6}$

Ferricenium salts containing the trichloroacetate anion have been amply investigated in recent years.[12,15, 21, 34, 38, 50-53] They are conveniently prepared either by allowing ferrocene to react with trichloroacetic acid in benzene medium or by treating an aqueous acidic ferricenium sulfate solution with trichloroacetic acid. Ferrocene oxidation in the former case is effected both in an autoxidation process with participation of dioxygen (eq. 1) and in a redox reaction involving an excess of the trichloroacetic acid, which affords trichloroacetaldehyde by-product (eq. 3).[15,52,53] The primary ferricenium trichloroacetate generated in either type of reaction stabilizes through increase of anion bulk by way of complexation with one or two additional molecules of the tri-chloroacetic acid (TCA) to give the mono- and bis(trichloroacetic acid)-solvated salts $\underset{\sim}{5}$, $Fc^+ CCl_3COO^- \cdot CCl_3COOH$, and $\underset{\sim}{6}$, $Fc^+ CCl_3COO^- \cdot 2CCl_3COOH$.

$$2\ Fc\ +\ 3\ CCl_3COOH \longrightarrow 2\ Fc^{\pm}\ CCl_3COO^-\ +\ CCl_3CHO\ +\ H_2O$$
$$\downarrow TCA$$
$$\longrightarrow \underset{\sim}{5},\ \underset{\sim}{6}$$

There has been some inconsistency in the literature as to the specific experimental conditions conducive to the formation of either $\underset{\sim}{5}$ or $\underset{\sim}{6}$. In a recent study aiming at a clarification of the controversy it was found[53] that the disolvate $\underset{\sim}{6}$ can reproducibly, and in satisfactory purity, be prepared from aqueous ferricenium sulfate in the presence of trichloroacetic acid in more or less stoichiometric proportions. The alternative preparation from trichloroacetic acid and ferrocene (3:1) in dioxygen-saturated benzene was found to be equally reproducible. Both approaches had been suggested previously by other authors.[38,51] On the other hand, at variance with an earlier report,[52] the mono-solvate $\underset{\sim}{5}$ could not be synthesized cleanly in benzene medium since the less soluble, concurrently generated disolvate $\underset{\sim}{6}$ invariably crystallized from the solution as the predominant or exclusive product even under widely varied conditions of reactant ratio and concentration.

The most satisfactory (and, in fact, the only reproducible) method of preparing 5, developed in the cited investigation,[53] originates from 6 and involves a simple recrystallization of the latter from water.   In aqueous solution, the mono-solvated trichloroacetate anion exists in equilibrium with the disolvated anion (eq. 4) and, presumably less hydrated, forms a salt (5) that possesses lower solubility in water and so crystallizes preferentially.   The same salt 5 can also be obtained, although

$$CCl_3COO^- \cdot 2\ CCl_3COOH \rightleftharpoons CCl_3COO^- \cdot CCl_3COOH + CCl_3COOH \qquad (4)$$

$$\Big\uparrow H_2O$$

$$CCl_3COO^- + H_3O^+$$

in an inferior state of purity, from the precursor 6, by controlled extraction of one molecule of trichloroacetic acid with diethyl ether.   The salt 6 used in the present study was prepared from trichloroacetic acid and ferrocene (3:1) in benzene,[53]   and the aqueous recrystallization method[53] was utilized for the conversion of 6 to 5.

Ferricenium trichloroacetate mono(trichloroacetic acid)-solvate (5) melts at 127–128°C, i.e. in the range reported by other authors[21,34,38]   for 6;   it is stable in the solid state. Although dominated by powerful absorption of the anion system,[53,54] the IR spectrum displays most of the prominent ferricenium bands, only the 850-cm^{-1} peak being obscured by anion absorption in this region (Table 2).   In the Raman spectrum (Table 3), the most characteristic band, as expected, emerges at 308 cm^{-1} (ν_4).

Ferricenium trichloroacetate bis(trichloroacetic acid)-solvate (6) melts with decomposition in the neighborhood of 140–144°C. The salt is stable in the absence of moisture and gives ferricenium bands similar to those of 5 in the IR and Raman spectra (Tables 2 and 3), a notable exception being the IR doublet appearing in moderate intensity at 486–472 cm^{-1}.   This doublet, ascribed to the skeletal modes ν_{21} and ν_{11} , is not normally exhibited in noticeable intensity by the ferricenium cation, and the cause of its significant enhancement in the spectrum of 6 is not known. Comparative X-ray diffraction and ^{57}Fe Mössbauer studies of both 5 and 6 may help shed some light on this anomalous spectroscopic feature.[55]

Ferricenium Tetrachloroantimonate (7) and Tetrachlorobismuthate (8)

Ferricenium salts comprising complex halogenopentelide anions,

notably those derived from arsenic, antimony, and bismuth, have
been extensively investigated.[56,57] While no simple ferricenium
haloarsenates have been reported,[58] both the tetrachloro-
antimonate(III) 7 and the tetrachlorobismuthate(III) 8 have been
described. The preparation of $Fc^+ SbCl_4^-$ (7) was first reported
by Cowell et al.,[56] their procedure involving a redox reaction
between ferrocene and hexachloroantimonate(V) anion. The latter
acted as the oxidant, Sb(V) being reduced to Sb(III) in this
process (eq. 5) (X = benzyldimethylphenylammonium). The

$$2 \text{ Fc} + X^+ SbCl_6^- \rightleftharpoons 2 \text{ Fc}^+ + X^+ + \left[SbCl_6^{3-} \rightleftharpoons SbCl_4^- + 2 Cl^- \right]$$

$$\longrightarrow Fc^+ SbCl_4^- + Fc^+ + X^+ + 2 Cl^- \qquad (5)$$

$$\underset{\sim}{7}$$

net result thus was the formation of one mole of the (less soluble)
ferricenium tetrachloroantimonate(III) (7) crystallizing from the
reaction mixture, with another mole of ferricenium salt remaining
in solution as the chloride. The salt 7 was found in that work[56]
to be infusible up to 300°C. IR and electronic absorption
spectra confirmed the ferricenium cation structure. In a later
publication by Landers et al.,[57a] 7 was briefly mentioned as a
product arising from ferrocene and antimony(III) choride (1:1)
dissolved in an indifferent solvent, such as acetone, but no preparative details or characterizing data were given. In our
laboratory, the method of Cowell et al.[56] proved reproducible
only insofar as it afforded, in several repeat runs, grayish-blue
ferricenium salt giving the typical electronic absorption band
at 617 nm as stated.[56] The IR spectrum, however, exhibited
not only the ferricenium bands but those of the ammonium cation
X^- as well, and microanalytical results were in reasonable
agreement with the composition of a double salt of the ferricenium
bis(benzyldimethylphenylammonium) nonachlorodiantimonate(III) type,
$Fc^+ [C_{15}H_{18}N]_2^+ [Sb_2Cl_9]^{3-}$. The product was not further investigated.

Utilizing, next, the approach suggested by Landers et al.,[57a]
i.e. ferrocene oxidation by O_2 in the presence of $SbCl_3$ in dry
acetone, we collected an ill defined oxygen-containing crude salt,
from which it was possible to isolate by fractionating crystallization under anhydrous and anaerobic conditions a crystalline
product constituting reasonably pure 7. Best results with
respect to composition were obtained with $SbCl_3$/Fc molar ratios of
2-3. Another oxygen-containing crude product, partially melting
below 100°C, resulted from analogous reactions performed in

anhydrous benzene under the conditions of Rheingold et al.,[57c,59] although multiple recrystallization eventually gave small amounts of reasonably pure 7 as well.

The formation of 7 from ferrocene, antimony(III) chloride, and dioxygen is proposed to follow the path and stoichiometry of eq. 6. The reaction is probably preceded by partial hydrolysis

$$4 \text{ Fc} + 6 \text{ SbC}\ell_3 + O_2 \longrightarrow 4 \text{ Fc}^+ + 4 \text{ SbC}\ell_4^- + 2 \text{ SbOC}\ell \tag{6}$$

of antimony salt brought about by traces of moisture, the resultant hydronium ion then participating in the oxidation of the metallocene (eq. 1) and the chloride ion complexing with $SbC\ell_3$ reactant to form $SbC\ell_4^-$; eq. 6 in effect represents a summation of these reaction steps.

Ferricenium tetrachloroantimonate as obtained in our work gives microanalytical results (Table 1) in acceptable agreement with the composition of 7, although oxygen contents of 0.3-0.5% are stubbornly retained even after many cycles of recrystallization. The salt melts in the vicinity of 155°C. This contrasts with the reported[56] infusibility of 7 below 300°C but is well in keeping with the behavior of some aliphatic and aromatic tetrachloroantimonates observed[60] to melt in the range of 150-175°C. The salt is stable when protected from moisture but decomposes rapidly in (neutral) solution. The IR spectrum displays the typical ferricenium bands (v_{19} appearing as a doublet), and the emergence of two intense low-frequency bands due to IR active antisymmetric Sb-Cℓ stretching modes (322 and 290 cm^{-1} ; both split into multiplets; additional maxima possibly hidden in the KBr absorption edge below 270 cm^{-1}) indicates the anion either to be tetrahedrally coordinated yet considerably distorted from T_d symmetry, or else to be polymeric, i.e. octahedrally coordinated with edge sharing brought about via chlorine bridges.[61]

The tetrachlorobismuthate, $Fc^+BiC\ell_4^-$ (8), was first reported in the afore-cited paper by Landers et al.,[57a] who prepared the salt from ferrocene and bismuth(III) chloride in dioxygen-containing acetone without giving experimental details. Although analytical or melting data for the product were omitted in that work, the authors confirmed the composition of 8 by X-ray crystallography,[57a,b] which showed a normal cation structure with eclipsed cyclopentadienyl rings and an infinite chain of irregular edge-sharing $BiC\ell_6^{3-}$ octahedra in the anion complex.

Using the cited[57a] approach, we obtained reasonably pure 8, crystallizing first from the reaction mixture. Additional product fractions isolated from the mother-liquors proved to be combinations of 8 with BiOCℓ, the latter having formed in a reaction

analogous to that of eq. 6.   These fractions tended to deposit
from mother-liquors as uniform, silver-gray crystalline 2:1
adducts possessing nearly the same melting point as 8 yet giving
an X-ray powder diffractogram clearly indicative of a physical
mixture of the two components.

Ferricenium tetrachlorobismuthate (8) melts rather sharply at
210-212°C. Although a composition in accord with 8 is indicated
by elemental analysis (Table 1), the salt obtained in this laboratory possesses an oxygen content of 0.4-0.5% as similarly
observed with 7.   The tetrachlorobismuthate is even less stable
in solution than the antimony analog, and we have therefore been
unable to purify the compound by recrystallization.   The IR
spectrum is remarkably similar to that of 7 over the range from
4000 to 400 cm^{-1}, displaying ν_{19} as a doublet.   In the low-
energy part of the spectrum a broad region of moderately
strong absorptions due to Bi-Cl stretching modes of the
irregular edge-sharing anion octahedra extends from 300 cm^{-1}
towards lower frequencies, with distinct maxima apparent at 283
and 270 cm^{-1}.   The ferricenium bands in the IR and Raman spectra
of both 7 and 8 are recorded in Tables 2 and 3.

Ferricenium tetrachloroferrate(III)   (9)

Ferricenium tetrahaloferrates(III), $Fc^+FeX_4^-$, where
X = Cl, Br, I, have all been reported in the literature.   Although
it has since been found[46] that the salt labelled as the tetra-
iodoferrate[23] is, in fact, the triiodide 4, both the tetra-
chloro- and the tetrabromoferrate are stable salts and have received
much experimental attention.   Indeed, the readiness with which the
tetrachloroferrate 9 is generated from ferrocene under many degradative conditions has given rise to a great number of preparative
approaches described in the literature and recently summarized.[62]
Present coverage of 9 is limited to topics of major preparative
and spectroscopic significance, and for details the cited publication[62] may be consulted.

The salt 9 is a typical end product of the reaction of
ferrocene with chlorinating agents.   The preparatively most
useful one of these reactions, developed by Motz and Pinnell,[63]
involves the treatment of the metallocene with refluxing sulfuryl
chloride;   leading to oxidation of ferrocene and concomitant
partial destruction of the substrate with generation of Fe^{3+} cation,
the latter in turn complexing readily with excess Cl^- in the
chlorination mixture to give the stable $FeCl_4^-$ anion species, it
ultimately affords 9 in addition to cyclopentadiene-derived polymers (eq. 7).

$$2\ SO_2Cl_2 + 2\ Fc \longrightarrow Fc^+FeCl_4^- + 2\ SO_2 + polymer \qquad (7)$$

9

A different, and more convenient, route to $\underset{\sim}{9}$ utilizes the oxidizability of ferrocene by iron(III) chloride in indifferent media first proposed by Nesmeyanov et al.[3] and further elaborated by Spilners.[64]   The reaction leads to a crude product assigned[64] a hexachlorodiferrate(II,III) structure (eq. 8), which reportedly[3,64] converts to pure $\underset{\sim}{9}$ upon recrystallization from ethanol.

$$\text{Fc} + 2\ \text{FeC}\ell_3 \longrightarrow \text{Fc}^+\text{Fe}_2\text{C}\ell_6^- \xrightarrow[-\text{FeC}\ell_2]{} \text{Fc}^+\text{FeC}\ell_4^- \qquad (8)$$

$$\underset{\sim}{9}$$

In our hands, the method of Motz and Pinnell,[63] while affording $\underset{\sim}{9}$ as stated, proved most inefficient, with degraded and insoluble material found to constitute most of the crude reaction product.   Nesmeyanov's procedure[3] gave high yields of the intermediate as described.   It proved necessary, however, to perform the recrystallization of the crude product from methanol acidified with HCℓ, or alternatively, from thionyl chloride to bring about conversion to pure $\underset{\sim}{9}$.   In the absence of acid, but with traces of moisture present, the tetrachloroferrate, equilibrating in alcoholic solution with the oxygen-bridged diferricenium μ-oxo-bis(trichloroferrate),[65] converted to the latter (eq. 9) to an extent sufficient to cause appreciable contamination of $\underset{\sim}{9}$ with the cocrystallizing μ-oxo-diferrate.   The contaminant is readily identified[65] by its characteristic low-frequency IR absorption consisting of two maxima at 365(s) and 321(m) cm^{-1}.

$$2\ \text{FeC}\ell_4^- + \text{H}_2\text{O} \rightleftharpoons \left[\text{C}\ell_3\text{Fe-O-FeC}\ell_3\right]^{2-} + 2\ \text{HC}\ell \qquad (9)$$

Ferricenium tetrachloroferrate (9) is infusible below 300°C and is distinguished by an excellent stability deriving from the thermodynamically favored, undistorted tetrahedral coordination achieved in the FeCℓ_4 anion complex.[66]   The IR spectrum, in accord with Spilners[64] but in substantial disagreement with Pavlík and Klikorka,[24,67] features the clean absorption pattern of the ferricenium cation (Table 2), and the characteristic anion band derived from the antisymmetric Fe-Cℓ stretching mode[66b,68] emerges in high intensity at 379 cm^{-1} The Raman spectrum, in substantial agreement with that of Gächter et al.,[35] exhibits the ferricenium shifts (Table 3).   We do not, however, observe the strong 213-cm^{-1} band[35] under our room-temperature conditions.   On the other hand, we find the symmetric Fe-Cℓ stretching band at 336 cm^{-1} to appear in appreciably higher intensity than reported;[35] this may indicate that the sample of these authors was impure, containing a major proportion of the oxodiferrate, which fails to exhibit a signal near 336 cm^{-1} while otherwise giving a very similar spectrum.

Ferricenium Tetrabromoferrate (10)

An early paper by Riemschneider and Helm[69] described the
formation of 10 by attack of molecular bromine on ferrocene.
The reaction involved both oxidation and competing degradation of
the metallocene complex, the ferric ion generated in the degradative process then complexing with Br^- to give the stable $FeBr_4^-$
anion.   Several research groups using identical or similar
experimental conditons[24,41,70] have subsequently confirmed the
earlier work, and an independent synthesis by treatment of an
acidic,ferric ion-containing ferricenium sulfate solution with
bromide anion has also been reported.[23]   The product salt 10
has variously been described as khaki,[70] brown,[23,67] and dark
brown[24] in color; one publication[70] lists a decomposition point
of 244°C.

In our hands the original procedure of Riemschneider and
Helm[69] proved more satisfactory for the preparation of 10 than
some of the more recent ones,[23,41]   giving crude tetrabromoferrate
in the same high yield and degree of purity as reported in the
early paper,[69] and recrystallization from suitable solvent
systems, while feasible, did not enhance the salt's purity.

Crude ferricenium tetrabromoferrate (10) as obtained by
Riemschneider and Helm's procedure[69] has a decomposition point of
240-242°C, well in accord with that reported by others.[70]
Recrystallized salt no longer decomposes sharply; it merely
sinters at 185-195°C and undergoes gradual decomposition above
240°-250°C.   The salt, like its analog 9, possesses excellent
stability in the solid state and, in solution, is superior in
stability to the majority of other ferricenium compounds
investigated.   The IR and Raman spectra show the usual ferricenium
pattern (Tables 2 and 3).   In addition, the strong IR band due
to the antisymmetric Fe-Br stretching mode of the tetrahedral
$FeBr_4$ complex appears at 285 cm^{-1}.

Ferricenium (Diamminetetraisothiocyanatochromate)   (11)

Several ferricenium salts are known in which the counterion
represents a complex iso- or heteropolyacid anion structure,
such as the molybdate, vanadate, tungstate or silicotungstate.[24,71]
These have not, however, been investigated in detail.   The reason
for this can be seen in variability of composition and properties
caused by even minor modifications of the preparative procedure,
which generally involves treatment, in aqueous phase, of the
polyacid with acidic ferricenium sulfate solution.   In our
laboratory, for example, ferricenium molybdate species varying
in iron content from 5 to 9% (thus containing mono-, di- and tri-
ferricenium octamolybdates as major components) were obtained by

this method, and other authors reported the preparation under very
similar conditions of salts formulated as molybdates with much
higher iron contents.[24,71]   A similar situation holds for salts
comprising heteropolyacid anions.   Furthermore, we found the
IR spectra without exception to be dominated by anion bands to such
an extent as to render these spectra useless for characterization
of the ferricenium complex.   Salts of the iso- and heteropolyacids
will therefore not be covered in the present communication.

A metal-containing anion complex more useful within the scope
of the current investigation is the "Reineckate",   ferricenium
diamminetetraisothiocyanatochromate(III) (11).   Ferricenium
Reineckate   reportedly[24] precipitates from aqueous acidic ferricenium sulfate solution by treatment with Reineckate anion and
possesses an elemental composition well in accord with the simple
1:1 stoichiometry, Fc^+ $[Cr(NH_3)_2(SCN)_4^-]$ (11).

In our hands, the tersely described[24]   procedure proved
reproducible after slight modification,   affording 11   in high
yield.   The crude salt, in agreement with the literature,[24]
possessed substantially the expected elemental composition, and
recrystallization did not enhance the degree of purity.

Ferricenium diamminetetraisothiocyanatochromate(III) (11)
is infusible up to 300°C and, while quite stable as a solid,
degrades rapidly when dissolved in dipolar aprotic solvents.
The IR and Raman spectra display the ferricenium bands essentially
unencumbered by anion signals (Tables 2,3).   The moderately strong
to very strong IR bands at 1195 and 781 cm^{-1} reported for 11   in
the literature[24]   have not been observed by us, and the peaks at
1125, 1114, and 1010 cm^{-1} reported to be medium-strong to strong
we find to possess vanishingly low to low intensity, suggesting
the presence of unoxidized ferrocene in the examined[24] sample.

Ferricenium   2,3-Dichloro-5,6-dicyanohydroquinonide   (12)

2,3-Dichloro-5,6-dicyanoquinone   (DDQ) has in recent years
found use as an electron acceptor component in charge transfer
reactions.   Interaction of the compound with strong donors,
such as ferrocene or some of its derivatives, results in complete electron transfer with formation of ionic products.
With ferrocenes as donors, this reaction affords ferricenium
salts now known[72] to comprise a diamagnetic anion of the hydro-
quinonide type, $[C_8HC\ell_2N_2O_2]^-$, the hydrogen probably being
abstracted from the solvent   (eq. 10; SH = solvent).   The first
description of the electron transfer reaction between ferrocene
and DDQ is due to Brandon et al.,[73]   who obtained the ferricenium salt 12 by treating the metallocene with an approximately

equimolar quantity of DDQ in benzene medium, although the salt's
structure was believed in that early work to be of the ferricenium
dichlorodicyanoquinone radical anion type, DDQ⁻ .   The recrystallized salt was reported to melt at 178° and possess semi-conductor
properties.   A more recent publication covers the subject
of semiconducting ferricenium-DDQH salts in some detail.[74]

$$ Fc \; + \; O = \!\!\! \underset{\underset{\text{DDQ}}{\overset{CN \quad CN}{}}}{\overset{Cl \quad Cl}{}} \!\!\! = O \quad \xrightarrow[-S \cdot]{SH} \quad Fc^{+} \; \; {}^{-}O \!\!-\!\!\! \underset{\underset{12}{\overset{CN \quad CN}{}}}{\overset{Cl \quad Cl}{}} \!\!\! -OH \qquad (10) $$

Using the method of Brandon et al.,[73] we obtained 12
reproducibly in yields of 80-90%, and recrystallization gave analytically pure salt as stated.[73]   No compositional changes, such
as anion dimerization of the type observed with tetracyanoquino-
dimethane,[75] were detected when the DDQ/Fc molar ratio was
doubled.

Pure ferricenium 2,3-dichloro-5,6-dicyanohydroquinonide (12)
melts at 179-180°C.   The salt is very stable in the solid state
and is reasonably stable in dry ethanolic solution.   Our own
magnetic measurements (μ_{eff} = 2.42μ_B at 295K) confirm the earlier
observation of anion diamagnetism.[72] The IR spectrum shows
no anomalies, the ferricenium bands being exhibited (Table 2)
in addition to the reported[73] anion absorptions.   In the Raman
spectrum one finds the characteristic ν_4 at the expected position
(Table 3).

EXPERIMENTAL

Solvents, reagents

All solvents were dried over Molecular Sieves 4A (3A used
for carbinols and acetonitrile) and were routinely nitrogen-
saturated.   Ferrocene was recrystallized from hexane. p-Benzoquinone, antimony(III) chloride, and trichloroacetic acid were
purified by sublimation at 60-80°C/0.1 torr. Iron(III) chloride,
anhydrous, sublimed, was a commercial product (Merck).   All
other reagents were laboratory grade; they were used as
received from various commercial sources or were recrystallized
as indicated.

Instrumental Operations

Melting points, uncorrected, were determined in sealed capillaries.   IR spectra were taken on KBr pellets or Nujol mulls;
particularly sensitive salts were mulled with the matrix under argon.
Room-temperature Raman spectra (Ar laser; excitation wavelengths
488,0 and 514,5 nm) were obtained on samples contained in rotating
glass capillaries (2mm inner diam.) sealed under argon.   X-ray
photoelectron spectra ($A\ell$ K_α) were recorded at room-temperature on
powdered sample material coated onto a gold probe; binding
energies were calculated from kinetic energy data calibrated against
E_b(C 1s) = 284.6 eV.   X-ray powder diffractograms were recorded
using CuK_α radiation.   Magnetic susceptibilities were determined
at or near 295K by the Gouy method and were corrected for compound
diamagnetism; effective magnetic movements, μ_{eff}, were calculated
from the corrected mole susceptibilities, χ_{corr}, by the formula
$\mu^2_{eff} = 8\chi_{corr}T$.

Analyses

Microanalyses were performed by commercial laboratories
(Galbraith Laboratories, Knoxville, Tennessee; Analytical Laboratories Elbach, Engelskirchen, Germany; Robertson Laboratory,
Florham Park, N.J.) and by the microanalytical laboratories of this
University and the University of Mainz.   C,H, and N analyses were
routinely executed in duplo for each sample, and triple determinations were made for iron and the halogens.   In many instances,
however, poor combustion (or decomposition) behavior of the samples
and resultant erratic findings called for an appreciably larger
number of analyses, oftentimes performed by different laboratories,
and the results were averaged for evaluation.   Representative
analytical findings for the pure salts 1-12 are listed in Table 1.
Analytical data for the crude product salts, although routinely
collected, will not be reported in the subsequent experimental
procedures when merely confirming expected compositions; however,
they will be found included in the text whenever deemed necessary
for identification of compositional deviations.

Preparation of Ferricenium Salts   1 - 12

All experiments were performed in duplicate or triplicate (more
often if necessitated by apparent discrepancies), and the procedures
described below (including product yields) are representative.
Unless stated otherwise, preparative operations were conducted at
ambient temperature.   In the few instances where yield data were
reported in the literature these will be cited in the text.
Products were routinely dried over P_4O_{10} in evacuated desiccators
(2-5 torr); drying at elevated temperatures was brought about in
an Abderhalden apparatus (0.2 - 0.5 torr).   In both cases, several
cycles of purging with dry nitrogen and subsequent evacuation
were routinely interposed.

Ferricenium hexafluorophosphate (1) (1)

(a) By oxidation of ferrocene in H_2SO_4. - Essentially in accord with the literature,[6,9-13] a solution of 186 mg (1.0 mmol) of ferrocene in 2 ml of 98% H_2SO_4 was allowed to stand for 20 min at room temperature. Upon dilution with ice water to a volume of 20 ml and filtration, the solution of ferricenium sulfate was treated with 326 mg (2.0 mmol) of solid NH_4PF_6 with vigorous shaking. As the ammonium salt dissolved, product began to precipitate. The mixture was left to stand for 1 h. The bluish-black hexafluorophosphate 1 was collected by filtration, washed with several small portions of ice water until the washings were essentially neutral, and dried at room temperature. Yield, 210 mg (63%; Lit.[13] 78%). Anal. found: C, 33.16; H, 3.01%. Recrystallization of the crude salt from water/acetone (2:1), washing with ether, and drying at 60°C gave analytically pure 1 as purplish-black, fine needles, 78 mg (24%), in addition to a second, less pure portion of the salt (15 mg) collected from the concentrated mother-liquor.

(b) From ferricenium tetrafluoroborate (2) generated in situ.[10]

- p-Benzoquinone, 108 g (1.0 mmol), was dissolved in a solution of 186 mg (1.0 mmol) of ferrocene in 1 ml of benzene. The solution, briefly warmed (50°C) and allowed to stand for 30 min at room temperature, was treated with 0.2 ml of boron trifluoride etherate with vigorous stirring. After 1 h at ambient temperature, ether (5 ml) was added, and the precipitated crude ferricenium tetrafluoroborate was collected and thoroughly washed with ether. It was then dissolved in a few ml of warm acetonitrile and re-precipitated as a mixed BF_4/PF_6 salt, 266 mg (~80%), by the addition, with shaking, of 200 mg of NH_4PF_6, followed shortly afterwards by the addition of ether (7 ml). Recrystallization from water/acetone (2:1) containing NH_4PF_6 (25mg) and, once again, from the same solvent in the absence of the ammonium salt, gave microcrystalline 1 of analytical purity; the salt was washed with a few ml of ice water and ether/methanol and was dried at 60°C. Ultimate yield, 39 mg (12%). Final recrystallization can also be accomplished from acetonitrile or nitromethane.

(c) By oxidation of ferrocene with silver hexafluorophosphate.[17]

- A deoxygenated solution of 372 mg (2.0 mmol) of ferrocene in 10 ml of benzene was added to 506 mg (2.0 mmol) of $AgPF_6$ under argon, and the blackish mixture was shaken for 24 h in the dark. The precipitate collected by filtration (830 mg), containing colloidally dispersed silver (Anal. found: C, 30.93; H, 2.52%), was recrystallized three times from water/acetone (2:1) to achieve acceptable purification. Ultimate yield, 45 mg (7%).

    Analytically pure 1 does not melt below 300°C (lit.[13,18]
160°C dec.); it dissolves in hydrophilic solvents and in such polar
media as acetonitrile and nitromethane.   In the solid state, the
salt is stable for periods of several years when protected from
moisture.   (Impure samples tend to decompose slowly with concomitant deterioration of solubility.)   Acidified aqueous solutions
display stability for more than 24 h; more rapid decomposition
occurs in neutral solvents even in the absence of oxygen.

Ferricenium Tetrafluoroborate   (2)

(a) By Oxidation of Ferrocene with p-Benzoquinone. - The following
procedure was adapted from Hendrickson et al.[10] and somewhat
modified.   The solution of 1.86 g (10 mmol) of ferrocene in 30 mℓ
of ether was added to a stirred solution of 2.16 g (20 mmol) of
p-benzoquinone and 7.26 mℓ (40 mmol) of 5.5M HBF_4 etherate.   The
mixture was allowed to stand for 30 min at room temperature.   The
settled fine black crystals of crude 2 were collected by filtration, washed well with ether and dried for 48 h at 55°C;
mp 110-113°C (decomp. ~180°C).   Yield, 2.70 g (98.9%). Anal.
found: C, 47.58; H, 3.78; Fe, 17.99%.   Repeated recrystallization from acetonitrile/ether at -20°C under N_2, monitored by
X-ray powder diffractometry, gave analytically pure 2 as fine,
purplish-black crystals in an ultimate yield of 30%.

(b) By Oxidation of Ferrocene with Silver Tetrafluoroborate. -
Ferrocene (1 mmol) was treated with $AgBF_4$ (1 mmol) under argon
in exactly the same fashion as described for 1 (method c).
The black precipitate essentially constituting a mixture of 2
with colloidal silver (235 mg) was rapidly recrystallized from
acetone/water (2:1); this was followed by two carefully controlled
recrystallizations from acetonitrile/ether as under (a), above.
Ultimate yield, 35 mg (13%).

(c) By Oxidation of Ferrocene with Nitromethane.[33] - To a stirred
suspension of 3.07 g (16.5 mmol) of ferrocene in 5 mℓ of nitromethane was added dropwise, over a period of 1 h, 5.45 mℓ (30 mmol)
of 5.5M HBF_4 etherate.   Stirring for another h was followed by
solvent removal under reduced pressure (ultimately 0.5 torr).
The residue was stirred first with 3 x 25 mℓ of hexane for extraction of unreacted ferrocene and then with 50 mℓ of ether, which
caused solidification of the initially semiliquid product.   The
residue was dried for 4 days at 50°C; mp 200-210°C (partial dec.).
Yield, 4.53 g (100.6%, indicating inclusion of excess[76] HBF_4).
Anal. found: C, 41.04; H, 3.70; Fe, 19.20%.   Recrystallization
as under (a) afforded pure 2 in 37% ultimate yield.

    While crude samples of 2 melt over the wide range of
100-200°C, sometimes with decomposition (lower melting points

observed for incompletely dried salt), the pure salt is infusible below 300°C. The stability and solubility properties are similar to those of 1.

Ferricenium Picrate (3)

(a) By Oxidation of Ferrocene with p-benzoquinone.[10,36] – A solution was prepared[36] from 10.0g (53.8 mmol) of ferrocene and 13.0 g (56.8 mmol) of picric acid in 180 mℓ of benzene/abs. ethanol (3:1) with gentle warming. To this was added, with vigorous shaking, 2.9 g (26.9 mmol) of p-benzoquinone dissolved in the same solvent (10 mℓ). Product crystallization set in immediately. The mixture was allowed to stand for 3 h at ambient temperature and another h at 10°C, and the separated greenish crystals of crude picrate were washed with benzene and dried at 50°C. Yield, 21.0 g (94%); mp 158–161°C (sintering at 130°C). Anal. found: C, 45.78; H, 2.92; N, 9.30%. Concentration of the filtrate and cooling afforded additional crude picrate, 1.2 g, mp 136–141°C (sintering at 110°C). A 1.0-g sample, twice recrystallized from water (under an argon blanket in the second operation), afforded analytically pure 3, 0.3 g (corresponding to 28% total yield of purified material), as fine black needles, mp 181–182°C (dec.).

Similar results, with crude yields in the range of 85–95%, were obtained under the experimental conditions of other authors[10] (ferrocene/benzoquinone/picric acid = 1:2:5). Crops isolated in these runs from the original mother-liquors after collection of the principal product fractions contained ferricenium dipicrate. This was obtained (3%) in an acceptably pure state from such a crop in one experiment as the middle fraction resulting from careful fractionating crystallization from ethanol under N_2 (monitored by X-ray powder diffractometry); blackish needles, mp 129–132°C (sintering at 100–105°C) (lit.[34] 119–120°C). Anal. calcd.for $C_{22}H_{15}FeN_6O_{14}$: C, 41.08; H, 2.35; Fe, 8.68; N, 13.06; O, 34.82%; found: C, 41.03; H, 2.37; Fe, 8.58; N, 12.93; O, 33.22%.[77] Recrystallization from water converted the dipicrate to 3.

(b) By Oxidation of Ferrocene in H_2SO_4.[24] – Ferrocene, 4.92 g (26.4 mmol), was dissolved in 15 mℓ of 98% H_2SO_4 and the solution allowed to stand for 20 min. It was then diluted with ice water (100 mℓ; authors[24] prescribe some 900 mℓ), filtered, and treated with a hot solution of 5.98 g (26.1 mmol) of picric acid in 100 mℓ of water. The greenish-black needles, allowed to separate over a 24-h period at 5°C, were filtered off, washed with a few ml of water, then ether, and were dried at 50°C. The crude product, 5.14 g (61.1%, based on picric acid), melted over the wide range 110–140°C and constituted a mixture of ferricenium dipicrate with a minor proportion of 3; Anal. found: C, 42.33; H, 2.64; N, 12.29%. Recrystallization from ethanol as before gave dipicrate, mp. 129-131°C. The original aqueous acidic mother-liquor, on standing overnight, gave a small second crop, 0.3 g, of brownish-black fine needles, mp 120-128°C, constituting a mixture of dipicrate with a very small percentage of tripicrate; Anal. found: C, 40.40; H, 2.50; Fe, 9.05; N, 13.96; O, 36.41%. Recrystallization from water gave 3, mp. 168-172°C (dec.), which, upon a further recrystallization from water under N_2, was analytically pure, mp 180-181°C (dec.).

In one of several repeat experiments, the second crop, collected as above, possessed a major content of ferricenium tripicrate. The material formed light-green fine needles, mp. 95-100°C. Anal. calcd. for $C_{28}H_{18}Fe\ N_9O_{21}$: C, 38.55; H, 2.08; N, 14.45%; found: C, 38.70; H, 1.93; N, 15.06%. Multiple recrystallization from water furnished 3 as before.

Experiments performed as above, yet with an excess of picric acid employed, afforded increased product yields but no significant compositional changes.

Pure picrate 3 melts with explosive decomposition in the vicinity of 180-182°C (lit. 166-167°C;[34] 170-180°C dec.[38]) and retains its composition and spectroscopic features over periods of several years. The salt dissolves readily in water and carbinols and somewhat less so in a number of other polar solvents. While aqueous solutions are stable for more than 24 h, slow degradation is observed in other media.

Ferricenium Triiodide (4)

The solution of 3.81 g (15 mmol) of iodine in 60 mℓ of abs. ethanol was rapidly added to a vigorously shaken solution of 1.86 g (10 mmol) of ferrocene in 60 mℓ of the same solvent. The mixture was warmed for 15 min at 50°C and left to stand for 6 h at room temperature in the dark. The black precipitate collected by filtration was washed with ethanol (2 x 10 mℓ) and ether (3 x 15 mℓ) and was dried at 45°. Yield of crude 4, 4.93 g (87%); mp 183-185°C (dec.). Anal. found: C, 20.62; H, 1.86; I, 67.03%. A small (60 mg) second portion of the same composition, mp 181°C (dec.), was obtained from the mother-liquor upon volume reduction in a rotating evaporator. Recrystallization of the crude triiodide from nitromethane in the dark gave analytically pure 4 crystallizing as shiny black prisms. The salt, 2.55 g (46%), was collected as three consecutive fractions by stepwise volume reduction and storage at 0°C; each fraction melted in the range of 183-186°C. A small sample recrystallized twice more from the same solvent, under N_2 had mp 188°C (dec.). Recrystallization from acetone under N_2 proved equally feasible.

Experiments performed similarly in benzene (a total of 330 mℓ
used for the quantities of reactants given) afforded crude 4, very
slightly contaminated with higher-order polyiodides, in 80-88%
yield; decomposition points were in the range 175-183°C.
Analytically pure triiodide resulted from repeated recrystallization
as before.

Ferricenium triiodide (4) melts with abrupt decomposition at
180-190°C.   Rigorously purified material undergoes this decomposition sharply, within one degree, in the region of 186-190°C (lit.
140°C (dec.)[5], 169-171°C (dec.),[41] 192°C (dec.)[43]).   The salt
dissolves readily in carbinols, ketones, and other polar solvents
and, when left in the dark, exhibits stability in these solutions
for periods of many days, being superior in this respect to most
other ferricenium salts investigated in this study.   In the solid
state, it has proved stable for more than five years.

Ferricenium Trichloroacetates   5 and 6

(a)   Ferricenium Trichloroacetate bis(trichloracetic acid)-solvate
6. -   A solution was prepared[51,53] of 1.86 g (10 mmol) of
ferrocene and 4.90 g (30 mmol) of sublimed trichloroacetic acid
in 100 mℓ of benzene.   Following dioxygen saturation by bubbling
into the liquid a brisk O_2 stream for 10 min., the darkened
solution was allowed to stand in a stoppered vessel in the dark.
Crude 6 crystallizing slowly from solution as fine bluish-black
needles (3.05 g), mp 141-143° (dec.), was collected after four days,
washed with a small volume of benzene, and dried for 5 h at 45°C/
0.5 torr.   Three more portions of 6, mp 138-140°C (dec.) (1.06 g),
and a final portion melting at 137-138°C (dec.) (0.10 g) were
obtained from the mother-liquor over a period of another 10 days
on stepwise volume reduction and exposure to air, bringing the
total yield to 4.21 g (62%).   Three repeat experiments gave
fractions (60-68%) melting in the range of 135-144°C (dec.);   the
highest mp observed was 143-144°C (dec.).

Using the method of ferrocene (10 mmol) oxidation in H_2SO_4
(4 mℓ) and treatment of the diluted (ice water, 36 mℓ) and filtered
solution with trichloroacetic acid (70 mmol),[51,53] we obtained
6 in 65% yield, mp 135-139°C (dec.).

The disolvate 6 forms monoclinic crystals melting typically,
with mild decomposition, in the region of 135-143°C; purest
samples melt at 143-144°C.   The solid salt has retained its mp,
composition, and spectrocopic properties over a period exceeding
two years when protected from moisture.   It dissolves readily in
water and, owing to its high acidity, degrades only slowly in this
solution, whereas rapid degradation is observed in most organic
solvents.   Therefore, recrystallization from nitromethane or

acetonitrile (under Ar), although feasible, is accompanied by
appreciable loss of material through degradation, and no purity
enhancement is achieved.   In contrast to most other ferricenium
salts, 6 is not entirely insoluble in nonpolar media, such as
benzene and ether;   hence, appreciable dissolution can be observed
on extended washing or digestion with such solvents.

(b) Ferricenium Trichloroacetate mono(trichloroacetic acid)-
solvate 5. - Hydrolytic conversion of 6 to 5 was brought
about[53] by heating 3.0 g (4.45 mmol) of the former salt with
25 mℓ of water for 20 min. at 95-100°C, and allowing the rapidly
filtered solution to stand at ambient temperature.   The crystallized bluish-black material (680 mg) was collected by filtration,
washed with water, and dried at 50°/0.5 torr; mp 127-128°C
(mild decompn.).   Two more portions of crystals, melting at
124-127°C (dec.), were isolated on stepwise volume reduction under
reduced pressure and cooling to 5°C, bringing the total yield of
5 to 1.14 g (50%).   Several repeat experiments afforded the salt
in 40-60% yield.

        The monosolvate 5 crystallizes in the monoclinic system;
typical melting points (with mild decomposition) are in the range
of 120-128°C, purest samples melting at 127-128°C (dec.).   The
salt, slightly less soluble in water than 6, otherwise shows much
the same solubility behavior as the latter.   In the solid state,
it has shown no signs of degradation after 5 years.   Aqueous
solutions are stable for more than 24 h, whereas solutions in
carbinols or acetonitrile degrade rather rapidly even in the
absence of air, and recrystallization from these solvents is
impracticable.

Ferricenium tetrachloroantimonate(III)   (7)

        To a solution of 1.86 g (10 mmol) of ferrocene in 40 mℓ of
acetone (dried by two passes through Molecular Sieves 4A) was
added 4.56 g (20 mmol) of SbCℓ$_3$ and the mixture shaken for
complete dissolution. After a brisk stream of dioxygen has been
bubbled into the solution for 5 min., the latter was rewarmed to
dissolve some crystallizing ferrocene and was left undisturbed
for two weeks in a stoppered flask in the dark.   The separated
solid was collected by filtration, washed with a few mℓ of acetone,
then ether, and was dried at 45°C overnight.   The bluish-black
antimonate salt, 950 mg, melted partly at 135°C; Anal. found:
Fe, 12.31; O, 0.92%.   On partial solvent removal from the
combined mother-liquor and washings to the point of beginning
solute crystallization and brief warming for redissolution, the
solution was left in the dark for another week.   This afforded
a second portion of crude antimonate, which was collected as
before.   Total crude yield, 2.41 g (53.6%).   A painstakingly
conducted fractionating crystallization from acetonitrile under

argon gave a less soluble, grayish-blue, powdery majority component constituting ferricenium oxychloroantimonates, partially melting at 148-150°C (Anal. found: Fe, 11,11; O, 0,89%), and a somewhat more readily soluble minority component constituting the tetrachloroantimonate 7, which was isolated as bluish-black crystalline material completely melting at 154-156°C and giving the analytical data recorded in Table 1 in addition to 0.41% oxygen. Ultimate yield of 7, 611 mg (14%).

Very similar results were obtained in analogous experiments employing $SbC\ell_3$/Fc ratios of 3.0-4.5, with crude product yields in the range of 60-70% and oxygen contents of the crude salts 1.0-1.5%. Exposure of the solutions to daylight during the crystallization periods did not substantially affect product yields, although the found Fe contents (9-11%) were generally lower, and the oxygen contents (up to 2%) higher, than in the products of the dark reaction.

Experiments performed in benzene medium instead of acetone ($SbC\ell_3$/Fc = 2.0) either in daylight or in the dark gave crude grayish-blue solid salt (35-50%), mp 65-95°C; Anal. found (typ.): Fe, 6.71; O, 1.22%; no IR absorption at 320-270 cm[1]. Fractionating crystallization from acetonitrile as described above afforded 7, mp 150-152°C, in 5% ultimate yield.

Ferricenium tetrachloroantimonate (7) as isolated by fractionating crystallization melts cleanly within the range of 150-156.°C (purest samples have mp 154-156°C; lit.[56] no mp < 300°C) and decomposes at about 180°C. In the solid state, protected from moisture, it has proved to be stable for at least two years. The salt dissolves in polar organic solvents, although with slow degradation; solutions in dipolar aprotic solvents degrade rapidly.

Ferricenium tetrachlorobismuthate(III)   (8)[57a]

The solution of 946 mg (3 mmol) of bismuth(III) chloride and 558 mg (3 mmol) of ferrocene in 15 mℓ of acetone (dried as described for 7 and not deoxygenated) was left to stand in the dark in a stoppered flask. Some fine whitish deposit of bismuth oxychloride appearing after 3 days was removed by filtration from the solution briefly prewarmed to redissolve some cocrystallizing 8. After another 5 days, a portion of bluish-black crystals of 8 had separated, which were collected and dried as described for 7. The product, 210 mg (13%), melted at 210-212°C and gave the analytical data listed in Table 1 in addition to 0.38% oxygen. A portion (100 mg) of silver-blue crystals collected on the subsequent day constituted an impure 8, mp. 208-210°C; (Anal. found: C, 19.95; H, 1.55; Cℓ, 22.21; Fe, 9.73%.) A final portion of solid material, isolated after a total of 24 days as silver-

gray crystals (165 mg), melted at 210-215°C; Anal. Calcd. for
2 FcBiCl₄·BiOCl (C₂₀H₂₀Bi₃Cl₉Fe₂O): C, 18.01; H, 1.51;
Fe, 8.37; 0, 1.20%; found: C, 18.42; H, 1.52; Fe, 8.07;
0, 1.29%.

    The tetrachlorobismuthate 8 crystallizes as fine bluish-
black, glossy needles, melting abruptly within the range of 208-
212°C and decomposing subsequently. While soluble in polar
organic media, the salt tends to degrade rapidly in these
solutions even under anaerobic conditions, and purification by
recrystallization has not proved practicable.

Ferricenium tetrachloroferrate(III)   (9)

(a) By Oxidative Degradation of Ferrocene with Sulfuryl Chloride[63]

The solution of 1.86 g (10 mmol) of ferrocene and 1.34 g (10 mmol)
of sulfuryl chloride in 50 ml of benzene was heated for 2 h at the
reflux temperature with moisture protection. After cooling to
ambient temperature, the separated brownish-black solid was
collected by filtration, washed well with benzene, and dried for
12 h at 80°C. Although the yield of this crude product was 1.35 g
(70.4%, based on ferrocene; lit.[63] 97%), more than 70% failed to
dissolve on hot extraction with 96% ethanol (4 x 10 ml). From the
combined ethanolic extracts, on stepwise partial solvent removal
under reduced pressure and storing at -10°C, a total of 310 mg
(16%) of bluish-black crystalline material constituting crude 9
was obtained. Anal. found: C, 32.48; H, 2.90; Fe, 30.81;
Cl, 35.34%. Recrystallization from aqueous-methanolic (2:98)
0.2M HCl gave analytically pure 9 as bluish-black platelets.

    Reducing the reflux time to 30 min. allowed for a larger
proportion of soluble salt to be formed, and ethanol extraction
afforded 28% of crude 9.

(b) By Oxidation of Ferrocene with Iron(III) Chloride[3] - Iron(III)

chloride 650 mg (4 mmol), suspended (and partially dissolved) in
ether (3 ml), was added to the vigorously shaken solution of
372 mg (2 mmol) of ferrocene in ether (6 ml). A precipitate
appeared immediately. After 1 h the greenish-blue solid was
filtered off, washed with ether, a small volume (~0.5 ml) of
abs. ethanol and, again, ether; it was then dried at 50°C overnight.
The crude material, 0.65 g (85%), was recrystallized from aqueous-
ethanolic HCl (0.2M) as before, to give two fractions totalling
190 mg (24.8%) of analytically pure 9. Recrystallization may
also be accomplished from freshly distilled thionyl chloride,
thorough washing of the crystallized salt with ether being required
for complete solvent removal.

The tetrachloroferrate $\underset{\sim}{9}$ forms bluish-black plate-like crystals infusible below 300°C and stable for periods of more than five years.   The salt is soluble in many polar organic solvents and very soluble in water, from which it can be recovered with only minor changes in composition.   Aqueous or alcoholic solutions are stable for several days, whereas more rapid degradation is observed in acetonitrile or nitromethane solutions.

Ferricenium tetrabromoferrate $(\underset{\sim}{10})$[69]

The solution of 5.19 g (65 mmol) of bromine in carbon tetrachloride (25 ml) was added dropwise, over a period of 30 min., to a boiling, magnetically stirred solution of 1.86 g (10 mmol) of ferrocene in the same solvent (25 ml) under a blanket of nitrogen. After cooling to ambient temperature, the brownish precipitate was filtered off, washed well with carbon tetrachloride and ether, and dried at 75°C; mp 240-242°C (dec.).   Yield, 2.69 g (95.8% ; Lit.[69] 94.4%).   The product was of acceptable analytical purity at this stage (analytical findings recorded in Table 1).   Recrystallization from Ar-saturated acetonitrile/abs. ethanol (1:1) (preferred over neat ethanol or the proposed[70] ethanol/benzene/ carbon tetrachloride (12:2:1) mixture) gave dark brown, microcrytalline $\underset{\sim}{10}$ sintering at 189-190°C.   Ultimate yield, 35%.

Ferricenium tetrabromoferrate $(\underset{\sim}{10})$, upon recrystallization, sinters at temperatures typically in the region of 185-195°C and undergoes a color change from dark brown to grey in the vicinity of 240°C, after which it very gradually decomposes.   High stability is indicated by the unchanged melting behavior, appearance and spectroscopic properties after three-years' storage at ambient temperature.   Aqueous, deoxygenated solutions are stable for more than a day; more rapid degradation is observed with alcoholic or acetonitrile solutions.

Ferricenium diamminetetraisothiocyanatochromate(III) $(\underset{\sim}{11})$

Essentially as reported,[24] an aqueous acidic ferricenium solution was prepared by dissolving 2.23 g (12 mmol) of ferrocene in 7.5 ml of 98% H_2SO_4, introducing a brisk dioxygen stream for 10 min., and diluting with 75 ml of ice water.   The solution of ferricenium sulfate was filtered and added slowly, with stirring, to the hot filtered solution of ammonium Reineckate (ammonium diamminetetraisothiocyanatochromate(III)), 10 g (28.2 mmol), in 150 ml of water.   The mixture was allowed to stand for several h at room temperature.   The reddish-blue, precipitated salt $\underset{\sim}{11}$ was collected by filtration, washed thoroughly with water, then ether, and was dried for 8 h at 100°C.   Yield, 4.37 g (72%). The salt gave acceptable analytical results (Table 1) at this stage.   When recrystallized (with great difficulty because of poor solubility) from nitrogen-saturated acetonitrile/methanol/(1:1),

it was obtained as a purplish-gray, microcrystalline solid possessing similar properties and composition as the crude compound.

The Reineckate 11, crude or recrystallized, is infusible up
to 300°C (color change to dark brown at ~240°C); it is practically
insoluble in water and slightly soluble in acetonitrile or aceto-
nitrile/methanol. In N,N-dimethylformamide or dimethylsulfoxide
the salt forms a purple solution, which soon degrades with concomitant reduction of the ferricenium cation. In the solid state
the salt is stable for periods of several years.

Ferricenium 2,3-dichloro-5,6-dicyanohydroquinonide   (12)

For the preparation of the DDQH⁻ salt 12 a slightly
modified literature procedure[73] was used. The solution of
456 mg (2 mmol) of 2,3-dichloro-5,6-dicyanobenzoquinone (DDQ;
recrystallized from benzene under N_2) in 5 mℓ of benzene was added
with continuing agitation to a warm solution of 372 mg (2 mmol) of
ferrocene in 4 mℓ of the same solvent. The mixture was warmed
for 10 min at 60°C with occasional shaking and was allowed to cool
to room temperature. The black, solid 12 that had separated
from the solution was filtered off, washed with warm benzene, then
ether, until the washings were colorless. The salt was dried for
12 h at 50°C; mp 179-180°C. Yield, 730 mg (88%; lit.[73] 97%).
Recrystallization from nitrogen-saturated acetonitrile gave a
material melting at 177-178°C; another portion recrystallized
from nitrogen-purged ethanol melted at 180-181°C.

An experiment performed as above except that the quantity
of DDQ was raised to 4 mmol gave crude 12 in 98% yield;
mp 167-170°C. Anal. found: C, 53.63; H, 2.88; Fe, 14.21;
N, 7.01%. Recrystallization from ethanol as before raised the
mp to 179-180°C.

Pure 12 crystallizes as fine black needles which typically
melt rather sharply within the range of 178-180°C (from ethanol),
although an occasional mp 181-182°C has been observed. The salt
is moderately soluble in nitromethane, acetonitrile, and alcohols
and dissolves readily in dipolar aprotic solvents, although these
solutions undergo rapid degradation. In the solid state the
salt has been found to be stable for more than five years when
protected from moisture.

SUMMARY AND CONCLUSIONS

The recent discovery of antineoplastic properties of certain
hydrophilic ferricenium salts prompted an extended reinvestigation,
described above, of the 12 most important known ferricenium
compounds with respect to synthesis, composition and spectroscopic

features. This study proved necessary in view of numerous inconsistencies found in the literature covering the subject. In the present effort, we have critically examined the published experimental procedures and (i) singled out those found to be reproducible, (ii) identified those found to be unreproducible, (iii) improved and completely described those found to be poorly or incompletely covered in the literature. We further reinvestigated compositional details for all 12 salts and established their IR and Raman spectral characteristics.

From the standpoint of biomedical use, which requires hydrophilicity or even complete solubility in aqueous media without undue degradative tendencies, the most promising ferricenium salts are the picrate 3, the two trichloroacetates 5 and 6, and the tetrahaloferrates 9 and 10, all of which are of well defined composition, are stable in the solid state, and readily dissolve in water to form acidic solutions in which the solutes degrade only slowly (with regeneration of ferrocene). Less hydrophilic ferricenium salts are the hexafluorophosphate 1 and the tetrafluoroborate 2; these species both dissolve moderately well in certain aqueous-organic media although forming solutions less stable than observed for the first group of salts. Other ferricenium compounds, such as the triiodide 4, the tetrachloroantimonate and -bismuthate 7 and 8, the Reineckate 11, and the hydroquiononide 12 are representative of a class of ferricenium compounds which, while stable in the solid state and in some instances (4, 12) also in certain organic solutions, do not possess sufficient hydrophilic character to dissolve in aqueous media and are, therefore, not of primary interest for biomedial investigations in the mononuclear, as well as in the polymer-anchored, state.

ACKNOWLEDGMENT

This investigation was supported by the CSIR and by Hoechst South Africa (Pty) Ltd., for which grateful acknowledgment is made. The author is obliged also to Professor H. Hoberg and Dr. K. Seevogel, Max-Planck-Institut für Kohlenforschung, Mülheim/Ruhr, for their friendly cooperation in obtaining the Raman spectra, and to his loyal assistants, Rokaya and Mohamed Loonat, for their skilled participation in the preparative work. Some preliminary experiments pertaining to this investigation were performed by the author while on study leave at the Institute of Organic Chemistry, University of Mainz, and it is his privilege to express once more his appreciation of the kind hospitality provided by Professor R.C. Schulz and the experimental assistance rendered by Mrs. U. Grimm.

REFERENCES

1.   G. Wilkinson, M. Rosenblum, M.C. Whiting, and
R.B. Woodward,   J. Am. Chem. Soc., 74 2125 (1952).

2.   P.Köpf-Maier, H. Köpf, and E.W. Neuse, submitted.

3.   A.N. Nesmeyanov, E.G. Perevalova and L.P. Yureva
Chem. Ber., 93, 2729 (1960).

4.   R. Prins, A.R. Korswagen and A.G.T.G. Kortbeek,
J. Organometal. Chem., 39, 335 (1972).

5.   M.M. Aly, Indian J. Chem., 11, 134 (1973).

6.   E. Shih Yang, M.-S. Chan and A.C. Wahl,
J. Phys. Chem., 79, 2049 (1975).

7.   M. Cais, P. Ashkenazi, S. Dani and J. Gottlieb,
J. Organometal. Chem., 122, 403 (1976).

8.(a) V.N. Babin, Yu. A. Belousov, L.R. Lyatifov,
R.B. Materikova and V.V. Gumenyuk, J. Organometal. Chem., 214,
C11 (1981). (b) V.N. Babin, Yu. A. Belousov, V.V. Gumenyuk,
R.B. Materikova, R.M. Salimov and N.S. Kochetkova, ibid., 214,
C13 (1981).   (c) N.S. Kochetkova, R.B. Materikova, Yu. A. Belousov
R.M. Salimov and V.N. Babin, ibid., 235, C21 (1982).

9.   W.L. Jolly, "The Synthesis and Characterization of
Inorganic Compounds", Prentice Hall, Englewood Cliffs, N.J., 1970,
p. 487.

10.   D.N. Hendrickson, Y.S. Sohn and H.B. Gray, Inorg.
Chem., 10, 1559 (1971).

11.   W.H. Morrison, Jr., and D.N. Hendrickson, Inorg. Chem.,
14, 2331 (1975).

12.   D.M. Duggan and D.N. Hendrickson, Inorg. Chem., 14,
955 (1975).

13.   I.R. Lyatifov, S.P. Solodovnikov, V.N. Babin and
R.B. Materikova, Z. Naturforsch., 34b, 863 (1979).

14.(a) T.E. Bitterwolf and A. Cambell Ling, J. Organometal.
Chem., 40, C29 (1972). Enhanced ease of oxidation of protonated
ferrocene in comparison to the unprotonated complex is also
apparent from the electrochemical behavior of ferrocene in different
solvents.[14b] The autoxidation of ferrocene in acidic medium is
catalyzed by iron ions.[14c]   (b) V.I. Ignatov, V.T. Solomatin and

A.A. Nemodruk, Zhurn. Anal. Khim., 33, 1268 (1978).
(c) J. Lubach and W. Drenth, Recl. Trav. Chim. Pays-Bas, 92, 586 (1973).

15. R. Prins and A.G.T.G. Kortbeek, J. Organometal. Chem., 33, C33 (1971).

16. F.H. Köhler, J. Organometal. Chem., 69, 145 (1974).

17. S.E. Anderson and R. Rai, Chem. Phys., 2, 216 (1973).

18. A.N. Nesmeyanov, R.B. Materikova, I.R. Lyatifov, T. Kh. Kurbanov and N.S. Kochetkova, J. Organometal. Chem., 145, 241 (1978).

19. M. Rosenblum, "Chemistry of the Iron Group Metallocenes", Part 1, Wiley, New York, 1965.

20. O. Traverso and F. Scandola, Inorg. Chim. Acta, 4, 493 (1970).

21. Y.S. Sohn, D.N. Hendrickson and H.B. Gray, J. Am. Chem. Soc., 93, 3603 (1971).

22. J. Holček, K. Handlíř and J. Klikorka, Proc. 7th Conf. Coord. Chem., Sept. 11-14, 1978, Smolenice, Bratislava, p.65.

23. P. Sohár and J. Kuszmann, J. Mol. Struct., 3, 359 (1969).

24. I. Pavlík and J. Klikorka, Coll. Czech. Chem. Commun., 30, 664 (1965).

25. J.A. Kramer, F.H. Herbstein and D.N. Hendrickson, J. Am. Chem. Soc., 102, 2293 (1980). J.A. Kramer and D.N. Hendrickson, Inorg. Chem., 19, 3330 (1980).

26.(a) In the spectra of 1 and all other ferricenium salts here discussed (excepting 4) we observe the 1114 cm^{-1} band due to ν_{10} (antisym. ring breathing) in such vanishingly low intensity as to warrant its exclusion from the list of prominent ferricenium bands. Similarly low intensities were observed[23,26b] for certain other Fc^+ salts. In contrast, Pavlík and Klikorka[24] reported the band as medium-strong to strong in the ferricenium spectra, which suggests that most of their samples were contaminated with ferrocene known[26c] to absorb strongly at the 1110-cm^{-1} position. (b) P.M. Maitlis and J.D. Brown, Z. Naturforsch., 20b, 597 (1965). (c) E.R. Lippincott and R.D. Nelson, Spectrochim. Acta, 10, 307 (1958).

27.   H.P. Fritz and R. Schneider, Chem. Ber., 93, 1171 (1960).

28.   D.O. Cowan, R.L. Collins and F. Kaufman, J. Phys. Chem., 75, 2025 (1971).

29.(a) A broad Raman band at 1580 cm^{-1} in the spectrum of the tetrafluoroborate 2 was assigned by Gächter et al.[29b] to an electronic Raman transition. This finding needs further corroboration. A similar unassigned Raman shift near 1540 cm^{-1} was recently reported for neutral polynuclear ferrocenes.[29c]   (b) B. Gächter, G. Jakubinek, B.E. Schneider-Poppe and J.A. Koningstein, Chem. Phys. Lett., 28, 160 (1974).   (c)   E.W. Neuse and L. Bednarik, Macromolecules, 12, 187 (1979).

30.   D. Hartley and M.J. Ware, J. Chem. Soc. A, 138 (1969).

31.   G. Stölzle, Ph.D. Thesis, University of Munich, 1961. A.N. Nesmeyanov, E.G. Perevalova, L.P. Yureva and K.I. Grandberg, Izv. Akad. Nauk SSSR, Ser. Khim., 1772 (1962).

32.   V.A. Nefedov and L.K. Tarygina, Zh. Org. Khim., 12, 2012 (1976).

33.   H. Schumann, Chem. Z., 107, 65 (1983).

34.   M. Aly, R. Bramley, J. Upadhyay, A. Wassermann and P. Woolliams, Chem. Commun., 404 (1965). See also M.M. Aly, D.V. Banthorpe, R. Bramley, R.E. Cooper, D.W. Jopling, J. Upadhyay, A. Wassermann and P.R. Woolliams, Monatsh., 98, 887 (1967).

35.   B.F. Gächter, J.A. Koningstein and V.T. Aleksanjan, J. Chem. Phys., 62, 4628 (1975).

36.   I.M. Kolthoff and F.G. Thomas, J. Phys. Chem., 69, 3049 (1965).

37.   R.C. Pettersen, Ph.D. Thesis, University of California at Berkeley, 1966.

38.   A. Horsfield and A. Wassermann, J. Chem. Soc. A, 3202 (1970).

39.   A.N. Nesmeyanov, E.G. Perevalova and O.A. Nesmeyanova, Dokl.Akad. Nauk SSSR, 100, 1099 (1955).

40.   J.C.D. Brand and W. Snedden, Trans. Faraday Soc., 53, 894 (1957).

41. A.N. Nesmeyanov, L.P. Yureva, R.B. Materikova and B.A. Hetnarski, Izv. Akad. Nauk SSSR, Ser. Khim., 731 (1965).

42. T. Bernstein and F.H. Herbstein, Acta Cryst., B 24, 1640 (1968).

43. M.A. Wassef and S.H. Abou el Fatouh, Indian J. Chem., 14A, 282 (1976).

44. I.M. Motoyama, M. Watanabe and H. Sano, Chem. Lett., 513 (1978)

45. P. Fornier de Violet and S.R. Logan, J. Chem. Soc. Faraday I, 76, 578 (1980).

46. E.W. Neuse and R. Loonat, submitted.

47. V.G. Tsvetkov and Yu. A. Aleksandrov, Zh. Obshch. Khim., 50, 2725 (1980).

48. Gächter et al.[35] reported an additional shift at 325 cm^{-1}, which they ascribed to ν_4. However, we have been unable to observe any shift due to ν_4 at frequencies higher than about 318 cm^{-1} in more than 25 different ferricenium salt spectra and, therefore, regard the 325-cm^{-1} signal as an overtone band of the moderately intense 164-cm^{-1} shift reported by these authors. Pure 4 gives a weak Raman band at 167-169 cm^{-1}, which we assign to the antisym. anion stretching mode (ν_3) that has gained Raman allowedness, perhaps by a minor deviation from I_3^- linearity. The much higher intensity of the 164-cm^{-1} shift in the spectrum of Gächter et al.[35] suggests that band to contain a major contribution from a stretching mode of the I_2, I_3^- or I_5^- systems known[49] to give a scattering signal in the vicinity of 170 cm^{-1}. The sample of these workers[35] may thus have been contaminated with additional I_2 giving rise to a I_5^- component in the anion.

49. M. Cowie, A. Gleizes, G.W. Grynkewich, D. Webster Kalina, M.S. McClure, R.P. Scaringe, R.C. Teitelbaum, S.L. Ruby, J.A. Ibers, C.R. Kannewulf and T.J. Marks, J. Am. Chem. Soc., 101, 2921 (1979). R.C. Teitelbaum, S.L. Ruby and T.J. Marks, ibid, 101, 7568 (1979).

50. J.C.A. Boeyens and E.W. Neuse, in preparation.

51. D.N. Hendrickson, Y.S. Sohn, D.M. Duggan and H.B. Gray, J. Chem. Phys., 58, 4666 (1973).

52. M. Castagnola, B. Floris, G. Illuminati and G. Ortaggi, J. Organometal. Chem., 60, C17 (1973).

53.   E.W. Neuse, R. Loonat and J.C.A. Boeyens,
Transition Metal Chem., in the press.

54.(a) The spectral details[53] suggest the absence of distinct
CCl_3COOH and CCl_3COO^- species in the anion.   Instead, approximate
equivalence of both acyl moieties is suggested, with negative charge
delocalized over both units.   The complex anion structure may thus
be of the type $[XHX]^-$ (X = trichloroacetate) reminiscent of the
structure of sodium hydrogen diacetate.[54b]   Preliminary X-ray
diffraction data[50] are in accord with a structural arrangement of
this type.   (b)   J.C. Speakman and H.H. Mills, J. Chem. Soc.,
1164 (1961).   See also H.N. Shrivastava and J.C. Speakman, ibid.,
1151 (1961).

55.(a) An X-ray crystal structure analysis of 6, was briefly
reported by Schlueter et al.[55b] The study showed no anomalies in
the ferricenium cation structure; no details were published of the
structural arrangement of the anion.   (b) A.W. Schlueter and
H.B. Gray, Am. Crystl. Assoc. Summer Meeting 1971, Abstr. Papers,
p.41.

56.   G.W. Cowell, A. Ledwith, A.C. White and H.J. Woods,
J. Chem. Soc. B, 227 (1970).

57.(a) A.G. Landers, M.W. Lynch, S.B. Raaberg, A.L. Rheingold,
J.E. Lewis, N.J. Mammano and A. Zalkin, Chem. Commun., 931
(1976).   (b) N.J. Mammano, A. Zalkin, A. Landers and A.L.
Rheingold, Inorg. Chem., 16, 297 (1977). The inter-ring distance
in the cation was found to average 340 pm; well in accord with
other ferricenium salt structures. (c) A.L. Rheingold, A.G. Landers,
P. Dahlstrom and J. Zubieta, Chem. Commun., 143 (1979).

58.(a) A ferricenium salt having a complex oxychloroarsenate(III)
counterion has been prepared and its crystal structure determined.[58b]
The cation structure possesses eclipsed ring ligands, and the average
inter-ring distance is 340 pm.   (b) M.R. Churchill, A.G. Landers
and A.L. Rheingold, Inorg. Chem., 20, 849 (1981).

59.   The ferricenium salt isolated by these workers[57c]
possessed a complex oxo- and chloride-bridged antimony cluster
ion.   The cation, with cyclopentadienyl rings eclipsed as in other
ferricenium salts, features an average inter-ring distance of 336 pm.

60.   G.Y. Ahlijah and M. Goldstein, J. Chem. Soc. A, 326
(1970).

61.   The two bands at 322 and 290 cm^{-1} are also observed
in the spectrum of tetraethylammonium tetrachloroantimonate, in which
the anion is appreciably flattened so as to assume C_{2v} symmetry.[60]
A third IR band, of low intensity, which is shown by this salt

near 335 cm^{-1} and represents the IR activated axial symmetric Sb-Cℓ stretching mode (a$_1$) in C$_{2v}$ symmetry,[60] has been observed by us in the spectra of some of the crude, but not of the multiply recrystallized samples of 7, suggesting a symmetry higher than C$_{2v}$ for the highly purified ferricenium salt.

62.   E.W. Neuse and B.S. Mojapelo, submitted.

63.   L.P. Motz and R.P. Pinnell, J. Organometal. Chem., 54, 255 (1973).

64.   I.J. Spilners, J. Organometal. Chem., 11, 381 (1968).

65.(a) J.C.A. Boeyens, P.C. Lalloo, E.W. Neuse and H.-H. H.-H. Wei, S. Afr. J. Chem., in the press. The compound had earlier beenprepared[65b] but erroneously assigned the ferricenium trichloroferrate(II) structure.   (b)   S.M. Aharoni and M.H. Litt, J. Organometal. Chem., 22, 179 (1970).

66.(a) The existence of regular, discrete FeCℓ_4 units is apparent from an X-ray structure analysis,[66b] which shows parallel layers of these anion units to alternate with parallel layers of Fc$^+$ cations. The cation possesses a normal sandwich structure with nearly parallel and eclipsed cyclopentadienyl rings, the mean inter-ring distance being 336 pm.   (b) E.F. Paulus and L. Schäfer, J. Organometal. Chem., 144, 205 (1978).

67.   We fail to detect the weak bands at 1450 and 496 cm^{-1} as well as the strong bands at 1594 and 1192 cm^{-1} listed by these workers;[24] furthermore, the features at 1114 and 412 cm^{-1} reported[24] as strong and medium-strong, respectively, appear as very faint signals in our spectrum.

68.   D.M. Adams, J. Chatt, J.M. Davidson and J. Gerratt, J. Chem. Soc., 2189 (1963).

69.   R. Riemschneider and D. Helm, Chem. Ber., 89, 155 (1956).

70.   S.M. Aharoni and M.H. Litt, J. Organometal. Chem., 22, 171 (1970).

71.   P.N. Gaponik, A.I. Lesnikovich and Yu. G. Orlik, Zhurn. Neorg. Khim., 22, 376 (1977).

72.   W.H. Morrison, Jr., S. Krogsrud and D.N. Hendrickson, Inorg. Chem., 12, 1998 (1973).

73.   R.L. Brandon, J.H. Osiecki and A. Ottenberg, J. Org. Chem., 31, 1214 (1966).

74.   E. Gebert, A.H. Reis, Jr., J.S. Miller, H. Rommelmann, and A.J. Epstein, J. Am. Chem. Soc., 104, 4403 (1982).

75.   L.R. Melby, R.J. Harder, W.R. Hertler, W. Mahler, R.E. Benson and W.E. Mochel, J. Am. Chem. Soc., 84, 3374 (1962).

76.   Schumann[33] reports a crude product composition of $2 \cdot \frac{3}{4}(CH_3CH_2)_2O$.   Analytical and IR data indicate decomposition of the solvate under our drying conditions with concomitant loss of the bound ether.   Excess incorporated fluoroboric acid, however, is much more tightly bound and less readily eliminated in the drying operation.

77.   Ferricenium dipicrate can also be prepared by slow cocrystallization of a 1:1 mixture of 3 and picric acid from ethanol and careful recrystallization from the same solvent.

POLYMER BOUND BIMETALLIC COMPLEXES AS SURFACE REACTANTS

ON SEMICONDUCTOR ELECTRODES

Thomas E. Bitterwolf

Department of Chemistry
United States Naval Academy
Annapolis, MD 21402 (U.S.A.)

INTRODUCTION

The photolytic splitting of water into H_2 and O_2 has attracted considerable interest because of its potential importance as a component in the development of a hydrogen fuel based economy. Several of the techniques which have been examined utilize the broad spectral absorbance of n- and p- type silicon semiconductors as the principal method of harvesting sunlight. The solar energy collected by the silicon semiconductors can be utilized immediately to drive the water splitting reaction by immersing the semiconductors in an aqueous medium.

Figure 1 illustrates the processes which occur at a pair of idealized electrodes[1]. For such a photoelectrolysis cell, the n-type semiconductor serves as the anode. Holes in the valence band of the n-type semiconductor oxidize water to oxygen at the surface. Transfer of an electron from water to the semiconductor destroys a hole. The hole is regenerated by the absorption of light which excites a valence band electron to the conduction band. At the p-type semiconductor, electrons, which are pumped by light to the conduction band, serve to reduce water to dihydrogen.

While the naked semiconductor itself can serve directly as the electrode, problems such as corrosive attack on the electrodes by the aqueous media make this impractical. One particularly attractive alternative to the use of naked semiconductor electrodes is to incorporate a film onto the surface of the semiconductor which contains a chemical species capable of carrying out the desired redox reaction. This film serves as a barrier protecting

137

the semiconductor surface from water while permitting electrons to be transferred to and from the semiconductor. The general theory and background of the use of derivitized electrodes has been discussed by Wrighton, et al.[2]

The compounds selected for use as semiconductor films must be capable of carrying out the desired redox chemistry and must be capable of undergoing repeated cycling through an oxidation/ reduction cycle. By purposeful design of the surface species it should be possible to control the potential of the surface reactions as well as their efficiencies. If appropriate surface species can be developed it should be possible to expand the range of redox chemistries which can be performed by photochemical cells.

Two options are available for the attachment of chemical species to semiconductor surfaces. In one method, the species to be attached to the surface, R, is derivitized with $SiCl_4$ to produce an $RSiCl_3$ or R_2SiCl_2 species which can be reacted with a treated semiconductor surface to produce a directly bound layer of reactant species. Alternately, the surface reactant can be incorporated into a polymer which can be coated onto the surface of the semiconductor as a thin film. Both methods for the intro-duction of surface species have recently been examined for [1.1] ferrocenophane modified silicon p-type semiconductors.[3]

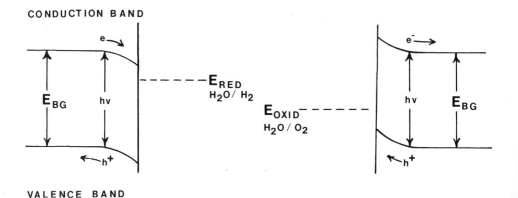

Fig. 1 Photolytic Splitting of Water on Semiconductor Electrodes

BIMETALLIC COMPLEXES FOR DIHYDROGEN EVOLUTION

The protonation of transition metal complexes in acids was first observed by Wilkinson and Birmingham in 1955[4]. Since that time a large number of compounds have been shown to protonate at the metal to yield metal-hydride species[5,6,7]. The basicities of metal complexes varies considerably, with compounds such as cyclopentadienyl bis(trimethylphosphine) cobalt undergoing complete protonation in methanol[8], while compounds such as ferrocene[9] and benzenechromiumtricarbonyl[10] require super acids such as trifluoroboric acid, HBF_3OH, or trifluoromethanesulfonic acid, HCF_3SO_3, for complete protonation.

During an extended study of ferrocene protonation[11], Bitterwolf and Ling examined the protonation of a series of ferrocene dimers, Ia-c[12]. ^1H NMR of the freshly prepared solutions revealed that both ferrocenyl moieties of the dimers were protonated in trifluoroboric acid. However, unlike the monomeric ferrocenes, these compounds formed solutions in trifluoroboric acid which were not oxidatively stable for greater than about fifteen minutes.

Reaction of [1.1] ferrocenophane, IIb, with trifluoroboric acid resulted in the immediate oxidation of the compound and the evolution of a gas which was shown by mass spectroscopy to be H_2. The oxidized [1.1] ferrocenophane, could be reduced with either tin(II)chloride or ascorbic acid to quantitatively recover the neutral dimer.

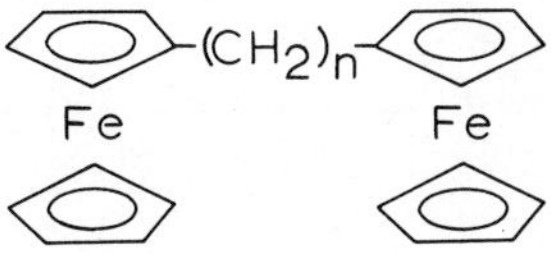

I

a. n = 0
b. n = 1
c. n = 2

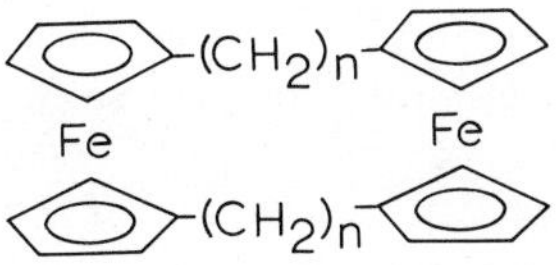

II

a. n = 0
b. n = 1

Scheme I

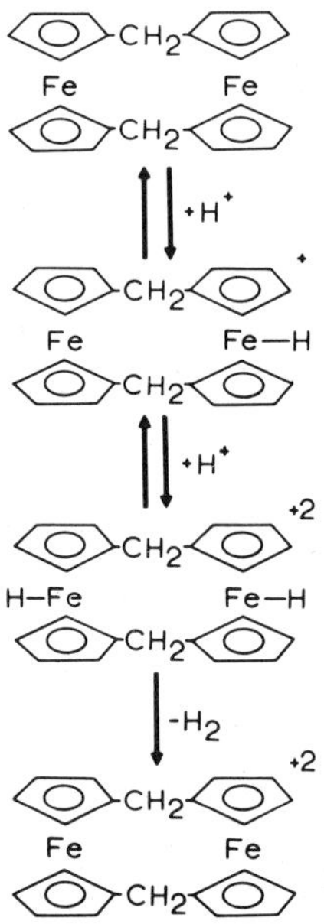

Fe   Fe
CH$_2$

$\uparrow\downarrow$ +H$^+$

Fe   Fe—H

$\uparrow\downarrow$ +H$^+$

H—Fe   Fe—H

$\downarrow$ -H$_2$

Fe   Fe

The reactions of ferrocene dimers with acids have recently been reinvestigated[13]. Detailed analyses of these reactions indicate that the ferrocene dimers react sequentially with acid to form a doubly protonated species which then undergoes reductive elemination of dihydrogen to leave the doubly oxidized ferrocene species. The probable mechanism for this process is illustrated in Scheme I. The exact mechanism whereby adjacent metal hydride centers interact to liberate dihydrogen is not yet understood.

The relative rates of dihydrogen liberation by ferrocene dimers was found to be:

$$\text{IIb} \gg \text{Ib} > \text{Ia} > \text{IIa}$$

This order suggests that the reaction rate is increased when the iron centers are held in close proximity, but is slowed when the ferrocene centers are directly coupled. This latter observation suggests that inductive electron withdrawl by one protonated ferrocene moiety reduces the basicity of the second ferrocene. For compounds IIb and Ib a methylene unit insulates the ferrocenyl groups reducing the effect of this inductive electron withdrawl.

   The ability of ferrocene dimers to generate dihydrogen from
acids and then be recovered by reduction stimulated interest in
the use of these dimers as surface agents for semiconductor
electrodes to mediate the cathodic reactions in water splitting.
[1.1]Ferrocenophane surface-modified electrodes can be produced
by reacting [1.1]ferrocenophane with n-butyl lithium to give the
anion, III, which can then be reacted with a silane to yield IV.
Reaction of IV with a prepared silicon surface binds the [1.1]
ferrocenophane directly to the surface- [1.1]Ferrocenophane surface
modified electrodes prepared in this way were shown to be capable
of liberating dihydrogen from acid solutions when exposed to light.
Underpotentials of 80 mV were observed, but saturation currents
amounted to only a few mA.

   Reaction of the [1.1]ferrocenophane anion, III, with
poly(p-chloromethylstyrene) gave a [1.1]ferrocenophane-containing
polymer, V, which was applied as a film onto the surface of a
p-type semiconductor to produce a photoelectrolysis device.   Studies
with this device demonstrated that H_2 could be generated from
trifluoroboric acid at a potential approximately 430 mV more
positive than for H_2 evolution from platinum under the same
conditions.

Scheme II

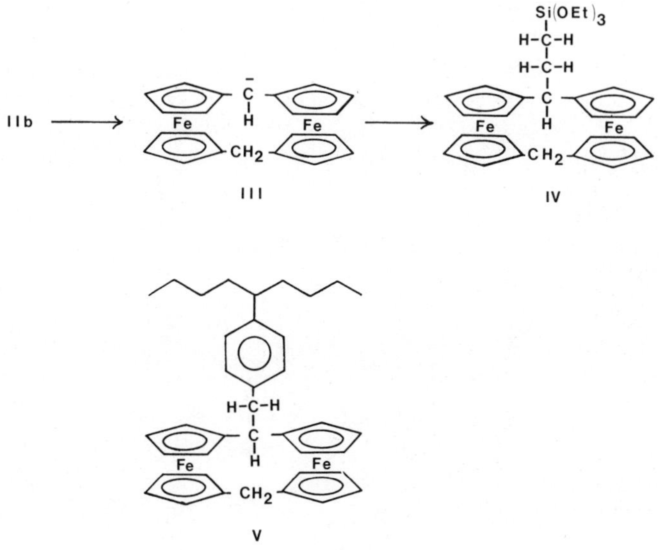

Saturation currents of 31 mA were recorded for this device.
Continuous irradiation for 120 hours resulted in no loss of
activity.   In dilute acids (1 $\underline{M}$ HCl, HBF_4, $HClO_4$) H_2 was
liberated at underpotentials of about 300 mV.

The demonstration of reductive elimination of dihydrogen
from ferrocene dimers and the incorporation of these compounds
into photoelectrolysis devices prompted us to consider the
possibility of preparing a series of bimetallic complexes to
examine their ability to act as dihydrogen liberating species.
In designing molecules which might have potential as electrode
surface species, it was decided to limit our consideration to
compounds with immediately adjacent metal centers.   Of particular
interest was the possibility of preparing dimers from organometallic species which were known to be more basic than ferrocene.
It was felt that by using more basic metal compounds it might
be possible to shift the dihydrogen evolution reaction to neutral,
or only slightly acidic solutions.

Arene chromiumtricarbonyl complexes have been shown to undergo
protonation at the chromium atom in either trifluoroboric acid
or trifluoromethanesulfonic acid[10].   Replacement of one of the
carbonyl groups with a triphenylphosphine has been shown to
increase the basicity of the chromium which permits the compound
to be fully protonated in trifluoroacetic acid.[14]

We have prepared a series of chromium dimer species, VIa-e,
in which the chromium atoms are held in close proximity.   Details
of the synthesis of VIc and VId have been reported previously[15].
Reaction of compounds VIa-d with trifluoroboric acid results in
the liberation of dihydrogen with the relative rate of the reaction
decreasing from VIa to VId.   Only VIa was found to liberate
dihydrogen from trifluoroacetic acid, but the reaction was slow.
Compound VIe does not react with trifluoroboric acid, but it does
react with hexafluoroantimonic acid to rapidly decompose the dimer
giving carbon monoxide, methane and a trace of dihydrogen.   The
methane apparently arises from cleavage of the bridge $N-CH_3$ bond
by the acid.

VI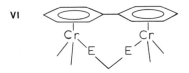

a. $(Me_2P)_2CH_2$ d. $(Ph_2As)_2CH_2$

b. Me_4P_2 e. $(F_2P)_2NMe$

c. $(Ph_2P)_2CH_2$

The rate of reaction of the chromium dimers parallels the
expected order of metal basicity, which in turn reflects the
electron donor ability of the phosphine ligands. The failure of
these compounds to liberate dihydrogen from trifluoroacetic acid
despite the fact that the parent arene chromiumdicarbonyltriphenyl-
phosphine monomer is protonated in this acid, indicates that
electron withdrawl upon protonation of the first metal center
reduces the basicity of the second center.

Unlike the compounds in the ferrocene series, it has not
proven possible to recover the starting chromium dimers from the
acid reaction mixtures by reduction. Workup of the reaction of
VIc with acid gives a mixture of biphenyl chromiumtricarbonyl
and biphenylchromiumdicarbonylbis(diphenylphosphido)methane. Both
compounds VIa and VIb yield biphenyl chromiumtricarbonyl upon
workup of the reaction mixtures. Preliminary electrochemical
studies show that both the mono- and dications of the chromium
dimers are stable in noncoordinating solvents such as methylene
chloride; thus it appears that the observed products are the result
of nucleophilic attack by solvent on the dicationic dimer species.

We have proposed that the dihydrogen liberation reaction
of the chromium dimers proceeds by a reaction mechanism which
is very similar to that of the ferrocenes. As illustrated in Scheme
III, protonation of both metal centers proceeds sequentially to
give a doubly protonated species. Dihydrogen reductive elimination
presumedly proceeds by interaction of the hydride centers.

Scheme III

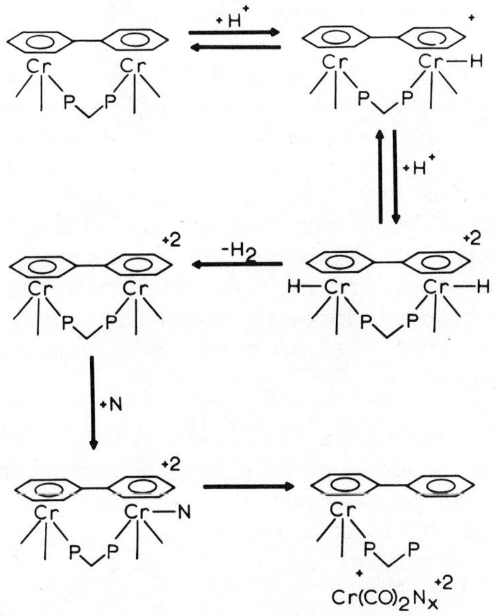

   We have recently prepared fulvalene bis(manganesedicarbonyl)
μ-bis(diphenylphosphido)methane, VII, by photolysis of bicymantrenyl
and bis(diphenylphosphido)methane in benzene. This compound
dissolves in trifluoroboric acid to give a brown solution which
slowly evolves dihydrogen. It has not been possible to characterize
the manganese containing reaction products recovered from this
reaction.

ORGANOMETALLIC POLYMER PRECURSOR STUDIES

   Inour efforts to develop methods which could be used to
incorporate biphenyl bimetallated dimers into polymers we have
investigated a variety of possible reactions. Regrettably those
reactions examined thus far have been uniformly unsuccessful,
however, additional routes remain which might prove successful.

   By analogy to the techniques developed by Müller-Westerhoff
and Nazzal[3] we have examined the possibility of generating an
anion of a simple biphenyl chromium dimer which could then be
reacted with an appropriate alkyl halide to yield a bound species.
Using compound VIc we have attempted to generate the anion VIII
by reaction of VIc with n-butyl lithium, lithium diisopropylamide,
lithium bis(trimethylsilyl)amide, or 18-crown-6 complexed potassium
hydride. After an appropriate reaction time methyl iodide was
added to the reaction mixtures and the reactions worked up. In
all cases, VIc was recovered in good yield with no evidence for
methylated products.

   It has been shown by several workers[16] that alkylbenzene
chromiumtricarbonyl compounds react with strong bases to generate
carbanions which can be reacted with D_2O or benzaldehyde to give
products. To examine the possibility of carrying out similar

reactions with a bimetallated compound we have prepared 4-methyl-
biphenyl bis(chromiumtricarbonyl), IX, and its bis(diphenylphosphido)methane derivative, X.

Reaction of IX with KOH in DMSO gives a dark red solution
which turns yellow upon addition of benzaldehyde. Work up of this
reaction gives a mixture of products which have not been successfully separated.

Reaction of IX with the more powerful bases, e.g. 18-crown-6
complexed potassium hydride or lithium bis(trimethylsilyl)amide
followed by reaction with methyl iodide gives only 4-methylbiphenyl
chromiumtricarbonyl, the chromium tricarbonyl group on the
unsubstituted ring having been lost in the reaction. The nature
of this rather unexpected reaction is now under study.

Reactions of X with trifluoroboric acid have established
that X liberates dihydrogen at about the same rate, or a little
faster, than VIc. No attempt has been made as of yet to generate
the anion of this compound with strong bases.

Attempts to metallate biphenyls which bear electron withdrawing substituents which might be useful for polymerization
reactions result in the formation of the monosubstituted
derivatives as the exclusive products. We are presently preparing
4-chloromethylbiphenyl and its methoxy analog to see if it is
possible to metallate both of the arene rings when the electron
withdrawing group is removed from the ring.

FUTURE STUDIES

The successful development of a photoelectrolysis device
by Müller-Westerhoff and Nazzal provides a strong motivation to
pursue the chemistry of the bimetallic complexes to optimize the
efficiency of the device. We are now engaged in the synthesis
of a series of dimer compounds with strongly basic metal centers
in an effort to lower the acidity requirements of the photoelectrolysis reaction.

Of considerable future importance will be the development
of compounds which are capable of mediating the anode processes
in water splitting. It is quite possible that a manganese dimer
modeled after the active site in the oxygenase enzyme might provide
a starting point for the development of these compounds.

ACKNOWLEDGEMENTS

We wish to thank The Research Corporation and the Naval Academy Research Council for their generous support of this research.

REFERENCES

1.   A.J. Nozik, D.S. Boudreaux, R.R. Chance and F. Williams, "Interfacial Photoprocesses : Energy Conversion and Synthesis", (M.S. Wrighton, ed.) American Chemical Society, Washington, D.C., 1980, Chapt. 9.

2.   (a) M.S. Wrighton, A.B. Bocarsly, J.M. Bolts, M.G. Bradley, A.B. Fischer, N.S. Lewis, M.C. Palazzotto and E.G. Walton, "Interfacial Photoprocesses : Energy Conversion and Synthesis" (M.S. Wrighton, ed.) American Chemical Society, Washington, D.C., 1980, Chapt. 15. (b) J.M. Bolts, A.B. Bocarsly, M.C. Palazzotto, E.G. Walton, N.S. Lewis and M.S. Wrighton, J. Amer. Chem. Soc., 101, 1378(1979). (c) A.B. Boscarsly, E.G. Walton and M.S. Wrighton, J. Amer. Chem. Soc., 102, 3390(1980). (d) A.B. Bocarsly, D.C. Bookbinder, R.N. Dominey, N.S. Lewis and M.S. Wrighton, J. Amer. Chem. Soc., 102, 3683(1980).

3.   (a) A.I. Nazzal and U.T. Müller-Westerhoff, Abstracts of the 186th National Meeting of the American Chemical Society, Aug. 1983, INOR 101. (b) U.T. Müller-Westerhoff and A.I. Nazzal, J. Amer. Chem. Soc., In press, 1983. (c) U.T. Müller-Westerhoff and A.I. Nazzal, U.S. Patent 4,379,740(1983).

4.   G. Wilkinson and J.M Birmingham, J. Amer. Chem. Soc., 77, 3689(1955).

5.   J.C. Kotz and D.C. Pedrotty, Organometal. Chem. Rev. A, 4, 479(1969).

6.   D.F. Schriver, Accounts of Chem. Res., 3, 231(1970).

7.   H. Werner, Pure and Appl. Chem., 54, 177(1982).

8.   H. Werner and R. Feser, Z. anorg. allg. Chem., 458, 301(1979).

9.   T.J. Curphey, J.O. Santer, M. Rosenblum and J.H. Richards, J. Amer. Chem. Soc., 82, 5249(1960).

10.   (a) A. Davidson, W. McFarlane, L. Pratt and G. Wilkinson, J. Chem. Soc., 3653(1963). (b) C.P. Lillya and R.A. Sahatjian, Inorg. Chem., 11, 899(1972). (c) G.A. Olah and S.H. Yu, J. Org. Chem., 41, 717(1976).

11. (a) T.E. Bitterwolf and A.C. Ling, J. Organometal. Chem.,
40, 197(1972). (b) T.E. Bitterwolf and A.C. Ling, J. Organometal.
Chem., 141, 355(1977). (c) T.E. Bitterwolf and A.C. Ling,
J. Organometal. Chem., 215, 77(1981). T.E. Bitterwolf, Tetrahedron
Lett., 22, 2627(1981), T.E. Bitterwolf and M.J. Golightly, I. Chem.
Acta., In press, 1983.

12. T.E. Bitterwolf and A.C. Ling, J. Organometal. Chem.,
57, C15(1973).

13. T.E. Bitterwolf, J. Organometal. Chem., In Press, 1983.

14. (a) B.V. Lokshin, V.I. Zdanovich, N.K. Baranetskaya,
V.N. Setkina and D.N. Kursanov, J. Organometal. Chem., 37,
331(1972). (b) D.N. Kursanov, V.N. Setkina, P.V. Petrovskii,
V.I. Zdanovich, N.K. Baranetskaya and I.D. Rubin, J. Organometal.
Chem., 37, 339(1972). (c) L.A. Fedorov, P.V. Petrovskii, E.I.
Fedin, N.K. Baranetskaya, V.I. Zdanovich, A.A. Tsoy, V.N. Setkina
and D.N. Kursanov, Koordinats Khim., 5, 1339(1979).

15. T.E. Bitterwolf, J. Organometal. Chem., 252, 305(1983).

16. (a) W.S. Trahanovsky and R.J. Card, J. Amer. Chem. Soc.,
94, 2897(1972). (b) J. Brocard, J. Libibi and D. Couturier, Chem.
Comm., 1264(1981). (c) A. Ceccon and G. Catelani, J. Organometal.
Chem., 72, 179(1974).

SYNTHESIS AND PROPERTIES OF CATIONIC CYCLOPENTADIENYL

IRON (II) MOIETY SUPPORTED ON POLYSTYRENE BEADS

Enrique A. Roman*[+], Gerardo J. Valenzuela*,
Ramon O. Latorrre** and John E. Sheats***
* Facultad de Quimica. Pontificia Universidad
  Catolica de Chile, Casilla 114-D, Santiago
  Chile
** Facultad de Ciencias Basicas y Farmaceuticas
   Universidad de Chile, departamento de quimica
   Casilla 653, Santiago, Chile
*** Rider College, Lawrenceville, NJ 08648
    U.S.A.

INTRODUCTION

In the study of arene modifications activated by a
12- electron organometallic moiety, [CpFe], the versatility of the neutral precursor compound 3, (figure
1), that has great nucleophilicity has been shown[1].
The reaction of strong bases with the cation $[\eta^5 CpFe\eta^6$
$C_6 Me_6]^+$, 1, by regioselective deprotontaion in the
$C_6 Me_6$ ring, provides the thermodynamically stable
neutral compound 3, $[\eta^5 CpFe\eta^6 C_6 (CH_3)_5 (=CH_2)]$. Compound 3 rapidly reacts under mild conditions with a
great variety of electrophilic substrates permitting
the modification of the permethylated arene to form
$C_6 Me_5 CH_2$-R, (R=C, Si, P, halogen, metal).

Insoluble reticulated resins, especially styrene -
DVB copolymer (2%), have many applications as solid
supports for transition metal catalysts, organic functional group protection and organic synthesis. Our
interest is to anchor cation 1 in a polymeric matrix,
and to carry on there the regioselective deprotonation
of the pendant organometallic group. In this way it
is possible to develop the chemistry of 3 in a supported phase. The first supported polymer prepared

with the cationic metallocene group 1 was a soluble
PVC polymer containing this redox catalyst 3,4.

In this work we report the synthesis and physical
characterization of a new supported cationic cyclopentadienyl iron (II) moiety in polystyrene-DVB beads,
containing besides another organometallic function
like $CpFe(CO)_2-$, $-(Ph)_2PRhCl(COD)$, $-Mo(CO)_5$, or
$-(Ph)_2PRhCl(CO)$. Catalytic studies of olefin hydrogenation and studies of photoactive behaviour of
supported polymers containing the $CpFe(CO)_2PPh_2-P$
group in the solid state are shown.

EXPERIMENTAL

The neutral cyclohexadienyl derivative 3 was prepared as described previously[1]. Chloromethylated
polystyrene-DVB (2%) beads were suspended in dry THF
in an argon atmosphere, and the solid compound 3 was
added. The mixture was refluxed for one day. The
supported polymer 5 was isolated by filtration, washed
with water and acetone, and finally dried. Polymer 5
is deprotontated in THF suspension with tert-BuOK at
room temperature in an argon atmosphere. In this way,
the supported polymer 6 was obtained, which contains
the methylated exo-cyclohexadienyl group bound to the
polymeric chain and also coordinated to the [CpFe]
moiety. This supported polymer 6 is the starting
reagent for several supported (heterobimetallic) polymers (Fig. 1). All of these polymers were prepared in
adequately dry solvents and inert atmosphere at room
temperature.

All of the polymers were characterized by elementary microanalysis for metals, chloride and carbon.
The infrared spectra for 8a, 8b and 11 were obtained
using KBr pellets or Nujol mulls.

The Mossbauer spectra were obtained with a conventional spectrometer with constant acceleration; the
intensity of the ^{57}Co source in Pd matrix was 20
milicurie. Natural iron was used as reference for the
isomer shift. The low temperature studies were carried out in an "Air Product Chemical CS-202" with a
DMX-20 interface model.

Catalyst 10 was tested in a Parr autoclave with 5
atmospheres of hydrogen pressure. The products of

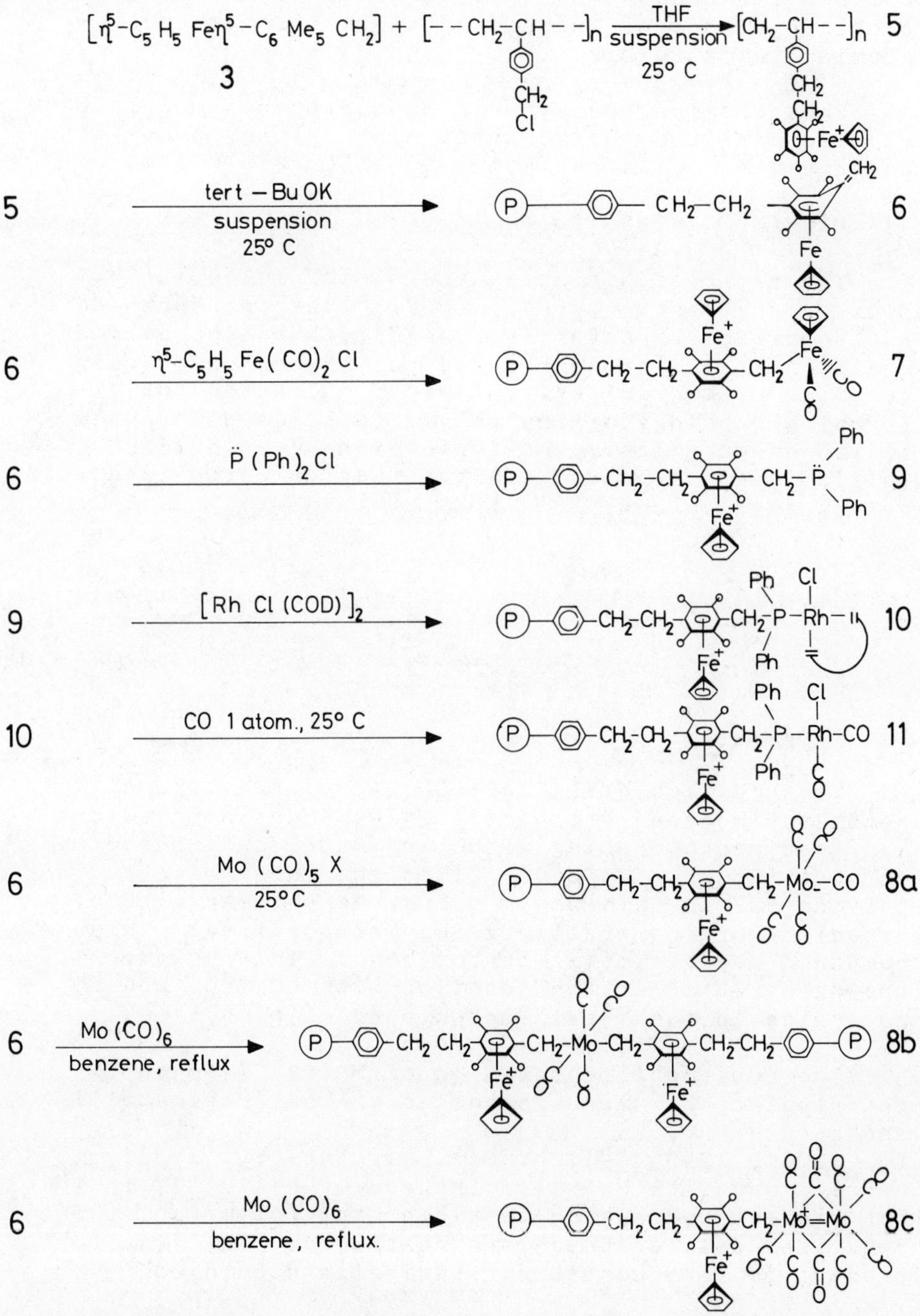
[η5-C5 H5 Feη5-C6 Me5 CH2] + [--CH2-CH--]n
3
THF
suspension
25° C
[CH2-CH--]n
5
tert-Bu OK
suspension
25° C
6
η5-C5H5 Fe(CO)2 Cl
7
P̈ (Ph)2 Cl
9
[Rh Cl (COD)]2
10
CO 1 atom., 25° C
11
Mo (CO)5 X
25°C
8a
Mo (CO)6
benzene, reflux
8b
Mo (CO)6
benzene, reflux.
8c

Fig. 1.

reaction were analyzed by gas chromatography using .
tricresylphosphate and carbowax 1000 columns. The
olefin cyclohexene (Aldrich) was percolated over
alumina prior to use.

Photochemical assays for supported polymer 14 were
carried out in a thermostated microcell of pyrex with
an external radiation from a tungsten 250 W lamp.

RESULTS AND DISCUSSION

The exo-olefinic methylene of the molecular com-
plex 3, [η^5CpFeη^5 C$_6$(CH$_3$)$_5$(=CH$_2$)], has an enhanced
nucleophilic character, so the attack on the polymeric
chain of the chloromethylated polystyrene proceeds
readily. The resultant polymer 5 maintains the external morphology of the polymeric beads with a new
redish brown color. The ionic exchange behaviour of
this polymer was studied with classical experiments in
an exchange column:

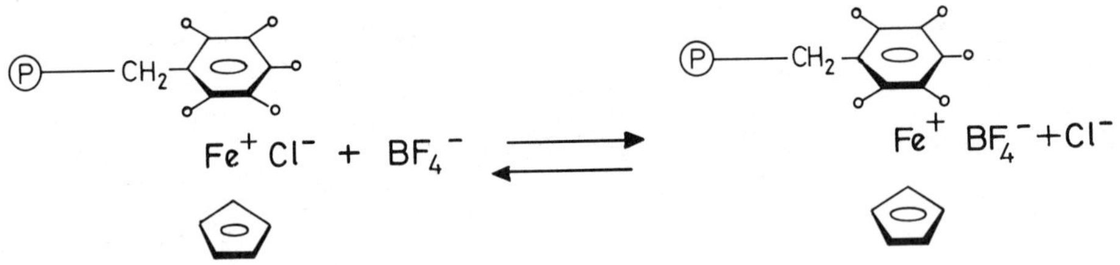

The addition of [η^5CpFe(CO)$_2$-] to the lateral
hexamethylbenzene group of polymer 5 was carried out
by first deprotonating with tert-BuOK, and second by
an addition of η^5CpFe(CO)$_2$Cl to the intermediate
polymer 6. In this way a supported polymer with a
binuclear organometallic group was obtained. This
compound shows a metal-carbon bond. This polymer 7
undergoes an insertion reaction with CO, CO$_2$ and SO$_2$
molecules in the metal-carbon bond (Fig. 2).

The nucleophilic addition of Mo(CO)$_5$I to the
intermediate polymer 6 produces the heterobimetallic
anchored group, 8a
(ν_{CO}=2050, 1940 cm-1, broad). If the reaction is
carried out with Mo(CO)$_6$, instead of Mo(CO)$_5$I, in a
hot benzene suspension that contains polymer 6, the
spherical polymer beads are destroyed. The resultant
product is very complex: terminal and bridged

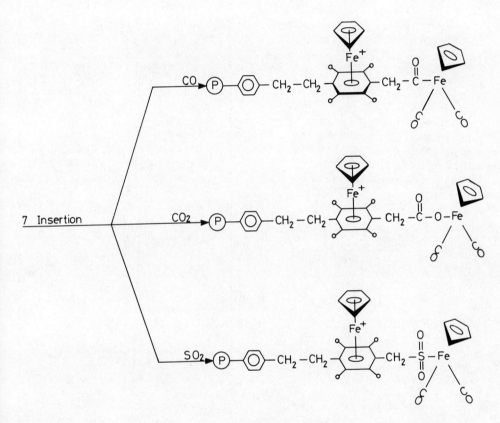

Figure 2.

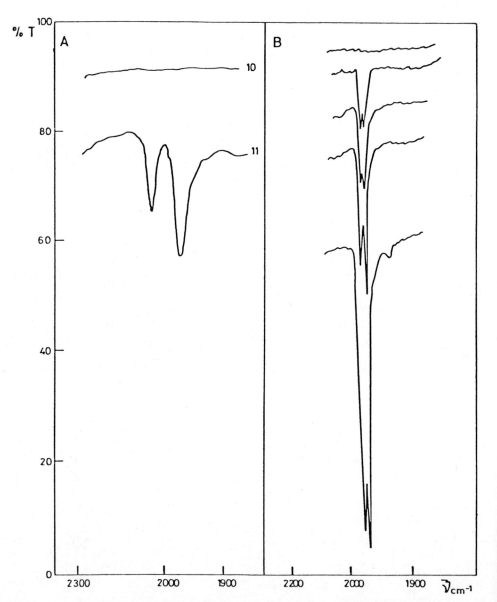

Fig. 3. Comparison of the carbonyl infrared spectra of (A) carbonyla-
         tion (1 atm) of polymer **10** in CH_2Cl_2 suspension; (b) stereo-
         selective carbonylation (1 atm) of $[CpFeC_6(CH_3)_5CH_2(PPh)_2-P-RhCl(COD)]PF_6^-$ in CH_2Cl_2 solution.

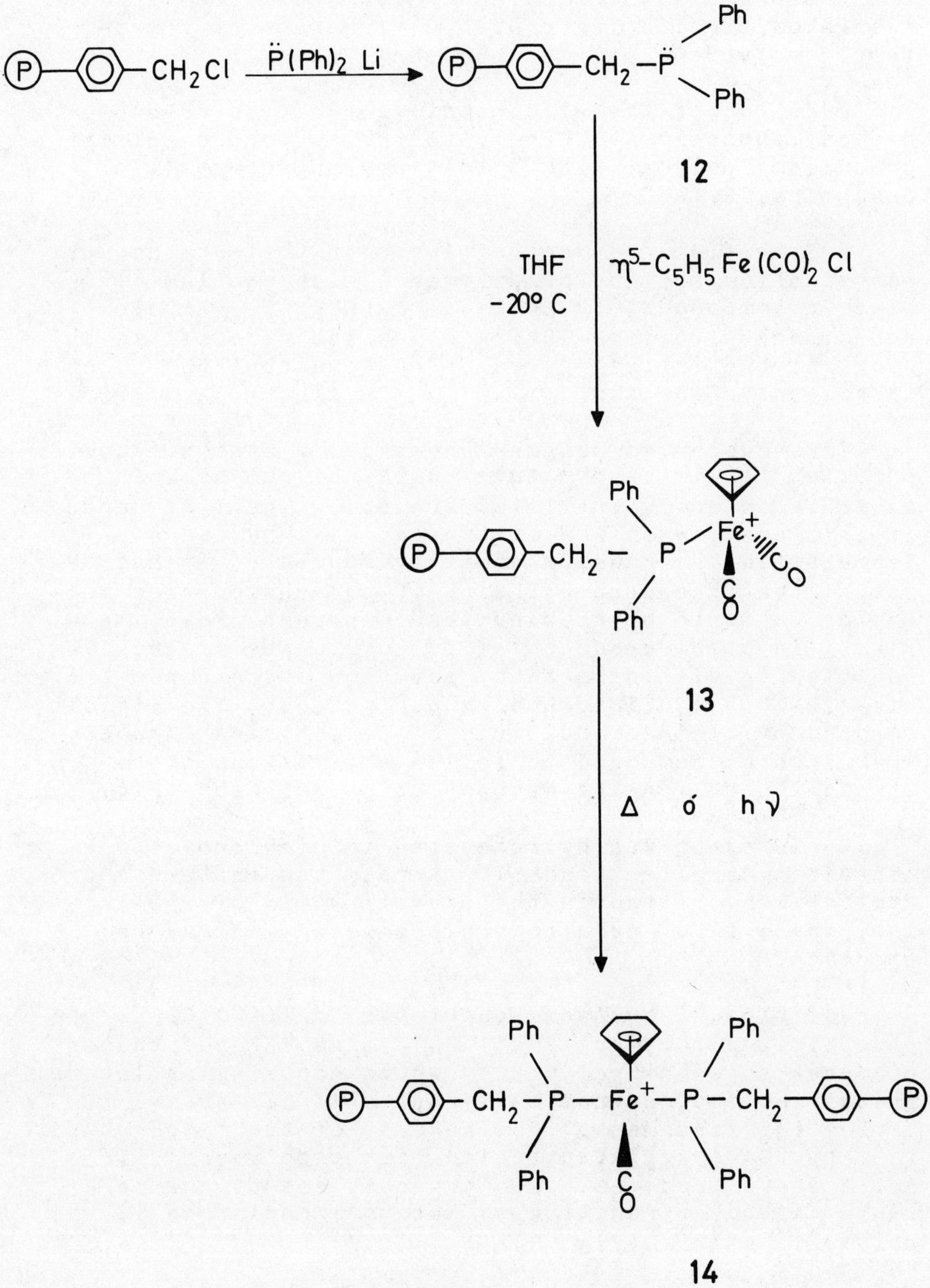
P
CH2Cl
P̈(Ph)2 Li
P
CH2–P̈
Ph
Ph
12
THF
-20° C
η5-C5H5 Fe(CO)2 Cl
Ph
P
CH2 –
P
Fe+
CO
C
O
Ph
13
Δ ó hν
Ph
Ph
P
CH2–P – Fe+ – P – CH2
P
Ph
C
O
Ph
14

Figure 4.

carbonyls appear (ν_{CO}= 2050, 1950 and 1860 cm^{-1}) and
the Mossbauer spectra at $30°K$ show many sites for the
iron atom, indicating probable formation of a supported molybdene polymer, 8b, 8c (Table 1).

Addition of the electrophilic PPh_2, as chloro-
diphenylphosphine ($ClPPh_2$), to the supported polymer 6
produces a polymer with a tertiary organometallic
phosphine, 9.

Polymer 10 was prepared in order to test the metal
coordination ability of polymer 9 with rhodium(I)
dimeric compounds, $[Rh(COD)Cl]_2$, (COD= 1, 5-cyclo-
octadiene). Polymer 10 preserves the external morphology of the initial polystyrene beads but the color
is golden yellow now.

This supported polymer undergoes a carbon monoxide
addition (1 atm of pressure) with the release of the
diolefin and polymer 11 is obtained. The I.R. bands
of this polymer show a mixture of cis and trans
isomers for the rhodium atom environment. It has been
shown[5] that the carbonylation of the unsupported molecular cationic heterobinucler compounds, Fe-d_6, and
Rh-d^8, in homogeneous phase follows a steroselective
addition in dichloromethane solution and affords the
cis-dicarbonyl chlororhodium derivative. The rigidity
imposed by the macromolecule of polystyrene induces
the anchored compound to follow a different stero-
chemical path for the carbonylation reaction (Fig. 3).

Cyclohexane was hydrogenated in dichloromethane
containing swollen beads 10 at room temperature and 5
atm for 24h. Although the overall yield of 15% is
low, these first results are promising and may be
improved.

Addition of the semisandwich $\eta^5CpFe(CO)_2Cl$ to the
phosphinated polymer 12 at $-20°C$ in a THF suspension
produces a yellow polymer 13 which contains as lateral
groups the thermodynamically unstable dicarbonyl [$\eta^5CpFe(CO)_2$] fragment. An increase of the temperature
or a photochemical excitation of the polymer 13 in
solid state produces a spontaneous loss of one carbonyl ligand, producing an internal chelation in the
polymeric matrix (Fig. 4).

Table 1. Infrared spectra.

Polymer	ν_{CO}	(cm^{-1})	(Nujol)
7	2.010	1.900	
8a	2.050	1.945	1.930
8b, 8c	1.945	1.860	
11	2.050	1.980	

This is an example of a ligand labilization imposed by the matrix rigidity, and at the same time, by the attack of the uncoordinated phosphine anchored to the same matrix. The carbonyl elimination from the yellow polymer 13 in the solid state by photochemical or thermal activation produces the brown polymer 14 with an external morphological change of the spherical beads. The IR spectrum shows a new medium intensity band at 1860 cm^{-1} separate from the principal band (ν_{CO}= 1940 cm^{-1}). This means that internal chelation is produced, with a lateral electronic transference from the [η^5CpFe(CO)] fragment, giving rise to a binuclear species of iron (I) with carbonyl bridges. This fact is observed commonly in [η^5CpFe(CO)$_2$] chemistry[1].

Mossbauer parameters for supported polymers 5, 8b, 10, at different temperatures are shown in Table 2. Mossbauer spectra of mixed metallocenic systems [η^5CpFeη^6 Arene]$^{1+,6}$ have principally been studied by Astruc and co-workers[1,6,7]. In the paramagnetic neutral molecules of that pattern, (Fe-d^7), there is a noticeable effect in the solid state, and from an interaction with the solid lattice when the molecule is in certain crystalline matrix hosts. This guest-host interaction removes the electronic ground state $^2E_{1g}$ by a Jahn-Teller distortion[6,8]. Now the Mossbauer spectra at different temperatures show two sites with different electronic environment; in our case when the cationic molecule, Fe-d^6, is covalently bonded to a reticular matrix of polystyrene, the results in QS values are greatly reduced (Table 2, Fig. 5).

The rigid polystyrene reticular matrix provokes a drastic decrease of the QS value with respect to the unanchored molecule. Mixed molecular metallocenes with Fe-d^6 in polycrystalline samples normally show QS values of 1.6-2.00 mm sec^{-1}. Now, the same anchored cationic metallocenes show a QS = 0.58 mm sec^{-1} at 300°K. Nevertheless, from the calculated relative intensities, only 20% of the supported polymer contains metallocenic centers with a [η^5CpFeC$_6$(CH$_3$)$_5$-CH$_2$] equivalent, (QS: 2.00 mm sec^{-1}), (Fig. 4). From Table 2, the iron nucleus and only mixed ferrocenic metallocenes with oxidation state I and III, have such small QS value. That is, the ferricinum (QS=0.00 mm sec^{-1}), [BFD(3,3)]$^+$, (QS= 0.568 mm sec^{-1}) and [η^5CpFe

Table 2.   Mössbauer parameters[a] of Fe-d^6 and Fe-d^7 metallocenic species.

Compound	T(°K)	QS (mm·s^{-1})	IS (mm·s^{-1})	% rel.intens.[b]	Ref.
[CpFeHMB]BF$_4$	293	2.00	0.45	–	This work
	297	2.00	0.46	–	This work
[CpFeHMB]	293	0.50	0.74	–	(1)
	42	1.54	0.90	–	(1)
[Cp$_2$Fe]DDQ	77	0.00	0.46	–	(9)
[BFD(3,3)](BF$_4$)$_2$	77	0.58	0.48	–	(9)
Polymer 5	300	0.58	0.32	80	This work
		2.00	0.45	20	
	30	0.50	0.45	57	
		1.40	0.30	29	
		2.10	0.35	14	
Polymer 8b, 8c	300	0.40	0.25	50	This work
		0.80	0.35	40	
		2.00	0.45	10	
Polymer 10	30	0.70	0.25	42	This work
		1.50	0.45	32	
		2.35	0.60	26	

a: Isomer shift relative to natural iron.

b: Relative intensity percentage of each doublet compared to the whole area of the spectrum.
BFD: biferrocenylene[bis(fulvalene)diiron].

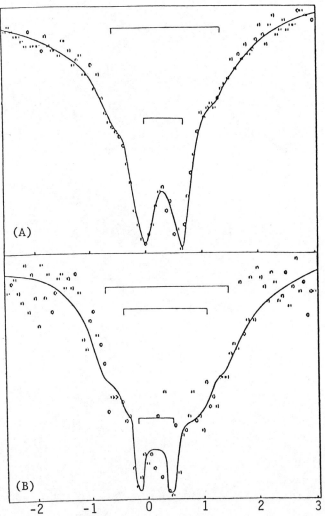

Fig. 5.   Mössbauer spectra of

$\textcircled{P}\!\sim\![(CH_2-(CH_3)_5-C_6-\eta^6)\overset{+}{F}e(\eta^5-Cp)]$,

5 :   (A) 300°K; (B) 30°K.

$^6C_6Me_6]$, (QS = 0.50 mm sec^{-1}). No QS values smaller
than 1.8 mm sec^{-1} had before been observed in Fe(II)
metallocenes with a d^6 electronic configuration. At
low temperatures (30°K) the same supported polymer 5
shows a new doublet with QS = 1.40 mm sec^{-1} in addition to the other two observed at higher temperatures, which show a slight difference in the QS now:
0.50 and 2.10 mm sec^{-1}.

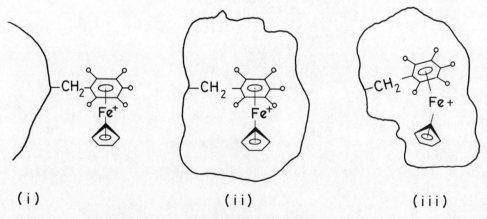

(i)   (ii)   (iii)

Fig. 6 Models of heterogenous sites of anchoring for a
        mixed metallocenic molecule in a grain of cross-
        linked chloromethylated polystyrene, creating
        different host-lattice interaction on the
        polymeric matrix.

The relative intensities of each doublet show that there exists a heterogeneous distribution of the metallocenic molecules anchored in the polymer with a wide variety of molecular environments.

From these remarks, and previous work in connection with the two principal factors that act on the QS values of these compounds, it is possible to infer that Mossbauer parameters of supported polymers that contain a particular metallocenic species will depend on (i) the vibrational state of the anchored molecule and its consequent "electronic lability"[8], strongly influencd by a guest-host lattice interaction; (ii) a temperature-effect that provokes a drastic change in the polymeric conformation, and therefore different metallic center distributions in the polymeric bead. Here it is necessary to establish that the $Fe-d^6$ supported molecule presents a ground state, $^1A_1$. Nevertheless with the molecular environment imposed by the rigid polymeric lattice, that electronic state may be disturbed due to an external perturbation.

It is qualitatively possible to infer that in a supported polymeric bead there exist at least three different types of active sites with different environments at the same time: (i) superficial distribution or no guest-host lattice interaction, (QS= 2.00 mm sec^{-1}); (ii) intermediate distortion (QS= 1.40 mm sec^{-1}) and (iii) high distortion in the network of the polymeric lattice (QS= 0.5 mm sec^{-1}) (figure 6).

For the polymer 8a case, that has bimetallic Fe(II) and Mo(I), (II) the QS value for the doublet with greater relative intensity is higher (QS= 0.7 mm sec^{-1}) than for polymer 5. The other two doublets in the Mossbauer spectrum represent less unprotected active sites in the polymeric matrix. Undoubtedly an "unprotection" of the sandwich implies an approaching to the ground state of the $CpFeC_6Me_6$ sandwich by the influence of mono and bimolybdenum centers (polymers 8a, 8b, 8c, Fig. 1).

CONCLUSIONS

The [$\eta^5C_5H_5Fe\eta^6(CH_3)_6$] redox catalyst fixation in PVC and in reticulated polystyrene through mild reaction

condition shows the nucleophilic properties of the
neutral [$\eta^5 C_5 H_5 Fe\eta^5 C_6(CH_3)_5(=CH_2)$] molecule through the
exocyclic methylene bond.

That fixation has allowed us (i) to prepare soluble
PVC polymers that have organometallic centers with
electrochemical activity. In the future it may be possible to anchor those "electronic reservoirs" in polymer
membranes, and on modififed electrode surfaces, possibly
causing selective and catalytic phenomena; (ii) to incorporate chiral heterobimetallic compounds into reticulated polystyrene in order to catalyze assymetric organic synthesis; (iii) to prepare a photosensitive supported polymer in the solid state for the temporary
coordination of unsaturated and photochemically activated species, such as olefins, dinitrogen, carbon
dioxide, hydride, etc. That phenomenon is possible due
to an anchored 16-electron intemediate, [P (Ph$_2$)P-
(CpFe)-P(Ph$_2$) P] (iv) to show that Mossbauer spectroscopy has, in this case, been a good tool for the
detection of the different types of polymeric matrix
site containing more or less protected sandwich
molecules.

REFERENCES

1. D. Astruc, J-R. Hamon, E. Roman, P. Michaud, J. Am.
    Chem. Soc. 103, 7502 (1981).

2. P. Hodge, D. Sherington, Eds., "Polymer Supported
    Reactions Synthesis", John Wiley & Sons, (1980).

3. E. Roman, G. Valenzuela, L. Gargallo, D. Radic, J.
    Polym. Sci. 21, 2057 (1983).

4. a) E. Roman, R. Dabard, C. Moinet, D. Astruc,
    Tetrahedron Letters 1433 (1979).

    b) D. Astruc, J-R. Hamon, G. Althoff, E. Roman, P.
    Batail, P. Michaud, J-P. Mariot, F. Varret, D.
    Cozak, J. Am. Chem. Soc., 101, 5445 (1979).

    c) A. Buet, A. Darchen, C. Moinet, J. Chem. Soc.
    Commun., 447 (1979).

5. E. Roman, V. Castro and M. Camus, manuscript in
    preparation.

6. M. Rajasekharan, S. Giezynsky, J. H. Ammeter, N.
    Oswald, P. Michaud, J-R. Hamon, D. Astruc, J. Am.
    Chem. Soc., 104, 2400 (1982).

7. J.C. Green, M.R. Kelly, M. P. Payne, E. A. Seddon,
    D. Astruc, J-R. Hamon and P. Michaud, Organo-
    metallics, 2, 211 (1983).

8. J. Ammeter, L. Zoller, J. Bachman, P. Baltzer, E.
    Gamp, R. Bucher, E. Deiss, Helvet. Chim. Acta., 64,
    1063 (1981).

9. C. LeVanda, K. Bechgaard, D.O. Cowan, U.T. Mueller-
    Westerhoff, P. Eilbracht, G.A. Candela and R. L.
    Collins, J. Am. Chem. Soc., 98, 3181 (1976).

MODIFICATION OF DEXTRAN THROUGH REACTION WITH
BISCYCLOPENTADIENYLTITANIUM DICHLORIDE EMPLOYING
THE INTERFACIAL CONDENSATION TECHNIQUE

Yoshinobu Naoshima[a,b], Charles E. Carraher, Jr.[a], George
G. Hess[a], and Masahide Kurokawa[b]

Wright State University[a]
Department of Chemistry
Dayton, Ohio 45435 and
Okayama University of Science[b]
Department of Chemistry
Ridai-ocho, Okayama 700, Japan

INTRODUCTION

Setting

Mankind inhabits a spherical ball with finite resources.
Initially mankind employed numerous renewable materials for shelter
and building as well as clothing and food. The 19th and 20th
centuries saw a turn toward non and slowly renewable resources
as building materials, household goods and clothing - most derived
from the basic feedstocks - coal and petroleum. The recognition
that the supply of petroleum and coal is decreasing and uncertain
coupled with increasing cost of both has signaled for a return
towards the use of renewable resources for technological applications. Renewable resource materials can offer both controlled
and known abundance and unique physical properties.

Table 1 contains a brief listing of renewable resource materials with regard to abundance. The two most abundant renewable
resources are lignin and carbohydrates. Lignin is a major polymeric component of woody tissue constituting about 25% of wood
(by dry weight). While lignin's structure is complex and variable
(with plant species, growth conditions, time of year) it is water
soluble and contains a variety of reactive, functional groups
including aromatic and aliphatic hydroxyls, ether, ketone and
aldehyde groups.

Table 1. Relative Availability of Selected Natural Products (1).

        Small scale - speciality uses
                Alkaloids
                Heme, Bile and Chlorophylls
                Phenolic Plant Products
                Steroids
                Tannins
        Medium scale with some having the potential for large scale
            usage
                Amino Acids
                Fungus
                Lipids
                Proteins (specific)
                Purines, Pyrimidines, Nucleotides, Nucleic Acids
        Large scale
                Bacteria
                Carbohydrates
                Drying Oils, Alkyd Resins
                Lignins
                Polyisoprenes
                Proteins (general)
                Terpenes and Terpenoids

        Carbohydrates are the most abundant (weight-wise) class
of organic compounds constituting three-fourths of the dry weight
of the plant world. They represent a great storehouse of energy
as a food and fuel for man and animals. About 400 billion tons
of sugars are produced annually through natural photosynthesis.
While carbohydrates come in a variety of sizes, the bulk of the
carbohydrates are mono and disaccharides with molecular weights
below 400 Daltons or are polymeric with molecular weights above
10^5. Table 2 contains a listing of select polysaccharides.

        The chemical and physical modification of polysaccharides
is one of mankind's oldest technologies typically focusing on
cotton since the 1850s. The majority of these modifications
is topochemical in nature occurring through reactions involving
hydroxyl groups available in the amorphous regions and on surfaces
of crystalline areas. Our group[1-5] has focused on the more thorough, homogeneous modification of polysaccharides.

Rationale

        Dextran was the polysaccharide chosen for preliminary study.
It is water soluble, permitting the evaluation of aqueous reaction
systems. Also, it is readily available on a large scale in a
wide variety of molecular weights, the latter permitting modification of dextran to be readily studied as a function of chain
length. Dextrans are a source of cellulose primarily found in

Table 2. Naturally Occurring Polysaccharides

Polysaccharide	Source	Monomeric Sugar Unit(s)
Agar	Red seaweed	Complex - contains beta-galacta-pyranose and 3,6-alpha-L-galactopyranose
Alginic Acid	Brown seaweed	Beta-D-mannuronic acid and Alpha-L-glucuronic acid
Arabinan	Plant cell walls	Heteropoly saccharide - most contain arabinofuranose
Arabinogalactans	Plants	Complex with galactose backbone
Arabinoxylans	All land plants	Xylose backbone - complex side mostly L-arabinofuranose
Carrageenan	Red seaweed	Complex - contains beta-galactopyranose linked to 3,6-anhydro-D-galactopyranose
Cellulose	Plants	D-Glucose
Chitin	Animals	2-Acetamido glucose
Dextran	Bacteria	D-Glucose
Galactan	Plant cell walls, intercellular material	Galactose
Galactoglucommannans	Softwood cell walls	Complex - glucose, mannose, galactose
Galactomannans	Seed mucilages	Mannopyranose and galactose
Glucomannans	Some seeds and bulbs, hardwood	D-Glucopyranose and D-mannopyranose
Glucuronoxylans	All land plants	Xylose backbone - complex side chains mostly 4-O-methyl-alpha-D-glucuronic acid
Glycogen	Animals	D-Glucose
Gum Arabic	Tropical trees	D-Galactopyranose chain with varying side chains
Gum Tragecanth	Plant resin	L-Arabinose and D-Galactose
Heparin	Lung, liver, arterial walls	Mostly glucuronic acid, iduronic acid and glucosamine
Inulin	Artichokes	D-Fructose
Laminaran	Brown seaweed	Beta-1-3-D-glucopyrabose
Levan	Grass, bacteria	Fructose
Lichenan	Iceland moss	Glucose

(continued)

Table 2. (continued)

Mannan	Plants, yeasts, algae	D-Mannose
Peptidoglycan, murein	Bacteria	Polysaccharide (-acetyl-glucosamine, N-acetylmu-ramic acid/strands cross-linked by peptides
Starch - amylose, amylopectin	Corn, potatoes	D-Glucose
Xyloglucans	Some seeds, hardwood cell walls	Complex - glucose, xylose, fucose, galactose
Yeast Mannan	Yeast	D-Mannose

yeast and bacteria consisting of branched storage polysaccharides of D-glucose. The polymers consist essentially of branched chains of alpha-1 $\rightarrow$ 6 linked D-glucopyranose residues. They differ from glycogen, starch and cotton in having a variety of backbone linkages which may be 1 $\rightarrow$ 2, 1 $\rightarrow$ 3, or 1 $\rightarrow$ 4, depending on the particular source. Below is a representation of 1 $\rightarrow$ 6 dextran with a 1 $\rightarrow$ 4 branch.

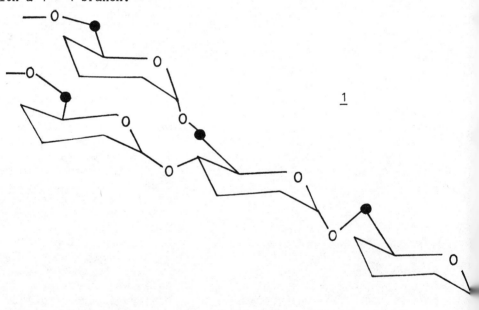

$\underline{1}$

An underlying assumption is that dextran is a representative polysaccharide source and that results derived from studying its modification can be directly applied to other sources of polysaccharides. If dextran can be successfully modified utilizing a specific reaction system, other polysaccharides can likewise be modified using the same reaction system if that particular polysaccharide is soluble in the given reaction phase employed. Thus, polysaccharide solubility is the key factor in determining which reaction system can be employed.

Our group had previously synthesized a number of metal containing polymers through reaction with a variety of diols including hydroquinone, ethylene glycol, 1,6-hexanediol and bisphenol A.[6-10] Further polyvinyl alcohol has been successfully modified employing analogous synthetic routes.[11-13]

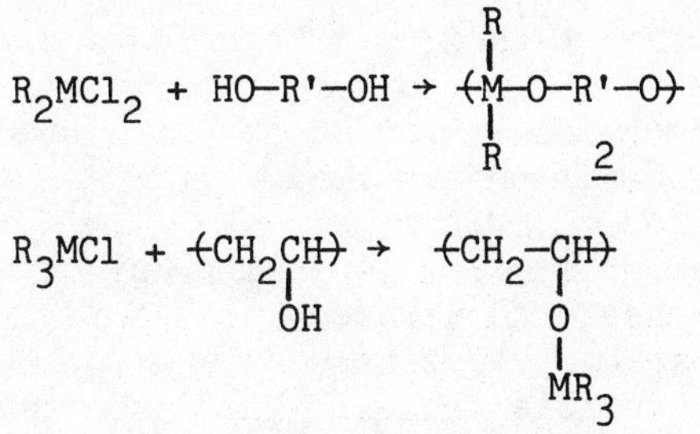

$$R_2MCl_2 + HO-R'-OH \rightarrow \left(\underset{\underset{R}{|}}{\overset{\overset{R}{|}}{M}}-O-R'-O \right) \quad \underline{2}$$

$$R_3MCl + \left(CH_2\underset{\underset{OH}{|}}{CH} \right) \rightarrow \left(CH_2-\underset{\underset{\underset{MR_3}{|}}{O}}{CH} \right)$$

$$\underline{3}$$

More recently we effected the modification of cellulose derived from cotton and dextran employing organostannane mono and dichlories.[1-5] The initial modification of dextran employing biscyclopentadienyltitanium dichloride, BCTD, is reported here.

BCTD was employed for at least the following reasons. First, the amount of Cp_2Ti moiety in the product is readily determined through thermolysis of the product leaving (quantitatively) titanium dioxide.

Second, it is readily available and relatively inexpensive. Third, BCTD has two color sites, yellow Cp-Ti and red Ti-Cl. Thus its inclusion in the product can be readily determined through observation of product color. Fourth, BCTD has been successfully condensed with a variety of hydroxyl-containing compounds. Fifth, BCTD is soluble in a variety of liquids including typical organic solvents as chloroform, hexane and carbon tetrachloride; dipolar aprotic solvents as acetone, DMSO and DMF, and in water forms chemically active moieties which behave as the Cp_2M^{+2} ion.[10,12,13]

There are three general condensation techniques: melt, solution and interfacial. The melt method is eliminated for at least three reasons. First, true solution of polysaccharide is not effected in the "neat" state through heating because degradation

occurs before melting. The goal of this project is the homogeneous
modification of a polysaccharide which requires the polysaccharide
to be in solution so modification can occur throughout the polysaccharide and not just at the surface of exposed hydroxyl moieties. Thus, "neat" melt condensation systems are unsuitable
for this particular study. Second, polysaccharides undergo degradation at temperatures of 250°C and above while glucose units experience ring opening at about 200°C. Moreover, BCTD begins to degrade
at about 120°C. Thus, reaction temperatures of above approximately
100°C should be avoided to minimize undesirable, thermally induced
side reactions. Third, from an energy conservation standpoint,
systems utilizing less energy are favored for industrial acceptance
if other factors are the same.

The two remaining reaction technique categories are the
solution and interfacial techniques, collectively referred to
as the low temperature condensation procedures. Both techniques
encompass a wide variety of reaction systems.

Interfacial polymerization systems generally employ an aqueous
phase containing the Lewis base, and a water immiscible organic
phase containing the Lewis acid. Reaction occurs at or near
the interface of the two immiscible phases. The interfacial
method provides a way to partially control the rate of reaction
by varying the stirring rate thereby controlling the interfacial
contact area. When the stirring is stopped, the interfacial
contact area is greatly reduced thus effectively stopping the
reaction.[14-16]

EXPERIMENTAL

Biscyclopentadienyltitanium dichloride (Aldrich, Milwaukee,
Wisconsin) and dextran (molecular weight 200,000-300,000; USB,
Cleveland, Ohio) were used as received. Reactions were carried
out in a one quart Kimex emulsifying jar placed on a Waring Blendor
(Model 1120) with a "noload" stirring rate of 18,500 rpm. The
freshly prepared aqueous phase, containing dextran was added
to stirred organic solutions containing BCTD. The product was
obtained as a precipitate employing suction filtration. Repeated
washings with carbon tetrachloride and water assisted in the
purification of the product.

Solubility studies were carried out by placing about 20
mg of sample with three milliliters of liquid. Elemental analyses
for titanium were conducted by heating known weights of samples
until a white product, titanium dioxide, was formed.

Infrared spectra were obtained using KBr pellets employing
Perkin-Elmer 457 and 1330 Infrared Spectrophotometers and a Digilab
FTS-20 C/D FT-IR.

Mass spectral analyses were performed using direct insertion
probes in Kratos MS-50 or MS-80 mass spectrometers, operating
in the EI mode, 8 KV acceleration, and 10 sec/decade scan rate
with a probe temperature of 350°C.

RESULTS AND DISCUSSION

Structure

The products may contain a variety of units including unreacted glucose units, and structures depicted as 4-7, with the percentage of Ti calculated to be 9.6% for 4, 14.2% for 5, 14.7% for
7, and 16.9% for 6 and 13.2% for 4, 17.1% for 5 and 19.3% for
6 if crosslinking through the Cp_2Ti moiety is not assumed and end
groups are OH.

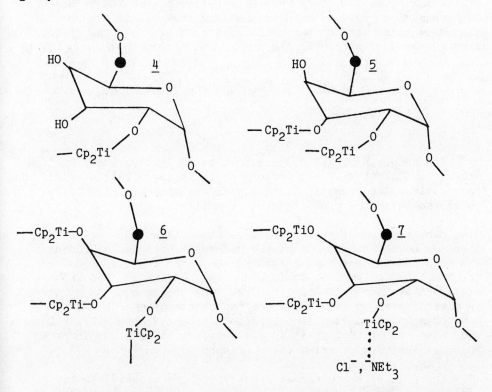

Analysis of the infrared spectra of the products is consistent
with the dextran being modified by formation of Ti-O-R ether
linkages through condensation at the hydroxyl sites. Bands characteristic of the dextran are present at 1650, 1480, 1440, 1360,
1275, 1240, 1160, 1100, 810 and 760 (all bands given in units
of cm^{-1}) and between 3000 to 3600 characteristic of unreacted

hydroxyl groups (Figure 1). Bands characteristic of the Cp_2Ti moiety are present at 1405, 1030 and 855.

Products employing triethylamine, TEA, as the added base also exhibit weak bands characteristic of TEA between 2840 and 2490. (These bands are absent when sodium hydroxide is employed as the added base, Figure 1). Further, as noted following, mass spectra of products derived using TEA show ion fragments attributed to TEA. The sample could have TEA as an end group (8) or as a separate titanium-containing product (9). Product 9 was synthesized separately. It is a green solid unstable in dilute nitric acid, dissolving on contact. The products are brown to yellow in color and are somewhat stable in nitric acid solutions (weight retention of about 70% after standing in 1M HNO_3 for 1 min. accompanied by a loss or decrease in bands associated with the presence of the TEA moiety). It is currently believed that TEA is present as an endgroup (8). A third possibility that TEA is trapped liquid is eliminated on the basis of a lack of evolution of odor from the product since TEA has a characteristic odor.

$$\underset{\displaystyle 8}{\text{Sugar}-\text{O}-\underset{\displaystyle Cp}{\overset{\displaystyle Cp}{\text{Ti}}}-\text{NEt}_3\text{,Cl}} \qquad\qquad \underset{\displaystyle 9}{Cp_2Ti(NEt_3)_2Cl_2}$$

The products range in color from being light brown to yellow brown, consistent with the inclusion of the Cp-Ti yellow-brown color site. The lack of solubility in any attempted liquid is consistent with the product being crosslinked.

Mass spectra of the products can be related more or less directly to the structures of the molecular components present. Tables 3 and 4 present mass spectral data for two compounds. The first two columns in Table 3 contain results derived from a product synthesized employing TEA. The presence of the TEA is clearly seen in Table 4, where the observed ion intensities are shown to be in excellent agreement with the John Wiley reference spectrum of TEA. This is also consistent with the infrared spectral analysis reported above (Figure 1).

The John Wiley reference spectrum of cyclopentadiene is compared in Table 4 with scans of the products, in which the ion intensities have been adjusted so that m/e 66 = 100%. From the data it is possible that cyclopentadiene is being evolved, but evidently other decomposition processes are more prevalent. The titanium atoms are not being volatilized since no intensity ratios characteristic of Ti can be found. This is consistent with other studies involving titanium polyesters and polyethers.

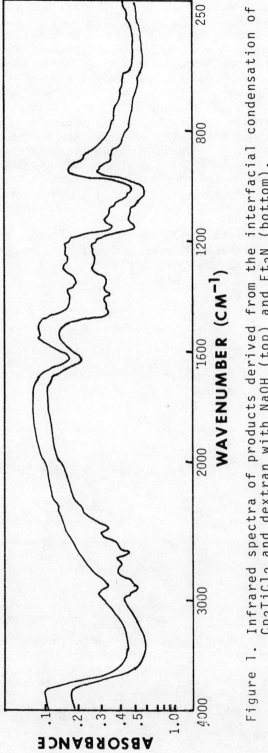

Figure 1. Infrared spectra of products derived from the interfacial condensation of Cp₂TiCl₂ and dextran with NaOH (top) and Et₃N (bottom).

Mass spectra of mono- and disaccharides are complex and temperature dependent. Thus it would be very difficult to predict what one might observe from the polymers synthesized here. Not shown in column 5 of Table 3 is the fact that the scan of the product from the NaOH base, evolved a scattering of low intensity ions out to ca. m/e 730; although the few of higher intensity can be seen. Compositions which may be assigned to the lower mass ions (<150) in all columns of Table 3 include a variety of C, H and O ratios which would be plausible for a thermally degraded modified dextran polymer. Additional mass spectra of these polymers will be required to establish what patterns should be observed in each case. The mass spectrum in column 3 of Table 3 has a large m/e 44, presumably CO_2. It would be interesting to know whether this mass is typical of these polymers in general.

Table 3.   Percent Relative Abundance of Ion Fragments Detected in Direct Insertion Probe Analysis of Condensation Products of Dextran and Cp_2TiCl_2.[a]

m/e	Early[b]	Middle[b]	350°C[c]	Ca.250°C[d]	350°C[e]
528					12
371		22			
368					11
262					12
149				14	
137				10	
123					14
121					10
119					11
111		12		12	16
109		18		10	23
107					11
105		12			9
101	15	9			
100	6				
98		19			23
97				26	36
96		24	5	11	20
95		38	7	19	42
93		12			15
91		22			17
87	6				
86	100	100			
85		28		25	23
84		15		17	29
83		43		39	51
82		38		17	25
81		45		42	50

(continued)

Table 3. (continued)

m/e	Early[b]	Middle[c]	350°C[c]	Ca.250°C[d]	350°C[e]
80					18
79		20			23
78		10			
77		10			10
73				30	20
71		45		42	40
70	4	26		22	18
69		54		100	75
68		23		22	33
67		40		24	59
66		10	11	12	
60				22	18
59		29			
58	24			19	
57		84		86	89
56	8	41		20	20
55		80		66	100
54		22			22
53		16			
44			100		
43			11		
39			15		
30			9		

[a]All ions $\geq$ 10% relative intensity, plus selected masses < 10%.

[b]Heated to 350°C and scanned repeatedly from m/e 50. "Early" denotes mass spectrum taken shortly after reaching 350°C. The middle scan was taken about 5 minutes later.

[c]Heated to 350°C, but scanned from m/e 30 to 200 with the MS-80.

[d]Heated to ca. 250°C, and scanned from m/e 50.

[e]Heated to 350°C, and scanned from m/e 50.

Most titanium-containing organometallic polymers decompose quantitatively to white titanium dioxide below 900°C. The products marked with the letter "a" were exposed to temperatures in excess of 2000°C but retained coloration. Thus percentage titanium values for these materials should be considered only as an upper value. Elemental analysis results (Table 5-7) where the final product was white agreed with a fairly high loading of the available hydroxyls such that between one (14% titanium, forms 4 and 5) and two (19%, form 6) of the three available hydroxyl groups are occupied by titanium-containing moieties. Elemental analysis was repeated except using perchloric acid degradations. Results (Tables 5-7) were similar to those obtained by thermolysis where the thermolysis products were white but lower and within reasonable limits for products that did not yield titanium dioxide on thermolysis. The results indicate a high degree of substitution. It is also possible that some titanium dioxide (% Ti = 60) is present derived from the base-catalyzed decomposition of Cp_2TiCl_2.

Table 4. Ion Fragmentation Patterns (Percent Relative Intensities) for Cyclopentadiene and Trimethylamine

Cyclopentadiene

Ions (m/e):	66	65	39	40	38	63	67	62
John Wiley	100	47	32	27	8	8	6	6
Et$_3$N, Middle[a,b]	100	65	e	e	e	35	422	5
NaOH, 350°C[c,b]	100	43	131	30	28	9	14	9
NaOH, ca.250°C[d,b]	100	38	e	e	e	4	203	7

Triethylamine

Ions (m/e):	86	58	101	56	100	87	70	72	57	54
John Wiley	100	26	16	8	7	6	4	3	2	2
Et$_3$N, Early[a]	100	24	15	8	6	6	4	3	3	-

[a]Cf. Table 3, footnote b.

[b]The intensities of the ions have all been divided by the same factor to make the major ion equal 100% (i.e. m/e 66 = 100%).

[c]Cf. Table 3, footnote c.

[d]Cf. Table 3, footnote d.

[e]Only masses > 50 were scanned.

In summary, infrared spectroscopy is consistent with the presence of both moieties derived from dextran and BCTD. Products derived employing TEA also show bands characteristic of its inclusion, at lower concentrations of TEA, yielding products containing structures of form 8. Elemental analyses for titanium are consistent with a high degree of substitution by BCTD. Mass spectroscopy is consistent with the presence of moieties derived from both the dextran and BCTD. When Et$_3$N is employed, ion fragments characteristic of Et$_3$N are found.

Synthesis

There exists a number of synthetic variables with regard to the interfacial technique.[14-16] These include molar ratio of reactants, stirring time, stirring rate, shape of reaction vessel and blade configuration, order and duration (time) of addition of reactants, reactant concentration, nature and amount of additives as base, pH, concentration of reactants, nature and amount of organic solvent, amount of aqueous phase and temperature. These synthetic variables are studied to ascertain the dependency of the reaction with regard to a particular reaction variable in hope of gaining insight as to the important mechanistic dependencies and to allow the construction of reaction conditions which allow for good product yield with low to high degrees of chain substitution. From previous studies several reaction variables appear to be both easily studied and to yield variable results allowing a brief evaluation of selected reaction dependencies. These reaction variables are length of stirring, monomer ratio, nature of base and monomer concentration.

Yield reaches a maximum about a stirring time of 30 seconds (Table 5). Thus 30 seconds was the reaction time employed for most of the other studies.

Table 5.   Results as a Function of Stirring Time.

Cp_2TiCl_2 (mmole)	Dextran (mmole)	Stirring Time(secs)	Yield[c] (%)	Yield (g)	Ti[d] (%)	Ti[e] (%)
3	2	10	0	0	–	–
3	2	15	7	0.05	25[a]	20
3	2	30	12	0.09	30[a]	19
3	2	45	4	0.03	19	–
3	2	180	0	0	–	–
2	3	30	42	0.21	30[a]	19
2	3	45	62	0.31	19[b]	17
3	3	30	0	0	–[b]	–
3	3	80	65	0.48	13[b]	–

Reaction conditions - Dextran and TEA or NaOH(b) (6.0 mmole) dissolved in 50 ml water added to stirred (18,500 rpm) solution of Cp_2TiCl_2 in 50 ml $CHCl_3$ at 25°C.

[a]Incomplete combustion of titanium to titanium dioxide

[c]Based on structure 6 assuming no crosslinking through the Cp_2Ti moiety.

[d]By thermolysis

[e]By use of $HClO_4$.

As with the analogous dextran and cellulose products modified through reaction with organostannane halides,[1-5] the current products are unstable in the presence of added base. This is responsible for the maximum observed as a function of stirring time and is demonstrated by the disintegration of previously isolated product when exposed to ca 0.1 molar NaOH aqueous solutions and rapid stirring for one minute. It also signals the importance of the employed added base nature and amount.

Table 6 contains results as a function of base amount for barium hydroxide, TEA and sodium hydroxide. For sodium hydroxide and barium hydroxide, reactions were typically run till precipitate was formed or for several minutes. If precipitate formed, stirring was stopped and the product collected. For systems employing TEA apparent product yield increased as the amount of TEA increased. The intensities of infrared bands associated with the presence of TEA also increased as the amount of TEA increased. Thus caution must be exercised in employing TEA such that excessive amounts of TEA are not incorporated into the products. At high TEA amounts (9.0 mmolar and above) the products become darker in coloration indicating that the product may contain products of form 9.

Table 6. Results as a Function of Added Base.

Triethylamine

Cp_2TiCl_2 (mmole)	Dextran (mmole)	Et_3N (mmole)	Stirring Time (secs)	Yield[b] (%)	Yield (g)	Ti[c] (%)	Ti[d] (%)
3	2	0	30	0	0	–	–
3	2	3	30	0	0	–	–
3	2	6	30	12	0.09	30[a]	19
3	2	9	30	31	0.23	27[a]	18
3	2	12	30	51	0.38	17	15
3	2	15	30	69	0.51	19	16
2	1	3	30	0	0	–	–
2	1	6	30	30	0.15	29[a]	21
2	1	9	30	52	0.25	23[a]	16
1.3	1	3	30	3	0.01	–	–
1.3	1	6	30	41	0.13	19	18
1.3	1	9	30	50	0.16	18	17
1	1	3	15	0	0	–	–
1	1	3	30	37	0.09	29[a]	21

(continued)

Table 6.   (continued)

Cp_2TiCl_2 (mmole)	Dextran (mmole)	NaOH (mmole)	Stirring Time (secs)	Yield[b] (%)	Yield (g)	Ti (%)
Barium Hydroxide						
1	1	0.5	30	0	0	-
1	1	1	30	16	0.04	22
1	1	1.5	30	0	0	-
Sodium Hydroxide						
2	3	9	30	0	0	-
2	3	6	30 to 120	0	0	-
2	3	4	30	0	0	-
3	3	6	30	0	0	-
3	3	6	80	65	0.48	13

Reaction conditions: Dextran and added base dissolved in 50 ml water and added to stirred (18,500 rpm) solutions of BCTD in 50 ml $CHCl_3$ at 25°C.

[a]Incomplete oxidation to titanium dioxide.

[b]Based on structure 6 with OH end groups.

[c]By thermolysis.

[d]By use of $HClO_4$.

Table 7 contains results as a function of mole ratio of reactants. Here yield (real and percentage) pass through a maximum as the amount of dextran is held constant and the amount of BCTD increased with the maximum occurring within BCTD/dextran ratios of 0.67 to 1.00 corresponding to reactive group ratios of 0.4 to 0.7. Percentage yield decreases while real yield increases as the amount of dextran is increased while holding the amount of BCTD constant.

In summary, the modification of dextran through condensation with BCTD occurs within a well defined set of reaction conditions. Added base is needed but excessive amounts of base act to hydrolyze formed product and, in the case of TEA, compete for BCTD. With lesser amounts of TEA, satisfactory product yields are obtained with some products containing units of form 8. Stirring time is also important with yield maximizing about 30 seconds stirring time for several systems studied.

Table 7.   Results as a Function of Molar Ratio of Reactants.

BCTD (mmole)	Dextran (mmole)	Molar Ratio $(Cp_2TiCl_2/$ Dextran)	Yield[c] (%)	Yield (g)	Ti[d] (%)	Ti[e] (%)
0.5	3.0	0.16	24	0.03	8	7
1.0	3.0	0.33	57	0.14	9	9
2.0	3.0	0.67	66	0.33	16	12
3.0	3.0	1.0	22	0.34	29[b]	21
4.0	3.0	1.3	1	0.01	19	18
6.0	3.0	2.0	0	0	-	-
3.0	0.5	6.0	68	0.24	23[b]	21
3.0	1.0	3.0	41	0.30	25[b]	15
3.0	4.0	.75	46	0.34	19	17

[a]Reaction conditions - Dextran and TEA (9.0 mmole) dissolved in 50 ml water is added to stirred (18,500 rpm no load) solutions of BCTD in 50 ml $CHCl_3$ with 30 secs stirring time at 25°C.

[b]Incomplete oxidation of titanium to titanium dioxide.

[c]Based on structure 6 with OH end groups.

[d]By thermolysis.

[e]By use of $HClO_4$.

REFERENCES

1.   C. Carraher and L. Sperling (Eds.), "Polymer Applications of Renewable-Resource Materials," Plenum Press, N.Y., 1983.
2.   C. Carraher, T. Gehrke, D. Giron, D. Cerutis, H.M. Molloy, J. Macromol. Sci.-Chem., A19, 1121(1983).
3.   C. Carraher, W. Burt, D. Giron, J. Schroeder, M.L. Taylor, H.M. Molloy, T.O. Tiernan, J. Applied Polymer Sci., 28, 1919(1983).
4.   C. Carraher, D. Giron, J. Schroeder and C. McNeely, U.S. Pat. 4, 312, 981, Jan., 1982.
5.   C. Carraher and T. Gehrke, Organic Coat. App. Polymer Sci., 46, 258(1982).
6.   C. Carraher, J. Chem. Ed., 58(11), 921(1981).
7.   C. Carraher, J. Sheats and C. Pittman (Eds.), "Advances in Organometallic and Inorganic Polymer Science," Dekker, N.Y., 1982.
8.   C. Carraher and S. Bajah, Br. Polymer J., 7, 155(1975).
9.   C. Carraher and G.F. Burish, J. Macromol. Sci.-Chem., A10(8), 1457(1976).
10. C. Carraher and S. Bajah, Polymer (Br.), 15, 9 (1974) and 14, 42 (1973).

11. C. Carraher and J. Piersma, J. Macromol. Sci.-Chem., A7(4),
     913 (1973).
12. C. Carraher and J.L. Lee, J. Macromol. Sci.-Chem., A9(2)
     191 (1975).
13. C. Carraher, J. Polymer Sci., A-1, 9, 3661(1971).
14. F. Millich and C. Carraher (Eds.), "Interfacial Synthesis
     Vols. I and II," Dekker, N.Y., 1977.
15. P.W. Morgan, "Condensation Polymers: By Interfacial and
     Solution Methods," Wiley, N.Y., 1965.
16. C. Carraher and J. Preston (Eds.), Interfacial Synthesis,
     Volume III," Dekker, N.Y., 1981.

Acknowledgements

     The mass spectra were obtained at the Midwest Center for
Mass Spectrometry, supported by NSF Regional Instrumentation
Facility Grant #CHE 78-18572. This work was supported in part
by the American Chemical Society - Petroleum Research Foundation
Grant #3682-Y3.

ELECTRON MICROSCOPIC STUDIES OF OSMIUM CARBOHYDRATE POLYMERS

C. C. Hinckley, S. Sharif[*], and
L. D. Russell

Department of Chemistry and Biochemistry
Department of Physiology
Southern Illinois University
Carbondale, IL 62901

INTRODUCTION

Osmium carbohydrate polymers[1], termed osmarins, are synthetic "osmium blacks." We have proposed that these materials be employed as antiinflammatory agents in the treatment of arthritis[2]. The suggestion is novel, and it is not immediately apparent that these materials should have this use. The substances are new, and the proposed application requires that several diverse criteria be simultaneously met. Though the criteria are separately precedented, they have not previously been addressed together with reference to a single substance.

In this paper, we use electron microscopy to consider osmarin toxicity. But first, the connection between osmium and arthritis is briefly reviewed. Then, the application of osmarins as antiinflammatory agents is discussed. Finally, results of electron microscopic studies are presented. High magnification micrographs of the polymers on an alumina support further the characterization of these substances. Micrographs of mouse liver cells are examined in which osmarins have been used as a stain for glutaraldehyde fixed tissue. These results enlarge understanding of osmarin-biomolecule associations. Micrographs derived from studies in which osmarin solutions were injected into the synovial spaces of arthritic pigs are presented. These data lead to suggestions concerning the toxicity of osmarins.

OSMIUM AND ARTHRITIS

Osmium tetroxide has been used for more than thirty years in the treatment of arthritis in humans, principally in Europe[3]. The technique is not used in the United States, and is controversial everywhere. OsO_4 is an extremely toxic material. One of its principal uses is as a fixative of biological tissue. The reaction of OsO_4 with biological material is extensive; virtually every component of such a system is attacked in the reaction. The arthritis treatment makes use of this reactivity. In the treatment, a solution of OsO_4 is injected into the synovial space of the diseased joint in order to achieve a "chemical synovectomy." Surprisingly, there are few reported toxic side effects.

When successful, the treatment is long-lasting. And osmium containing deposits are found in the treated joints long after the injection. Recently, a group of Swiss scientists examined seventy three persons who had been successfully treated[4]. They found osmium containing material in joints as long as five years after treatment. They suggested that the deposits may have a long term function.

OSMARINS AND ARTHRITIS

When employed as an antiarthritic, we have proposed that osmarin solutions be injected into the synovial space of affected joints[2]. The aim is to produce beneficial osmium containing deposits, while at the same time, avoiding the trauma of OsO_4 injections. Results of preliminary studies have been uniformly promising. Continuing research, aimed at producing materials which may be safely and effectively used, must address a number of issues. These are associated with the characterization of the polymers, their toxicity, and their mode(s) of action.

   Characterization. Osmarins are prepared through the reaction of an osmium(VI) acetate and glucose[1]. The composition of the polymers and their detailed properties are dependent upon the conditions of preparation. They have been characterized by elemental analyses, gel filtration, solution density and viscosity measurements, gel electrophoresis, and ultracentrifugation. The picture that emerges from this is that the molecules of the polymers are spherical, polyanionic, polydisperse, and may have molecular

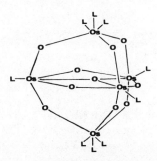

Figure 1. A proposed osmarin structure based upon OsO_2.

weights of a few thousand daltons to a few hundreds of
thousands of daltons. We have suggested that the
polymers be considered as carbohydrate solubilized OsO_2,
and the structure pictured in Figure 1. is based upon
that idea.

This diagram represents the kind of characterization we seek. Those things which we would like to
know, and do not, include the way(s) in which the osmium
ions are bound in the backbone of the polymer, the
distribution and concentration of Os(III) which is
thought to be present with Os(IV), the ways in which the
carbohydrate ligands are bound to the backbone, and
finally, the ways in which the backbone of the polymer
fits together to yield the spherical structure of the
molecules. In this paper we present micrographs which
show this spherical shape. But, the internal structure
of the polymers is unknown.

Toxicity. Osmarins have very low acute toxicities.
No lethal dose has been determined. Doses of 1 g/Kg can
be administered (IP) to mice with no mortality. This
finding is encouraging, but not final. Behind the broad
biological issues which toxic effects illuminate, are
chemical issues which must be addressed. If the
materials are not toxic, why are they not? At the
origin of this question is the delineation of osmarin
associations with biomolecules, and the determination of
the fate of osmarin molecules absorbed in biological
systems. Osmarins bind extensively with biological
tissue. They are absorbed by phagocytic cells. It is
important to know whether or not the molecules are
changed in these processes. Electron microscopy
provides useful information. If the polymers are
dissociated, and dispersed, then images of the polymers

will not be found. If, on the other hand, structures
similar to those of the polymers are observed, then the
possibility exists that the polymers are absorbed
unchanged.

　　　Mode of action. There are two possible modes of
action which may rationalize the beneficial effect of
osmarin deposits. The first is based upon the reported
effect of osmarin injections upon the synovial cells of
rabbits[2]. Osmarins are apparently more toxic to this
tissue than to others. They may, therefore, provide a
means to achieve a "chemical synovectomy" having fewer
damaging side effects.

　　　The second mode of action is based upon the oxygen
chemistry of osmarins. Osmarins exhibit an extensive
oxygen species chemistry that includes oxygen[5],
superoxide[6], peroxide[2], and, presumably, all oxy-radical
species derived from these. The involvement of
superoxide and related oxy-radicals in inflammation is
well established[7]. Osmarin deposits in a joint could
reduce oxy-radical concentrations, diminishing the
damage caused by these species.

EXPERIMENTAL

　　　Preparation of osmarins. The preparation and
characterization of osmarins have been reported[1].
Material used to obtain the micrographs in Figures 2.
and 3. was prepared by a slightly modified proceedure.
For this preparation, 2.83 g (11 mmoles) of OsO_4 was
dissolved in 60 ml of methanol to which was added 56 ml
of an 1.0N KOH methanol solution. This mixture was left
overnight in a refrigerator ($-20°C$). The green solution
contains dipotassium tetramethyldioxoosmium(VI). To
this solution, 250 ml of glacial acetic acid was added.
The resulting blue solution contains potassium
triacetatodioxoosmium(VI), TAKO. A solution of 8.04 g
(44 mmoles) of glucose was prepared by first dissolving
the sugar in 50 ml of water and adding this to 250 ml of
glacial acetic acid. The glucose and TAKO solutions were
then combined and allowed to react at room temperature
overnight. Solvent was removed from the product
solution by rotary evaporation. The resulting black oil
was redissolved in 50 ml of water and passed in 25 ml
aliquots through a Sephadex G-25 column (total volume:
600ml). Water was removed from the black product bands
by rotary evaporation, and the solid dried over
phosphorus pentoxide in a vacuum desicator. Yield: 4.7
g. Elemental analysis: Os-30.6%, C-19.4%, H-3.0%,
K-6.0%.

Alumina supported osmarin. Osmarin (20 mg) was dissolved in 2.0 ml of water, and alumina (0.5 g, 200 mesh) was added to the solution. Binding of the osmarin to the alumina occurred in a few seconds as the mixture was shaken until no color remained in the liquid. The water was decanted from the blackened solid, which was then washed several times with acetone. The final solid/acetone mixture was placed in a 50 ml beaker and 20 ml of acetone added. The beaker was placed in a cleaning bath and sonicated for three minutes. A few drops of solution containing suspended particles were pipetted and dropped one at a time onto a cabon coated copper grid. The acetone was allowed to evaporate from each drop, leaving the osmarin coated alumina particles on the grid. The grids were placed in a transmission electron microscope (Hitachi 500B), and the thin edges of the particles examined under high magnification (480,000X). Each field chosen was photographed three times at different focal settings.

Preparation of osmarin stained mouse liver sections. Normal mouse liver was cut into 1 mm blocks and placed in a buffered 3% glutaraldehyde solution for two hours^8. The glutaraldehyde fixed blocks were then placed, after washing, into a 3% osmarin solution for 1.5 hours. Stained tissue blocks were dried by sequencial soaking in graded ethanol/water solutions, and then embedded in plastic. Thin sections obtained from the blocks were placed on copper grids, treated briefly with drops of lead citrate and uranyl acetate solutions, and then dried.

Preparation of tissue sections from osmarin treated animals. Tissue samples from osmarin treated animals were prepared for sectioning by a method which closely parallels the above, with the exception that OsO_4 was used as the second fixative. After gluteraldehyde treatment, the tissue blocks were fixed in 4% OsO_4 solution. They were dried, embedded in plastic, and sectioned. Lead citrate and uranyl acetate were used as stains.

RESULTS

Figure 2. is a micrograph of an osmarin preparation supported on alumina. The dark spots on the surfaces are due to the osmarin. Micrographs of alumina not treated with osmarin are free of these features.

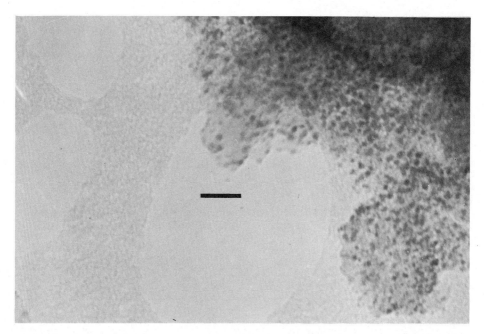

Figure 2. An electron micrograph of an osmarin on an
         alumina support. The bar indicates 100 Å.

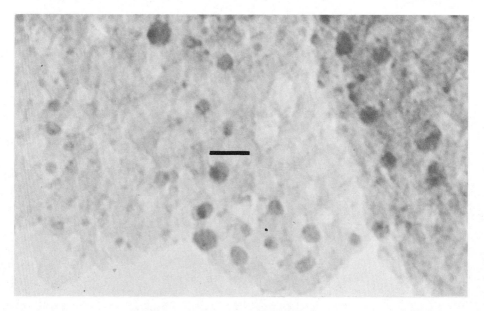

Figure 3. Aggregates are formed when the alumina
         supported osmarin in Figure 2. is treated with
         lead acetate solution. The bar indicates 100 Å

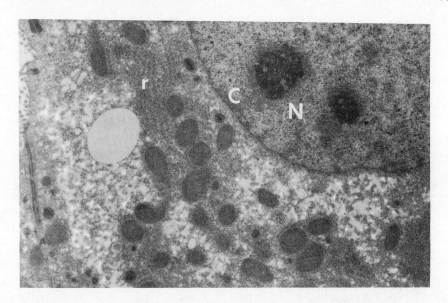

Figure 4. A section of mouse liver stained with osmarin
(N, nucleus; C, chromatin; r, ribosomes).

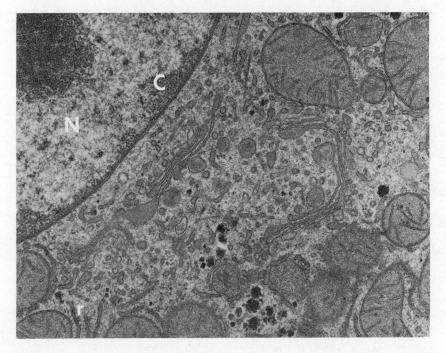

Figure 5. A conventionally prepared mouse liver section
(N, nucleus; C, chromatin; r, ribosomes).

Measurements taken from the micrograph indicate
molecular diameters up to thirty angstroms. On this
micrograph there are 77 molecules by count in five
representative 1 cm^2 areas. Of these, molecular
diameters range from below 10 to near 30 Å, with 21
below 12 Å, 31 between 12-18 Å, 18 between 18-24 Å, and
7 above 24 Å. The average diameter is 15.7 Å.

When alumina supported osmarin is treated with lead
acetate, larger aggregates are found on the surface.
Figure 3. is a micrograph taken of a sample of the
osmarin/alumina preparation pictured in Figure 2. after
treatment with a lead acetate solution. The diameters
of the aggregates range from 30 to 100 Å. In this case,
the number of aggregates is only 1 or 2 per cm^2. Of
these, about half are below 30 Å, while most of the
remainder are between 30-70 Å.

Figure 4. is an electron micrograph of a mouse
liver section stained using an osmarin solution. Figure
5. is a micrograph of a conventionally prepared liver
section for comparison. In this case, OsO_4 is used as a
second fixative, and the osmium blacks in the fixed
tissue are formed <u>in situ</u> through the reaction of OsO_4
with cell constituents. The sharp contrast which allows
the identification of the various organelles is
characteristic of the technique.

Ultrastructure in the osmarin stained section is
much less clearly defined, though still descernable.
Osmarin binding in the cell is general. Membranes are
visible, though not sharply outlined. Ribosomes and
chromatin material in the cell nucleus are stained,
indicating osmarin binding. The micrograph indicates
generalized binding of osmarins with proteins and
nucleic acids. This cell was fixed with glutaraldehyde
before being placed in the osmarin solution, and so, was
not alive. Osmarins are distributed throughout the
cell. The distribution of osmium deposits in cells which
have absorbed osmarin while alive is substantively
different. The distribution is not uniform. Localized
deposits are found.

When injected into a joint, osmarins bind to all of
the tissue surfaces facing the synovial space. Figure
6. is a photograph of a pig tarsal joint which has been
stained with osmarin. The joint recieved 1 ml of a 5%
osmarin solution. After one month, the pig was
sacrificed and tissue specimens obtained. Many cells
within stained tissue contain osmarin deposits not found
in unstained tissue.

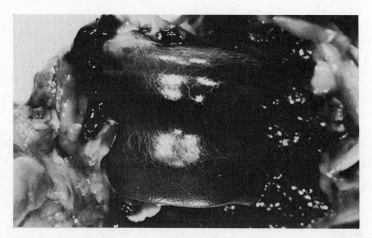

Figure 6. An osmarin stained joint surface. The white
line pattern on the joint surface is due to
growth of the joint after the staining.

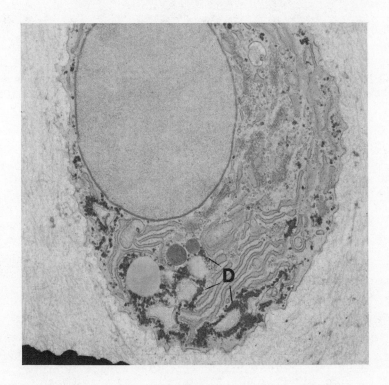

Figure 7. An electron micrograh of a chondrocyte
found in osmarin stained cartilage. The dense
deposits(D) are not found in cell in unstained
cartilage, and are presumed to be osmarin.

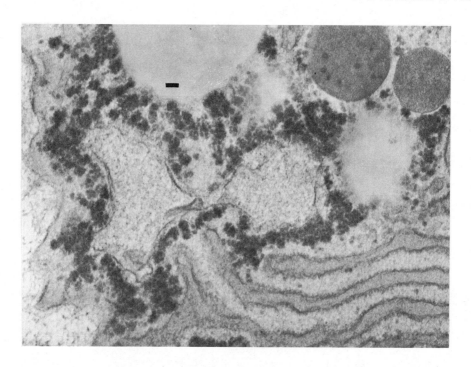

Figure 8. A higher magnification micrograph of the
deposits pictured in Figure 7.   The bar indicates 100 Å.

        Figure 7. is an electron micrograph of a
chondrocyte found in osmarin stained cartilage.   Dense
deposits are localized outside the cell nucleus, a
feature common to all those observed cases in which
osmarin was absorbed in living cells without killing
them.   Figure 8. is a micrograph of the osmarin deposits
of this cell taken at higher magnification.   The
magnification is approximately half that of Figure 2.
The deposits are composed of structures of approximately
the same size as the osmarin molecules pictured in
Figure 2.

DISCUSSION

        The micrograph in Figure 2. shows that the osmarin
preparation is polydisperse in character, and that the
molecules are spherical is shape.   This confirms
solution studies of these materials reported earlier.
The average diameter for this preparation, 15.7 A, is
substantially less than that derived from solution
studies[1], but this micrograph is of anhydrous molecules
not in solution.

The micrograph in Figure 3., likewise, confirms earlier findings. Gel electrophoresis and ultracentrifugation studies of osmarin solutions, to which lead ions had been added, demonstrated that aggregation accompanies complex formation[9]. The formation of insoluble complexes between osmium blacks and heavy metal ions is general, and is the chemical basis of the use of heavy metal stains for osmium fixed tissue sections.

Osmarins bind with proteins. This result, first obtained in solution[2], is generalized in the micrograph of Figure 4., and extended to include nucleic acid - containing structures in cells. It is not known whether or not the binding with chromatin material and ribosomes is with the nucleic acids or with the associated proteins. Osmarin was found to bind throughout the cell in this experiment, and the finding raises the question of osmarin toxicity. If binding with biological material is this widespread, how can osmarins not be toxic? The answer appears to be that in living cells, osmarin distributions are not this extensive.

The chondrocyte pictured in Figure 6. was exposed to osmarin while alive. An osmarin solution was injected into the synovial space of a living pig's tarsal joint. The cell contains dense deposits not seen in untreated tissue, and presumed to be osmarin. The deposits are localized outside the nucleus of the cell. The cell appears to be able to restrict the distribution of the foreign material. This supports earlier similar observations in liver and kidney tissue of osmarin treated mice[2]. The efforts of a cell to restrict osmarin distribution does not always succeed, osmarin is toxic to synovial tissue and killed chondrocytes are found in small numbers in osmarin stained cartilage.

Though osmarins are certainly associated with proteins in living systems, and possibly subject to some ligand exchange, they appear to be absorbed in cells as largely intact polymers. This is consistent with their confined mobility in cells, and with high magnification micrographs which show osmarin sized particles in intracellular deposits (Figure 8.).

CONCLUSIONS

Electron microscopy confirms that osmarin molecules

are generally spherical in shape, and that the preparations are polydiperse. In situations where the molecules are not bound with proteins or other biomolecular polymers, complex formation with heavy metal ions is accompanied by aggregation.

Osmarins bind generally with intercellular material, and may have limited application as stains in electron microscopic studies of biological specimens. The binding of osmarins with nucleic acids is a new observation which may lead to other uses for the materials.

The findings of this work suggest that the low toxicity of osmarins is due to the success of living systems in confining the polymers to nonessential regions. The polymers possess sufficient reactivity to destroy intercellular chemical organization. They are either prevented from entering the cell, or when absorbed, in the case of phagocytic cells, are distributed away from essential components. These defenses are not always successful, and some tissues are sensitive to osmarins. In a sense, osmarins may be said to have latent toxicity. Were it not for successful defenses, osmarins would be uniformly toxic. This latent toxicity may be exploitable. Other substances known to effect cellular organization may increase the toxicity of these polymers, providing the means thereby, to make constructive use of an otherwise distructive tendency.

ACKNOWLEDGMENTS

This research was supported by the departments listed, and the office of Research Development and Administration of Southern Illinois University. We are grateful for the assistance provided by the SIU School of Medicine and the Center for Electron Microscopy.

REFERENCES

1. C. C. Hinckley, P. S. Ostenburg and W. J. Roth, Polyhedron, 1, 335(1982).
2. C. C. Hinckley, J. N. BeMiller, L. E. Strack and L. D. Russell,"Platinum, Gold, and Other Metal Chemotherapeutic Agents: Chemistry and Biochemistry,"(S. J. Lippard, ed.)American Chemical Society Symposium Series, No. 209, 1983, Chapt. 21.
3. M. Nissila, Scand. J. Rheumatology, Suppl. 29, 1(1979).

4. I. Boussina, R. Lagier, H. Ott and G. H. Fallet, Scand. J. Rheumatology, 5, 53(1976)

5. C. C. Hinckley, M. A. Islam, J. N. BeMiller, L. D. Russell and L. E. Strack, "Oxy Radicals and Their Scavenger Systems. Vol. II: Cellular and Medical Aspects," (R. A. Greenwald and G. Cohen, eds.) Elsevier Science Publishing Co., 1983, p. 388.

6. J. Skosey, D. Chow, F. Lichon and C. Hinckley, "Oxy Radicals and Their Scavenger Systems. Vol. II: Cellular and Medical Aspects," (R. A. Greenwald and G. Cohen, eds.) Elsevier Science Publishing Co., 1983, p. 264.

7. I. Fridovitch, Science, 201, 875(1978).

8. M. A. Hayat, "Principles and Techniques of Electron Microscopy, Vol. I," Van Nostrand, London, 1970.

9. W. J. Roth, "Osmium Biomolecule Chemistry, ", Dissertation, Southern Illinois University, 1981.

BIOLOGICAL ACTIVITIES OF MONOMERIC AND POLYMERIC DERIVATIVES OF CIS-DICHLORODIAMINEPLATINUM II

Charles E. Carraher, Jr.[a], Claredine Ademu-John[a],
David J. Giron[b] and John J. Fortman[a]
Departments of Chemistry[a] and Microbiology and
Immunology[b]
Dayton, Ohio 45435

INTRODUCTION

Malignant neoplasms are the second leading cause of death
in the USA. Prior to the work of Rosenberg and co-workers,
only totally organic compounds were seriously investigated as
potential antineoplastic agents. Rosenberg and coworkers in
1964 found that bacteria failed to divide, but continued to
grow giving filamentous cells in experiments involving platinum
electrodes and ammonia/ammonium chloride buffers. The cause
of this inhibition of cell division was eventually traced to
the formation of minute (about 10 ppm) amounts of cis-dichloro-
diamineplatinum II (cis-DDP; c-DDP) trans-dichlorodiamineplatinum
II (trans-DDP) and cis-tetrachlorodiamine platinum IV (cis-TCP),
with the activity of the c-DDP being the greatest. It was found
that c-DDP is active against a wide range of tumors in man and
animals leading to the licensing of c-DDP under the name Platinol.
Platinol is also referred to as cis-platin in the literature.

1 - cis-DDP

Relatively few studies have been carried out to determine
the variety and proportion of platinum-containing compounds
in live animals (including humans) derived from c-DDP though
extensive chemical studies have been undertaken of c-DDP in non-
biological systems. This is because the major impetus involves
biological, not physical or chemical, characterization of these

compounds though of late much effort has focused on the structures of c-DDP - DNA-like complexes (for instance 3). Conversely, more is known about the biological properties of c-DDP, including the mode of transport and distribution within select tissues and organs, than for any other metal-containing group of compounds.[4]

TOXICITY

The positive attributes of c-DDP and its derivatives are found to be coupled with a number of negative side effects including gastrointestinal, immunosuppressive, hematopoietic, auditory and renal dysfunction with the latter two being the most serious.

Nephrotoxicity, which is the major dose limiting side effect of cis-DDP, is also cumulative and eventually irreversible. This is not surprising since platinum is a heavy metal and soon after application, a large percentage of the drug is filtered from the blood by the kidneys and excreted with the urine. Renal damage is usually observed by the elevation of blood urea nitrogen and creatinine or by a decreased creatinine clearance. Histological studies have shown this to be due to local acute tubular necrosis, affecting primarily the distal convoluted tubules and collecting ducts, dilation of convoluted tabules and the formation of casts.[9,10]

It would be difficult to effectively summarize the actual frequency of heavy metal toxicity caused by cis-DDP. Literature provides a wide range of clinical data. Studies such as those done by Kovach and coworkers[11] (Table 1) and others (Table 2) provide a somewhat limited view for major side effects enocuntered under given conditions. Kovach conducted his study on patients after rapid administration of an initial dose of 50 mg/m^2 body surface area.

A number of approaches have been investigated to reduce the toxicity of cis-DDP and its derivatives. The initial effect was the hydration technique of Cvitokovic et al.[12] where the patient's fluid uptake is greatly increased prior to the administration of the cis-DDP. The cis-DDP is then administered along with mannitol, a diuretic. This technique allowed the administration of about ten times the normal dosage, while lowering damage to the kidneys. It also signals a real problem related to retention of the drug. Typically well over 80% of the water soluble cis-DDP is flushed through the body and excreted within 16 hours after administration, and it is this high dosage of cis-DDP delivered to the body's circulatory system that is responsible for the majority of negative biological effects.

Table 1. Toxic Effects of cis-DDP.[11]

Toxic Effect[a]	Evaluable Patients	Patients with Toxic Effects
	No.	
Vomiting	32	31
Hearing Loss		
subclinical[b]	26	8
clinical	28	1
Renal Function Impairment[c]	28	6
Myelosuppression[d]		
Leukopenia	31	4
Thrombocytopenia	31	2
Anemia	28	2

[a]Observed after an initial dose of 50 mg/m^2

[b]Hearing loss about 400 Hz

[c]Evidenced by increase in creatinine or blood urea nitrogen and decrease in creatinine clearance

[d]Leukopenia - leukocyte count < 4000/mm
Thrombocytopenia - platelets > 100,000/mm
Anemia = hemoglobin decrease > 2g/dl

Another early approach, which met with moderate success, involved repeated application of the drug in an attempt to build up the amount of platinum compound retained. It is believed that small amounts of cis-DDP and its derivatives become associated with the body's protein and other cellular components where the retained platinum has a body half life in excess of a week.

Most of the more recent investigations involve two approaches. First, the use of cis-DDP in conjunction with other drugs as adriamycin, vinblastine, bleomycin, actinomycin D and cyclophosphamide. Second, the synthesis of new compounds exhibiting equal or enhanced activity and lowered toxicity. This latter approach has been taken by the research groups associated with Allcock[14-18] and Carraher[19-26] with the inclusion of platinum diamines in polymers. (A brief review of monomeric derivatives is given in reference 27.)

Table 2.   Toxicity in Patients Treated with <u>cis</u>-DDP Alone and in
Connection with Other Antitumoral Drugs.[13]

Toxic Effect	cis-DDP (%)	cis-DDP vincristine, bleomycin, prednisone, and actinomycin D (%)	Totals (%)
Nausea	184(100	66(100	250(100)
Vomiting	184(100)	66(100)	250(100)
Renal toxicity	8(4.3)	2(3)	10(4)
Hearing loss	40(21.7)	2(3)	42(16.8)
Increased uric acid	50(27)	2(3)	52(20.8)
Leukopenia	31(16.8)	19(28.7)	50(2)
Thrombocytopenia	18(9.7)	0	18(22.4)
Anemia	50(27)	6(9)	56(22.4)
Peripheral neuropathy	2(1)	0	2(0.8)

Toxicity Minimization

From studying the structural features of compounds which
are the most active antitumoral agents, several features have
emerged. The most active compounds are a) neutral, b) contain
two inert and two labile ligands and c) must have the ligands
<u>cis</u> to each other.

Where both the <u>cis</u> and <u>trans</u> compounds were tested, the
<u>cis</u> isomers are found to be more reactive toward tumor inhibition.

The <u>trans</u> effect predicts that a chloride ligand is more
readily replaced (more labile) for the <u>trans</u> isomers than the
<u>cis</u> isomer.[28] Thus <u>trans</u>-DDP hydrates about four times faster
than the <u>cis</u> isomer[28] and undergoes ammination ten times faster.[29]
This greater chemical reactivity might imply a lowered reaction
specificity for the <u>trans</u> isomer. Thus, even though the two
isomers are of approximately equal toxicity, their therapeutic
levels differ vastly.[29-31] This difference might be due to
the reaction of the <u>trans</u> isomer with various constituents of
the body prior to reaching the tumor site. Distribution and
excretion studies showed <u>cis</u>-DDP to be excreted much faster
initially.[32] However, within five days the levels were comparable
at about 20% retention. Even at this point the distribution

of the two compounds differed radically, platinum levels from
trans-DDP remaining high in plasma at all times, while levels
from cis-DDP fall off markedly.[32,33] This might be explained
by suggesting that trans-DDP reacts with some constituents in
the blood, remaining there for some time, while cis-DDP reacts
somewhat later, thus being readily filtered from the blood by
the kidneys.

The biological and chemical activity of the platinum drug
is also dependent on the nature of labile moieties. Thus, if
the leaving groups are too reactive, the drug may chelate prior
to reaching the tumor site. If the leaving groups are not labile
enough, the drug will not chelate with the proper cellular material.

The nature of the amine ligand also affects the biological
and chemical activity of the platinum drug.

First, aliphatic amines are found to offer better activity
than aromatic amines.[31,32] Second, nonlabile amines are preferred.
Third, solubility, hydrolysis rate and mobility are decreased
as the organic portion of the amine is increased.[32,34]

Neutrality is important for drugs which exert their effects
within the cell. Due to the non-polar nature of the cell membrane,
substances would have to be relatively neutral to pass through
it. One exception to this rule has been seen in the case of
"platinum blues".[35] These are an unusual class of antitumoral
drugs that are believed to exert their effects by a unique mechanism that does not involve their migration through the cell
wall.

The two biologically inert ligands should also be neutral.
Amine systems (as opposed to oxygen or sulfur-containing Lewis
bases) are more likely to give active compounds. These ligands
are considered inert only in the sense that they do not undergo
displacement within the cell. Under the same biological conditions, the two labile ligands will function as leaving groups.
Ligands of intermediate lability, such as chloride or bromide,
are generally active but oxalate and malonate have also been
effective.[36]

A point that is generally not appreciated regards the multi-
plicity of platinum-containing moieties once the cis-DDP, or
other platinum-containing drug is administered. While it is
known that the vast majority (>80%) of cis-DDP is readily flushed
through the human body, the exact form of the stored, tumoral-
active platinum-containing drug is not known. Thus, while there
exists a great deal of "hand-waving" regarding the elements
necessary to give a biologically acceptable platinum compound,

the "established" structural requirements must be considered
as first approximations and nonlimiting. For instance, the
idea that the platinum drug should be neutral is counter to
the experience of researchers dealing with charged complexes
of platinum-containing compounds called the "platinum blues".
It is true that the platinum blues appear to inhibit tumoral
growth through other routes than does cis-DDP but even this
point is not sure.

Fortunately, nature has provided a ready synthetic route
to the synthesis of both the cis or trans isomers at the exclusion
of the other isomer. The trans effect is in operation for many
Group VIII B square planar compounds including Pt, Pd and Ni.
Thus reaction of tetrahaloplatinum II salts with nitrogen containing compounds gives exclusively the cis-isomeric product. The
trans-isomeric product is obtained from reaction of the tetraamino-
platinum II with halide ions.

$$\begin{bmatrix} Cl & Cl \\ & Pt & \\ Cl & Cl \end{bmatrix}^{-2} + 2RNH_2 \rightarrow \begin{matrix} RH_2N \quad NH_2R \\ Pt \\ Cl \quad Cl \end{matrix}$$

2

$$\begin{bmatrix} H_3N & NH_3 \\ & Pt & \\ H_3N & NH_3 \end{bmatrix}^{+2} + 2Cl^- \rightarrow \begin{matrix} Cl \quad NH_3 \\ Pt \\ H_3N \quad Cl \end{matrix}$$

3

MODE OF ACTIVITY

Several attempts have been made to explain how cis-DDP
and its active derivatives work within the body.

One rationale is that many antineoplastic drugs act as
blocking agents against nucleic acid or protein synthesis.
Due to the general lack of biological differences between tumor
cells and normal cells, those which multiply at the greatest
rate will also typically experience the greatest kill rate.
Thus, areas characterized by rapid cell division such as bone
marrow, whether cancerous or not, are usually highly affected
by these drugs.

As previously mentioned, active species will typically
contain two cis-oriented leaving groups which in many cases
are chloride ligands. Due to the relatively high chloride ion
concentration of the circulating blood (103 mM)[37], these leaving
groups will generally remain in position causing the molecule

to remain electrically neutral until it enters the cell wall.
As would be expected, chloride ligands would do this to a greater
extent than bromide or iodide leaving groups which are not present
in sufficient concentrations in the blood to retard ligand transfer. Once within the cell, where the chloride ion concentration
is much lower (4 mM), hydrolysis readily occurs to give a number
of aquated species including the diaquo complex (6) in which
each chloride ligand is replaced by a water molecule. At 37°C
the half life for the completion of this reaction is 1.7 hours
with an activation enthalpy of approximately 20 Kcal/mole.[38]

In addition to the diaquo complex, the aquated species
can also exist in the form of the dihydroxy complex (9) or the
aquo hydroxy complex (8). The actual form is dependent upon
the pH of the surrounding liquid. Furthermore, although these
aquated species predominately exist in monomeric form, they
will also readily oligomerize to form hydroxo bridged centrosym-
metric dimers (10) and hydroxo bridged trimer (11).[35,38] The
proportion of each form present in solution is also pH dependent.

At low pH's the major monomeric form will be the diaquo complex
at high pH's, the dihydroxy complex. Oligomerization most readily
occurs at the neutral pH of 7.

The activity of the monomer and each oligomer has been studied by injecting each directly into animals (bearing in mind that the high chloride concentration of extracellular fluids may affect each form differently). Each injected form was more toxic than the original (chlorinated) cis-DDP, with the monomer being the least toxic injected form. High doses of dimer and trimer were instantly lethal and neither dimer nor trimer displayed anticancer activity at any dose level whereas the monomer was active at dose levels below the toxic dose.[38]

These results may explain the cytotoxicity of normal cells containing accumulated concentrations of cis-DDP, such as the kidney where the dimerization rate increases as the square of monomer concentration.[38] These reactions may also explain the reduced toxicity (but maintained efficacies) observed when cis-DDP is given as a divided dose or as a slow infusion as opposed to it's being given rapidly or as a single dose.

Injection of cis-DDP is followed by rapid excretion. In man it has a half life of 1.5 hours with only 5% remaining in the body after 24 hours.[38] It is believed that a great proportion of that remaining is of the more toxic aquated and oligomeric forms, hence its toxicity; but this also indicates that very little of the cis-DDP used actually accounts for its favorable inhibitory activity against various tumors.

In actuality the possible aquated derivatives of cis-DDP is great with the proportion dependent on such variables as time, temperature, pH and concentration of associated reactants (Cl$^-$ and NH$_3$). Structures illustrating some of these forms are given in Figure 1. Charged forms (d, e, f, h, i, j, k and l) will not penetrate cell walls preventing them from exhibiting biological inhibition analogous to cis-DDP. Even so, such forms may stimulate cell walls in a manner analogous to the "platinum blues". Thus more complete studies are needed to more clearly define the structure(s) and proportion of the cancer-inhibiting agent(s).

It has been shown in vivo and in vitro that DNA is the primary biological target for cis-DDP and its active derivatives. Many groups have studied the resulting DNA-Pt complex obtained from both cis and trans derivatives in order to better understand their actual mode of binding and the differences that select only the cis derivatives as antitumoral agents.[3,38]

Marquet and Butour[39] studied possible binding modes of cis-complexes to DNA by saturating salmon sperm DNA with cis-DDP. All interactions were found to result in chelate type complexations. Their reaction kinetics were determined by chlorine displacement during hydrolysis.

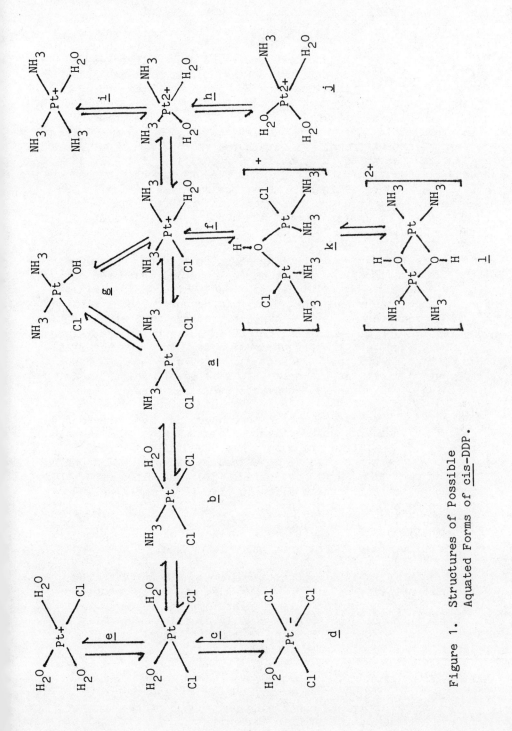

Figure 1. Structures of Possible
Aquated Forms of cis-DDP.

The first interaction characterized by proton liberation
was attributed to binding at the N-7 position and O-6 positions
of guanine, a mode of binding which resulted in considerable
destabilization of the double helix. The second interaction
which involved a pH drop from 6.0 to 4.5, also involved only
4% of the total platinum used and was attributed to binding
at the N-1 position of guanine and N-3 of cytosine forming interstrand crosslinks which held the helix firmly. The third and
final interaction involved 50% of the total platinum used with
no proton liberation and was attributed to binding at N-7 of
adenine and the amino group at the C-6 position of adenine.
Thus DNA saturation with cis-DDP involved 4% interstrand and
96% intrastrand crosslinking. Previous studies[40,41] had shown
that thymine molecules do not react with cis-DDP. Similar results
were obtained by Robins[42] using cis-dichloroethylenediamine
platinum II with DNA.

Due to the rapid rate of excretion of cis-DDP, DNA saturation
is unlikely to occur in vivo, thus the mode and proportion of
binding may vary slightly under more realistic conditions.
For instance, there have been reports of monofunctional binding
of cis derivatives to DNA[36,37] which correlate with studies
done involving proteinase that result in DNA-protein crosslinked
complexes. However, the frequency of these are low and this
mode of binding has been considered of little or no importance
to the cytotoxicity of the drugs.[37] This monofunctional mode
of binding is also the type characteristic of trans isomers.

Whether in vitro or, more importantly, in vivo, the binding
of platinum compounds to DNA involves the breakage of a certain
number of bonds. It has been seen that the initial binding
to guanine bases causes local denaturation exposing additional
sites for chelation and crosslinking. Electron micrograph studies
have shown that this denaturation involves the unwinding and
considerable shortening of the DNA helix.[43]

DNA repair does occur but can be inhibited by nonlethal
concentrations of caffeine or trimethylxanthine.[44]

In summary, the bulk of evidence is consistent with the
anticancer activity of cis-DDP being due (at least in part)
to its binding to DNA molecules for the systems investigated.
Much remains to be done to more clearly describe such activity
as a function of type of cancer, form of platinum-containing
species, location of cancer, etc.

RATIONALE

The use of polymers containing platinum compared with platinum
delivery through use of smaller molecules is clearly debatable

but there is sufficient evidence to justify preliminary studies
of such materials. The use of biological and synthetic polymers
as tumor-suppressants is well-known and has, in some cases,
proved to be advantageous (for instance 45-48).

Specifically, advantages for synthesizing polymeric derivatives of cis-DDP include a) restricted biological movement,
b) controlled release, c) increased probability of critical
attachment and d) delivery of increased amounts of drug.

Heavy metal toxicity, related to the presence of large
quantities of cis-DDP derivatives in the circulatory system
(such as renal failure), is well established. Chain lengths
of about 100 and greater are typically prevented from easy movement
through biological membranes.[4] Thus the location of the platinum
drug can be somewhat restricted, for instance avoiding the kidney,
decreasing damage to the kidney and associated organs.

Studies by our group have established that most metal-containing polymers undergo hydrolysis when wetted. DMSO solution
of polymeric derivatives of cis-DDP also undergo hydrolysis
when added to water. Thus, polymeric derivatives of cis-DDP
can act as controlled release agents, releasing therapeutic
quantities of the active drug.

If attachment - interstrand, intrastrand or otherwise -
to DNA is essential for antitumor activity and multiple attachments
are required or advantageous (i.e. more than one attached platinum
compound per DNA) then the fact that the platinum is itself
an integral portion of a polymer is advantageous since the probability that successive attachments will be made on the same
strand and adjacent strands is high after the first attachment.

POLYMER SYNTHESIS

Cis-DDP is synthesized by the action of ammonia on the
tetrachloroplatinate. This reaction, due to the trans effect,
yields exclusively the cis isomer[2], (the tetraamine platinate
reacted with chloride ions yields exclusively the trans isomer,

$$PtCl_4^{2-} + NH_3 \longrightarrow \begin{array}{c} NH_3 \\ \diagdown \\ \diagup \\ NH_3 \end{array} Pt \begin{array}{c} Cl \\ \diagup \\ \diagdown \\ Cl \end{array}$$

12

$$PtCl_4^{2-} + RNH_2 \longrightarrow \begin{array}{c} RNH_2 \quad Cl \\ Pt \\ RNH_2 \quad Cl \end{array}$$

<u>13</u>

<u>13</u>). Similarly, the tetrachloroplatinate will react with any amine to yield the corresponding <u>cis</u>-dichloroderivative.

Allcock and co-workers[14-18] studied reactions of the tetra-chloroplatinate with synthetic polymers in the formation of polymeric derivatives of <u>cis</u>-DDP. He used the approach of incorporating <u>cis</u>-DDP onto a preformed polymer, poly[bis(methylamine)-phosphazine] (<u>14</u>)., which is a water soluble polymer with coordination sites on both the side groups and the nitrogens. The compound reacts with K_2PtCl_4 in the presence of 18-crown-6-ether to give a coordination complex <u>15</u> which has shown antitumoral activity against P388 lymphocytic leukemia cells in mice and the Erlich Ascites tumor cells. It would be of interest to see if reaction with the amine nitrogens also occurred.

$$\begin{array}{c} RHN \quad NHR \quad RHN \quad NHR \\ N = P \quad N = P \end{array} + K_2PtCl_4 \longrightarrow \begin{array}{c} RHN \quad NHR \quad RHN \quad NHR \\ N = P \quad N = P \\ Pt \\ Cl \quad Cl \end{array}$$

<u>14</u>   <u>15</u>

A second approach has been to synthesize a polymer with the <u>cis</u>-dichloroplatinum moiety included in the polymer's repeat unit. This approach was used by Carraher and co-workers[19-21] in the synthesis of poly[<u>cis</u>-dichlorohexamethylenediamineplatinum (II)] (<u>16</u>) which involved the action of K_2PtCl_4 on 1,6-hexamethyl-enediamine when stirred in aqueous solution for 48 hours. The <u>cis</u>-dibromo and <u>cis</u>-diiodo derivatives have also been synthesized using the appropriate tetrahaloplatinate.

$$K_2PtCl_4 + NH_2(CH_2)_6NH_2 \xrightarrow{H_2O} \begin{array}{c} Cl \quad Cl \\ Pt \\ NH_2 \quad NH_2(CH_2)_6 \end{array}$$

<u>16</u>

More recently these syntheses were extended to include amide-containing Lewis bases as urea, thiourea, pyrimidines, purines and hydrazines (<u>17-19</u>).[22-26]

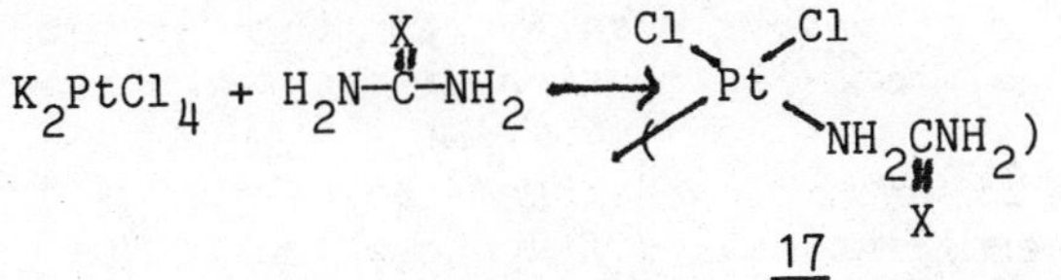

$$K_2PtCl_4 + H_2N-\overset{\overset{X}{\parallel}}{C}-NH_2 \longrightarrow$$

17

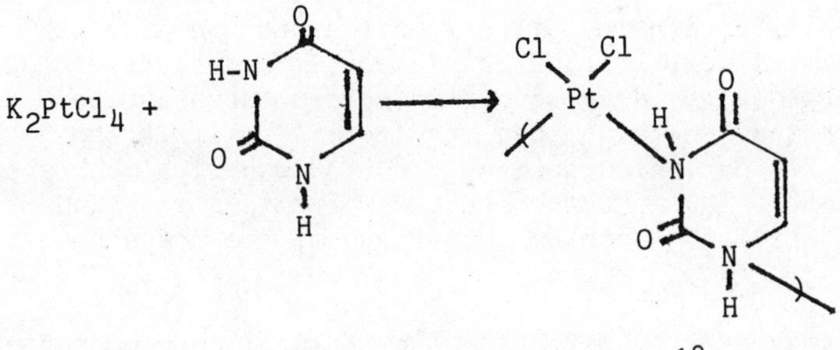

$$K_2PtCl_4 + \longrightarrow$$

18

$$K_2PtCl_4 + NH_2NHR \longrightarrow$$

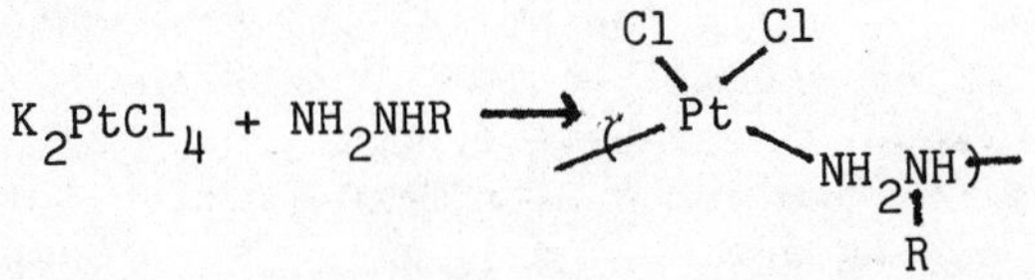

19

BIOLOGICAL CHARACTERIZATION

Several phosphazene carrier polymers have undergone preliminary (culture and cell line) testing. These polymers inhibited
mouse P388 lymphocytic leukemia cells and showed tumor inhibition
in the Ehrlich Ascites tumor regression test.[14-18]

The following comments emphasize results derived from studying
linear polymers of forms 16-19 derived from the condensation
of the tetrahaloplatinates with nitrogen-containing reactants.
The "windows of activity" established for the monomeric derivatives
of c-DDP appear to hold for the polymeric derivatives. Activity
is typically aliphatic > aromatic and Cl > I against L929,
WISH and HeLa cell lines.

Almost without exception, the polymeric derivatives show
good inhibition to all tested tumor cell lines (L929, HeLa,
Detroit, WISH) exhibiting an activity cut-off generally within
the 30 to 60 ug/ml region. The polymers show a wide variety
of behavior against RNA viruses as Polio Type I virus and the
MM strain of encephalomyocarditis where some polymers inhibit
viral replication while others show no effect toward viral replication. The majority (ca 60 to 75%) of the polymers inhibit
viral replication at concentration levels below that where tumoral
cell lines are inhibited. It is believed that some cancers

are virally induced, thus it may be possible for the virus(es) to be rendered inactive without destruction of the tumor itself.

Giron, Espy and Carraher employed transformed mouse (T3T) fibroblast cells in determining cell differentiation values for a number of the platinum II polyamines.[49] The cells were transformed, infected using the oncogenic virus SV40. The ratio of % cell death-transformed cells/% cell death-normal cells is called the cell differential ratio and is a measure of the tendency of the agent, drug to differentiate between cancer-like (transformed) cells and healthy (normal) cells. The cell differential ratios varied from about 15 to 0.5 with most showing values around 2. Thus a number of the polyamines show good cell differentiation, favoring inhibition of the transformed T3T cells.

The polymeric derivatives show low toxicities compared with c-DDP itself. An often employed upper dosage of c-DDP (single dose) is about 4×10^{-4} g/kg for humans which is increased to 4×10^{-3} g/kg when flushing is employed. For rats the LD50 (ipr-rat) is 1.2×10^{-3} g/kg.[50] Mice have been injected on an alternate day schedule for one month with DMSO-H_2O (10-90% by volume) solutions containing 2×10^{-2} g/kg per dose without apparent harm with the polymer derived from 1,6-hexanediamine.

In tests related to extending the lifetimes of mice injected with a lethal dose of a cancer, many polymers extended the lifetimes of the mice.

There appears to be a synergistic effect when the nitrogen-containing portion is also an antitumor agent. This activity may be due to the presence of both reactants through controlled release of each moiety or an enhanced activity of a larger fragment of the whole chain.

SUMMARY

Synthesis and physical characterization of polymeric derivatives of c-DDP has been well established. Biological characterization has begun with promising preliminary results obtained with regard to lowered toxicity to live animals, favorable cell differentiation and good inhibition toward RNA viruses and a wide number of tumor cell lines.

REFERENCES

1. B. Rosenberg, L. Van Camp and T. Krigas, Nature (London), 205, 698 (1965).
2. F.R. Hartley, "The Chemistry of Platinum and Palladium," John Wiley, N.Y., 1973.

3. S.J. Lippard (Ed.), "Platinum, Gold, and Other Metal Chemo-
    therapeutic Agents," A.C.S., Washington, D.C., 1983.
4. C. Carraher, "Biologically Active Macromolecules," (C.
    Gebelein and C. Carraher, Eds.), Plenum Press, N.Y., 1984.
5. B. Rosenberg, Cancer Chemother. Rep., Pt. 1, 59, 589 (1975).
6. S. Stadnicki, R. Fleischman, U. Schaeppi and P. Merriman,
    Cancer Chemother. Rep., Pt. 1, 59, 467 (1975).
7. J. Ward, D. Young, K. Fauvie, M. Wolpert, R. Davis and
    A. Guarino, Cancer Treatment. Rep., 60, 1675 (1976).
8. J. Ward and K. Fauvie, Tox. and Appl. Pharm., 38, 535 (1976).
9. M. Rozenwieg, D. VonHoff and M. Slavik, Ann. Int. Med.,
    86, 803 (1977).
10. I.W. Krakoff, Cancer Treat. Rep., 63, 1523 (1979).
11. C.L. Litterset, I. Torres, and A.M. Guarino, J. Clin. Hem.
    Onc., 7, 169 (1977).
12. J.S. Kovack, C.A. Moertel and A.J. Schutt, Cancer Chemother.
    Rep., 57, 357 (1973).
13. D. Hayes, E. Cvitokovic, R. Golby, E. Scheiner and I. Krakoff,
    Proceedings of American Assoc. Cancer Res., 17, 169 (1976).
14. C.E. Merrin, Cancer Treat. Rep., 63, 1579 (1979).
15. H. Allcock, R. Allen and J. O'Brien, Chem. Comm. 717 (1976).
16. H. Allcock, Science, 193, 1214 (1976).
17. H. Allcock, Polymer Preprints, 18, 857 (1977).
18. H. Allcock, Organometallic Polymers (C. Carraher, J. Sheats
    and C. Pittman, Editors), Academic Press, N.Y. (1978),
    pgs. 283-288.
19. H. Allcock, R. Allen and J. O'Brien, J. Amer. Chem. Soc.,
    99, 3984 (1977).
20. C. Carraher, C. Admu-John and J. Fortman, unpublished results.
21. C. Carraher, D.J. Giron, I. Lopez, D.R. Cerutis and W.J.
    Scott, Organic Coatings and Plastics Chemistry, 44, 120
    (1981).
22. C. Carraher, W.J. Scott, J.A. Schroeder and D.J. Giron,
    J. Macromol. Sci.-Chem. A15(4), 625 (1981).
23. C. Carraher, Organic Coatings and Plastics Chemistry, 42,
    428 (1980).
24. C. Carraher, T. Manek, D. Giron, D.R. Cerutis and M. Trombley,
    Polymer Preprints, 23(2), 77 (1982).
25. C. Carraher and A. Gasper, Polymer Preprints, 23(2), 75
    (1982).
26. C. Carraher, Biomedical and Dental Applications of Polymers
    (G. Gebelein and F. Koblitz Editors), Plenum Press, N.Y.,
    1981, Chpt. 16.
27. C. Carraher, M. Trombley, I. Lopez, D.J. Giron, T. Manek
    and D. Blair, unpublished results.
28. R. Speer, H. Ridgway, L. Hall, D. Stewart, K. Howe, D.
    Lieberman, D.A. Newman and J. Hill, Cancer Chemother. Rep.,
    Pt. 1, 59, 629 (1975).

29. M. Tucker, C. Colvin and D. Martin, Inorganic Chem., 3, 1373 (1964).
30. C. Colvin, R. Gunther, L. Hunter, J. McLean, M. Tucker and D. Martin, Inorganic Chimica Acta, 3, 487 (1968).
31. T. Conners, M. Jones, W. Ross, P. Braddock, A. Khokharard, M. Tobe, Chemico-Biological Interactions, 5, 415 (1972).
32. M. Cleare and J. Hoeschele, Platinum Metals Rev., 17, 2 (1973.
33. M. Cleare and J. Hoeschele, Bioinorg. Chem. 2, 187 (1973).
34. C. Litterset, T. Gram, R. Dedrick, A. Leroy and A. Guarino, Cancer Res., 36, 2340 (1966).
35. M. Tobe and A. Khokhar, J. Clinical Hematology and Oncology, 7, 114 (1977).
36. L.J. Lippard, Science, 218, 1075 (1982).
37. T.A. Connors, J.J. Cleare and K.R. Harrap, Cancer Treat. Rep., 63, 1499 (1979).
38. L.A. Zwelling and K.W. Kohn, Cancer Treat. Rep., 63, 1439 (1979).
39. B. Rosenberg, Cancer Treat. Rep., 63, 1433 (1977).
40. J.P. Macquet and J.L. Butour, J. Clin. Hem. Onc., 7, 469 (1977).
41. L.L. Munchausen and R.O. Rahn, Biochem. Biophys. Acta, 414, 242 (1975).
42. J.P. Macquet and T. Theophanides, Biochem. Biophys. Acta, 402, 160 (1975).
43. A.B. Robins, Chem. Biol. Interactions, 7, 223 (1973).
44. C. Carraher, W.J. Scott and D.J. Giron, "Biologically Active Macromolecules" (C. Gebelein and C. Carraher, Eds.), Plenum Press, N.Y., 1984.
45. H.W. Vandenberg, H.N. Fraval and J.J. Roberts, J. Clin. Hem. Onc., 7, 349 (1977).
46. J. Davidson, P. Faber, R. Fischer, S. Mansy, J. Persie, B. Rosenberg and L. Van Camp, Cancer Chemother. Rep., Pt. 1, 59, 287 (1975).
47. B. Rosenberg, Naturwissen schaften, 60, 399 (1973).
48. M. Cleare, J. Clinical Hematology and Oncology, 7, 1 (1977).
49. T. Conners and W. Ross, Advs. in Antimicrobial and Antineoplastic Chemotherapy, 3, 771 (1972).
50. D. Giron, M. Espy and C. Carraher, unpublished results.
51. H. Christensen and T. Luginbyhl (Eds.), "The Toxic Substance List - 1974 Edition," U.S. Department of HEW, Rockville, Maryland, 1974.

STRUCTURAL CHARACTERIZATION OF <u>CIS</u>-PLATINUM II DERIVATIVES OF

POLYETHYLENEIMINE AS A MODEL CARRIER FOR NATURAL DRUG CARRIERS

Charles E. Carraher, Jr.[a], Claredine Ademu-John[a] John
J. Fortman[a] and David J. Giron[b]
Departments of Chemistry[a] and Microbiology and
Immunology[b]
Wright State University
Dayton, Ohio 45435

INTRODUCTION

While the synthesis of biologically active compounds is
widespread, the major area lacking definitive results concerns
the site specific delivery determined by the use of controlled
drug release, natural polymers, suitably tailored synthetic polymers and polymer length and shape. The previous paper describes
the use of monomeric and linear polymeric derivatives of <u>cis</u>-
dichlorodiamine platinum II in medical applications emphasizing
cancer chemotherapy.

It is our intention to employ site specific natural molecules
such as proteins, nucleic acids and liposomes as carriers of
the <u>cis</u>-dichloroplatinum II moiety. Our initial step in this
investigation is the use of synthetic, readily available, inexpensive macromolecules which possess known structures. The evaluation
of factors such as tendency to crosslink, tendency to form <u>cis</u>
products, effect of monomer ratios on load (amount of substitution
on the preformed polymer) would assist in the design of reaction
conditions with natural, less fully characterized, expensive
potential drug carriers. This chapter describes an initial attempt
in this area.

RATIONALE

The preference of including platinum-containing moieties
within polymers compared to delivery via smaller molecules is
clearly debatable. Smaller molecules are usually more soluble
and can move more easily within the blood stream and also traverse
the cell membrane with greater ease. However, this freedom in

213

movement allows for a greater frequency of heavy metal toxicity.
Polymeric derivatives, on the other hand, have shown restricted
biological movement and controlled release mechanisms which have
been effective in the reduction of these toxicities. Polymeric
derivatives may also show an increased probability of attachment
to critical sites over that observed in the case of smaller molecules.

Metal-containing polymers are not foreign to the human body
since it synthesizes its own hemoglobin. Nitrogen-containing
polymers occur naturally in all living organisms in the form
of proteins and nucleic acids. As previously noted, the "targeting"
of drugs to specific sites within the body is presently being
studied and it has been suggested that naturally occurring macromolecules may act as "magic bullets" in delivering drugs. However,
the use of naturally occurring macromolecules in research can
be costly and availability may be limited. Thus, more readily
available synthetic macromolecules are of value in acting as
model systems in the study of site specific drug delivery agents.

Polyethyleneimine (PEI) is one such synthetic polymer which
is both readily available and relatively inexpensive. Previous
work done by Carraher and co-workers[1-3] has already established
the activity of PEI as a Lewis base in reactions involving such
organometallic reactants as mono and dihalo stannous compounds
(i.e., R_3SnX and R_2SnX_2). The resulting structure from the dihalo-
stannanes is complex and includes unreacted polyethyleneimine,
crosslinked moieties and the conformationally favored five-membered
ring moieties.   In addition, PEI has been shown to exhibit antitumoral activity.[4] Thus its association with the dichloroplatinum
moiety forming units analogous to cis-DDP, may provide a synergistic
effect due to the presence of two potential antitumoral agents.

$$R_2SnCl_2 \; + \; \{CH_2-CH_2-NH\} \; \rightarrow \;$$

$$\underline{1}$$

$$\begin{array}{c} \{CH_2-CH_2-N\} \\ | \\ \underline{2} \qquad R-Sn-R \\ | \\ \{CH_2-CH_2-N\} \\[4pt] \{CH_2-CH_2-N-CH_2-CH_2-N\} \\ Sn \\ \underline{3} \qquad Cl \quad Cl \end{array}$$

An additional purpose of this study is to create biologically
active polymeric material that may in some way enhance the antitumoral nature of the platinum-containing agent.

EXPERIMENTAL

The following chemicals were used without further purification: potassium tetrachloroplatinate (J and J Materials, Neptune City, NJ; 99.9% purity) and polyethyleneimine (molecular weight - 40,000 to 60,000; Polysciences, Inc., Warrington, PA).

The synthesis of each product occurred in a glass Kimax flask. An aqueous solution of each amine was mixed with an aqueous solution of K_2PtCl_4 (at the appropriate concentrations). The reaction mixtures were magnetically stirred for a specified time, typically several hours.

The resulting polyethyleneimine-platinum (PEI-Pt) complex precipitated from the reaction mixture and was removed by vacuum filtration, washed repeatedly with deionized, doubly distilled (D/D) water (used throughout), transferred onto a glass petri dish and air dried at room temperature. The product was typically a light to medium brown colored powder.

Attempts were made to dissolve the products in a wide variety of liquids including dipolar aprotic solvents such as dimethyl sulfoxide (DMSO), dimethyl formamide (DMF), hexamethyl phosphoramide (HMPA) triethyl phosphate (TEP) and acetone. Solubility tests were conducted by placing approximately 1 mg of product in 3 ml of liquid, shaking vigorously and observing the mixture over a period of one week.

Elemental analyses involved the determinations of the platinum and chloride content of the products. The percentage platinum was obtained by thermal decomposition of the products. Each produce (0.05 to 0.10 g) was heated in a crucible over a bunsen flame for at least 8 hours or until a silver gray residue of platinum was obtained and further heating resulted in no further weight loss of the substance. The percentage chloride was determined by sodium fusion followed by the precipitation of silver chloride. Weighed samples of each product were ionized by sodium fusion. The resulting mixture was diluted with deionized, doubly distilled (D/D) H_2O, acidified with nitric acid and centrifuged. A 1M $AgNO_3$ solution was added to the decantant to give a precipitate of silver chloride. The precipitates were collected and purified by dissolving in 6M NH_4OH and reprecipitating with 6M HNO_3. Due to the light sensitive nature of silver chloride, all precipitates were air dried in a darkened hood. The percentage chloride was calculated from the weight of silver chloride obtained.

Infrared spectra were obtained using potassium bromide pellets employing a Perkin-Elmer 1330 Infrared Spectrophotometer and a Digilab FT-1R model FTS-20 C/D.

Ultraviolet spectra were obtained of DMSO soluble portions of products using a Cary 14 Spectrophotometer.

Mass spectral analyses were performed using a direct insertion probe connected to a Kratos MS-50 mass spectrometer, operating in the E1 mode, 8KV acceleration and 10 sec/decade scan rate with a probe temperature of 350° to 550°C.

Molecular weight determinations were made using a Brice-Phoenix 3000 Universal Light Scattering Photometer with a mercury light source at 546 mm. Due to the low solubility of products in DMSO, determinations were carried out using concentrations lower than the usual 1% solutions. Refractive indices were measured on a Fisher Scientific Refractometer (Model 5018).

RESULTS AND DISCUSSION

The products of reaction between polyethyleneimine and potassium tetrachloroplatinate may contain a variety of repeat units including unreacted ethyleneimine, internal cis-bidentate chelation, crosslinked cis-bidentate chelation, crosslinked trans bidentate chelation and monodentate coordination. The following evidence is consistent with the majority of the product containing internal cis bidentate chelation.

Infrared and Ultraviolet Spectral Analyses

Infrared spectra of the PEI-Pt produce (Figure 1) contain bands characteristic of the presence of both reactants and new bands characteristic of the formation of the platinum-nitrogen (Pt-N) moiety. The three bands at 3500, 3195 and 3105 cm^{-1} are due to primary and secondary amine stretching and the two bands at 1615 and 1580 cm^{-1} are due to primary and secondary amine deformation.[5] Additional bands due to PEI occur at 1460, 1360, 1050, 1020 and 755 cm^{-1}. The band at 318 cm^{-1} is characteristic of Pt-Cl stretching and two small Pt-N bands are evident at 520 and 450 cm^{-1}. The number of Pt-N stretching vibration bands is often used to identify the geometrical configuration of the compound. Trans platinum diamines exhibit only one Pt-N band whereas cis-platinum derivatives exhibit two bands. However, the second band is weak and sometimes missed, thus while the presence of two bands indicates the cis geometry, the presence of one band is not firm evidence for the trans geometry. The two bands exhibited by the PEI-Pt product indicate the presence of the cis geometry.[6,7]

Further evidence for the cis configuration is found in the ultraviolet spectra of the products. Trans derivatives of DDP exhibit only one UV band in the 200 to 300 nm region, whereas cis derivatives exhibit two bands.[8,9] Figure 2 is a representation of the spectrum over the region of 200 to 300 nm showing the presence of the two bands. The product exhibits two bands within this UV region (Table 1) consistent with a cis structure for the product as depicted by 5 and 6.

Mass Spectral Analyses

Mass spectra were obtained on a PEI-Pt product (Sample 6, Table 3). Pyroprobe-MS analyses were performed at a slow heating rate from ambient to 400°C. Few ion fragments were observed. The sample was reexamined using ballistic heating to 400°C. Ion fragments suitable for identification were obtained. Table 2 contains a summary of

Table 1. UV Spectral Bands of Selected Platinum Compounds

	wavelength, nm	
Cis-DDP	240	273
trans-DDP	-	268
cis-HMDA-PtCl$_4$[a]	250	310
PEI-Pt	256	263

[a]Product from K_2PtCl_4 and 1,6-hexanediamine.

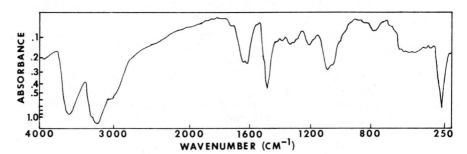

Figure 1.   Infrared spectrum of the product derived from polyethyleneimine and potassium tetrachloroplatinate.

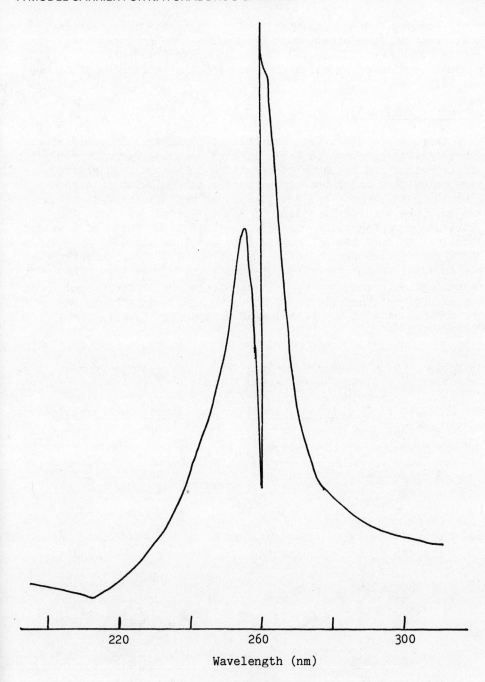

Wavelength (nm)

Figure 2.   Ultraviolet spectrum of product derived from poly-
            ethyleneimine and potassium tetrachloroplatinate
            (note scale change at 260 nm).

ion fragments and assignments for ion fragments with relative
(with respect to most abundant ion; m/e - 36) ion itensities
greater than one. These ion fragments are consistent with a
product containing moieties derived from the tetrachloroplatinate
and PEI.

Elemental Analyses

Samples of PEI-Pt were synthesized using variations in reagent
concentration as shown in Table 2, thus allowing for differences
in the extent and sequence of chelation. The obtained yields
were calculated using the structure of unit 5 (which is also
equivalent to forms 6 and 7). Two major variations of platinum
are possible with Cl/Pt ratios of 2:1 (5-7) and 3:1 (8). The
platinum percentages are consistent with a high degree of platinum
substitution onto the PEI chains but cannot be utilized to differentiate between the forms due to the closeness of the percentage
platinum values for the platinum-substituted forms 5-8. The
percentage chlorine, however, is significantly different and
the results obtained (Table 2) for both % Cl and Cl/Pt ratios
are consistent with the presence of a majority of units containing
Cl/Pt ratios of 2:1.

Table 2. Ion Fragments Derived from the Chelation Product of
          Polyethyleneimine and K_2PtCl_4.

m/e	Intensity (%)	Associated Ion Fragment
35	13	Cl
36	100	HCl
37	4	Cl
38	34	HCl
39	2	C_2HN
40	3	C_2H_2N
41	7	C_2H_3N
42	3	C_2H_4N
43	2	CH_2H_5N
44	6	C_2H_6N
67	1	C_4H_5N
80	1	$C_4H_4N_2$
85	4	$C_4H_9N_2$
87	1	$C_3H_{11}N_2$

Solubility

The actual proportion of units 5 and 6 present is difficult to determine. Chains containing units of form 5 would be more soluble than those containing units of form 6 since forms such as 6 constitute crosslinking. The number of crosslinks needed for the onset of insolubility is unknown for these compounds; but for many other polymers it is as little as 2 to 5%. The solubility of moderate amounts is consistent with the presence of large amounts of unit 5 which is reasonable since this involves the formation of five-membered ring structures and such structures are conformationally favorable for square planar platinum compounds.

Molecular Weight

Molecular weight determinations were carried out on solubilized portions of the PEI-Pt products.

The molecular weight for PEI was given as approximately 50,000. Total occupancy of a single PEI complex with a molecular weight of approximately 2×10^5. Weight average molecular weight values ranging from approximately 1×10^5 to 9×10^5 were obtained (Table 3). These results are consistent with a high degree of substitution of the $PtCl_2$ moiety onto the PEI backbone. They are also consistent with small numbers of PEI chains being linked together. This is consistent with the major chain unit being of form 5.

Structural Summary

The presence of moieties derived from both reactants is indicated by infrared spectroscopy and mass spectral analyses. The bidentate chelation of platinum onto the PEI backbone is indicated by the elemental analyses results where the Cl to Pt ratios are typically between 1.8 and 2.0. The cis structure is indicated by: (a) the trans effect; (b) IR spectra (presence of two Pt-N bands characteristic of a cis structure); and (c) the UV spectrum (presence of two bands characteristic of a cis structure). The lack of substantial crosslinking is indicated by the solubility of major portions of the product and molecular weight results which are consistent with highly substituted single PEI chains.

In conclusion, the physical characterization of the PEI-Pt product is consistent with its containing a majority of structural units of form 5.

Table 3. Results of Chelation of Polyethyleneimine with K_2PtCl_4.

Sample Desig-nation	K_2PtCl_4 (mmole)	PEI (mmole)	Yield (g)	Yield[b] (%)	Pt (%)	Cl (%)	Ratio Cl/Pt	Soluble[c] (%)-28 hours	Soluble (%)-10 hrs	Molecular Wt.[d] ($\overline{M}_w$)
8	1.00	4.00	0.09	25	41.6	15.2	2.0		18	
7	1.00	3.00	0.39	110	46.6				29	
6	1.00	2.00	0.37	105	48.3	16.0	1.8		17	1.4×10^5
1	1.00	2.00	0.39	111	46.9	16.8	2.0	35		2.0×10^5
2	2.00	2.00	0.39	112	48.2	15.3	1.8	22	16	4.0×10^5
3	3.00	2.00	0.41	117	55.7	18.9	1.8	38		0.9×10^5
4	4.00	2.00	0.42	119	55.3			53	21	9.3×10^5

[a] Reaction conditions: PEI (in 10 ml water) and K_2PtCl_4 (in 10 ml water) are mixed together and stirred for 48 hours.

[b] Based on repeat unit 5.

[c] Solubility at room temperature; 0.02 g of product in 20 ml DMSO.

[d] Soluble portion determined employing light scattering photometry.

REFERENCES

1. C. Carraher, D.J. Giron, W.K. Woelk, J.A. Schroeder and
    M.F. Feddersen. J. Applied Polymer Sci., 23, 1501 (1979).
2. C. Carraher, J.A. Schroeder, W. Venable, C. McNeely, D.J.
    Giron, W.K. Woelk and M.F. Feddersen, "Additives for Plastics,
    Volume 2." (R. Seymour, Ed.), Academic Press, N.Y., 1978.
3. C. Carraher and M.F. Feddersen, Angew. Makromolekulare Chemie,
    54, 119 (1976).
4. E.M. Hodnett, Polymer News, 8, 323 (1983).
5. C.N.R. Rao, "Chemical Applications of Infrared Spectroscopy,"
    Academic Press, N.Y., 1963.
6. A.A. Grinberg, M. Serator and M.I. Gel'fman, Russian J.
    Inorg. Chem., 13, 1695 (1968).
7. G. Barrow, R. Krueger and F. Basolo, J. Inorg. and Nuc.
    Chem., 2, 340 (1956).
8. H. Ito, J. Fujita and K. Saito, Bull. Chem. Soc. Japan,
    42, 1286 and 2677 (1969).
9. J. Chatt, G.A. Gamien and L.E. Orgel, J. Chem. Soc. London,
    486 (1958).

ORGANOTIN PIEZO- AND PYROELECTRIC POLYMER FILMS

R. Liepins

University of California
Los Alamos National Laboratory
P.O. Box 1663
Los Alamos, NM 87545

and

M. L. Timmons and N. Morosoff

Research Triangle Institute
P.O. Box 12194
Research Triangle Park, NM 27709

and

J. Surles

University of North Carolina
School of Pharmacy
Chapel Hill, NC 27514

INTRODUCTION

Since the early work on the piezoelectric effect in polymeric materials, many types of polymers have been shown to be piezoelectrically active.[1] Of the synthetic polymers, poly-(vinylidene fluoride) (PVF_2) in β and γ crystal forms, has the highest piezoelectric constant, $d_{31} = (2 - 5) \times 10^{-11} CN^{-1}$. The piezoelectricity in polymers may also be roughly related to the dielectric constant.[2] The dielectric constant for PVF_2 is 8.0 for the β and γ crystal forms. In the β and γ crystal forms the dipoles can be oriented to a sufficiently high degree, whereas in the α crystal form the chain has a small dipole moment, the overall dipole moment in a unit cell is zero, and no piezo-electric activity is found.

     In amorphous polymers, such as poly(methyl methacylate),
the piezoelectricity is explained on the basis of the freezing
mechanism.[3]   The semipermanent dipoles are oriented under the
influence of a high dc field applied above the glass transition
temperature (T_g) and maintained during cooling toward T_g.

     The work reported here involved designing polymers that in
addition to being piezoelectrically active also possessed
antifouling properties and could be formulated into a paint.
The polymer systems investigated consisted of various
tributyltin and trimethyltin methacrylate (TBTM and TMTM,
respectively) homo-, co-, and terpolymers. These polymers are
known to be antifouling, and alignment of the strong tin-oxygen
dipoles by poling procedures was expected and later found to be
possible also by medium hydrogen bonding solvents without the
poling procedures.

EXPERIMENTAL
Monomeric Materials
Bis(tri-n-Butyltin)oxide (TBTO). This material was used in the
synthesis of the tin-containing methacrylates as received from
Alfa Division of Ventron Corp.

Tributyltin Methacrylate (TBTM). TBTO (312.9g, 0.525 mol) and
ligroin (750mL) are added to a 2-L, one-necked flask equipped
with a stir bar and liquid addition funnel; the freshly
distilled methacrylic acid (MAA (93.10 g 1.0825 mol) is then
placed in the addition funnel. The flask is cooled to ca. $4^{\circ}C$
in an ice bath and the MAA is added at a rate of ca. 3 mL/min so
that the flask temperature will not rise above $15^{\circ}C$. After
addition the ice bath is removed and the mixture is stirred for
about 1 h.

     The reaction mixture is poured into a separatory funnel and
washed extensively with H_2O to remove the excess MAA. (Note:
Excess MAA is used because TBTO is more difficult to remove;
also, do not wash with $NaHCO_3$ solution because it causes product
decomposition). The ligroin solution is dried (on 3-Å molecular
sieves or with $MgSO_4$), filtered, and concentrated on a Rotovac
(by removing 600mL of ligroin). The solution is then stoppered
and placed in a freezer to induce crystallization. The liquid
is poured off the crystals, the crystals are warmed to room
temperature, more ligroin is added, and the solution is
stoppered and refrozen. This freeze-thaw procedure should be
carried out until the yellowish product yields colorless
crystals on freezing (2-4 cycles; mp $18^{\circ}C$). The yield of
crystallized product is typically ca. 95%.

Trimethyltin Methacrylate (TMTM). This monomer was synthesized
according to the general procedures of Koton et al.[4,5] and

Shostakovskii et al.[6] from trimethyltin hydroxide (used as
received) and MAA in benzene. The product was recrystallized
from a 50-50 mixture of petroleum ether and carbon tetrachloride
to yield fine white needles with a mp of 117°C as determined in
a differential scanning calorimeter at a heating rate of
20°C/min. The melting points reported in the literature ranged
from 100 to 122°C.[6-8] The yield of the recrystallized material
was 69%.

Polymeric Materials

Poly(tri-n-Butyltin Methacrylate). The glassware used was
always dried overnight at 100°C. Azoisobutyronitrile (AIBN) was
used as the initiator at a ratio of 1:1000 mol of monomer.
Toluene was used as the solvent. The monomer-toluene mixture
was degassed on a vacuum line (10^{-14} torr) in three freeze
(liquid N_2)-thaw cycles. The polymerization vessel was then
sealed, warmed to RT, and placed in an oven at a constant
temperature of 80°C for at least 16 h. The flask was shaken
occasionally during this time. The polymer was isolated from
methanol as a yellowish, somewhat tacky, rubbery mass. Films
cast from toluene were elastic but not very strong. The yield
was above 95%.

Copolymers of TBTM and TMTM. The general procedure was the same
as that for the polymerization of poly(TBTM). Both benzyol
peroxide and AIBN were used as the initiators, with the AIBN
giving the higher yields. The amount of toluene was always
equal to the weight of the monomers. The copolymers were
isolated from their viscous toluene solutions by pouring the
solution slowly into a 10 times larger volume of ligroin. The
isolated material was dried in a vacuum oven at 50°C/16 h.
Typically, a 29 mol% TBTM/MMA copolymer was obtained in a 90%
yield with an inherent viscosity in dimethylformamide as high as
1.2.

Sample Preparation. Films for piezo-and pyroelectric activity
measurements were cast from dilute toluene or cyclohexanone
solutions having typical concentrations of 10% or less. The
films were cast in a clean-air hood on specially cleaned, large
($\sim$ 15.25 by 25.4 cm) glass plates in at least three layers, each
< 10 µm thick. The individual layers were only partially dried
and were soft to the touch before the next layer was cast. The
final, multilayered films were dried overnight in a vacuum oven
at 323 K (50°C).

The films were then floated off the glass plates in tanks
filled with distilled, deionized water, and were dried again in
a vacuum oven at 50°C for about 16 h. Typical film thicknesses
for activity evaluations were 30 ± 5 µm. Electrodes of tin were
vacuum evaporated onto the films. Electrode thicknesses ranged

from 750 to 1100 Å. Experimental film samples were cut into
rectangles of typically 1.3 x 12.7 cm. The electrodes were
offset at opposite ends and wires were attached with a silver
epoxy (Emerson and Cummings, Inc., Eccobond Solder 57C).

Poling Technique. Films were routinely poled for 30 min. at 800
- 850 kV/cm, at 15 - 20 K above the T_g of the particular polymer
film. The films supported fields of 1000 kV/cm at room
temperature for short periods of time (long-term studies were
not performed). The length of time required to pole films
adequately was first established for a specified length of time,
measuring the activity, and then depolarizing the films by
heating to ~ 20 K above their T_g with the leads shorted for 1 h.
This cycle was repeated for increasing lengths of time until
essentially constant activity was reached. Each cycle started
from zero activity after depolarization. Using this procedure
it was found that the maximum activity was usually developed
after 20 - 25 min. of poling at 800 - 850 kV/cm, 20 K above the
T_g.

Piezoelectric Activity Measurements. The measurement system was
similar to that of Hayakawa and Wada.[1] A thin strip of film is
clasped between jaws, one of which is immovable whereas the
other is attached to the impedance head of a B&K instruments,
Model 8001 joined to an electromechanical shaker. A sinusoidal
current drives the shaker's movable coil producing linear
oscillations of the impedance head and attached jaw.

The piezoelectric output from the film is measured as an
open-circuit voltage by an electrometer (Kiethley Model 610B) as
the unity gain amplifier with an input impedance of $> 10^{12}$ ohms.
Care was taken to ensure that the film's operating frequency was
within the amplifier's unity gain frequency response. The
output from the electrometer can be connected either to an
oscilloscope for visual display or to a lock-in amplifier for
accurate voltage measurement. The reference signal to the
lock-in amplifier was provided by the oscillator of the system.

RESULTS AND DISCUSSION
Polymer Synthesis

Organotin esters of acrylic and methacylic acids have been
polymerized by a variety of free-radical initiators as well as
thermally in bulk, solution or aqueous emulsion phases.[9-12]
Because of the superior paint formulation properties of TBTM
polymers a series of co- and terpolymers were prepared.

Under identical polymerization conditions for charged
21-mol%, TBTM-comonomer compositions and a copolymerization time
of 16 h the isolated yield of MMA copolymers was 90%, of acry-

lonitrile copolymer, 40%, of methacrylonitrile copolymer, 30% and of styrene copolymer, 25%. Numerous attempts at copolymerizing α-methylstyrene were unsuccessful. This failure apparently is an example of steric hindrance due to the tributyltin group because copolymerization of a α-methylstyrene were unsuccessful. This failure apparently is an example of steric hindrance due to the tributyltin group because copolymerization of a α-methylstyrene with MMA was successful.

Five TBTM terpolymer compositions were also prepared and are summarized in Table 1. Most interesting was the fact that methacrylic acid in the TBTM/MMA terpolymerizations acted as a polymerization "sensitizer." Polymerizations with methacrylic acid were complete (100% conversion) in less than 1 h and the inherent viscosity of the terpolymers in dimethylformamide was always above 0.7.

Table 1. TBTM Terpolymer Compositions[a]

Composition, Charged (mol%)						Yield
TBTM	MMA	MAA	α-MeSty	AN	MAN	(%)
40	30	30	---	--	---	100
40	50	10	---	--	---	100
40	30	---	---	--	30	5
30	35	---	35	--	---	1
30	---	---	31.5	38.5	---	51

[a]MMA = methylmethacrylate; α-MeSty = α-methylstyrene; MAA = methacrylic acid; AN = acrylonitrile; MAN = methacrylonitrile.

Corresponding copolymers without the methacrylic acid, would require at least 16 h for an 80% conversion. Terpolymerization of α-methylstyrene in the compositions tried was not possible. The use of methacrylic acid in a terpolymerization of α-methylstyrene may be more promising.

Tin Content in TBTM/MMA Copolymers. Qualitatively, TBTM, hence the tin content, could be monitored by IR spectroscopy-specifically the tin-ester carbonyl absorption at 6.1 μm. Thus we observed that the tin-ester carbonyl absorption increased in the copolymer with increasing conversion. Chemical analysis for tin in a series of different TBTM/MMA copolymers, polymerized to about the same conversion (80-85%), also confirmed the observation that TBTM is the slower polymerizing comonomer. The data are summarized in Table 2. Thus in the initial stages of polymeri-

Table 2.   Tin Content in TBTM/MMA Copolymers

Composition, Charged (mol%)	Sn Found (%)	Sn Found (mol%)	Sn Difference (mol%)
10	7.5	9	1
21	14.5	18	3
30	17.5	25	5
35	19.0	29	6
40	21.0	32	8

zation the copolymer is low in the tin comonomer and the tin concentration approaches that of the charged concentration only if the polymerization is conducted essentially to completion. Furthermore, this difference in tin concentration between that charged and that found in the copolymer is more pronounced with increasing TBTM concentration in the charged composition.

Glass Transition Temperature Versus Composition. Because of the importance of the relationship between stability of the dipole alignment (after poling), piezoelectric activity, and T_g in amorphous polymers, we determined the T_gs on most of the materials prepared. They were determined on vacuum-dried films cast from various solvents and measured in a Perkin-Elmer differential scanning calorimeter at a heating rate of 20°C/min. Some representative copolymer T_gs are listed in Table 3. By extrapolating the available TBTM/MMA data to 100 mol% TBTM we obtained a T_g of -70°C for TBTM.

Table 3.   Glass Transition Temperature of Various Polymers

Polymer	T_g ($^{\circ}$C)
MMA	105
TBTM/Sty (50:50 wt%)	55
TBTM/MMA (18 mol% TBTM)[a]	71
TBTM/MMA (29 mol% TBTM)[a]	56
TBTM/MMA (41 mol% TBTM)[a]	32
TBTM/AN (30 mol% TBTM)	20
TMTM/MMA (50:50 mol%)	75

[a]These particular compositions were established analytically; the others represent the charged compositions.

X-Ray Diffraction Analysis

The crystallite size and/or perfection of the crystals can
be estimated from the breadth of reflections. Crystallite size
and perfection may play a role in the piezoelectric behavior of
a polymer and would be expected to be affected by the choice of
solvent and the rate at which the polymer is cooled down during
poling.

Because of the considerable piezoelectric activity in the
30-mol% TBTM/MMA copolymers, we investigated the crystallizing
and/or orientation effects of these materials. Guinier x-ray
diffraction photographs and microdensitometer traces of the
photographs were obtained on the copolymer films, unpoled and
poled, while being formed. The poled and unpoled films
indicated the existence of crystals. A more detailed study is
needed to elucidate the extent of "ordering" of the amorphous
phase.

X-ray diffraction photographs and microdensitometer traces
of the photographs were also obtained on the 50:50 mol%
(charged) TMTM/MMA copolymer. Again the photographs indicated
the existence of crystals in the unpoled copolymer film.
However, a comparison of the 50:50 mol% TMTM/MMA and 30 mol%
TBTM/MMA copolymer microdensitometer traces revealed apparent
difference in the "ordering" of the amorphous phases in the two
polymers. The broad peaks in the microdensitometer traces of
the two copolymers had 2θ maxima at the following positions:

Position of 2θ maxima (deg)

30 mol% TBTM/MMA	11	17	30 (poled only)
50 mol% TMTM/MMA	--	16	--

The fact that the 11° maximum was not observed in the 50
mol% TMTM/MMA unpoled copolymer, coupled with the fact that the
maximum at $16-17^{\circ}$ was observed, is a strong indication that we
are seeing certain "orientation" effects in these largely
amorphous materials. The x-ray reflection at $2\theta = 30^{\circ}$ is weak
and seen only in the poled TBTM/MMA copolymer. Interestingly,
recently, vinylidene cyanide/vinyl acetate copolymer was also
shown to be piezoelectrically active and from the weak and broad
x-ray reflections at $2\theta = 15^{\circ}$ and 30° was interpreted as possessing crystals in the copolymer.[13]

Piezoelectric Activity

Piezoelectric activity data here will be reported only on

the 25- to 30mol% TBTM/MMA copolymers because they represent the best compromise between reasonably long-term piezoelectric activity at room temperature, antifouling properties, and paint formulation characteristics.

Films cast from hydrocarbon solvents, such as toluene, supported poling fields up to 1000 kV/cm. Poling increased g_{31} activity from barely detectable in the unpoled film to better than 10% of poled PVF_2 activity:

	d_{31} Activity (C/N)	
Casting Solvent	25 mol% TBTM/MMA	PVF_2*
Cyclohexanone	1.7×10^{-12}	20×10^{-12}
Toluene	0.53×10^{-12}	---

	g_{31} Activity (Vm/N)	
	30 mol% TBTM/MMA	
Toluene	$16.1 \text{ z } 10^{-3}$	150×10^{-3}

*Kureha KF Piezofilm; sample obtained from NBS

These d_{31} and g_{31} values are not the highest obtainable in this class of materials, rather, these polymer compositions represent the compromise of activity, anti-fouling properties, and paintability characteristics. The activity in these materials increased with increasing TBTM content and we have not established the maximum activity obtainable. The g_{31}-constant of the TBTM polymer films is comparable to that of ceramics.

Pyroelectric Activity

Initial pyroelectric activity evaluation on 30-mol% TBTM/MMA films with a pyroelectric detector was done at NASA/Langley Research Center, VA.[14] Unfortunately, the films supplied were 50 μm thick or about 10 times thicker than desirable for a good pyroelectric response.

To obtain a figure of merit for a pyroelectric detector, one must also consider thermal noise and intrinsic noise of the system due to dielectric loss. These considerations lead to a figure of merit, called the normalized detectivity, D*:[15]

$$D^* = \frac{pn}{2c\rho} \left[\frac{1}{t\omega\varepsilon\tan\delta\kappa T} \right]^{1/2}$$

where ρ is the pyroelectric coefficient, η is the efficiency with which the incident radiation is absorbed, t is the detector thickness, ε is the dielectric constant, tan δ is the dielectric dissipation factor, k is Boltzmann's constant, and T is the absolute temperature. The normalized detectivity D*, obtained with our films, was 1.5×10^5 cmHz$^{1/2}$/W. This value is low, for reasons indicated before, and compares with D* for PVF$_2$ as follows: 1×10^8 cmHz $\frac{1}{2}$/W.

Solvent Induced Orientation Effects

Films cast from certain solvents, most notably cyclohexanone and methyl ethyl ketone, possessed piezoelectric activity in the unpoled state, whereas films of equal composition but cast from other solvents, for example, toluene, possessed no activity in the unpoled state:

	d_{31} Activity (C/N)	
Casting Solvent	25 mol% TBTM/MMA	PVF$_2$*
Cyclohexanone	1.5×10^{-12}	0
Toluene	Near Noise Level	---

*Commercial sample of an unpoled film.

Ketone solvents have <u>medium</u> hydrogen bonding characteristics, which we believe are important. The presence of activity is attributed to a unique combination of an inherent chain orientation in cast thin (< 10 μm) films[16-18] and an oxygen-tin dipole orientation by the ketone solvents as they evaporate during film deposition. The ketone molecules would be expected to coordinate with the tin-oxygen dipole because of their medium-strong hydrogen bonding parameter value. Also, it is extremely difficult to remove the last traces of cyclohexanone from a polymer film; and thus any remnants would act as a solvent plasticizer and facilitate the dipole orientation. Further study is needed especially because this dipole orientation eliminates the possibility of charge injection during film poling, a mechanism that has been frequently cited as possibly causing piezoelectric activity in polymeric films.

CONCLUSION

A 30-mol% TBTM/MMA copolymer possessed good piezoelectric activity, antifouling properties, and paint formulation characteristics. The polymers could be made piezoelectrically active in thin film by preferentially dipole-orienting solvents or by poling procedures. Unpoled 25-mol% TBTM/MMA film, cast from cyclohexanone, possessed the d_{31} piezoelectric activity of 1.5 x 10^{-12} C/N. Whereas pole 30-mol% TBTM/MMA film, cast from toluene, possessed a g_{31} activity of 16.1 x 10^{-3} Vm/N. For pyroelectric activity evaluation a very thick (50 μm) 30-mol% TBTM/MMA film in a pyroelectric detector had a normalized detectivity, D*, of 1.5 x 10^5 cmHz$^{1/2}$/W. To our knowledge this is the first evaluation piezo- and pyroelectric activity in organotin polymers and of orientation effects induced by solvents.

ACKNOWLEDGEMENT

    The authors express their appreciation to Mr. C. Walker, the program manager for the Naval Sea Systems Command, and Dr. M. G. Broadhurst, D. G. T. Davis, Mr. S. Edelman, and Mr. S. Roth, all of the National Bureau of Standards for valuable discussions, recommendations and sample evaluation. Our grati- tude goes also to Charles Hicks of the Naval Ocean Systems Center, San Diego, and to Dr. James Robertson of NASA/Langley Research Center for evaluation of some of our samples for piezo- and pyroelectric activity. We appreciate greatly the discussions and assistance of Prof. V. T. Stannett of North Carolina State University, Rayleigh, N.C, and of Dr. J. J. Wortman and Mr. T. R. Howell of the Research Triangle Institute, Research Triangle Park, NC.; and Mrs. J. A. Montemarano and Dr. E. Fisher of the David Taylor Naval Ship Research and Development Center for their recommendations of potential paint formulations. This work was supported by the Naval Sea Systems Command, Contact No. N00024-76-6246.

REFERENCES

1. R. Hayakawa and R. Wada, "Piezoelectricity and Related Properties of Polymer Films," Adv. in Polymer Sci., Vol. 11, 1973, p. 1.

2. N. Murayama, K. Nakamura, H. Obara, and M. Segawa, Ultra- sonics, 15 (January 1976).

3. M. G. Broadhurst and G. T. Davis, "Piezo- and Pyroelectric Properties of Electrets," NBSIR 75-787, October 1975.

4. M. J. Koton, T. M. Kiseleva, and F. S. Florinskii, Vysokomol. Soedin., 2, 1639 (1960).

5. M. J. Koton and F. S. Florinskii, Zh. Obshch. Khim. 32(9), 3057 (1962); English translation, J. Gen. Chem. 32I(9), 3008 (1962).

6. M. F. Shostakovskii, S. P. Kalinina, V. N. Kotrelev, D. A. Kochkin, G. I. Kuznetsova, L. V. Laine, A. I. Borisova, and V. V. Borisenko, J. Polym. Sci., 52 223 (1961).

7. C. M. Langkammerer, U.S. Pat. 2,253,128 (August 19, 1941).

8. M. J. Koton, T. M. Kiseleva, and R. M. Paribok, Dokl, Akad, Nauk SSSR, 125 1263 (1959).

9. F. G. A. Stone and W. A. G. Grahm, "Inorganic Polymers," Academic, New York, 1962, Cap. IV, Part III.

10. M. F. Lappert and G. H. Leigh, "Developments in Inorganic Polymer Chemistry," Elsevier, Amsterdam, 1962, Cap. 8, Sec. 4.

11. K. A. Andrianov, "Metalorganic Polymers," Interscience, New York, 1965.

12. J. Montemarano, L. P. Marinelli, and T. M. Andrews, J. Polym. Sci., 32 523 (1958); U.S. Pat. 3,016,369 (1962).

13. S. Miyata, M. Yoshikawa, S. Tasaka, and M. Ko, Polymer Journal, 12, (12), 857 (1980).

14. J. B. Robertson, NASA/Langley Research Center, VA, unpublished results (1977).

15. W. R. Peters, Proc. Carnahan Crime Countermeas., 78, 141 (1977).

16. W. A. Prest, Jr. and D. J. Luca, J. Appl. Phys., 50, (10), 6067 (1979).

17. Y. Cohen and S. Reich, J. Poly. Sci.; Poly. Phys. Ed., 19, 599 (1981).

18. A. N. Cherkasov, M. G. Vitovskaya, and S. B. Bushin, Vysokomol. Soedin., Ser. A, 18, 1628 (1976).

NUCLEAR MAGNETIC RESONANCE AS A PROBE OF ORGANIC CONDUCTORS

Larry J. Azevedo

Sandia National Laboratories

Albuquerque, New Mexico 87185

ABSTRACT

The uses of nuclear magnetic resonance as a probe of organic conductors are discussed with emphasis on the study of electronic properties in the superconducting Bechgaard salts. Several examples are given depicting how spin lattice relaxation rate measurements over a wide range of frequencies, temperature and applied pressure can uncover the important electronic properties of these systems.

INTRODUCTION

The uses of magnetic resonance techniques as a microscopic probe of matter in the condensed state are well known. In this paper several examples of the application of magnetic resonance techniques are presented which are useful in the microscopic study of organic conductivity. This study will concentrate on the so-called Bechgaard salts,[1] $(TMTSF)_2X$ where $X = PF_6$, ClO_4, AsF_6, etc. and TMTSF is tetramethyltetraselenafulvalenium. These salts exhibit a wealth of interesting properties of both fundamental and technological interest including high conductivity, insulating behavior, and an antiferromagnetic spin density wave ground state. All of these states are accessible by an appropriate choice of applied pressure, magnetic field and temperature. The structure[2] consists of planes of TMTSF

* This work performed at Sandia National Laboratories supported by the U. S. Department of Energy under Contract Number DE-AC04-76-DP00789.

molecules separated by anion sheets leading to strong electronic overlap in two dimensions and much weaker overlap in the third dimension (the c-axis). As is shown below, the electronic properties reflect the quasi two-dimensional structure.

The discovery[3] of superconductivity in these materials, the first in an organic, has motivated many experimental and theoretical studies. One of the problems in the study of these systems is the extreme experimental conditions necessary to observe superconductivity: pressures above 6 kbar (except for the ClO_4 salt which exhibits superconductivity at ambient pressure) and temperatures of 1K or below. Due to these limitations many standard experimental techniques such as x-ray crystallography, light scattering, conductivity, magnetic susceptibility, etc., are either impossible or at least very difficult to perform. One additional problem is the small size and fragile nature of crystals of these materials. Fortunately, magnetic resonance techniques are well suited to these systems if one is careful in experimental design and technique. By observing all of the nuclear sites in the crystals, one can determine where the charge carriers are located and details about their microscopic motion. Through frequency dependent relaxation measurements one can deduce, on a microscopic basis, the effective dimensionality of the conduction process; the magnetic state of these materials can be studied; and the anisotropic properties of the superconductivity can be measured.

BACKGROUND

There are several properties of the Bechgaard salts which are suited for study by NMR techniques. The ground state (in the limit of zero temperature) of most of the Bechgaard salts is an insulating magnetic state characterized by a spin density wave (SDW). The SDW spin ordering is antiferromagnetic with an exchange temperature on the order of 200 K.[4] The properties of the SDW state have been fully characterized by microwave spin wave experiments[4] and susceptibility studies.[5] Upon application of pressure, the SDW transition temperature is suppressed with the result that a superconducting ground state appears at temperatures of the order of 1 K. This behavior is illustrated in Fig. 1 which shows the temperature/pressure phase diagram of $TMTSF_2PF_6$.

A fundamental question is the mechanism by which the SDW state is suppressed by the application of pressure, resulting in a superconducting ground state. In other words, we would like to understand the competition of the SDW and superconductivity (SC). Another important study is the nature of the electronic motion in the normal metallic state. Previous

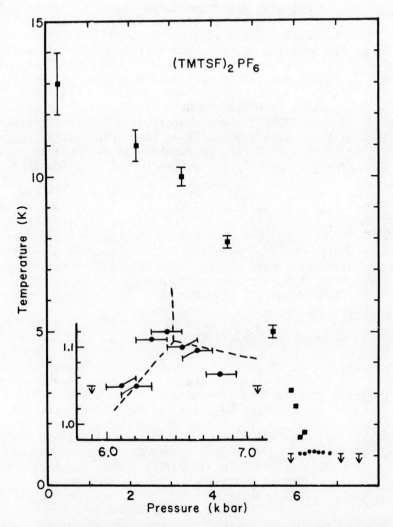

Fig. 1. The temperature/pressure phase diagram of the super-
         conductor (TMTSF)$_2$PF$_6$. The inset shows the details
         of the phase diagram in the superconducting regime.
         Squares depict the SDW/metallic transition.   The
         system is a metal in the high temperature regime and
         a SDW insulator in the low temperature regime (below
         the squares). The circles depict the superconducting
         transition, below the circles the sample is super-
         conducting. The lowest attainable temperature was
         1.04 K.

studies of other organic conductors such as $Qn(TCNQ)_2$ (quino-
linium tetracyano-quinodimethan) have shown that the electronic
motion is highly one-dimensional along a particular crystalline
axis.[6] This large anisotropy in the conductivity leads to unusual power law behavior of the susceptibility,[7] heat capacity,[8]
and magnetization[7] which is not observed in higher dimensional
systems. In many cases the low-dimensional nature of the
electronic system leads to thermal and magnetic properties
that are not present in three-dimensional systems. For instance,
in a one-dimensional metal the singularity in the susceptibility
response function leads to a finite nuclear spin-lattice relaxation rate in the limit of zero temperature instead of the usual
Korringa relaxation rate which is proportional to temperature.

NMR can provide detailed knowledge of the conductivity
anisotropy in the Bechgaard salts. Are there sufficient overlap
of electronic wavefunctions to lead to anisotropic three-
dimensional electronic motion or are the Bechgaard salts one or
two-dimensional? The high magnetic field (> 60 KOe) properties
of the Bechgaard salts are particularly unusual: a field-driven
phase transition[9] triggers a magnetic state that might be related to the SDW. Magnetic resonance has been used to address
these problems.

Competition Between the SDW and SC

The observation of low field electron spin resonance (ESR)
serves as an excellent probe to study the pressure/temperature
phase diagram. The SDW and SC states have pronounced effects
on the nuclear and electron magnetic resonance properties of
the Bechgaard salts. For instance, in the SDW state the antiferromagnetic pairing between spins causes the g = 2 ESR signal
to vanish. A spin-wave resonance appears at high frequencies
(15-30 GHz).[4] By monitoring the low-field (resonance field
10 Oe, frequency 28 MHz) ESR signal with a Q-meter circuit
using continuous-wave absorbtion techniques, one can follow the
metallic/SDW phase transition as a function of pressure or
temperature. This technique has the important advantage that
the crystalline sample is merely mounted inside a rf coil and
no leads are attached to it. Because of the small size of
crystals (typically a few mg in mass) the sample coil must be
very small; sometimes 100 microns or less in diameter. The
metallic, SDW, and SC states are monitored as follows. The
metallic state is characterized[10] by a narrow ESR line (1 Oe)
whose intensity is independent of temperature indicative of
the Pauli susceptibility of a metal. Passage into the SDW state
is observed by the disappearance of the ESR signal. The superconducting state is monitored by its shielding effect on the
applied rf in the magnetic resonance coil. When the sample
passes into the superconducting state the effective filling

factor of the coil changes due to the Meissner currents and is
detected by the Q-meter. Using this technique the superconducting transition temperature can be measured as a function of
pressure, temperature, and applied magnetic field. Also, the
lower and upper critical field are measured as functions of
temperature, sample orientation, and field.

Figure 1 shows the phase diagram for the superconductor
$(TMTSF)_2PF_6$ as measured by the ESR technique. The low pressure
(< 6 kbar), low temperature (< 12 K) region is characterized by
the SDW state. The transition between SDW and metallic states
is narrow, typically 100 mK in width. Above 7.0 kbar and below
1.1 K superconductivity is observed. This region of the phase
diagram is shown in detail in the inset of Fig. 1. There is a
narrow pressure regime (on the order of 0.3 kbar) where, with
only a lowering of temperature one passes from the metallic
state, to the SDW state and then to the SC state. This is the
first microscopic observation of a transition from a magnetic
state directly to a superconducting state by a change of temperature in an organic superconductor. Theoretical calculations
by K. Yamaji[11] show that there is no coexistence of the superconducting and SDW states and that the phase boundary is
characterized by a first-order phase transition. Our electron
spin resonance results confirm the calculations of the phase
boundary between superconductivity and SDW states, but to date
have not determined whether the transition is first order.
Yamaji's determinations of the complete pressure/temperature
phase diagram are based on microscopic calculations of the
free energy in the metallic, SC and SDW states. The details of
these measurements and comparison with theoretical calculations
by K. Yamaji[11] are presented elsewhere.[12]

The Dimensionality of the Conduction

The low-dimensional electronic structure of these materials
has led to speculations that the superconductivity may be one-
dimensional in nature.[13] That is, experimental evidence has
been presented claiming that fluctuations into a one-dimensional
superconducting state occur at temperatures as high as 40K
even though the three-dimensional superconducting transition
is on the order of 1K. This question is addressed through a
study of the diffusion rate of the electronic motion in the
different crystalline directions. The ideal probe is nuclear
magnetic resonance.

Measurement of the frequency (or field) dependence of the
nuclear spin lattice relaxation rate, T_1^{-1}, yields details
of the electronic motion. The spin-lattice relaxation rate of
a nuclear spin which is hyperfine coupled to an electron spin
may be written as[14]

$$T_1^{-1} = \Omega^z \, F_z(\omega_N) + \Omega^+ \, F_+(\omega_e) \tag{1}$$

where Ω^z and Ω^+ are electron-nuclear coupling constants and $F_z(\omega_N)$ and $F_+(\omega_N)$ are the Fourier transforms of the longitudinal and transverse electron spin correlation functions and the nuclear and electron Larmor frequencies, respectively. If an electron is diffusing in d dimensions, the time dependence of its spin-spin correlation function is proportional[14] to $t^{-d/2}$ which leads to characteristic frequency dependences for T_1^{-1}. Basically, the diffusional motion is a random walk in d dimensions. The results are:

$$\text{1D:} \quad (k_B T_1 T \chi)^{-1} = A + \frac{\Omega^+}{(2D_1 \omega_e)^{1/2}}; \quad D_1 > \omega_e > D_2, D_3 \tag{2}$$

$$\text{2D:} \quad (k_B T_1 T \chi)^{-1} = A + \frac{\Omega^+}{2\pi(D_1 D_2)^{1/2}} \, \ln \frac{4\pi^2 D_2}{\omega_e}; \tag{3}$$

$$D_1, D_2 > \omega_e > D_3$$

$$\text{3D:} \quad (k_B T_1 T \chi)^{-1} = A + \frac{\Omega^+}{2\pi(D_1 D_2)^{1/2}} \, \ln \frac{4\pi^2 D_2}{D_3}; \tag{4}$$

$$D_1, D_2, D_3 > \omega_e \quad .$$

Here T is the temperature, χ the reduced susceptibility ($\chi = N(g\mu_B)^2$, D_j is the diffusion rate in the jth direction, and $\omega_e = g\mu_B H/h$ is the electronic Larmor frequency. The constant A is given by

$$A = (\Omega^z + \Omega^+)f_2 + \Omega^z \, f_1(\omega_N) \tag{5}$$

where f_1 is the diffusive part of the Fourier transform of the spin correlation function and f_2 is a constant.

By covering a large range in frequencies (or times) one can uncover the dimensionality of the conduction process by the observation of the frequency dependence of the nuclear spin-lattice relaxation rate. One added advantage is that the crossover between one and two or two and three-dimensional motion may be observed if the time scale of the experiment is in the appropriate range. For instance, if the inter-chain diffusion rate is large enough so that the electron has a good chance to move to neighboring chains during a period of nuclear precession, then the motion is essentially two-dimensional as far as the nuclear spin in question is concerned. If, however, one goes to a lower frequency, thus probing longer times, then a crossover can be observed when the electronic motion in three dimensions is fast compared to the inverse nuclear frequency. These effects have been observed in many systems such as $Qn(TCNQ)_2$[6] and TTF-TCNQ.[15] Another advantage is that one can determine directly the diffusion rates for transport in different crystalline directions through an analysis of the frequency dependence of T_1^{-1}.

The results of relaxation studies are shown in Figs. 2-4 where the ^{1}H, [77]Se, and ^{13}C spin-lattice relaxation rates in

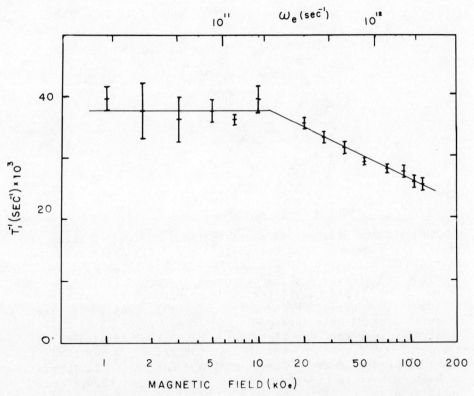

Fig. 2. The proton spin-lattice relaxation rate vs. logarithmic field for the PF_6 salt in the metallic state at T = 4kbar and P = 6.9 Kbar. The appropriate electronic frequencies are shown at the top of the figure. The logarithmic field dependence in the high field regime is indicative of relaxation to a charge diffusing in two dimensions.

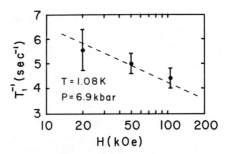

Fig. 3.   The selenium spin lattice relaxation rate vs. logarithmic field in the PF$_6$ salt.

(TMTSF)$_2$PF$_6$ are plotted vs. magnetic field. Figure 2 illustrates the abovementioned crossover[16] from relaxation by a two-
dimensional diffusion process (at fields above 12 KOe) to relaxation mediated by a three-dimensional process (at fields
below 12 KOe). An analysis of these results shows that above
fields of 12 kOe the transport is essentially two-dimensional
since the logarithmic dependence of T$_1^{-1}$ vs. field is
observed. Note that the field value of 12 kOe corresponds to
a time of $\tau = 2 \times 10^{-12}$ sec. Hence, for times greater than
τ the transport is three-dimensional and for times less than
τ the transport is two-dimensional. One can also show, from
applying eqns. (1-3) to the data of Fig. 2, that the anisotropy
of the transport in the two poorly conducting directions is
about 25, in agreement with transport measurements[17] and that
the diffusion rates along the b and c axes (the highly conducting direction is the a axis) are D$_2$ = 5.5 x 10^{12} and D$_3$ = 2.2 x
10^{11} sec^{-1}, respectively.[16] Since one-dimensional diffusive behavior is not observed, a measure cannot be obtained of D$_1$.

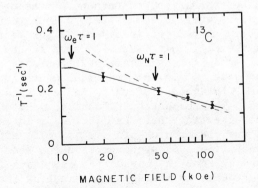

Fig. 4.   The carbon-13 spin lattice relaxation rate vs. logarith-
          mic field in the PF_6 salt. The dotted line depicts a
          fit of eqn. (2) asuming one-dimensional diffusion of
          the electron instead of two-dimensional motion (solid
          line).

        The same high-field logarithmic field dependence is observed
for the other nuclear spins which are hyperfine coupled to the
electron spin as shown for the ^{77}Se in Fig. 3[18] and ^{13}C in
Fig. 4.[19] The only difference is the values of the hyperfine
coupling constants. Due to lower signal intensities the cutoff
to three-dimensional diffusional motion at 12 kOe could not be
observed but the same high-field dependence on relaxation rate as
the proton data provides proof that the analysis in terms of
eqns. (1-3) is correct.

High-Field Properties of the Bechgaard Salts

        Above 60 kOe in the metallic state the magnetic and elec-
tronic properties of these salts display a phase transition.
A large magnetoresistance is observed for the applied field
parallel to the lowest conducting direction.[20] Above a
certain critical field, oscillations in the magnetoresistance
have been observed in the PF_6 salt.[20] This effect has been

interpreted[20] as the de Haas-Shubnikov (dHS) effect with the
conclusion that the Fermi surface must be closed in at least
two dimensions. The transition, or turnon, of the oscillations
is a function of temperature, field, field orientation, and
pressure and has been mapped out in several experiments.[9,20]
The transition in the ClO_4 salt is easily seen in the magnetic
resonance properties as an enhancement of the proton spin
lattice relaxation rate,[21] similar to measurements in the
PF_6 salt,[9] and the "wiping out" of the ^{77}Se resonance[21,22]
as shown in Fig. 5. The resonance is found to vanish above
the turn-on field suggesting that the applied field establishes
magnetic order in the electronic spin system. The effects on
the proton resonance is not as pronounced due to the remote
location of the proton from the conduction electrons in the
unit cell but nevertheless a strong enhancement of the proton
T_1^{-1} is observed[21] consistent with the idea of induced
magnetic order. From an analysis of the proton and selenium
data in the ClO_4 salt, an estimate of the induced field at
the selenium site can be mde. This induced field is on the
order of several kOe at least.

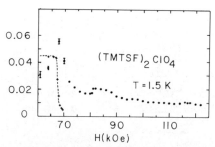

Fig. 5.   The proton spin lattice relaxation rate (●) and the
          selenium nuclear susceptibility (+) vs. applied
          magnetic field for the ClO_4 salt at 1.5 K in the
          highly conducting state. Note that at the turn-on
          field (60 kOe) discussed in the text that the proton
          relaxation rate is strongly enhanced and the selenium
          signal vanishes.

SUMMARY

As shown above, magnetic resonance is a ideal local probe
of organic conductors. Resonance is of special use in the
study of small crystalline samples where one needs high pressures, low temperatures, and high magnetic fields. In the
examples shown above, magnetic resonance studies have measured
the diffusion rates for conduction in different crystalline
directions, uncovered the dimensionality of the electronic
motion, demonstrated that the state at high field is magnetic
in origin, and been used to map out accurately the pressure/
temperature phase diagram of the organic superconductors.

ACKNOWLEDGEMENTS

The work reviewed here was done in close collaboration
with J. E. Schirber (Sandia), J. M. Williams (Argonne), and
E. M. Engler (IBM/San Jose). I also wish to acknowledge the
expert technical assistance of D. Stuart and D. Overmyer.

REFERENCES

1.   K. Bechgaard, D. O. Cowan, A. N. Bloch and L.
Heinriksen, J. Org. Chem., 40, 746 (1975).
2.   K. Bechgaard, K. Carneiro, F. B. Rasmussen, M. Olsen,
G. Rindorf, C. S. Jacobsen, H. Pedersen and J. C. Scott, J.
Amer. Chem. Soc., 103, 2440 (1981); N. Thorup, G. Rindorf,
H. Soling, and K. Bechgaard, Acta. Cryst. 37, 1236 (1981).
3.   D. Jerome, A. Mazaud, M. Ribault, and K. Bechgaard,
J. Phys. (Paris) Lett., 41, L95 (1980).
4.   J. Torrance, H. J. Pedersen and K. Bechgaard, Phys.
Rev. Lett., 49, 881 (1982).
5.   J. C. Scott, H. J. Pedersen, K. Bechgaard, Phys. Rev.
Lett., 45, 2125 (1980); K. Mortensen, Y. Tomkiewicz, T. D.
Schultz and E. M. Engler, Phys. Rev. Lett., 46, 1234 (1981).
6.   F. Devreux, Phys. Rev., B13, 4651 (1976).
7.   L. M. Bulaevskii, A. V. Zvarykina, R. B. Lyubovskii
and I. F. Shchegolev, Zh. Eksp. Teor. Fiz., 62, 725 (1972)
(Sov. Phys. JETP, 35, 384 (1972).
8.   L. J. Azevedo and W. G. Clark, Phys. Rev., B16, 3252
(1977).
9.   L. J. Azevedo, J. E. Schirber, R. L. Greene and E. M.
Engler, Physica, 108B, 1183 (1981).
10.   L. J. Azevedo, J. E. Schiber, R. L. Greene and E. M.
Engler, Mol. Cryst. Liq. Cryst., 79, 123 (1982).
11.   K. Yamaji, J. Phys. Soc. Japan, 51, 2787 (1982); K.
Yamaji, J. Phys. Soc. Japan, 52, 1361 (1983).

12. L. J. Azevedo, J. E. Schirber, J. M. Williams, to appear in Phys. Rev. B, Rapid Comm.

13. A. Fournel, C. More, G. Roger, J. P. Sorbier, J. M. Delrieu, D. Jerome, M. Ribault and K. Bechgaard, J. Physique Lett., 42, L-448 (1981); H. K. Ng, T. Timusk, J. M. Delrieu, D. Jerome, K. Bechgaard and J. M. Fabre, J. Physique Lett., 43, L-513 (1982).

14. F. Devreux and M. Nechschein, in Quasi-One-Dimensional Conductors I, Proc. of the Int. Conf., Dubrovnik, 1978, edited by S. Barisic, et. al., Lecture Notes in Physics, Vol. 95 (Springer, New York, 1979), page. 145.

15. G. Soda, et. al., J. Physique (Paris), 38, 931 (1977).

16. L. J. Azevedo, J. E. Schirber, J. C. Scott, Phys. Rev. Lett., 49, 826 (1982).

17. R. L. Greene, P. Haen, S. Z. Huang, E. M. Engler, M. Y. Choi and P. M. Chaikin, to be published.

18. L. J. Azevedo, J. E. Schirber, and E. M. Engler, Phys. Rev. (Rapid Comm.), B27, 5842 (1983).

19. L. J. Azevedo, J. E. Schirber and E. M. Engler, to appear in Phys. Rev. B, Rapid Comm.

20. J. F. Kwak, J. E. Schirber, R. L. Greene and E. M. Engler, Phys. Rev. Lett., 46, 1296 (1981).

21. L. J. Azevedo, J. M. Williams, and S. J. Compton, to appear in Phys. Rev. B, Rapid Comm.

22. T. Takahashi, D. Jerome and K. Bechgaard, J. de Physique, C3, 805 (1983).

POLYMER-METAL COMPLEXES FOR SOLAR ENERGY CONVERSION

Masao Kaneko and Akira Yamada

The Institute of Physical and Chemical Research

Hirosawa, Wako-shi, 351 Japan

1) INTRODUCTION

Solar energy conversion is an attractive and important research subject, which aims at the development of permanent and clean energy. Since solar irradiation is intermittant, it is desirable to store the converted energy as a form of fuel in order to use it efficiently. For this purpose, photochemical conversion is most suited to produce fuels directly from the irradiation. It is encouraging to know that photosynthesis in green plants, which actually supports all the life on the earth, is based on photochemical processes. In the solar energy conversion research field, utilization of polymers or molecular assemblies has begun to attract much attention.[1-7]

Two main processes in solar energy conversion are photoinduced charge separation occurring within very short excitation times, and the reactions of the separated charges at catalysts or electrodes to give either products or electricity. In these processes, unidirectional transportation of the charges has to be realized in order to prevent energy-consuming back reactions. It is almost impossible to achieve such an anisotropic electron flow in a homogeneous solution system where all the components and intermediates can react at random. For this reason, the utilization of hetergeneous reaction systems provided by polymers or molecular assemblies has become the center of interest.

Metal complexes are the most promising candidates for the photoreaction centers and catalysts in solar energy conversion systems. In this review article, polymer-metal complexes are described for use in the photochemical conversion of solar energy and for solar cells.

2) SOLAR ENERGY AND ITS CONVERSION

Solar spectrum on the earth ranges from 250 to 2400 nm with its maximum at 500 nm.   Since the visible light between 400 and 800 nm occupies about half of the spectrum, the utilization of this region is the important subject of solar energy conversion.   The total irradiation energy reaching the earth is 3×10^{24} J/year, which is about ten-thousand times as much as the energy consumption of mankind. The magnitude of this energy is illustrated by the fact that the earth's estimated total petroleum reserve corresponds to only five days of the irradiation energy supplied to the earth by sunshine.

Photosynthesis by green plants in nature is a typical example of photochemical solar energy conversion.   It can be understood as an electron flow from H_2O to the fixed CO_2 which is driven by the adsorbed visible-light energy (Fig. 1) to produce carbohydrates. The water molecule works as a reducing agent, and the light energy as a kind of electron pump at the photosystems II and I (PSII and I) where chlorophyl molecules (Mg complex of porphyrin containing a long alkyl chain) play the main role for harvesting light and charge separation.   In the reaction centers of PSII and I, a highly ordered arrangement of the components as donor-chlorophyl-acceptor brings about photoinduced charge separation with almost 100% efficiency.

A photochemical conversion model was proposed as Fig. 2.[1,8] At the photoreaction center (P_2-$P_1 \equiv P$), the excitation energy should be smaller than 3.1 eV in order to utilize visible light. T_1 and T_2 (acceptor and donor, respectively), are the electron mediators to realize charge separation. C_1 and C_2 are catalysts to facilitate production of reduction and oxidation products, respectively. They can be electrodes when electricity should be obtained instead of reaction products.   In order to acquire free energy by this conversion scheme, the redox potential of the oxidant R_1 (E_{R1}) should be lower than that of the reductant (E_{R2}).   The free energy acquired (ΔG) can then be expressed by eq. 1;

$$\Delta G = -nF (E_{R1} - E_{R2}) \qquad (1),$$

where n is the equivalent of the reaction and F is the Faraday constant (23.1 kcal/volt·equivalent).

When water (R_2) is oxidized at C_2 giving O_2, and H^+ (R_1) is reduced at C_1 giving H_2; the total reaction is water photolysis. The free energy of 57 kcal/mol H_2O is obtained in this reaction, because $E_{O_2/H_2O} = 0.82$ V (vs. NHE at pH 7) and $E_{H^+/H_2} = -0.41$ V. For the water photolysis, at least 1.23 V has to be acquired at the photoreaction center.

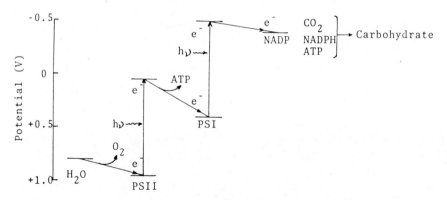

Fig. 1   Simplified scheme of electron flow in photosynthesis

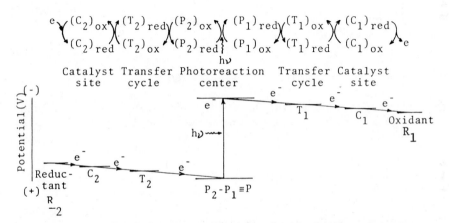

Fig. 2   A model system of photochemical conversion

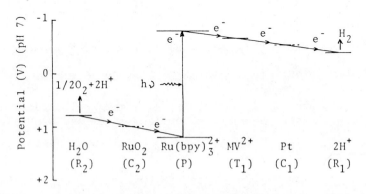

Fig. 3   A possible scheme of photochemical water splitting

For the photoreaction center, metal
complexes are promising candidates. Semiconductors also can work efficiently for
charge separation as described later.
Tris(2,2'-bipyridine)ruthenium(II) complex
($\underline{1}$; Ru(bpy)$_3^{2+}$) is attracting much attention,
since it is suitable as a photoreaction
center of water photolysis. It absorbs
visible light (λ_{max}; ca. 450 nm), and its
ground and excited states are fitted to the
redox reactions of water.$_1$

$$\underline{1};\quad Ru(bpy)_3^{2+}$$

Since water photolysis is a multi-electron process, the
mechanism of the electron transfer should be considered. When
water is oxidized by a stepwise process, the first step requires
a very high positive potential (eq. 2). However, when water is
oxidized by a 4-electron process in which no intermediate is isolated, a much lower redox potential suffices (eq. 3).

$$H_2O \longrightarrow HO\cdot + e^- + H^+ \quad (E_o' = 2.33\text{ V})^{*a)} \quad (2)$$

$$2\,H_2O \longrightarrow O_2 + 4e^- + 4H^+ \quad (E_o' = 0.82\text{ V}) \quad (3)$$

Also for proton reduction, the 2-electron process needs a much
larger redox potential than a stepwise mechanism (eqs. 4 and 5)

$$H^+ + e^- \longrightarrow H\cdot \qquad (E_o' = -2.52\text{ V}) \quad (4)$$

$$2H^+ + 2e^- \longrightarrow H_2 \qquad (E_o' = -0.41\text{ V}) \quad (5)$$

In order to split water by a stepwise mechanism, the potential
difference of 4.90 V is needed for the first step (eq. 6). This
corresponds to the ultraviolet light of 252 nm. When water is
split by a 4-electron process, however, a potential difference of 1.23 V is required (eq. 7). This corresponds to a wavelength of 1000 nm.

$$H_2O \longrightarrow HO\cdot + H\cdot \quad \text{(Photons required = 252 nm x 1)} \quad (6)$$

$$2H_2O \longrightarrow O_2 + 2H_2 \quad \text{(Photons required =1000 nm x 4)} \quad (7)$$

For the water photolysis with Ru(bpy)$_3^{2+}$, multi-electron processes
are required, since the redox potentials concerned are as follows.

$$Ru(bpy)_3^{3+} + e^- \longrightarrow Ru(bpy)_3^{2+} \quad (E_o' = 1.27\text{ V}) \quad (8)$$

$$Ru(bpy)_3^{3+} + e^- \longrightarrow Ru(bpy)_3^{2+*} \quad (E_o' = -0.83\text{ V}) \quad (9)$$

---

*a) All the values are vs. NHE at pH 7.

A catalyst is needed to achieve the multi-electron process. Metals or metal oxides are suitable candidates. They can be used as highly active colloids or sols when supported on polymers. A water photolysis system was proposed which is composed of $Ru(bpy)_3^{2+}$, methyl-viologen ($\underline{2}$; MV^{2+}) as T_1, platinum colloids as C_1 supported on poly(vinyl alcohol), and ruthenium dioxide sols as C_2 supported on poly(styrene-co-maleic anhydride). This conversion system, represented as Fig.3, has a problem in its reproducibility. The problems involved in this system seem to be, (1) rapid back electron transfer from MV^+ to $Ru(bpy)_3^{3+}$, (2) slow electron donation from H_2O to $Ru(bpy)_3^{3+}$ via RuO_2, and (3) back reaction of the formed O_2 with MV^+. The problems are therefore dynamic rather than thermodynamic. Although a design based on thermodynamics is required in order to construct conversion systems, dynamic considerations also are important to achieve uni-directional electron flow.

$$H_3C-N\overset{+}{\diagdown}\text{—}\diagup\overset{+}{N}-CH_3$$
$$Cl^- \qquad Cl^-$$

$$\underline{2};\ MV^{2+}$$

3) PHOTOINDUCED CHARGE SEPARATION , ELECTRON RELAY, AND CATALYSIS BY METAL-COMPLEXES OF POLYMER SYSTEMS

Polymers can provide micro- and macroheterogeneous reaction fields for the photoinduced charge separation, electron relay, and catalysis. Individual steps in photochemical conversion utilizing polymer systems are described in this chapter.

3-1) Photoinduced charge separation by polymer pendant $Ru(bpy)_3^{2+}$

Polymer pendant $Ru(bpy)_3^{2+}$ was first prepared from poly(styrene) containing 2,2'-bipyridine groups.[10,11] As its analogue, poly-(4-vinylpyridine) pendant bis(2 2'-bipyridine)Ru(II) was prepared, but it is susceptible to photoaquation.[12,13] 4-Methyl-4'-vinyl-2,2'-bipyridine (Vbpy) was prepared from 4-methylpyridine, and it was polymerized and copolymerized with various comonomers to give polymer pendant bpy. These polymers were reacted with cis-$Ru(bpy)_2Cl_2$ to give polymer pendant $Ru(bpy)_3^{2+}$[14-17] (Table 1). The polymer complexes showed slightly higher absorption maxima around 460 nm than the monomeric $Ru(bpy)_3^{2+}$. Their emission maxima were also higher (ca. 610 nm) than the monomeric complex, and their emission intensities were lower than $Ru(bpy)_3^{2+}$.

It is interesting that the solubility of the polymer complex depends on the comonomer. The monomeric complex, $Ru(bpy)_3^{2+}$, is very soluble in water, but almost insoluble in benzene or chloroform. The copolymer complex prepared from styrene was insoluble in water, but quite soluble in benzene or chloroform. This specific solubility of the complex ($\underline{4}$) was very useful for utilizing it as a solid phase or membrane in water as described later.

Table 1.   Polymer Pendant $Ru(bpy)_3^{2+}$

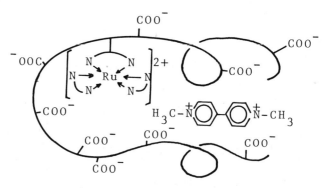

$\{CHCH_2\}_x$ ---------- $\{CHCH_2\}_y$ --- $\{A\}_z$

2+ $N{\frown}N$ = 2,2'-bipyridine

No	Complexes	Comonomer (A)	x	y	z	Ref
<u>3</u>	$Ru[PVbpy](bpy)_2^{2+}$	None	0.190	0.810	–	14,16
<u>4</u>	$Ru[P(St-Vbpy)](bpy)_2^{2+}$	Styrene	0.033	0.033	0.934	14
<u>5</u>	$Ru[P(AA-Vbpy)](bpy)_2^{2+}$	Acrylic acid	0.024	0.037	0.939	15
<u>6</u>	$Ru[P(Vpyr-Vbpy)](bpy)_2^{2+}$	N-vinyl-pyrrolidone	0.022	0.150	0.828	15

Table 2.   Lifetime, Stern-Volmer Constant and Rate Constant of
Electron Transfer from $Ru(bpy)_3^{2+*}$ to MV^{2+} in DMF/H_2O
= 9/1 at 30°C.

Complex	τ(ns)	$k_{sv}(M^{-1})$	$k_q(M^{-1}s^{-1})$
$Ru[P(St-Vbpy)](bpy)_2^{2+}$	434	245	5.65×10^8
$Ru(bpy)_3^{2+}$	350	234	6.69×10^8

Fig.4   Schematic presentation of the electrostatic
effect of polyanionic domains

The reaction of the excited states of the polymer complexes with MV^{2+} was studied. This reaction has been noted as an electron mediator for hydrogen evolution in a water photolysis system. The rate constant of the electron transfer from $Ru(bpy)_3^{2+*}$ to MV^{2+} can be obtained from the Stern-Volmer plots[18] of the emission quenching by MV^{2+} (eq. 10) and the lifetime (τ) of the excited state. The rate constant (k_q) is calculated from the eq. 11.

$$I_o/I = 1 + k_{sv}[MV^{2+}] \qquad (10)$$

I_o, I; emission intensities from $Ru(bpy)_3^{2+*}$ in the absence and presence of MV^{2+}

k_{sv}; Stern-Volmer constant

$$k_q = k_{sv}/\tau \qquad (11)$$

The lifetime of $Ru(bpy)_3^{2+*}$ was measured by a single photon counting method, and the results were shown in Table 2. The copolymer complex (4) showed almost the same reactivity as $Ru(bpy)_3^{2+}$ in its excited state.[14]

The polymer complexes, 3 and 5, also showed almost the same reactivity as $Ru(bpy)_3^{2+}$, but the reactivity of the polymer complex (5) prepared from the acrylic acid copolymer was specifically dependent on pH. The k_q of the electron transfer from the excited 5 to MV^{2+} increased sharply with the increase of pH at around 5 which corresponds to the pK_a of the poly(acrylic acid). In the alkaline region the k_q value was 10 times as high as that of the $Ru(bpy)_3^{2+}$. The high k_q value of 5 must be due to the electrostatic attraction of the positively charged MV^{2+} by the polyanionic domains provided by the dissociated carboxylates. The addition of NaCl to the alkaline solution diminished the k_q of 5 to almost the same magnitude as $Ru(bpy)_3^{2+}$, supporting the electrostatic concentration effect explanation. The electrostatic effect provided by polyanionic domains is schematically presented in Fig. 4.

The photochemical H_2 evolution using these polymer complexes together with a sacrificial reducing agent (Scheme 1), and the dark O_2 evolution with the trivalent polymer complex (scheme 2) were studied as model half reactions of water photolysis.[15,19]

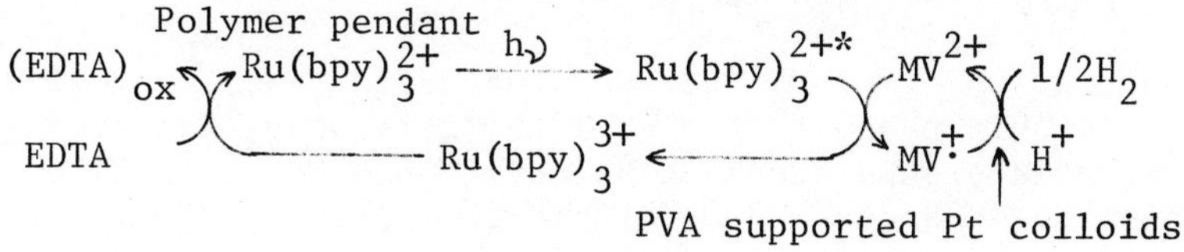

Scheme 1

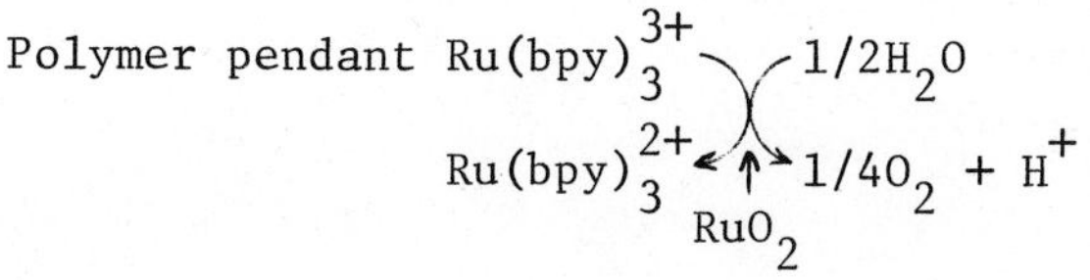

Scheme 2

The ionic domains of polyelectrolytes also affect the photoinduced charge separation between
low molecular weight compounds. The
quenching rate of $Ru(bpy)_3^{2+*}$ by
dibenzyl-sulfonated viologen (7;BSV)
was enhanced about 4 times in the
presence of poly(vinylsulfate) due
to the adsorption of both the compounds onto the polymer.[20]   The
binding of both the Ru complex and
viologen brought about   a   very
efficient quenching of $Ru(bpy)_3^{2+*}$ by
viologen units (see 8).[21]

3-2) Photoinduced electron relay utilizing solid phase of polymers

Heterogeneous phases for photochemical reactions such as the
solid phase or the solid-liquid interface are worth notice. especially from a practical point of view, the use of the solid phase
attracts much attention.   For the purpose of designing solid phase
conversion systems, polymer compounds are the most useful materials.
The utilization of polymer membranes for solar cells will be described in section 5.   In this section, the fundamental aspects of
photoinduced electron relay in the solid phase and at the solid-
liquid interface are briefly described.

It was found that the photoinduced electron relay in eq. 12

$$EDTA \xrightarrow{\ e^-\ } Ru(bpy)_3^{2+*} \xrightarrow{\ e^-\ } MV^{2+} \qquad (12)$$

occurs in the solid phase using cellulose onto which the three components were adsorbed.[22-24]   The electron relay occurs via the
excited state of the Ru complex, and the cellulose turned blue on
irradiation due to the formation of $MV^{+\cdot}$.

The reducing power of $MV^{+\cdot}$ formed at a solid phase can be
transfered to liquid phase.   The chelate resin beads containing
iminodiacetic acid groups adsorbed $Ru(bpy)_3^{2+}$ and MV^{2+}.   The
irradiation of these beads induced rapid formation of $MV^{+\cdot}$ in the
solid phase.[25]   When the beads are present in water containing
air, hydrogen peroxide was formed in the aqueous phase (eq. 13).

$$MV^{+\cdot} + O_2 \longrightarrow MV^{2+} + O_2^- \longrightarrow H_2O_2 \qquad (13)$$

Photoinduced charge separation at solid-liquid interface was studied using polymer pendant $Ru(bpy)_3^{2+}$ [22,25-27] (Table 3).   In order to accumulate $MV^{\ddagger}$ in the relay system of EDTA $\longrightarrow Ru(bpy)_3^{2+} \longrightarrow$ MV^{2+}, the EDTA and $Ru(bpy)_3^{2+}$ should be in the same phase (No.1 and 2 of Table 3).   Otherwise, the EDTA and MV^{2+} should be in the same phase (No.5 under stirring condition of Table 3).   These results can be understood by considering the rate constant of each of the electron transfer steps.   The rate constants of the electron transfer occurring in the aqueous solution are shown in Fig. 5 for reference.[28]   Since EDTA works as a sacrificial reducing agent, the back reaction from EDTA does not need to be considered. Since the back electron transfer from $MV^{\ddagger}$ to $Ru(bpy)_3^{3+}$ is very rapid (k_b), it is important to carry out the electron transfer from EDTA to $Ru(bpy)_3^{3+}$ as fast as possible to facilitate accumulation of $MV^{\ddagger}$. It is desirable therefore that both EDTA and $Ru(bpy)_3^{2+}$ be present in the same phase.   Hydrogen was formed in No.2 of Table 3 when platinum catalyst and protons are present in the solution.[26]

Photoinduced charge separation was also reported in $Ru(bpy)_3^{2+}$ [29] immobilized in a film of the cinnamate ester of poly(vinyl alcohol) and in tetraphenylporphyrin dispersed in resins.[30]

3-3) Polymerized molecular assemblies

Many model reactions have been done on molecular assemblies such as micelles and liposomes.[2,4,6]   They provide a heterogeneous reaction field in water for the photochemical conversion.   The interface at micelle and water, for instance, exhibits a strong coulomb field to facilitate photochemical charge separation.[31-33] Since these assemblies have transient stability, their stabilization with synthetic polymers is attracting attention.   A stable liposome was prepared by the ultra-violet initiated polymerization of liposomes composed of 9 containing α-diazo-β-trifluoropropionyl-oxy group in the hydrophobic section of phosphatidylcholine.[34,35]

Polymerized liposomes were also prepared by polymerizing lipid molecules containing vinyl groups at various positions (10,11 etc.) after they had formed liposomes.[3,36,37]

Table 3. Photoinduced electron transfer at solid-liquid interfaces.

$$D = EDTA, \quad P = [Ru(bpy)_3]^{2+}, \quad A = MV^{2+}$$

☐ = Solid phase

⌐⌐⌐⌐ = Liquid phase (No.2; MeOH,

Others; H_2O)

⟶ = Electron transfer occurred

No.	Reaction system	$MV^{\dotplus}$ accumulated	Reference
1	$\boxed{D \longrightarrow P \longrightarrow A}$	Yes	22, 25
2	$D \longrightarrow P \dashrightarrow A$	Yes	26
3	$D \quad\boxed{P \longrightarrow A}$	No	
4	$D \quad\boxed{P \dashrightarrow A}$	No	26
5	$P \nearrow^{\,D}_{\,A}$	Under standing; No Under stirring; Yes	27

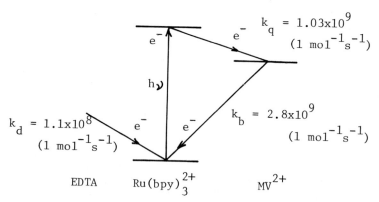

EDTA $Ru(bpy)_3^{2+}$ MV^{2+}

$k_q = 1.03 \times 10^9$ $(1\ mol^{-1}s^{-1})$

$k_b = 2.8 \times 10^9$ $(1\ mol^{-1}s^{-1})$

$k_d = 1.1 \times 10^8$ $(1\ mol^{-1}s^{-1})$

$h\nu$

Fig.5 Rate constants of the electron transfers among
EDTA, $Ru(bpy)_3^{2+}$, its excited state, and MV^{2+}.[28]

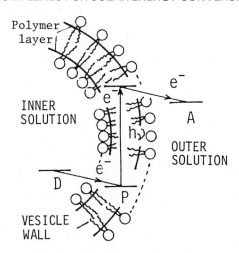

Fig.6 Photoinduced charge separation in polymerized liposome

$$n\text{-}C_{15}H_{31}COO(CH_2)_2$$
$$n\text{-}C_{15}H_{31}COO(CH_2)_2$$ $\text{NCO(CH}_2)_2\overset{+}{N}$ ◯—◯ $\overset{+}{N}CH_2CH{=}CH_2, \ 2Br^-$ __11__

Excitation of $Ru(bpy)_3^{2+}$ placed on the outside of the polymerized liposome composed of __10__ or __11__ resulted in the formation of long-lived reduced viologens on the inside. The polymerized liposome should assist charge separation as illustrated in Fig. 6 when the photoreaction center (P) is incorporated into the vesicle wall. When $Ru(bpy)_3^{2+}$ and EDTA were present in the outer solution and MV^{2+} was present in the inner solution, irradiation induced the accumulation of $MV^{\pm}$ at the quantum yield of 0.26.[3] The presence of PtO_2 in the inner solution brought about H_2 formation. The Pt colloids present in the inner solution of a polymerized liposome were stable and were retained for more than one month.[38]

Polymeric microemulsions with particle diameters of 20 - 40 nm were prepared by polymerizing microemulsions consisting of cetyltrimethylammonium bromide, styrene, and hexanol.[39] These stabilized emulsions also can provide heterogeneous reaction fields for photochemical charge separation.

3-4) Polymer supported metal colloids as catalysts

As described in section 2, the utilization of multi-electron processes is important for solar energy conversion. Metal colloids are promising candidates for this purpose. They are in many cases used after being stabilized with synthetic polymers.

In the model reaction system for photochemical generation of H_2 composed of EDTA, $Ru(bpy)_3^{2+}$, MV^{2+}, and Pt colloids supported on poly(vinyl alcohol)(PVA), smaller particles of Pt gave a higher activity.[40]   The maximum activity was observed at a Pt particle size of 3 nm using PVA or poly(vinylpyrrolidone) as a protective polymer.[41]   The efficiencies of poly(acrylamide), poly(ethtylene-imine), poly(vinylmethylether), PVA, poly(acrylic acid), poly-(vinylpyrrolidone), hydroxyethyl cellulose, gum arabic,   and gelatin as protective polymers for Pt colloids were studied[42] in the same model photochemical H_2 generation system discussed above. The ability to prevent coagulation of the colloids was important. Poly(vinylpyrrolidone) showed the highest efficiency for the H_2 production.   An optimal concentration of Pt colloid was observed for H_2 production.[43]   This is because the Pt colloids also catalyze the hydrogenation of MV^{2+}, which destroys its activity as an electron mediator.

The polymer supported metal colloids still have a problem in their long-term stabilities.   Dispersion of the colloids in solid polymer brings about long-term stability.   Platinum colloids were dispersed by two methods in solid of poly(hydroxyethylmethacrylate) (PHEMA).[44]   HEMA was polymerized in the presence of H_2PtCl_4 and crosslinking agent.   The resulting gel polymer was treated with citric acid to give Pt colloids dispersed in solid PHEMA.   Another method is to reduce H_2PtCl_4, which has been adsorbed in PHEMA gel, with citric acid.   These Pt colloids dispersed in PHEMA gel were stable catalysts in a photochemical system composed of sacrificial reducing agent (EDTA or triethanolamine), $Ru(bpy)_3^{2+}$, and MV^{2+}. They showed a constant activity for more than 96 h's use; their turnover number reached more than 2000, and they could be used repeatedly without any activity loss.

The kinetics of H_2 evolution from colloidal Pt supported on PVA were studied, and the possible role of Pt as a storage pool for hydrogen atoms or electrons was discussed.[45]

4) WATER PHOTOLYSIS BY POLYNUCLEAR METAL COMPLEXES

Polynuclear metal complexes are promising candidates as cat-alysts for multi-electron processes, because they can work   as electron or charge sinks.   It was found that the aqueous system composed of colloidal prussian blue (PB) and $Ru(bpy)_3^{2+}$ splits water under visible light to give H_2 and O_2 simultaneously.[46,47] The PB is a high molecular weight, mixed-valent, polynuclear iron cyanide complex whose composition was reported as $Fe_4^{3+}[Fe^{II}(CN)_6]_3$ (Fig. 7).[48]   Our recent study has revealed, however, that in the aqueous solution it is present as $KFe^{3+}[Fe^{II}(CN)_6]$.[48]   The H_2 and O_2 evolved were analyzed by gas chromatography and mass spectro-metry.   The photolysis of water containing D_2O and $H_2^{18}O$ showed that both gases were generated by water decomposition.[46]

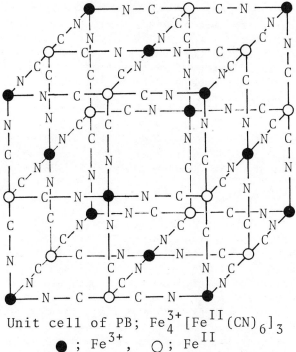

Unit cell of PB; $Fe_4^{3+}[Fe^{II}(CN)_6]_3$

● ; Fe^{3+},   ○ ; Fe^{II}

Fig.7 Crystal structure of PB[48]

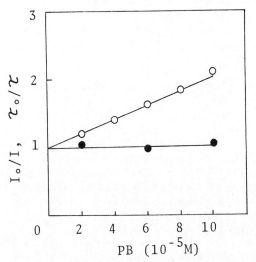

Fig.8 Stern-Volmer plots for the quenching of $Ru(bpy)_3^{2+*}$
by PB based on emission intensity (○) and life-
time (●). $Ru(bpy)_3^{2+}$; 20 μM, KCl 0.5 M, pH 2.

The average diameter of the PB colloidal particles was 23 nm according to electron microscopy. Excitations of both PB and $Ru(bpy)_3^{2+}$ were required for the photolysis to occur. The model half reaction involving EDTA, $Ru(bpy)_3^{2+}$ and PB produced H_2 under visible-light irradiation. The trivalent Ru complex $Ru(bpy)_3^{3+}$, oxidized water in the presence of PB to give O_2. The reaction scheme represented by eqs. 14 - 18 was proposed to account for these results.

$$Ru(bpy)_3^{2+*} + PB \longrightarrow Ru(bpy)_3^{3+} + PB \cdot e^- \qquad (14)$$

$$Ru(bpy)_3^{3+} + PB \cdot e^- \longrightarrow Ru(bpy)_3^{2+} + PB \qquad (15)$$

$$Ru(bpy)_3^{2+*} + PB \cdot e^- \longrightarrow Ru(bpy)_3^{3+} + PB \cdot 2e^- \qquad (16)$$

$$PB \cdot 2e^- + 2 H^+ \xrightarrow{700 \text{ nm}} PB + H_2 \qquad (17)$$

$$2 Ru(bpy)_3^{3+} + H_2O \xrightarrow{PB} 2 Ru(bpy)_3^{2+} + 1/2 O_2 + 2 H^+ \quad (18)$$

The excited state of the Ru complex is quenched by PB oxidatively (eq. 14) to give $Ru(bpy)_3^{3+}$ and reduced PB. Although the back electron transfer would prevail (eq. 15), the colloidal PB particle could be reduced by two electrons (eq. 16). The $PB \cdot 2e^-$ thus formed would reduce two protons on excitation to give H_2 (eq. 17). When four molecules of $Ru(bpy)_3^{3+}$ are accumulated, they would oxidize water with PB catalyst to give O_2 (eq. 18). The excitation of PB (λ_{max} = 700 nm) is due to intervalent charge transfer between the adjacent Fe^{2+} and Fe^{3+} ions through the antibonding π orbitals of the bridging CN ligand.[50] Such adjacent CN antibonding π orbitals in the excited state of PB would participate in the reduction of two protons.

In the present photolysis system the presence of neutral salt such as KCl is necessary. The concentration of KCl showed its optimum point at 0.5 M for the photolysis.[47] The neutral salts such KCl and RbCl whose hydrated cations are smaller than the pore size of the PB lattice (0.35 nm) were effective for the photolysis, but NaCl, LiCl, and $BaCl_2$ whose hydrated cations are larger than 0.35 nm were ineffective. To allow the reduction of PB, a cation(Cat^+) has to enter the PB lattice in order to compensate the negative charge of the electron (eq. 19).

$$(Cat^+)Fe^{III}[Fe^{II}(CN)_6] + e^- + Cat^+ \longrightarrow (Cat^+)_2Fe^{II}[Fe^{II}(CN)_6] \quad (19)$$

It has been shown that the presence of K^+ or Rb^+ is necessary for the electrochemical reduction of PB coated on a Pt electrode.[51].

Quenching of the excited state of $Ru(bpy)_3^{2+}$ by PB was studied using the Stern-Volmer plots based on emission intensities and lifetimes (Fig. 8). [*] The zero slope of the plots based on the lifetime of $Ru(bpy)_3^{2+}$ [*] indicates that only static quenching occurs in this system, and no dynamic quenching occurs. It could be concluded from this result that the Ru complex and PB form a complex in the aqueous solution. Complex formation was supported by NMR spectroscopy. Since PB is negatively charged the complex formation must be due to electrostatic forces. The Ru complex would be adsorbed on the surface of PB particles. The stepwise binding of the two compounds was analyzed according to the Poisson statistics, and the binding constant was estimated.[52] According to this calculation, the average size PB particle (23 nm) binds about 3000 molecules of $Ru(bpy)_3^{2+}$ under typical reaction conditions, $[Ru(bpy)_3^{2+}] = 10~\mu M$ and $[PB] = 1$ mM.

Although O_2 inhibits the rate, the photolysis occurs even under air indicating that the active sites for H_2 formation are fairly protected against O_2 attack. When considering that it is not easy for the O_2 molecule (0.28 x 0.4 nm) to enter into the PB lattice, the reduction sites would be located inside the PB particle. The water oxidation sites would be located on the surface of the colloid, because it should be the coordinated water that undergoes oxidation. The reaction scheme can then be illustrated as Fig. 9. The polynuclear PB colloid thus provides a heterogeneous reaction zone for the photochemical processes.

The potential diagram is showm in Fig. 10. The conversion system might be regarded as a kind of artificial model for photosynthesis since it consists of two excitation steps.

5) SOLAR CELLS

The development of solar cells to produce electricity directly from sunshine is an important subject. The principles of almost all the solar cells are based on semiconductor photophysical processes. Examples are shown in Fig. 11 (a) - (c). The typical example of the p-n junction (a) is the silicon solar cell used commercially. The Schottky junction (b) is formed by p- or n-semiconductor and metal. The use of polymeric materials is attracting notice for constructing the elements of (a) and (b), because polymers are easily molded, light weight, suited for mass production, and therefore can be inexpensive. Doped poly(acetylene) has evoked great interest as a polymeric semiconductor.[53,54] The liquid junction (c) is of interest, because the junction is easily formed simply by dipping a p- or n-semiconductor in an aqueous solution of electrolytes. This junction is important since it can give photochemical products in the liquid phase. It will be described in section 5-2.

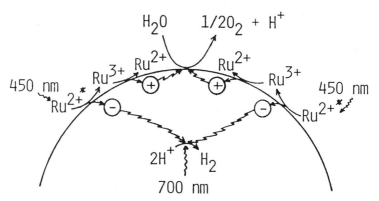

Fig. 9. Illustrated scheme for water photolysis with PB and
Ru(bpy)$_3^{2+}$

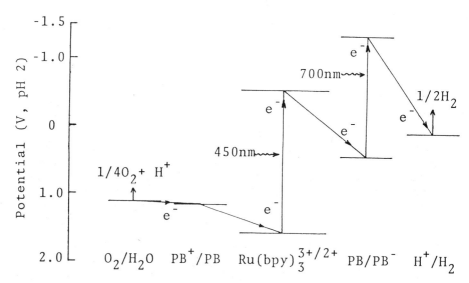

Fig. 10. Potential diagram for water photolysis with Pb
and Ru(bpy)$_3^{2+}$.

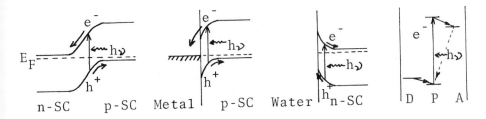

(a) pn junction   (b) Schottky   (c) Liquid   (d) Photo-
                       junction   junction   induced
                                                         reaction

Fig. 11 Photoinduced charge separation based on semiconductor
        (SC) photophysical processes (a - c) and photochemical
        reaction (d). E_f = Fermi level

    Besides these semiconductor junctions, a new photodiode must
be formed based on the photochemical electron transfer reaction (d).
It will be described in section 5-3.

5-1) Photovoltaic cells composed of polymer membranes containing dispersed metal complex

    Sandwich type photochemical cells have been
composed of a thin layer of sensitizer such as
metal-phthalocyanine (M-Pc; 12).   The construction of the cell is mostly; transparent conductive
electrode (e.g., nesa glass)/sensitizer layer/
semitransparent thin layer of metal.   The most
of these photovoltaic effects are due to the
Schottky junction formed at the interface of the
sensitizer and metal.[55,56]

12; M-Pc

    Dispersion of sensitizer in synthetic polymers increases the
conversion efficiency.[57,58] A photovoltaic cell was constructed
which was composed on a thin film (0.5 - 5μ) of poly(vinylcarbazole)
(PVK) containing dispersed x-H_2Pc (Fig. 12.)[57] The conversion efficiency of the cell; nesa glass/x-H_2Pc, PVA/Al, reached a maximum
value of 3.8% at the H_2Pc concentration of ca. 60 wt% under 670 nm
light irradiation. A much higher efficiency of poly(vinylidene-
fluoride) (PVDF) versus PYK suggested the microenvironmental effect
of the polar PVDF. A flexible cell; Al/sensitizer/InSnO$_2$ coated
on polyester film, showed an efficiency of 0.2% under sunlight.[60]

    A MIS (Metal-Insulator-Semiconductor) type cell; nesa glass/
NiPc/poly(ethylene)(PE)/Al, was constructed.[61]   The insulating
PE layer caused the open circuit voltage (V_{oc}) to be 0.65 V higher
higher than that in a cell without PE.

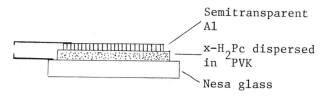

Fig.12   Side view of a sandwich type cell composed of polymer
         membrane containing dispersed H_2Pc

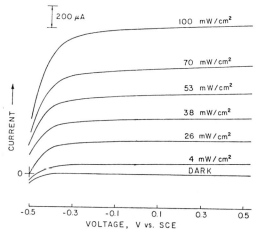

Fig.13   Current-voltage characterisitics of n-GaAs (area:0.09 cm^2)
         coated with polymer pendant Ru(bpy)$_3^{2+}$ (4) in contact with
         0.1 M Fe^{2+}/Fe^{3+} redox electrolytes (pH 2). 100 mV s^{-1}.

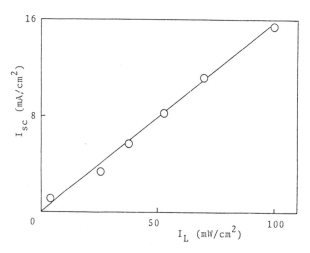

Fig.14   Dependence of I_{sc} on the light intensity (I_L) for n-GaAs
         coated with polymer pendant Ru(bpy)$_3^{2+}$ in Fe^{2+}/Fe^{3+} soln.

These photovoltaic cells have the great advantage of easy
fabrication, but their low conversion efficiency under white light
irradiation is a problem.   This is due to low carrier generation
efficiency and to low carrier mobility.   Although they can be used
for photosensors, these problems must be solved before they can be
used for solar energy conversion.

5-2) Liquid junction photochemical cells composed of semiconductors
     coated with polymer membranes

The UV irradiation of n-type titanium dioxide (TiO_2) electrodes
in aqueous electrolytes induced photocurrent, and, at the same time,
brought about water decomposition to give H_2 at the   Pt counter
electrode and O_2 at the TiO_2.[62]   In these liquid junction type
photochemical cells (Fig. 11 (c)), stable semiconductors are
limited to those whose bandgaps are large, e.g., TiO_2 (3.0 eV),
and ZnO_2 (3.2 eV), etc.   They adsorb only the UV light whose
wavelength is lower than 400 nm.   The narrow bandgap   semiconductors of n-type such as n-Si (1.1 eV), n-GaAs (1.35 eV), and
n-CdS (2.4 eV) are all unstable in water during irradiation.
Stabilization of these semiconductors is therefore important.
The degradation of these narrow bandgap semiconductors in water
is due to the holes formed near the semiconductor surface.   These
holes induce the formation of an inactive oxide layer (n-Si), or
the solubilization of the oxide (n-GaAs).   In order to prevent
these processes, the holes should be transported efficiently   to
the electrolytes in solution, and the semiconductor should be
protected against water.   For these purposes, polymer membranes
have attracted attention.

A poly(pyrrole) film coated on n-Si by electropolymerization
prevented degradation of the silicon under photoanodic conditions,
and induced stable photocurrent.[63]   The poly(pyrrole) film also
protected n-CdS against photocorrosion, and the combination of the
coated n-CdS with Pt counter electrode brought about water photolysis when RuO_2 was dispersed in the poly(pyrrole) film as O_2-evolv-
ing catalyst.[64]

The polymer pendant $Ru(bpy)_3^{2+}$ (4), described before,   also
stabilized n-GaAs in aqueous electrolytes under irradiation when
coated as a thin membrane.[65]   The I-V characteristics   of the
coated n-GaAs in contact with 0.1 M Fe^{2+}/Fe^{3+} redox electrolytes
are shown in Fig. 13.   While bare n-GaAs degraded soon after
irradiation, coated n-GaAs generated a stable photocurrent depending on the light intensity. The linear dependence of the short
circuit photocurrent (I_{sc}) on the light intensity (I_L) (Fig. 14)
indicates that the transportation of the holes from the semiconductor surface to the redox electrolytes through the polymer membrane
is not the limiting factor for the photochemical events.   Good
conversion efficiencies were obtained as shown in Table 4 with a

maximum efficiency of 12% and the fill factor of 0.71 for the Fe^{2+}/ Fe^{3+} electrolytes.

Although it is not certain at present if the redox reaction of the pendant $Ru(bpy)_3^{2+}$ contributes to the transportation of the holes through the membrane, the incorporation of effective redox species into the coated layer is an important subject in this research thrust.

5-3) New photoresponsive elements composed of membranes of polymer-metal complex

The operating principles of most photoresponsive elements are based on semiconductor photophysical processes (Fig. 11 (a) - (c)). But a new type of photoresponsive element may be constructed based on photoinduced electron transfer as shown in Fig.11 (d). For this purpose the arrangement of the photoreaction components is of great importance to realize unidirectional electron flow. The easiest approach to solve this problem is to utilize polymer membranes containing each of the components.

The photoinduced electron transfer between $Ru(bpy)_3^{2+}$ and MV^{2+} is one candidate to construct a photoresponsive element. But, since the back electron transfer from MV^+ to $Ru(bpy)_3^{3+}$ is very rapid, simply dipping the electrode into the solution of both the components does not give any photoresponse at the electrode. When the Ru complex is coated on the electrode as a thin membrane. However, the photoresponse at the electrode can be obtained.[66-69] The coating was made by adsorbing $Ru(bpy)_3^{2+}$ into the membrane composed of a polyanionic polymer such as poly(styrene sulfonate)[66] or Nafion[68] coated on carbon electrode. or by coating polymer pendant $Ru(bpy)_3^{2+}$ (4).[67,69]   Carbon electrodes were rendered photoreponsive by this modification.

The redox response of a basal plane graphite electrode (BPG) coated with the polymer pendant $Ru(bpy)_3^{2+}$ membrane was measured and its cyclic voltammogram is shown in Fig. 15 (a). The reversible redox reaction of $Ru(bpy)_3^{2+/3+}$ occurs at a slightly more positive potential than the $Ru(bpy)_3^{2+}$ present in solution. About 35% of the coating gave the response. The coated electrode responded to the redox reaction of MV^{2+} present in solution (Fig. 15 (b)). The coated electrode gave photoresponse in the presence of MV^{2+} in solution under visible light irradiation producing photocurrent densities of hundreds nA cm^{-2}. The direction of the photocurrent was dependent on the applied electrode potential: a lower electrode potential gave a cathodic photocurrent, and a higher potential gave an anodic photocurrent. The mechanism is shown in Fig. 16. MV^{2+} is considered to be adsorbed in the polymer layer and to react there. At lower potentials, the reaction of $Ru(bpy)_3^{3+}$ at the electrode would prevail, giving cathodic photo-

Table 4.   Performance parameters for the liquid junction
            n-GaAs coated with polymer Ru complex.

Redox couple	V_{oc} (V)	I_{sc} (mA/cm^2)	P_{max} (mW/cm^2)	FF	η (%)
$Fe^{2+/3+}$	1.09	10.89	8.41	0.71	12.0
	(1.10)	(15.11)	(11.54)	(0.69)	(11.5)[a]
$Fe^{2+/3+}$-EDTA	0.47	8.33	1.07	0.27	1.5
I^-/I_3^-	0.72	11.67	2.33	0.28	3.3

P_{max}; Maximum power output,   FF; Fill-factor,

η ; Optical-to-electrical conversion efficiency

Light intensity; 70 mW/cm^2

a) AM1 illumination (100 mW/cm^2)

(a)   (b)

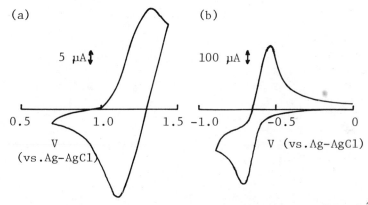

Fig. 15. (a) Cyclic voltammogram of Ru[P(St-Vbpy)](bpy)$_2^2$
             coated BPG on 0.2 M CF$_3$COONa solution (ph 7)
             at the scan rate of 200 mV/s.
         (b) Cyclic voltammogram of MV^{2+} in the same
             solution as (a) at the Ru[P(st-Vbpy)(bpy)$_2^{2+}$
             coated BPG.

(a)   (b)

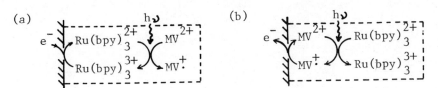

Fig. 16 The mechanism of the photocurrent generation at the
         coated electrode in the presence of MV^{2+} in solution

Table 5. Photocurrents induced by a BPG electrode coated
with either a bilayer or a monolayer (mixture)
membrane of polymer pendant Ru(bpy)$_3^{2+}$ and
viologen in 0.2 M CF$_3$COONa aqueous solution
(pH 2) under argon.

Applied potential	Photocurrent (nA cm^{-2})	
(V vs. Ag-AgCl)	Bilayer system[a]	Monolayer system[b]
-0.2	-238	- 59
0.2	- 72	- 13
0.7	12	159

a) BPG/Polymer Ru(bpy)$_3^{2+}$,0.45 μm/PMV^{2+},0.17 μm

b) BPG/mixture,0.62 μm

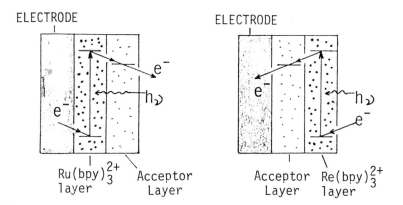

ELECTRODE   ELECTRODE

Ru(bpy)$_3^{2+}$ Acceptor   Acceptor   Re(bpy)$_3^{2+}$
layer   Layer   Layer   layer

Fig.17   New photoresponsive elements based on photochemical
reactions of multilayer coated polymers

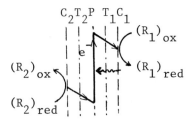

Fig.18   Photochemical conversion element composed of multilayers of thin polymer membranes

currents (a), and at higher potentials, the reaction of MV^+ would prevail giving anodic photocurrents (b). The presence of oxygen caused exclusive cathodic photocurrent generation, and the magnitude was enhanced about 100 times.

When MV^{2+} also was coated on the top of the complex layer as a polymer pendant compound to give a bilayer coated electrode, the selectivity for the cathodic photocurrent generation increased. These results are shown in Table 5 in comparison with those given by a monolayer coated electrode where both the components were coated as a mixture. The bilayer system showed selective cathodic photocurrent generation about 50 times as high as that given by the monolayer coated electrode.[69]

Thus, new photoresponsive elements based on photochemical reactions can be constructed by utilizing multilayer polymer membranes containing photoreactive components (Fig. 17). In order to increase the conversion efficiency, improved photoreaction components and more sophisticated arrangements of them must be devised.

6) CONCLUSIONS AND FUTURE SCOPE

Solar energy conversion systems utilizing polymer-metal complexes have been described. Polymer-metal complexes are attractive materials for constructing photochemical conversion systems and solar cells.

Metal complexes are indispensible as photoreaction centers, electron mediators or catalysts for solar energy conversion. An adequate combination and arrangement of the reaction components are essential requirements for constructing photochemical conversion systems. Polymers will undoubtedly be useful for this purpose and also for industrial production.

The molecular arrangement of the necessary components in a polymer matrix could be realized in the future as now exists in nature for photosynthesis. But, when considering the relevant difficult problems to be solved and also the practical aspects concerned, the nearest approach at present would be to utilize multilayers of ultra-thin polymer membranes as shown in Fig. 18. Such an element can be used for the production of both fuels and electricity. For fuel production, systems composed of microparticles containing the necessary functions in multiple polymer layers would be promising.

The authors are grateful to Prof. C. U. Pittman, Jr. for his valuable suggestions and discussions.

REFERENCES

1.   M. Kaneko and A. Yamada, Adv. Polym. Sci., 55 (1983) (in press).

2.   M. Grätzel, Acc. Chem. Res., 14, 376 (1981).

3.   J. -M. Lehn, Commentarii, 3, 1 (1982).

4.   T. K. Foreman, W. M. Sobol and D. G. Whitten, J. Am. Chem. Soc., 103, 5333 (1981).

5.   M. Calvin, Can. J. Chem., 61, 873 (1983).

6.   J. S. Connoly, Ed., "Photochemical Conversion and Storage of Solar Energy", Academic Press, 1981.

7.   J. H. Fendler, J. Photochem., 17, 303 (1981).

8.   M. Kaneko and A. Yamada, Kobunshi, 28, 85 (1979); Symp. Unsolved Problems on Polym. Chem. (Soc. Polym. Sci. Jpn.), 1976, p. 21.

9.   K. Kalyanasundaram and M. Grätzel, Angew. Chem., 91, 759 (1979).

10.   M. Kaneko, S. Nemoto, A. Yamada and Y. Kurimura, Inorg. Chim. Acta, 44, L289 (1980).

11.   M. Kaneko, A. Yamada and Y. Kurimura, ibid. 45, L73 (1980).

12.   J. M. Klear, J. M. Kelly, D. C. Pepper and J. G. Vos, ibid. 33, L139 (1979).

13.   T. Shimidzu, K. Izaki, Y. Akai and T. Iyoda, Polym. J., 13, 889 (1981).

14.   M. Kaneko, A. Yamada, E. Tsuchida and Y. Kurimura, J. Polym. Sci. Polym. Lett. Ed., 20, 593 (1982).

15.   M. Kaneko and A. Yamada, 4th Intern. Conf. Photochem. Conversion and Storage of Solar Energy, 1982, p. 227.

16.   M. Kaneko and A. Yamada, Polym. Prepr. Jpn., 31, 1665 (1982).

17.   M. Furue, K. Sumi and S. Nozakura, J. polym. Sci. Polym. Lett. Ed., 20, 291 (1982).

18.   O. Stern and M. Volmer, Phys. Z., 20, 183 (1919).

19.   M. Kaneko, N. Awaya and A. Yamada, Chem. Lett., 1982, 619.

20.   R. E. Sassoon and J. Rabani, Israel J. Chem., 22, 138 (1982).

21.   T. Matsuo, T. Sakamoto, K. Takuma, K. Sakura and T. Ohsako, J. Phys. Chem., 85, 1277 (1981).

22.   M. Kaneko, J. Motoyoshi and A. Yamada, Nature, 285, 468 (1980).

23.   M. Kaneko and A. Yamada, Makromol. Chem., 182, 1111 (1981).

24.   M. Kaneko and A. Yamada, Photochem. Photobiol., 33, 793 (1981).

25.   Y. Kurimura, M. Nagashima, K. Takato, E. Tsuchida, M. Kaneko and A. Yamada, J. Phys. Chem., 86, 2432 (1982).

26.   M. Kaneko, M. Ochiai, K. Kinosita, Jr. and A. Yamada, J. Polym. Sci. Polym. Chem. Ed., 20, 1011 (1982).

27.   E. Tsuchida, H. Nishide, N. Shimidzu, A. Yamada, M. Kaneko and Y. Kurimura, Makromol. Chem. Rapid Commun., 2, 621 (1981).

28. P. Keller, A. Moradpour, E. Amouyal and H. B. Kagan,
Nouv. J. Chim., 4, 377 (1980).
29. W. Kawai, Kobunshi Ronbunshu, 37, 303 (1980); 38, 451
(1981).
30. K. Kojima, T. Nakahira, Y. Kosuge, M.Saito and S. Iwabuchi,
J. Polym. Sci. Polym. Lett. Ed., 19, 193 (1981).
31. S. C. Wallace, M. Grätzel and J. K. Thomas, Chem. Phys.
Lett., 23, 359 (1973).
32. Y. Tsutsui, K. Takuma, T. Nishijima and T. Matsuo, Chem.
Lett., 1979, 617.
33. P. -A. Brugger, P. P. Infelta, A. M. Braun, and M. Grätzel,
J. Am. Chem. Soc., 103, 320 (1981).
34. C. M. Gupta, R. Radhakrishnan and H. G. Khorana, Proc.
Natl. Acad. Sci. USA, 74, 4315 (1977).
35. C. M. Gupta, R. Radhakrishnan, G. E. Gerber, W. L. Olsen,
S. C. Quay and H. G. Khorana, ibid., 76, 2595 (1979).
36. A. Akimoto, K. Dorn, L. Gross, H. Ringsdorf and H. Schupp,
Angew. Chem., 93, 108 (1981).
37. P. Tundo, D. J. Kippenberger, M. J. Politi, P. Klahn and
J. H. Fendler, J. Am. Chem. Soc., 104, 5352 (1982).
38. K. Kurihara and J. H. Fendler, American Chemical Society,
Symp. Inorg. Organometal. Polym., 1983, p.203.
39. S. S. Atik and J. K. Thomas, J. Am. Chem. Soc., 103,
4279 (1981); 104, 5868 (1982).
40. J. Kiwi and M. Grätzel, Nature, 281, 657 (1979).
41. N. Toshima, M. Kuriyama, Y. Yamada and H. Hirai, Chem.
Lett., 1981, 793.
42. N. Toshima, Y. Yamada and H. Hirai, Polym. Prepr. Jpn.,
30, 416, 1500 (1981).
43. P. Keller, A. Moradpour, E. Amouyal and H. Kagan, J. Mol.
Catal., 7, 539 (1980).
44. M. Akashi, T. Motomura, N. Miyauchi, K. F. O'Driscoll
and G. L. Rempel, Polym. Prepr. Jpn., 30, 1496 (1981).
45. M. S. Matheson, P. C. Lee, D. Meisel and E. Pelizetti,
J. Phys. Chem., 87, 394 (1983).
46. M. Kaneko, N. Takabayashi and A. Yamada, Chem. Lett.,
1982, 1647.
47. M. Kaneko, N. Takabayashi, Y. Yamauchi and A. Yamada,
Bull. Chem. Soc. Jpn. (1984) (in press).
48. H. J. Buser, D. Schwarzenbach, W. Petter and A. Ludi,
Inorg. Chem., 16, 2704 (1977).
49. M. Kaneko and A. Yamada, unpublished data.
50. M. B. Robin, Inorg. Chem., 1, 337 (1962).
51. K. Itaya, A. Ataka and S. Toshima, J. Am. Chem. Soc.,
104, 4767 (1982).
52. M. Kaneko and A. Yamada, to be submitted.
53. H. Shirakawa, E. J. Louis, A. G. MacDiarmid, C. K. Chiang
and A. J. Heeger, J. Chem. Soc., Chem. Commun., 1977, 578.
54. C. K. Chiang, M. A. Druy, S. C. Gau, A. J. Heeger, E. J.
Louis, A. G. MacDiarmid, Y. K. Park and H. Shirakawa, J. Am. Chem.
    Soc., 100, 1013 (1978).

55. A. K. Ghosh, D. L. Morel, T. Feng, R. F. Shaw and C. A. Rowe, Jr., J. Appl. Phys., 45, 230 (1974).

56. F. J. Kampus and M. Gouterman, J. Phys. Chem., 81, 690 (1977).

57. R. O. Loufty and J. H. Sharp, J. Chem. Phys., 71, 1211 (1979).

58. R. O. Loufty, J. H. Sharp, C. K. Hsiao and R. Ho, J. Appl. Phys., 52, 5218 (1981).

59. N. Minami and K. Sasaki, Chemical Society Jpn., Autumn Meeting. 1981, p. 1502.

60. T. Moriizumi and K. Kudo, Appl. Phys. Lett., 38, 85 (1981).

61. K. Misoh, S. Tasaka, S. Miyata, H. Sasabe, A. Yamada and T. Tanno, Polym. Prepr. Jpn., 31, 384 (1982).

62. A. Fujishima and K. Honda, Nature, 238, 37 (1972).

63. R. Noufi, A. J. Frank and A. J. Nozik, J. Am. Chem. Soc., 103, 1849 (1981).

64. A. J. Frank and K. Honda, J. Phys. Chem., 86, 1933 (1982).

65. K. Rajeshwar, M. Kaneko and A. Yamada, J. Electrochem. Soc., 130, 38 (1983).

66. M. Kaneko, M. Ochiai and A. Yamada, Makromol. Chem. Rapid Commun., 3, 299 (1982).

67. M. Kaneko, A. Yamada, N. Oyama and S. Yamaguchi, ibid., 3, 769 (1982).

68. N. Oyama, S. Yamaguchi, M. Kaneko and A. Yamada, J. Electroanal. Chem., 139, 215 (1982).

69. M. Kaneko, S. Moriya, A. Yamada, H. Yamamoto and N. Oyama, Electrochim. Acta, 29, 115 (1984).

CONDUCTING ORGANOMETALLIC POLYMERS AND ORGANOMETALLIC

COMPLEXES CONTAINING M(CO)$_3$ (DIENE) GROUPS

Hajime Yasuda, Ippei Noda, Yoshitsugu Morita,
Hitoshi Nakamura, Seiichi Miyanaga, and
Akira Nakamura

Department of Macromolecular Science
Faculty of Science, Osaka University
Toyonaka, Osaka 560, Japan

I. INTRODUCTION

A wide variety of organometallic polymers such as homo- and
copolymers of vinylmetallocenes,[1] vinyl-η^5-cyclopentadienyl-
(carbonyl)metals[2] and vinyl-η^6-benzene(carbonyl)metals[3] have been
synthesized in this decade seeking for novel conducting materials,
thermo-oxidatively stable materials, functional catalysts, adhesives and molding compounds for inorganic compounds, etc.
This paper focuses on the preparation of the novel type
conducting organometallic complexes.
Since mixed valence (bisfulvalenediiron)$^+$(TCNQ)$^-$ (1) is known
to show an extremely high electrical conductivity ($\sigma=10^2$ Scm^{-1}) in
the crystalline state,[4] the synthesis of mixed valence [Fe^{II}Fe^{III}]
poly(vinylmetallocene) has attracted special attention for the
development of practically useful conducting flexible films or
moldings. A TCNQ charge transfer complex of poly(3-
vinylbisfulvalenediiron) (2) indeed shows a fairly good conduct-
ivity(6-9 x 10^{-3} Scm^{-1}) as reported by Pittman.[5]

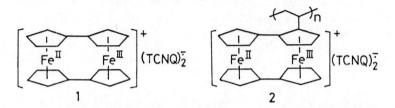

1   2

However, mixed-valence biferrocene, poly(vinylbiferrocene)[6] and doped poly(vinylferrocene)[7], poly(ferrocenylmethyl acrylate) exhibit a quite low conductivity ($10^{-6} \sim 10^{-9}$ Scm^{-1}) contrary to expectation. This is presumably due to the poor stacking or nonuniform spacing of the ferrocene or biferrocene groups in the solid state and/or the occurrence of undesirable oxidation reactions during the doping reaction with halogen, Lewis acids or protic acids. Actually poly(ferrocenylmethyl methacrylate) immediately decomposed by the addition of I_2 to give an insoluble black powdery compound both in the solid state and in solution. The M-C bond cleavage of ferrocene also occurred rapidly when I_2 was added to the ether or the CH_2Cl_2 solution at ambient temperature. Therefore the appropriate molecular design of the organometallic polymers which suppresses or controls the oxidation reaction is essential to obtain desirable conducting organometallic polymers. A systematic survey is necessary to find out the organometallic complexes which are sufficiently resistant to undesirable oxidative decomposition reactions.

    The main purpose of the present work is to synthesize soluble electrical conducting polymer materials or photoconductors which are readily cast into flexible thin films. Though there exist several highly conducting quasi-one-dimensional compounds composed of [IrCl$_2$(CO)$_2$], [Pt(CN)$_4$] etc, casting into film is generally very difficult for these polymers.

II. MOLECULAR DESIGN OF ORGANOMETALLIC POLYMERS

  1. Polymerization of Fe(CO)$_3$(3-allyl-1,4-η^4-1,3-pentadiene)

    We have examined the preparation of soluble linear polymers containing M(CO$_3$)(η^4-diene) groups (M=Fe, Ru) at the beginning of this work. Preparation of some of related polymers such as poly[Fe(CO)$_3$(η^4-2,4-hexadiene-1-ylacrylate)] has been reported.[8] The use of a M(CO)$_3$(η^4-diene) fragment has the following advantages.
  1) A series of (η^4-diene)metal complexes with different kinds of ligands and metals is readily available.
  2) The unit is generally inert to air, moisture and sun light under the normal conditions.
  3) This unit is known to react with HCl, HBF$_4$, or other acids without accompanying any undesirable decomposition reactions.
    As a typical example, preparation and polymerization of Fe(CO)$_3$(3-allyl-1,4-η^4-pentadiene) (3) was examined, starting from 3-vinyl-1,5-hexadiene which was obtained from pentadienyl potassium and allyl bromide.[9,10] Cationic polymerization of 3 with BF$_3$·OEt$_2$ (2 mol% of monomer) or SnCl$_4$ in CH$_2$Cl$_2$ at -20°C for 10 h followed by quenching in MeOH gave a pale-yellow powdery polymer (4) in 35 and 15% yield, respectively.[11] Anionic and radical initiators were not effective for polymerization of 3. The degree of polymerization was determined to be ca. 20 based upon the molecular weight measurement of the organic polymer

$$\text{(1)}$$

fraction ($\overline{M}n=1,900$) obtained after the removal of $Fe(CO)_3$ group from 4 with $(NH_4)_2Ce(NO_3)_6$. Though the present preparation method is straightforward, it has a conclusive disadvantage that the re-producibility of the polymer yield is very poor when appropriate precautions are not taken. This is due to the contamination of complex 5 formed by the isomerization of 3 during distillation. Preparation of pure samples of 5 could be attained by reaction of 3-vinyl-1,5-hexadiene or reaction of 3 with excess $Fe_3(CO)_{12}$ at high temperature (110–140°C) for a long period (40 h). The

$$\text{(2)}$$

complex 5 possesses the remarkable stability due to the conjugation of the three double bonds. Moreover, complex 5 behaved as an inhibitor of polymerization; i.e., when 3-6 mol% of 5 was added to 3, no polymerization occurred with any catalysts tested in this work (AIBN, BuLi, $TiCl_4/AlEt_3$, $BF_3 \cdot OEt_2$, $SnCl_4$). Stoichiometric reaction of 5 with these catalysts or the chain transfer of the growing end of the polymer to 5 should occur in preference to the propagation reaction. Quite similar behavior is known also for $Fe(CO)_3$(1,3,5-hexatriene).

    Above results gave following information:

1) $Fe(CO)_3(\eta^4$-diene) group is relatively stable toward cationic initiators such as Lewis acids.

2) Anionic initiators readily react with $Fe(CO)_3$ fragment of 3 to form $RFe(CO)_3^-M^+$(M=Na,Li) prior to initiation of the polymerization. Refer to reaction of iron carbonyl complexes with bases reported in literature.[12] Anionic initiators are unsuitable also for the polymerization of $Fe(CO)_3(\eta^4$-2,4-hexadienylacrylate) as reported by Hirao et al.[8]

3) Radical polymerization of 3 is possible in principle based upon the Q, e values (0.73, -1.47, respectively) determined by copolymerization with styrene.[11] However, the homo-polymerization of 3 with AIBN was unsuccessful. Oxidation of the $Fe(CO)_3$ group or chain transfer of the growing end

radical to monomer or solvent precludes the propagation
reaction in polymerization.

4) Vinyl group of the monomer must be suitably apart from the
$M(CO)_3(\eta^4$-diene) group to inhibit the isomerization reaction
leading to a stable conjugated $M(CO)_3$(triene) species. Some
examples of the desirable monomers are shown in eq. 3.

$$X=CH_2,O,S \quad n=2,3,4$$

2. Preparation and Polymerization of Vinylethers Containing Fe(CO)₃(diene) or Ru(CO)₃(diene) Group.

The chemical behavior of above mentioned $Fe(CO)_3(3$-allyl-η^4-
pentadiene) indicates that the use of a cationic initiator is
most appropriate for the polymerization of monomers containing
$M(CO)_3$ group. One of the representative monomers suited for this
purpose will be vinylethers involving $M(CO)_3$ group where the
vinyloxy group is apart from the metal center. Based upon this
presumption, we have examined the preparation of iron(tricarbonyl)-
[3-(2-vinyloxyethyl)-1,3-pentadiene)] (9) according to the
procedure given in eqs. 4 and 5.

When the reaction was carried out in a non-polar solvent
(hexane, benzene or ether), ca. 6:1:3 mixture of 6, 7, and 8 was
obtained, while the reaction in THF gave the desired compound 7

Table 1. Yield of poly[Fe(CO)$_3$(3-vinyloxyethyl-1,3-pentadiene)] (10)†

| Polymerization | Initiators | | |
Temperature(°C)	BF$_3$·OEt$_2$	SnCl$_4$	TiCl$_4$
−20	53	18	16
0	62	24	19
30	24	32	24

$\dagger$Polymerization; in CH$_2$Cl$_2$ for 20 h, cat.; 2 mol%.

preferentially (98% purity) in 70% yield. The formation of 7 is the result of the base-catalyzed isomerization of 6. The reaction of 7 with Fe$_2$(CO)$_9$ in refluxing isooctane for 10 h followed by distillation (95 °C/0.5 mm Hg) gave the (E)-isomer of complex 9 in 93% yield. The reaction of 6 with Fe$_3$(CO)$_{12}$ also gave the same complex in 73% yield as yellow oil. The (E)-structure was confirmed by the ^{1}H-NMR spectrum (Fig. 1) with reference to the (Z)- and (E)-isomers of (1,3-pentadiene)Fe(CO)$_3$.[13]

The reaction of 6 or 7 with Ru$_3$(CO)$_{12}$ underwent in refluxing[14] isooctane to give a monomer 11 in 65% yield (115°C/0.2 mm Hg). Though the preparation of Ru(CO)$_3$(diene) is generally far more difficult than the preparation of Fe(CO)$_3$(diene), the desired monomer could be obtained without C-C bond cleavage or the formation of ill-defined cluster compounds. The use of 3,4-dialkyl-substituted diene is crucial for obtaining such a diene complex of Ru in good yield.[14]

BF$_3$·OEt$_2$ catalyzed cationic polymerization of 9 took place successfully and a THF- or benzene-soluble linear polymer (10) was obtained in good yield as shown in Table 1. Relatively large amounts of catalyst and high polymerization temperature were required compared to the conventional cationic polymerization of alkyl vinylethers. Polymerization of 11 was also achieved with BF$_3$·OEt$_2$ under the same polymerization conditions as those for 9. Molecular weights of polymers 10 and 12 determined by a gel permeation chromatograph (column HSG 40-15, eluent THF) was $\overline{M}$n=21,000 and 15,000, respectively, and Mn/Mw for 10 and 12 were 1.5 and 2.4 respectively. Microstructure analysis of these polymers was carried out in terms of ^{13}C NMR spectrum (25.1 MHz). ^{1}H NMR spectrum is useless because the CH$_2$ or CH proton signals overlap Me or H signals of M(CO)$_3$(1,3-pentadiene) group (Fig. 1). Based upon the ^{13}C chemical shift values, the ratios of meso and racemo sequences for 10 and 12 were determined to be 86/14 and 62/38, respectively, with reference to the literature (Fig. 2).[15] Thus, polymer 10 was found to be highly isotactic. The low stereoregularity of 12 may be derived from the large covalent radius of Ru atom.

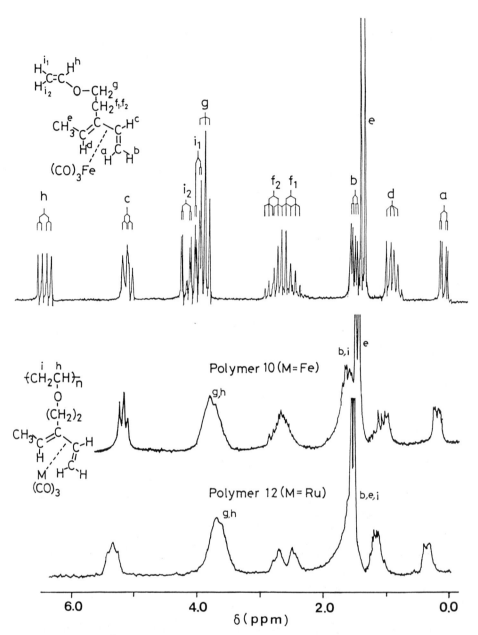

Fig. 1. ^1H NMR spectra of monomer 9 and polymers 10, 12 in $CDCl_3$ at $30°C$(100 MHz).

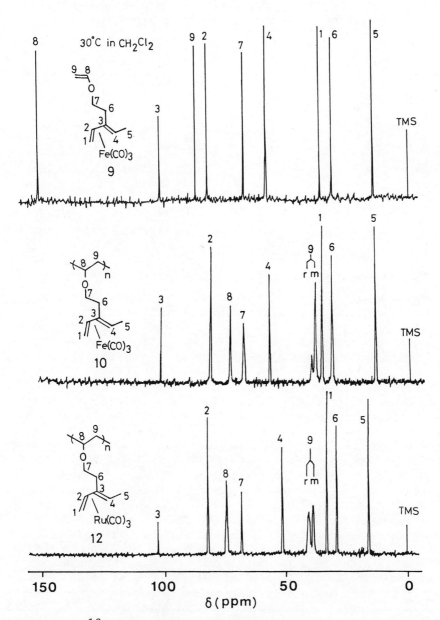

Fig. 2. ^{13}C NMR spectra of monomer 9 and polymers 10, 12 in CDCl$_3$ at 30°C(25.1 MHz).

Table 2. Apparent conductivities for undoped and doped
homopolymers, copolymers and polymer blendings

Complexes	Ratio of dopant to metal complex	σ_{DC} (S cm^{-1})
Polymer 10	Undoped	$< 1.0 \times 10^{-10}$
Polymer 10	0.18 (I_2/Fe)	3.2×10^{-3}
Polymer 10	0.34 (PF$_6$/Fe)	6.6×10^{-4}
Polymer 12	Undoped	$< 1.0 \times 10^{-10}$
Polymer 12	0.35 (I_2/Ru)	1.4×10^{-3}
Polymer 13	0.15 (I_2/Fe)	1.3×10^{-4}
Copolymer(13 /polyMMA=1/1.2)	0.10 (I_2/Fe)	1.3×10^{-5}
Copolymer(13 /polyMMA=1/5.8)	0.19 (I_2/Fe)	1.4×10^{-5}
Poly(styrene) + Complex 22	Undoped	$< 1.0 \times 10^{-10}$
Poly(styrene) + Complex 22	0.45 (I_2/Fe)	5.0×10^{-6}
Poly(MMA) + Complex 22	0.37 (I_2/Fe)	6.7×10^{-6}

III. CONDUCTING ORGANOMETALLIC POLYMERS

Polymers 10 and 12 can be readily cast into flexible thin films
by evaporation of the CH_2Cl_2 solution. Any chemical change or
denature was not observed on the film upon storage in the air at
ordinary temperature for > 20 days. When doping with I_2 was
conducted on the film of 10 (thickness 0.1 - 0.15 mm) by sublimation or dipping in ether solution of iodine, the direct current
conductivity (σ_{DC}) at 25°C (initially $< 10^{-10}$ Scm^{-1}) increased
rapidly in the range of I_2/Fe=0.02~0.2 (mol/mol) to reach the
level of $\sigma_{DC}=10^{-3}$ Scm^{-1}, one of the highest ever observed value
for organometallic polymers(Fig. 3). The color of the film
turned into black on doping. The conductivity leveled off at
I_2/Fe=0.2 and the addition of a large amount of I_2 (I_2/Fe=2.0~4.0)
resulted in a decrease in conductivity. Doping with NoSbF$_6$
in nitromethane/CH_2Cl_2 is also effective. The ratio of I_2/Fe or
PF$_6$/Fe given in Table 2 was determined by gravimetry and expressed
by molar ratio of total amount of dopant to complex. The maximum
room temperature conductivity of polymer 12 (I_2/Ru=0.35) is nearly
the same as that of polymer 10, suggesting that difference in
microtacticity does not affect greatly   the apparent conductivity. The effect of the crystallinity of the polymer is
unambiguous from the present work. When the doped polymer 10 or
12 was held in vacuum or on exposure in the air (5 days), initial
polymer could be recovered in good yield.

Based upon the experiences obtained from polymer 10 and 12,

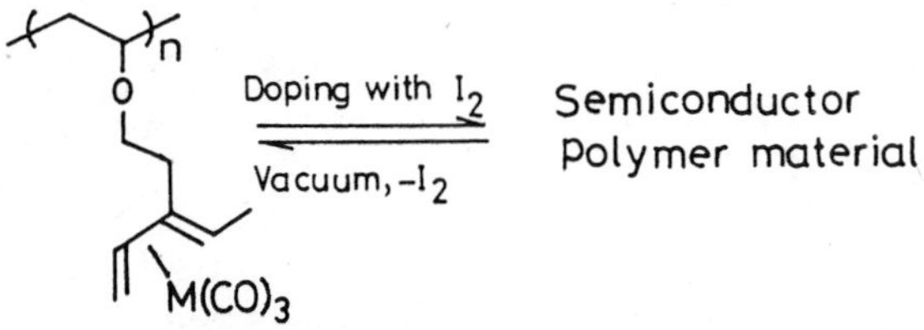

Doping with I_2 / Vacuum, $-I_2$ → Semiconductor Polymer material

the measurement of conductivity was applied to poly[(Fe(CO)$_3$(η^4-2,4-hexadienylacrylate)] (13) which was prepared according to the known method.[8] Polymer 13 also showed an enhanced conductivity, in the region of I_2/Fe=0.03~0.2 when sublimation of I_2 was conducted on the film (thickness 0.15 mm).(Table 2) The apparent conductivity measured in this work is nearly equal to the volume conductivity as evidenced by the measurement of volume and surface current with 3-terminal electrodes. In contrast to above polymers, ferrocene-containing polymers such as poly(ferrocenyl-methyl acrylate), poly(ferrocenylmethyl methacrylate) (14) and poly(1-ferrocenylethyl methacrylate)(15) showed only low conductivity when I_2 was doped on the film, in line with the low conductivity (σ=10^{-8} Scm^{-1}) observed for doped poly(vinyl-ferrocene).[7]

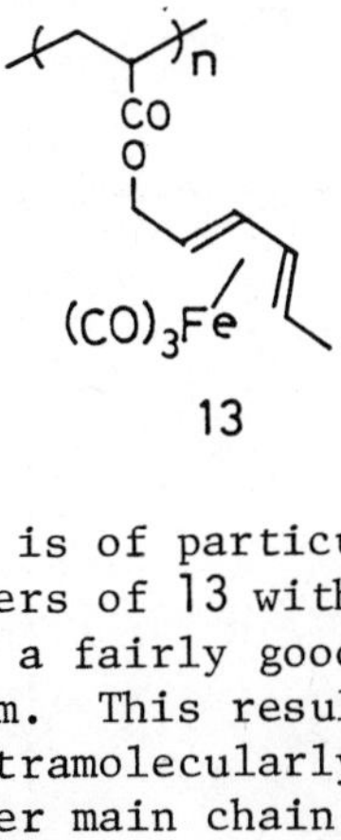

13

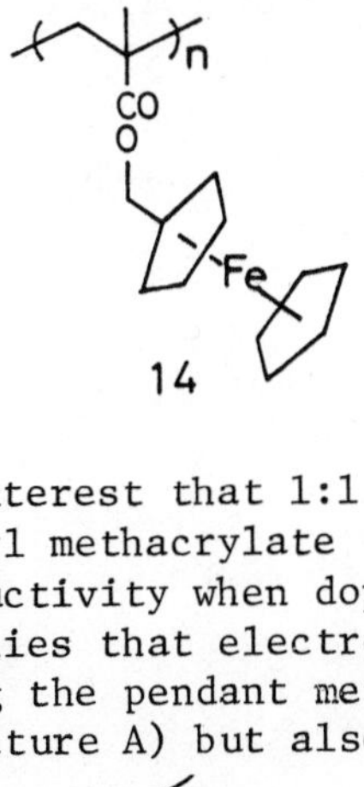

14

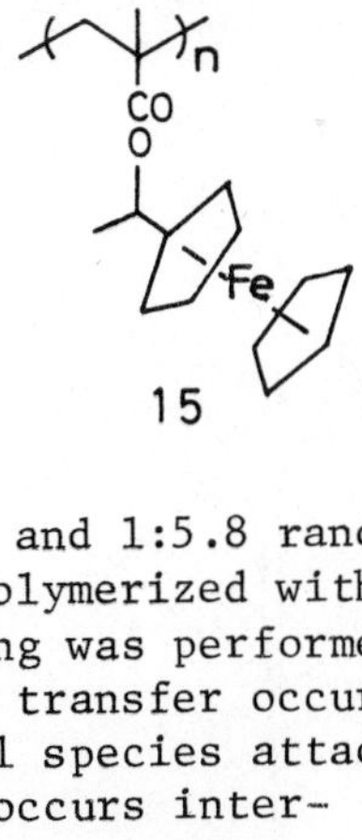

15

It is of particular interest that 1:1.2 and 1:5.8 random copolymers of 13 with methyl methacrylate (polymerized with AIBN) exhibit a fairly good conductivity when doping was performed on the film. This result implies that electron transfer occurs not only intramolecularly along the pendant metal species attached to a polymer main chain (structure A) but also occurs inter-

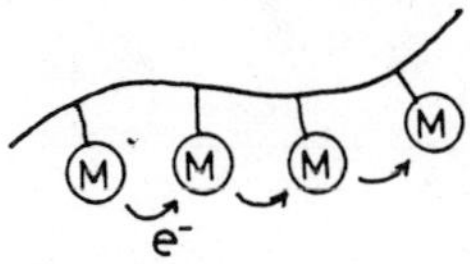

A

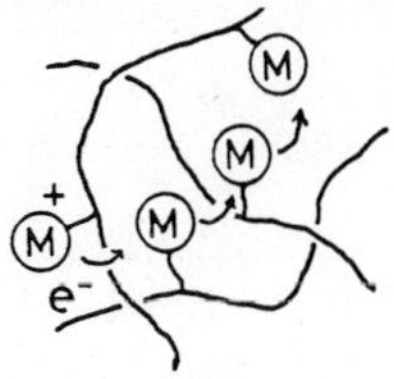

B

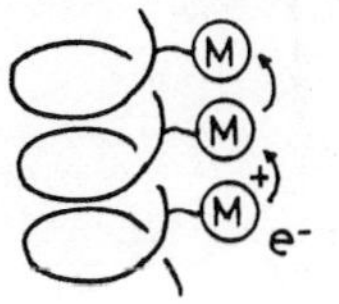

C

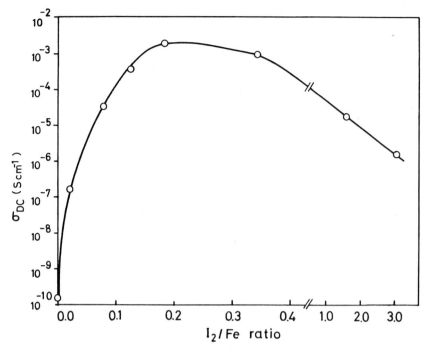

Fig. 3.   Dependence of dopant(I_2) concentration on the
conductivity of polymer 10.

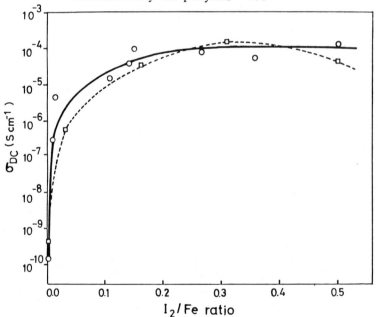

Fig. 4.   Apparent conductivity of complex 19 (—O—) and
and complex 21 (--□--) as a function of dopant
concentration.

molecularly through nonuniformly spaced metal species in different
polymer molecules (structure B) since the average distance of
adjacent two metal species (3–10Å) in structure A is estimated to
be longer than that in structure B. If the present homo- or
copolymer has helix coil structure, a through–space interaction
parallel along the helix will be also possible (structure C).

Above result suggests that conducting polymer material will be
available even by blending or mixing $M(CO)_3(\eta^4\text{-diene})$ complexes
with conventional organic polymers. Actually $Fe(CO)_3$bis(cyclo-
octadienyl) dispersed in poly(styrene) or poly(methyl methacrylate)
exhibits enhanced conductivity of the level of 10^{-6} Scm^{-1} at the
ratio of monomer unit/Fe=3–5 after doping with iodine (I_2/Fe=0.15).
Thus, the blending of organometallic complexes in organic
polymers is a valuable procedure to get semiconductor devices.
The advantage of this procedure lies in
1) Ease in preparation of samples. Mixing in solution followed
   by evaporation of solvent is favorable over the blending in
   the solid state.
2) Wide variety of combinations is possible because many kinds
   of metal complexes and organic polymer are readily available.
3) Facility in molding into desired shape of sample.
The disadvantage of the blending method lies in the
following:
1) Blending of metal complexes in poly(styrene) or poly(MMA) can
   be applied only for solid $M(CO)_3$(diene) complexes. Homogeneous mixing of oily $Fe(CO)_3$(butadiene) or $Fe(CO)_3$(isoprene)
   with poly(MMA) is difficult due to the occurrence of phase
   separation during the evaporation of a highly concentrated
   polymer solution.
2) Homogeneous dispersion of solid metal complexes is also
   difficult because microcrystals or sometimes large crystals
   of metal complexes precipitate on the film during casting
   into film.
Presence of low percentages of metal complexes (< 5 mol% per
monomer unit of polymer) in organic polymers is generally insufficient to get good semiconductor materials.

IV. CHEMICAL CHANGE DURING DOPING ON THE POLYMER

To elucidate the unique chemical structure of metal species
which is responsible for the electron conduction, chemical
characterization of a doped polymer was carried out. When polymer
10 was mixed with 2 equiv. of I_2 in solution (CH_2Cl_2), the polymer
was converted into diamagnetic black materials. The initial IR
absorption assigned to the CO stretching vibration of 10 (2060,
1987, 1970 cm^{-1}) shifted to higher wave number (2110, 2050 cm^{-1})
after doping, suggesting that the $Fe(CO_3)$(diene) group of 10
converted to $FeI(CO)_3(\eta^3\text{-allyl})$ species (17). Similar reaction
is known for the reaction of $Fe(CO)_3(\eta^4\text{-butadiene})$ with HX (X=Br,
Cl, BF_4, PF_6) to give $Fe^{II}(CO)_3X(\eta^3\text{-butenyl})$ where X=Br, Cl or

$[Fe(CO)_3(\eta^3\text{-butenyl})]^+X^-$ (20).[16][17] The reaction of polymer
10 with anhydrous HCl at -20°C also gave the polymer (16) containing $FeCl(CO)_3(\eta^3\text{-2-butenyl})$ species. The CO absorptions of
16 appeared at 2080, 2035, 2005 cm^{-1} which are close to 2100, 2030
cm^{-1} of complex 19. Thus, oxidation of polymer 10 with I_2 or HCl

(6)

(7)

occurred successfully to give polymers containing π-allylic Fe(II)
species, without accompanying significant decomposition to Fe(III)
species. The CO ligands on the Fe atom play an important role
to suppress the oxidative decomposition.

It is important to note here that apparent conductivity
(10^{-7} Scm^{-1}) of polymer 17 thus prepared in solution is much lower
than the conductivity of 10 doped on the film.

V. CONDUCTING BEHAVIOR OF ORGANOMETALLIC COMPOUNDS

To get a closer look at the role of π-allylic metal species
resulted from doping reaction, model reaction was carried out
using compressed pellets of pure $FeX(CO)_3(\eta^3\text{-allyl})$ (X=Br, Cl).
The complex $FeBr(CO)_3(\eta^3\text{-allyl})$ 19 itself is an insulator ($\sigma=10^{-9}$
Scm^{-1}) while the sublimation of I_2 onto the pellet (thickness
0.5 mm) brings about good conductivity reaching to the semiconductor region (Fig. 4). Complex 21 containing two $FeCl(CO)_3(\eta^3\text{-}$
allyl) group in a molecule was also used to observe the proximity
effect of adjacent metal species. The complex 21 thoroughly
purified by recrystallization from cold acetone showed an
extremely low conductivity but the σ_{DC} was increased rapidly with
an increase in amount of dopant (Fig. 4). The maximum conductivity is nearly equal to that of complex 19. Thus, remarkable
proximity effect was not observed for the complex 21. The
$[Fe(CO)_3]_2$ (1,3,7,9-decatetraene) was prepared by reaction of
1,3,7,9-decatetraene with $Fe_2(CO)_9$[11] or by the known method.[18]

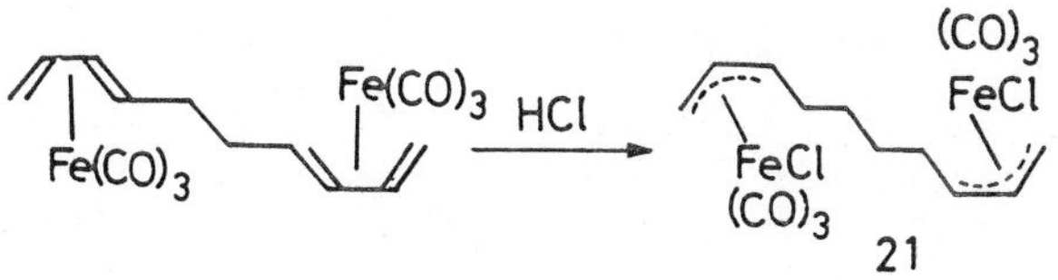

$$(8)$$

21

Doping of $FeCl(CO)_3(\eta^3\text{-allyl})$ will induce the formation of $(Fe^{II})^+$ or $(Fe^{III})^+$ species such as $Fe^+Cl(CO)_3(\eta^3\text{-allyl})$ or Fe^+ClI-$(CO)_3(\eta^3\text{-allyl})$ and conduction will take place through mixed valence metal species $[Fe^{II}Fe^{III}]^+$ or following the hopping mechanism.[19] Though more precise measurement is necessary to understand the exact conducting mechanism, the present system seems to involve both ionic and electronic conductions under the atmosphere. Observed gradual decrease of current with time indicates the involvement of the ionic conduction and a rapid increase in conductivity with the addition of a small amount of dopant supports the electronic conduction as estimated from the Drude's equation.[20] Iodine acts as an electron acceptor and produces positive charge carriers.

If the mixed-valence species exert important role, the electron transfer through $[Fe^0Fe^I]^+$ or $[Fe^IFe^{II}]^+$ seems also possible for the doped $Fe(CO)_3(\eta^4\text{-diene})$ complexes, because the maximum conductivity was observed at the small ratios of I_2/Fe for doped polymers 10, 12 and $Fe(CO)_3[\text{bis(cyclooctadienyl)}]$ (22).[21] The reaction of 22 with a dopant (I_2) occurred not only on the surface of a compressed pellet but also inside of the pellet. One of the candidates for the cationic species to give $[Fe^IFe^{II}]^+$ is $[Fe(CO)_3(\eta^3\text{-allyl})]^+$ because $Fe(CO)_3(\eta^4\text{-diene})$ is known to give $[Fe(CO)_3(\eta^3\text{-allyl})]^+X^-$ (X=BF_4, PF_6, ClO_4) on protonation with HBF_4 or HPF_6.[16] Therefore, following reaction seems possible for the solid state reaction of 22.

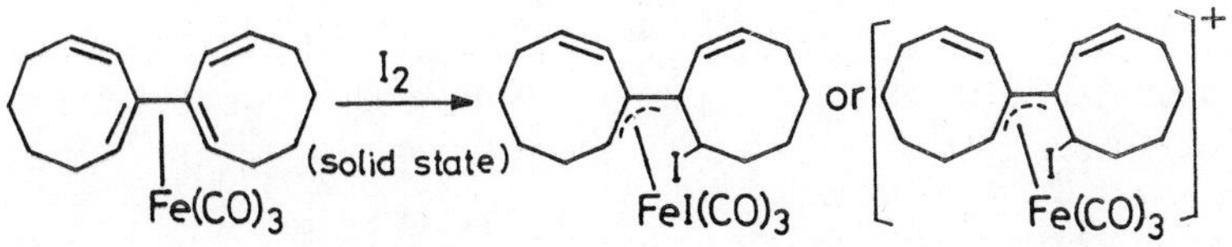

$$(9)$$

22

VI. CONDUCTIVITY OF CONVENTIONAL ORGANOMETALLIC COMPLEXES

The discovery of a remarkably high conductivity of doped $(CO)_3Fe^0(\text{diene})$ species raises the hope for various types of mixed-valence metal complexes as semiconductor materials. In the case of iron complexes, the formation of the following mixed-

$$[Fe^0 Fe^I]^+ \qquad [Fe^I Fe^{II}]^+ \qquad [Fe^{II} Fe^{III}]^+$$
$$[Fe^0 Fe^I]^- \qquad [Fe^I Fe^{II}]^- \qquad [Fe^{II} Fe^{III}]^-$$

valence metal complexes may be possible when Fe^0 or Fe^{II} species combines with $[Fe^I]^+$ or $[Fe^{III}]^+$ species to form the p-type conductor. Since Fe^0 (for example Fe(CO)$_5$) or Fe^{II} species are known to react or interact with electron donors such as Na or K,[12] the preparation of the n-type conductor seems also possible when the n-type doping is performed on the appropriate organometallics or organometallic polymers.

In relating this assumption, the apparent conductivity of Fe^0 or Fe^{II} species mixed with $[Fe^{II}]^+$ or $[Fe^{III}]^+$ species was measured using ferrocene, Fe(CO)$_3$(1,4-diphenyl-1,3-butadiene) as neutral species and $[FeCp_2]^+[BF_4]^-$, $[FeCp_2]^+[FeCl_4]^-$, $[Fe(CO)_4(\eta^3\text{-allyl})]^+$ $[BF_4]^-$ as cationic species. The conductivity of a 1:1 mixture was however rather low   ($10^{-6} \sim 10^{-8}$ S cm^{-1}) suggesting that the mixing of Fe^+ species with neutral Fe species is insufficient to obtain the mixed-valence metal species which show good conductivity. More systematic investigations are required to obtain the fundamental information on the electric properties of organometallic complexes.

Therefore, the effect of doping on the conductivity was examined with many conventional transition metal complexes. Some of the complexes tested are summarized in Table 3. Low valent

Table 3.   Room Temp. Conductivity of Organometallic Complexes After Doping with Iodine onto the Pellet Disc

Complexes		σ_{DC} (S cm^{-1})
FeBr(CO)$_3$(η^3-C$_3$H$_5$)	FeCl$_2$(PPh$_3$)$_2$	$> 10^{-6}$
Fe(CO)$_3$[bis(cyclooctadienyl)]	PdI$_2$(PPh$_3$)$_2$	
[FeCl(CO)$_3$]$_2$(C$_{10}$H$_{16}$)		
CoI$_2$(dppe)	CoCl$_2$(Bu^{t}NC)$_4$	
FeBr$_2$(PPh$_3$)$_2$	FeClCp(CO)$_2$	$10^{-6} \sim 10^{-7}$
NiCl$_2$(PPh$_3$)$_2$	NiClCp(PPh$_3$)$_2$	
Ni(CNC$_6$H$_4$CH$_3$)$_4$	PdBr$_2$(PPh$_3$)$_2$	
[FeCp(CO)$_2$]$_2$	FeICp(CO)$_2$	$< 10^{-9}$
Fe$_2$(CO)$_9$	Ru$_3$(CO)$_{12}$	
Mo(CO)$_6$	W(CO)$_6$	
Cr(CO)$_6$	PtCl$_2$(PBu$_3$)$_2$	
NiCl$_2$(dppe)	NiI$_2$(dppe)	

iron carbonyls or cyclopentadienyliron carbonyls do not always show good conductivity. Especially $Fe_2(CO)_9$, $Ru_3(CO)_{12}$ and $M(CO)_6$ showed an extremely low conductivity even when doping was conducted onto the pellet since these are inert to dopants. Some of the cobalt complexes showed good conductivity on doping. Our preliminary experiments indicate the possibility to find better conducting organometallics using low valent transition metals. In this way, many new potentially conducting organometallic complexes and polymers may be screened without the expenditure of excessive time and funds. Further search for more practical organometallic conductor devices is now in progress.

VII.   CONCLUDING REMARKS

The key point of the present work lies on the method to generate stable cationic species, M^+, which form the mixed valence metal species $[M][M']^+$ or specially oriented metal complexes. Mixed valence species using organometallic complexes are preferred over pure inorganic complexes in view of synthesizing flexible polymer materials, because a variety of techniques to combine the organometallic compounds to organic polymers has been already established by many workers. The synthesis of polymer supported conducting organometallics or conducting organometallic polymers promises their wide application because of the good solubility in common solvents and the facility in the preparation of films or blocks of desired shape.

REFERENCES

1) Y. Sasaki, L. L. Walker, E. L. Hurst, and C. U. Pittman, Jr., J. Polymer Sci., Polym. Chem. Ed., 11, 1213 (1973).
2) C. U. Pittman, Jr., G. V. Marlin, and T. D. Rounsefell, Macromolecules, 6, 1 (1973).
3) M. D. Rausch, G. A. Moser, E. J. Zaiko, and A. L. Lipman, Jr., J. Organometal. Chem., 23, 185 (1970).
4) U. T. Mueller-Westerhoff, and P. Eilbracht, J. Amer. Chem. Soc., 94, 9272 (1972).
5) C. U. Pittman, Jr. and B. Surynarayanan, J. Amer. Chem. Soc., 96, 7916 (1974).
6) C. U. Pittman, Jr. and Y. Sasaki, Chem. Lett., 1975, 383.
7) D. O. Cowan, J. Park, C. U. Pittman, Jr., Y. Sasaki, T. K. Mukherjee, and N. A. Diamond, J. Amer. Chem. Soc., 94, 5111 (1972).
8) C. U. Pittman, Jr, O. E. Ayers, and S. P. McManus, J. Macromol. Sci-Chem., A-7, 1563 (1973).
9) H. Yasuda, Y. Ohnuma, M. Yamauchi, H. Tani, and A. Nakamura, Bull. Chem. Soc. Jpn., 52, 2036 (1979).
10) H. Yasuda, M. Yamauchi, Y. Ohnuma, and A. Nakamura, Bull. Chem. Soc. Jpn., 54, 1481 (1981).

11) Y. Morita, M. Yamauchi, H. Yasuda, and A. Nakamura, Kobunshi
    Ronbunshu., 37, 677 (1980).
12) a) W. Hieber and G. Brendel, Z. Anorg. Chem., 289, 324 (1957).
    b)Y. Takegami, Y. Watanabe, H. Masada, and I. Kanaya,
    Bull. Chem. Soc. Jpn., 40, 1456 (1967). c) H. W. Sternberg,
    R. Markby, and I. Wender, J. Amer. Chem. Soc., 79, 6116 (1957).
13) J. E. Mahler and R. Pettit, J. Amer. Chem. Soc., 85, 3955(1963).
14) I. Noda, H. Yasuda, and A. Nakamura, Organometallics, 2, 1207
    (1983); idem, J, Organometal. Chem., 205, C9 (1981).
15) a)L. F. Johnson, F. Heatley, F. A. Bovey, Macromolecules,
    3, 175 (1970). b) K. Matsuzaki, T. Kanai, T. Kawamura, S. Matsumot
    and T. Uryu, J. Polymer Sci., Polym. Chem. Ed., 11, 961 (1973).
16) a) G. F. Emerson, J. E. Mahler, and R. Pettit, Chem & Ind.,
    1964, 836. b) R. F. Heck and C. R. Boss, J. Amer. Chem. Soc.,
    86, 2580 (1964).
17) H. D. Murdoch and E. Weiss, Helv. Chim. Acta., 45, 1927 (1962).
18) R. W. Jotham, S. F. A. Kettle, D. B. Moll, and P. J. Stamper,
    J. Organometal. Chem., 118, 59 (1976).
19) a)H. A. Pohl and E. H. Engelhardt, J. Phys. Chem., 66, 2085
    (1962). b) J. H. T. Kho and H. A. Pohl, J. Polymer Sci., Part
    A-1, 7, 139 (1969).
20) A. R. Blythe, "Electrical Properties of Polymers", Cambridge
    University Press (1979).
21) I. Noda, H. Yasuda, and A. Nakamura, J. Organometal. Chem.,
    250, 447 (1983).

POLYMERS CONTAINING POLYNUCLEAR COBALT AND IRON

CARBONYL COMPLEXES

J. C. Gressier[*], G. Levesque[*], H. Patin[+], F. Varret[°]

[*] Laboratoire de physicochimie et photochimie organiques
    Université du Maine 72017 Le Mans France
[+] Laboratoire de Chimie des Organométalliques
    Université de Rennes I 35040 Rennes France
[°] Laboratoire de spectroscopie Mössbauer
    Université du Maine 72017 Le Mans France

Organometallic polymers are of interest for a large variety of applications including catalysis, semiconductors, UV absorbers, antifouling agents. The most widely studied have been prepared either from metal-complexes of vinyl monomers (1), or by coordination between a metal atom and convenient ligands linked on the chain (2) ; but to our knowledge no work has been devoted to cluster-containing polymers.

For several years, some of us have developed the preparation of monomers bearing thiocarboxylic functions. A number of dithioesters, dithioacids and thioamides useful either for polymerisation or for polycondensation have been synthetised (3) and the corresponding (co)polymers widely studied (4).

On the other hand, metallic complexes have been obtained by action of metal carbonyls on dithioesters, xanthates, trithiocarbonates and thioamides (5). Some of these clusters offer interesting applications for the obtention of polymerisable molecules.

The conjunction of these two axes has led us to undertake the synthesis and study of macromolecular compounds bearing polymetallic groupings. The interest of these compounds is very large because the monomers are potential catalysts (6) and by attaching homogeneous catalysts to organic polymers or to inorganic supports, it is possible to combine the interests of homogeneous catalysis (selectivity, mild operating conditions) and of heteregeneous catalysis (duration, easy separation of the products from the reactants).

This article describes the synthesis of clusters obtained by reacting dicobalt octacarbonyl and diiron nonacarbonyl with polymerisable carboxydithioate and thiocarboxamides.

Reactions on pre-formed polymers bearing the required functionalities have been also realised.

TRICOBALT CARBONYL CLUSTER

Cluster 3b is prepared by action of $Co_2(CO)_8$ on dithioester 1b in THF (7) (Fig. 1). Desulfurization occurs and several byproducts, especially cobalt sulfides are formed. Pure monomer is obtained with 40 % yield after chromatography on TLC plates (silicagel, hexane) and recrystallisation from methanol. A similar route leads to cluster 4b from monomer 2b (yield 45 %).

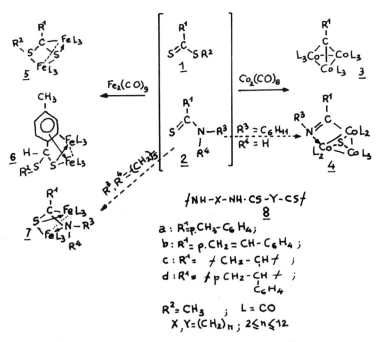

Fig. 1 Monomers and structural units

Free radical homopolymerisation of monomers 3b and 4b was unsuccessful and ionic initiators destroy them immediatly but copolymerisation with styrene (St) or methyl methacrylate (MMA) and water soluble comonomers such as [tris (hydroxymethyl) methyl] acrylamide (tris A) lead to copolymers for which the proportion of cluster is 1 to 1,5 % (mole %) (Table 1).

Another route to obtain organometallic polymers is the chemical modification of pre-formed polymers. Copolymers of 1b with styrene, MMA, and tris A have been treated by $Co_2(CO)_8$ in THF. A clusterisation yield of more than 80 % and high content in cluster units on the macromolecular chain is easily obtained.

The same reaction with copolydithioesters bearing the unit 1c obtained by chemical modification of copolyacrylonitrile (4b) leads to similar results.

However, the action of $Co_2(CO)_8$ on the copolymers of vinyl thiobenzamide 2b and polythiocarboxamides 8 do not lead to the excepted clusters but to cross-linked polymers. The same result is observed with other polymers bearing two or more thioamide functions. It appears that $Co_2(CO)_8$ does not react selectively with one function but acts as a cross linking reagent; a high proportion of cobalt under an undetermined form remains present in the chain (the IR spectra show no absorptions characterictic of the ligands).

At first glance, the chemical modification of polymers may appear to be the most convenient route to prepare supported cobalt clusters: the yields are high and the separation of reacted polymers is easy. However, the purification of these polymers is difficult because the clusterisation reaction involves at least two desulfurisation steps and sulfur containing organocobalt species and colloïd cobalt sulfide remains adsorbed on the chain. This is proved by excess of cobalt found by comparison to the theoretical value. The cobalt adsorbed on the polymer is approximately 10 % of the total amount (90 % of cluster)(Table I).

Since tricobaltcarbonyl clusters are active catalysts for the hydroformylation of olefins (8), in the purpose of studying the catalytic activity of such clusters anchored on polymers it seems more convenient to prepare them by the copolymerisation route which offers a high degree of purity and preserves the integrality of the cluster units. A wide variety of organo or water soluble copolymers can be obtained according to the nature of the backbone and an easily adjustable amount of cluster units can be introduced in the polymers.

Direct metallation remains useful for the clusterisation of resins bearing dithioester functions (4b).

Table I. Characteristics of cobalt cluster copolymers

Cluster unit in the copolymer	Obtention mode (a)	Starting Material (b)	Analysis Co% (w/w)	Cluster unit in copolymer (mole %)	$\overline{M}_n$; $\overline{M}_w$ (GPC)
	I	Styrene/3b (99/1)	5.45	3.7	6400, 14720
	I	Methyl methacrylate/3b(99/1)	1.78	1.0	21300, 44700
3d	I	Tris A/3b (99/1)	1.65	1.6	- (c)
	II	Styrene/1d (85/15)	18.91	13.0 (d)	5320, 12600
	II	Homopolymer 1d	20.33	37.8	4300, 8650
	I	Styrene/4b (99/1)	2.26	1.3	7200, 16800
4d	I	Methylmethacrylate/4b (99/1)	3.13	1.8	24600, 41300
	I	Tris A/4b (99/1)	1.60	1.6	- (c)
3c	II	Styrene/1c (91.3/8.7)	11.08	8.5	Cross-linked

(a) I copolymerisation ; II chemical modification

(b) in case I : in brackets, percentage of the comonomer and cluster monomer in the comonomers mixture

   in case II : in brackets, percentage (mole%) of each units in the starting copolymer

(c) Water-soluble

(d) dithioester unreacted : 2.0% (mole%) ; CoS absorbed on polymer 7.4% (w/w) give the best correlation with analytical data

IRON CARBONYL CLUSTERS

The same routes have been used in order to obtain polymers bearing hexacarbonyl diiron clusters.

Diiron nonacarbonyl reacts with dithioesters 1 and leads to clusters 5 with 40 % yield after purification (9). Small amounts of orthometalated products 6 are also obtained but they have not been studied because of their instability and low yields (10 %); complexation of thioamides shows that only NN disubstituted thiocarboxamides lead to the stable clusters 7 with 15 % yield. These results are consistent with Alper's data (10).

As found for cobalt cluster monomers, iron carbonyl clusters 5b and 7b do not homopolymerise but they are stable in the conditions of free radical initiation. Their copolymerisation leads to different copolymers of various solubility according to the nature of the comonomers (route I): cluster content between 1 and 2 % (mole %) (Table II). Clusterisation of pre-formed polymers (route II) is also possible except for "tris A" copolymers: the OH groups and the hydrophilic character of the polymer are not compatible with $Fe_2(CO)_9$ which is oxidised mainly to ferric oxide; with non-hydroxylic polymers, high content in cluster can be obtained ($\sim$ 40 %).

A detailed analysis of these new organometallic polymers exhibits a different Fe/S ratio for a similar polymer according to the choice of route I or II. In all cases, an excess of iron is found but in smaller quantity for copolymers obtained by route I (Table II).

<u>Mössbauer analysis</u>

The Mössbauer spectra of iron carbonyl compounds provide important information concerning the electronic structure and symmetry of the central iron atom. The isomer shift value (δ) is proportionnal to the change in the electron density at the iron nucleus. The quadrupole splitting value (ΔQ) deals with the symmetry of the electron cloud around the iron nucleus. For these reasons many studies have been carried out on mono and polynuclear complexes (11) but since the work of Pittman at al. (12) few Mössbauer measurements have been made on polymer-supported iron complexes (13).

We have used a conventional triangular mode spectrometer with a source ^{57}Co and an instrumental linewith $\sim$ 0.22 mm/s. The crystallised cluster 5a is chosen as a model of polymeric cluster unit; the same compound dissolved in polymethyl methacrylate (PMMA) and two coPMMA prepared by the two ways and containing the same percentage of cluster units, (1.1 %, mole %) have been analysed by this technique.

In all cases the characteristic doublet of ferric oxide is

Table II. Analytical data of the iron carbonyl cluster copolymers

| Type of unit in the copolymer | Route (a) | Starting material (b) | Analysis | | % cluster unit in the copolymer (mole %) (d) | % Fe as Fe_2O_3 (e) | $\overline{Mn}$, $\overline{Mw}$ (GPC) | n° |
			% Fe w/w (c)	Fe/S (c)				
5a	II	St/1c (91/9)	5.17	2.7	3.5 (41 %)	35	cross-linked	
5d	I	St/5b (98.9/1.1)	1.95	2.0	1.4	15	2250-3320	
5d	I	MMA/5b (99.1)	1.51	2.1	1.1	23 (*)	10060-18560	9
5d	I	Tris A/5b 98.5/1.5)	1.06	2.1	1.5	16	5500-7000 (f)	
5d	II	id (homopolymer)	20.58	2.0	39.0 (39 %)	30 (*)	2500 –	10
5d	II	St/1d (83/17)	9.57	3.0	6.7 (40 %)	40 (*)	4630-21640	11
5d	II	MMA/1d (88/12)	8.18	2.8	6.4 (54 %)	33	40000 –	
5d	II	MMA/1d (96/4)	2.76	2.8	1.8 (45 %)	31 (*)	– –	12
5d	II	Tris A/acrylamide /1d 31/66/3	0.74	3.2	0.4 (13 %)	67 (*)	– (f)	
7d	I	St/7b (99/1)	1.36	4.1	1.1	17	8200-16300	
7e	II	St/1c (98.2/1.8)	0.80	8.4	0.3 (17 %)	63	–	

(a) : I : copolymerisation ; II : chemical modification

(b) : in brackets composition (mole %) of the comonomers mixture

route II : in brackets composition (mole %) of the starting copolymer

(c) : included correction for unreacted dithioester or thioamide

(d) : calcul'd on Fe % and S % included correction for ferric oxide content : in brackets, % conversion of initial

function into cluster when route II is used

(e) : calcul'd on analytical or Mössbauer data (*)

(f) : water soluble

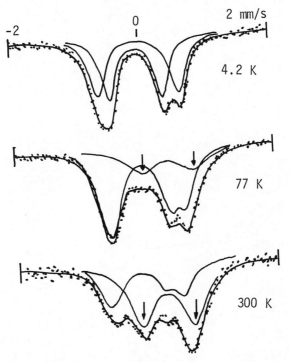

Fig. 2 : Mössbauer spectra of copolymer MMA − 5d (9) synthetized
         by copolymerisation

found at 300 K (14) (shown by arrows on the figure 2) and the amount
of impurity is calculated for each sample by means of a mathematic
treatment of the Mössbauer spectra.

Copolymers obtained by chemical modification have the higher
content (30 to 40 % of Fe under Fe_2O_3 form) while the others do not
exceed 15 %. Moreover, from the values of the isomeric shift (δ)
and of the quadrupole splitting (ΔQ), it is clear that the electronic state of the iron atom is the same for the crystallised complex
5a, the solution of 5a in PMMA and the copolymers obtained by copolymerisation, while important changes for δ and ΔQ, associated with
a broadening of the lines indicate several modes of coordination
on the copolymer obtained by chemical modification (Table III)

These different structures can arise from different routes:
coordination of $Fe_2(CO)_9$ on the polymer followed by degradation and
oxydation into ferric oxide, direct adsorbtion of ferric oxide
produced during the reaction, presence of different cluster units,

Table III. Mössbauer data at 4.2 K

route (a)	compound n°	cluster unit in the polymer mole %	2 sites (b)	isomer shift mm/s (c) (δ)	quadrupole splitting mm/s (ΔQ)	line width mm/s	relative area	average isomer shift mm/s	average quadrupole splitting mm/s
solid state	5a	92.0	yes	0.066 / 0.032	1.262 / 0.831	0.258 / 0.270	0.49 / 0.51	0.050	1.042
I	9	1.1	yes	0.084 / 0.037	1.287 / 0.843	0.30 / 0.27	0.51 / 0.49	0.061	1.069
	10	39.0	yes	0.086 / 0.045	1.294 / 0.869	0.35 / 0.30	0.49 / 0.51	0.065	1.08
II	11	6.7	poor	0.070	0.96	0.51	-	0.070	0.96
	12	1.8	no	0.072	1.01	0.51	-	0.072	1.01

(a) : I : copolymérisation ; II : chemical modification ; III : solution of complex 5a in PMMA

(b) : yes : resolved ; poor : poorly resolved ; no : unresolved (broad lines)

(c) : isomer shift relative to metallic iron at room temperature

(d) : half-height width of Lorentzian lines

for instance orthometalated products of type 6 which eventually
decompose. Removal of ferric oxide, either by dissolution-precipita-
tion techniques, or by solution centrifugation was unsucessful
which show that ferric oxide is probably colloïdal. However the presence of ferric oxide in polymers prepared by route I can be assumed
to be the result of partial destruction of the complexes by interactions with free radicals.

Our results show once more that obtention of pure organometallic
polymers is difficult and the best route to prepare them is the synthesis of cluster-monomers followed by copolymerisation under inert
atmosphere and degassed solvents.

The thermal dependance of the Mössbauer absorption, proportionnal to the Lamb-Mössbauer factor (log f) provides relevant informa-
tions about the stiffness of the lattices (15). In the solid state
log f is a linear function of the temperature according to the Debye
theory and the thermal dependance f(T) can be calculated as a function of the Debye temperature Θ_D.

Figure 3 shows the Mössbauer experimental absorption area matched at a log scale to the theoretical functions f(T) given by the
Debye model (fitted Θ_D in brackets). The departure from the straight
line indicates a softening of the cluster surrounding

The higher Θ_D is obtained for ferric oxide and is consistent
with the fact that an ionic lattice is stiffer than an inorganic
one (cluster 5a).

A significant decrease of Θ_D for the organometallic copolymer
is explained by low frequency vibrations occuring in the polymer
chain ; a weak departure from the linear law at 285 K is interpreted as the first glassy transition of PMMA.

The lowest Θ_D found for the dissolved cluster in PMMA agrees
with the fact that the cluster is weakly bonded to the chain, the
inflexion of the linear plot at 200 K is interpreted by modification of the geometry of the interchain spaces to which the electronic state of the cluster is very sensitive. This transition should
be specific for the localization of the cluster in the interchain
spaces.

So the great sensivity of Mössbauer spectroscopy allows the
detection of distorsions at low temperature, previously to the
glassy transition, while a Mössbauer probe linked to the chain
should exhibit the proper transition of the polymer support.

The study of organometallic polymers is only at the beginning.
Various clusters can be prepared in the form of polymerisable monomers and this way offers large possibilities for chiral polymers;

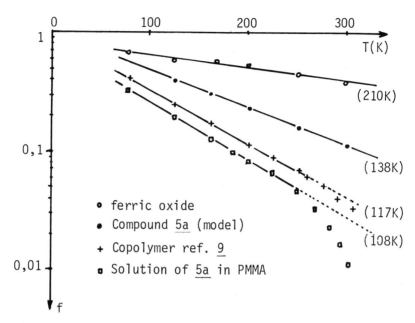

Fig. 3 : Mossbauer experimental absorption area (A) matched at a log
scale to the theoretical functions f(T) given by the Debye
model (fitted Θ_D in brackets). The departure from the
straight line indicates a softening of the cluster
surroundings.

also the electrochemical behaviour of these clusters is a field of
interest (16). The use of Mössbauer spectroscopy for analytical
purpose and for the study of transitions in polymer phase merits
a large development. With no doubt next years will see an increase
of research in these fields.

REFERENCES

1. C. E. Carraher Jr, J. E. Sheats, C. U. Pittman Jr: "Organometallic Polymers", Academic Press, New York, 1978

2. A. Seriya, J. K. Stille, J. Am. Chem. Soc., 103, 5096 (1981
and references therein;
M. Kaneka, E. Tsuchida, J. Polym. Sc., Macromol. Rev. 16,
397 (1981)

3. R. Haraoubia, G. Levesque, J. C. Gressier, Makromol. Chem.
176, 2143, 1975

R. Haraoubia, C. Bonnans-Plaisance, G. Levesque, Makromol. Chem., 182, 2409 (1981)

R. Haraoubia, C. Bonnans-Plaisance, J. C. Gressier, G. Levesque, Makromol. Chem. 183, 2383 (1982)

4. a) J. C. Gressier, G. Levesque, Europ Polymer J. 16, 1093, (1980); ibid 16, 1101, (1980); ibid 16, 1157 (1980); ibid, 16, 1175 (1980); ibid, 17, 695 (1981)

b) R. Haraoubia, J. C. Gressier, G. Levesque, Makromol. Chem. 183, 2367 (1982)

5. H. Patin, G. Mignani, M. T. Van Hulle, Tetrahedron Lett. 26, 2441 (1979)

H. Patin, G. Mignani, C. Mahe, J. Organomet. Chem. 208,C39, (1981)

H. Patin, C. Mahe, M. Benoist, J. Y. Le Marouille, J. Organomet. Chem, 218, C 67 (1981)

H. Patin, G. Mignani, A. Benoist, J. Y. Le Marouille, D. Grandjean, Inorg. Chem. 20, 4351 (1981)

A. Benoist, J. Y. Le Marouille, C. Mahe, H. Patin, J. Organomet. Chem. 233, C 51 (1982)

A. Benoist, A. Darchen, J. Y. Le Marouille, C. Mahe, H. Patin, Organometallics, 2, 555 (1983)

6. E. L. Mutterties, J. Organomet. Chem. 200, 177, (1980)

7. J. C. Gressier, G. Levesque, H. Patin, Polym. Bull. 8, 55 (1982)

8. C. U. Pittman, R. C. Ryan, Chem. Tech., 8, 170 (1978)

9. J. C. Gressier, G. Levesque, H. Patin, F. Varret, Macromolecules, 16, 1577, (1983)

10. H. Alper, A. S. W. Chan, J. Chem. Soc. Chem. Commun 1973 724

H. Alper, C. W. Foo, Inorg. Chem. 14, 2928 (1975)

11. P. Gutlish, R. Link, A. Trautwein, "Mössbauer Spectroscopy and Transition Metal Chemistry" Springer Verlag, W. Berlin, 1978

12. C. U. Pittman Jr., J. C. Loi, D. A. Vanderpool, M. Good, R. Prado, Macromolecules 3, 706, (1970)

13. G. L. Corfield, J. S. Brooks, S. Plimley, Polym. Prepr., Am. Chem. Soc., Div, Polym. Chem. 22(1), 3, (1981)

14. A. M. Van Diepen, J. J. A. Popma, J. Phys. <u>37 C6</u>, 755, (1976)

15. D. W. Macomber, W. L. Hart, M. D. Rausch, R. D. Priester and C. U. Pittman, J. Am. Chem. Soc., <u>104</u>, 884 (1982)

16. A. Darchen, C. Mahe, H. Patin, J. Chem. Soc., Chem. Comm. 243, (1982)
      A. Darchen, C. Mahe, H. Patin, N.J. de Chem. <u>6(11)</u> 539 (1982)
      L. E. Klir, C. Mahe, H. Patin, A. Darchen, J. Organomet. Chem. <u>246</u>, C 61 (1983)
      A. Darchen, C. Mahe, H. Patin, J. Organomet. Chem. <u>251</u>, C 9 (1983)

DECOMPOSITION OF IRON CARBONYLS IN SOLID POLYMER MATRICES:

PREPARATION OF NOVEL METAL-POLYMER COMPOSITES

R. Tannenbaum, E. P. Goldberg and C. L. Flenniken

Department of Materials Science and Engineering
University of Florida, MAE 217
Gainesville, FL 32611

INTRODUCTION

The aim of this study was the synthesis and
characterization of polymer composites in which microscopic
metal or metal oxide particles were incorporated by thermal or
photolytic decompositions of solid solutions of organometallic
complexes in polymers. In contrast to conventional composites
or filled polymer compositions, which are prepared by dispersion
of particles or fibers in a polymer matrix, this study has
typically involved the preparation of metal carbonyl solid
solutions in polymers and the phase separation decomposition of
the organometallic complex to form uniform metal or metal oxide
dispersions of very small particle size (50-500Å).

This approach to the preparation of metal-polymer
composites may afford unique opportunities for investigating
fundamental aspects of nucleation and growth of metal and metal
oxide clusters in the solid state as well as the properties of
such microparticles as a function of cluster size, concentration
and environment. For example, very dilute systems may afford
more versatility than conventional matrix isolation techniques
for studying the catalytic and electronic properties of nascent
metal atoms and small clusters.

Applications for metal-polymer composites made by this
phase-separation method are also potentially interesting and
include: (1) photorecording by formation of metallic phases in
polymer matrices according to photo-induced spacial patterns,
(2) magnetic recording and magneto-optic switching,

(3) conducting and semi-insulating polymers, (4) polymers with special mechanical properties and (5) metal-polymer catalyst compositions.

BACKGROUND

Nascent metallic species created by UV photolysis or by thermal or electron beam decomposition in a polymer matrix should be highly reactive[1,2]. There are two major pathways for chemical reaction available for these species: (a) metal ions/radicals can attack the polymer, a process that can lead to various polymer degradation, crosslinking, or metal attachment processes, and (b) metal species can aggregate to form very small clusters 10-100Å in diameter. Both processes may occur and may be significantly influenced by reducing or oxidative atmospheres. The chemistry of the system, i.e. the relative importance of process (a) or (b), is determined by the structure of the solid polymer matrix and by the rate of the decomposition process as compared to the diffusion time of the active metal species in the solid polymer matrix.

Metal-polymer composites are conventionally prepared by physically mixing metal or metal oxide powders with a polymer, and fabricating mixtures by extrusion or molding into materials with large metal or metal oxide particles dispersed in the polymeric matrix. Particle size and shape, volume loading of filler, fabrication technique, specific gravity of the metal versus the polymer, and compatibility (surface interactions) of the polymer with the metal are important factors affecting the properties of such conventional metal-polymer composite systems. Since these compositions are mechanically blended, it is a major technological problem to control the uniformity of the dispersion and the size of the aggregates.

The concept presented here for producing composites was designed to overcome difficulties connected with blending techniques to create uniform dispersions of metals or metal oxides by in-situ phase-separation in a solid polymer matrix under controlled conditions. This approach is somewhat analogous to "precipitation hardening" in metal alloys. However, in our approach, organometallic complexes were dissolved in polymer solutions to form homogeneous mixtures. These solutions were then cast and the resulting films exposed to thermal, photolytic or electron beam energy. These treatments decomposed the organometallic complex to form fine uniform dispersions of metal or metal oxide particles in the polymer matrix. The general scheme for this method is outlined in Figure 1.

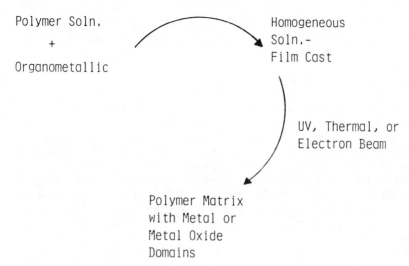

Figure 1.   Concept for Producing Metal-Polymer Composite Films

        The concept for preparing uniform dispersions of metals by
phase separation in solid polymer matrices is suggested by some
metal colloid literature. For example, thermolysis of
transition-metal carbonyls in fluids under an inert atmosphere
is a well known technique for the preparation of pure metal
powders and a widely used process for the preparation of
ferromagnetic cobalt particle dispersions[3,4]. This method was
successfully applied to the preparation of small iron particles
by Griffith et al.[5] and Smith et al.[6] in polymer solutions. The
macromolecules reportedly act as dispersing and stabilizing
agents for the iron. Stable colloidal dispersions (50-150 Å) of
zero valent iron were obtained by thermolysis of $Fe(CO)_5$ in
dilute polymer solutions. Exposure of dispersions to the
atmosphere decreased the observed magnetic moment due to
formation of an oxide film on the particles. Chlorinated
solvent based dispersions were observed to generate chlorine
which destroyed the γ-Fe_2O_3 spinel film passivity and promoted
reaction with water to give β-FeOOH. A "locus control"
formalism of particle nucleation and growth was proposed to
describe the formation of these colloidal dispersions[6]. It was
suggested that the polymer served as a catalyst for the
decomposition of the metal carbonyl and induced particle
nucleation. Selected polymer systems were termed "active"
because the initial and overall rate of decomposition of

$Fe(CO)_5$, as determined by the rate of CO evolution, was much
faster in the presence of the active polymers than in solvent
alone.

A similar study by Hess and Parker[7] describes stabilization
of colloidal cobalt particles in polymer solutions. Though a
number of addition and condensation polymers can act as
dispersants, the most successful are linear addition polymers of
high molecular weight having relatively nonpolar backbones.
Stabilization of the cobalt particles results by adsorption of
the polymer to the metal particles to form a film which
separates the particles sufficiently to keep Van der Waals
forces below thermal energy levels. The solvents for these
systems should be less polar than the most polar groups in the
polymer to minimize competition with the polymer-metal particle
interaction.

DECOMPOSITION OF IRON CARBONYLS IN SOLID POLYMER MATRICES

As model systems for studying the preparation of metal-
polymer composites by solid state phase separation, we selected
$Fe(CO)_5$ as the organometallic complex and the following
polymers: bisphenol polycarbonate (PC), polyvinylidene fluoride
(PVF_2), polystyrene (PS), aromatic polysulfone (PSF) and
polymethylmethacrylate (PMMA) (Figure 2). Some criteria for
selecting these polymers were:

1. Commercially available soluble thermoplastic polymers to
    facilitate sample preparation and analysis.
2. Mutual solubility of polymer and organometallic complex in
    common organic solvents.
3. Polymers exhibiting high melting and glass transition
    temperatures, preferably above the decomposition temperature
    of the metal carbonyl to be used.
4. Polymers with varied molecular structures, morphologies, and
    mechanical properties to assess interactions with active
    metal species and evaluate effects upon physical
    properties.

Polycarbonate (LEXAN®, 131-111, $\bar{M}_v$ = 32,200), polysulfone
(UDEL®, $\bar{M}_v$ 30,100), and polystyrene (Polyscience, Inc., $\bar{M}_v$ =
159,400) are soluble in methylene chloride whereas
polymethylmethacrylate (Polyscience, Inc., $\bar{M}_v$ = 323,000) is
soluble in benzene and polyvinylidene fluoride (Polysciences,
Inc., $\bar{M}_v$ = 119,000) is soluble in dimethylformamide. We knew
from preliminary studies[8,9] that in the PC system both thermal
and photolytic decompositions of $Fe(CO)_5$ lead to significant
polymer-Fe interactions which may be accompanied by severe

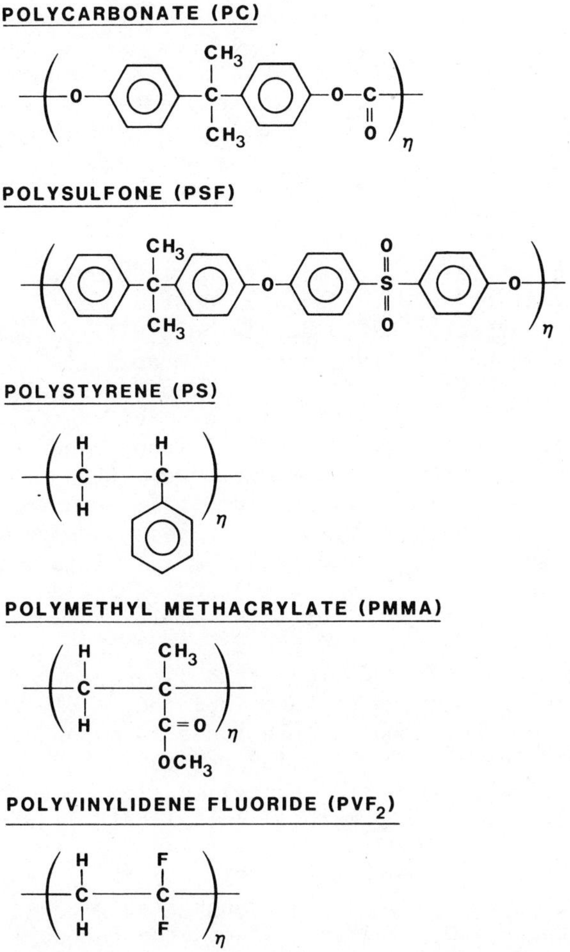

POLYCARBONATE (PC)

POLYSULFONE (PSF)

POLYSTYRENE (PS)

POLYMETHYL METHACRYLATE (PMMA)

POLYVINYLIDENE FLUORIDE (PVF$_2$)

Figure 2.   Model Systems Selected for Studying the Preparation
of Metal-Polymer Composites.

degradation of the polymer. The PC system may in fact be of
interest for photo-resist applications.

PVF$_2$ composites were initially studied in some detail in
this laboratory by S. Reich and found to produce ferrimagnetic
materials upon thermolysis due to formation of submicroscopic
α-Fe and γ-Fe$_2$O$_3$ particles.[10] The Fe-PVF$_2$ composites may be of
interest for magnetic recording applications. The decomposition
of Fe(CO)$_5$ in PVF$_2$ leads to some fluorine abstraction by Fe and
slight crosslinking of the polymer yielding insoluble but tough,
strong, magnetic polymer composite films. The PC and PVF$_2$
systems represent extremes of Fe(CO)$_5$-polymer chemistry. Other

polymer media produce various degrees of Fe-Fe versus Fe-polymer bond formation. Investigation of polymers of differing molecular structures and the analysis of the products obtained with each polymer was undertaken to help establish mechanistic pathways for decompositions of $Fe(CO)_5$ and other organometallic complexes in polymeric matrices. Understanding the chemistry of $Fe(CO)_5$ relevant to the systems described above was also regarded as important.

One aspect of $Fe(CO)_5$ chemistry pertinent for consideration here is the interaction with isolated double bonds. Molecular orbital theory suggests that a double bond will coordinate with the metal in a manner similar to carbon monoxide[11-13]. The substitution of one CO by a double bond would therefore produce a complex containing the C=C group: i.e. $Fe(C=C)(CO)_4$. Indeed, in recent literature, there are several reports concerning the formation of olefin-iron tetracarbonyl complexes. The first example was the preparation of acrylonitrile iron tetracarbonyl from acrylonitrile and $Fe_2(CO)_9$.[14] The same complex may also be obtained with $Fe(CO)_5$ upon irradiation.[15,16] This suggests that $Fe(CO)_5$ is first converted to $Fe_2(CO)_9$ and then reacts with the double bond. The x-ray data for this complex shows that the iron atom is bonded to the C=C group rather than to the -C≡N or the nitrogen atom[17]. Butadiene-$Fe(CO)_4$ and butadiene-$[Fe(CO)_4]_2$ have also been isolated.[18] In these complexes, one of the carbon double bonds is π-bonded to the Fe atom. The butadiene-$Fe(CO)_4$ complex reacts with HCl to form the 1-methyl-π-allylchloroiron tricarbonyl complex, possibly through intermediate formation of a methyl-π-allyl-$Fe(CO)_4$ cation.[18]

The most common feature of $Fe(CO)_5$ chemistry is its interaction with conjugated or non-conjugated dienes, with the formation of the $Fe(CO)_3$ moiety. Diene-iron carbonyl complexes were first prepared by two entirely different synthetic methods. In 1930, Reihlen and coworkers[19] obtained butadiene-iron tricarbonyl by reaction of butadiene with iron pentacarbonyl, and in 1953 Reppe and Vetter[20] reported the formation of organoiron compounds (since shown to be diene-iron carbonyl complexes) following reaction of acetylene with iron carbonyls. The diene-iron carbonyl complex was further investigated by Hallam and Pauson in 1958, and was found to resist hydrogenation and not to undergo Diels-Alder-type reactions.[21] Spectroscopic and chemical evidence led to the suggestion that the butadiene molecule remained essentially intact in the organoiron complex. Furthermore, it was shown that an analogous compound, 1,3-cyclohexadiene-iron-tricarbonyl, could be prepared in a similar manner from 1,3-cyclohexadiene and $Fe(CO)_5$. It was therefore concluded that the diene system

adopted a <u>cis</u> arrangement of double bonds within the complex
rather than <u>trans</u>. The following structure was proposed:

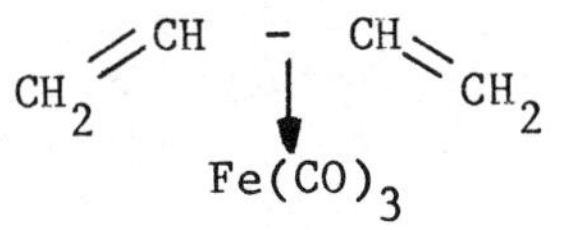

The butadiene structure is assumed to be nearly planar,
with the iron atom lying below this plane approximately
equidistant from the four carbon atoms of the diene system. The
nature of the diene-iron bonding was presumed to involve
interaction of the Fe atomic orbitals with π molecular orbitals
of the diene system as a whole. The structure is therefore more
analogous to π-bonding in ferrocene[22] rather than σ-type
interactions implied by Reihlen.[19] The conjugated diene system
was considered to be essential for the formation of iron
derivatives of this type.

Interesting correlations exist between the electronic
structure of the carbonyl groups for various diene-iron
tricarbonyl complexes. Cationic ligands shift the C=O IR
absorption frequencies to higher values.[23,24] In butadiene-iron
tricarbonyl itself, there are two regions of carbonyl
absorption,[25] a narrow intense band at 2053 cm^{-1} and a broader
band which is resolved into two maxima at 1985 and 1975 cm^{-1}.
In analogous derivatives the gross structure of those bands is
retained. However, positions shift according to the nature of
the ligand.[26]

The majority of diene-iron tricarbonyl complexes have been
made by direct reaction of dienes with one of the three common
iron carbonyls [$Fe(CO)_5$, $Fe_2(CO)_9$ and $Fe_3(CO)_{12}$], either by
simply heating the reagents together or by photochemical
reaction. Using $Fe(CO)_5$, a temperature of 120-160°C is required
and reactions are conducted in sealed tubes. Equivalent
reactions using $Fe_3(CO)_{12}$ in an inert solvent such as benzene or
ethylcyclohexene give satisfactory yields of products in shorter
times at somewhat lower temperatures (60-120°C). With $Fe(CO)_5$
and $Fe_3(CO)_{12}$, the products isolated are, with very few
exceptions, diene-iron tricarbonyls. Yet another important
feature of the chemistry of $Fe(CO)_5$ relative to interactions
with polymer matrices is the formation of complexes with
symmetric ring systems.

The chemistry of $Fe(CO)_5$ in the presence of halides is also
of great importance for our consideration. Iron pentacarbonyl

reacts with organic halides if two conditions are met. First, the halide must be activated by at least one, and preferably two, groups such as cyano, phenyl or halogen (in decreasing order of effectiveness). Second, there must be at least two halogens on the same carbon atom or in very close proximity to each other. For example, dichlorodiphenylmethane $(C_6H_5)_2CCl_2$, reacts readily with $Fe(CO)_5$[27] according to the following reaction:

$$2Fe(CO)_5 + 2(C_6H_5)CCl_2 \longrightarrow 10\ CO + 2FeCl_2 + (C_6H_5)_2C=C(C_6H_5)_2$$

Furthermore, 1,2 dichloro-1,1,2,2-tetraphenylethane, even though it does not have geminal halogen atoms, reacts readily with $Fe(CO)_5$ according to the following reaction:

$$Fe(CO)_5 + (C_6H_5)_2ClC-CCl(C_6H_5)_2 \longrightarrow 5CO + FeCl_2 + (C_6H_5)_2C=C(C_6H_5)_2$$

Molecular models of this halide reveal that the halogens (in the rotamer with the closest proximity of halogens) are as close as in a geminal dihalide and it is perhaps not surprising that reaction occurred. In no case does reaction occur with mono-halogenated compounds, even when the halide is as strongly activated as that in triphenylmethyl chloride.

Another type of reaction leads to the formation of iron complexes of the general form $FeCl_2L_2$ or $FeCl_3L$, where L stands for ligands such as formamide, N-methylformamide, aniline, benzamide, acetamide, triphenylphosphine and triphenylarsine. All three iron carbonyls [$Fe(CO)_5$, $Fe_2(CO)_9$ and $Fe_3(CO)_{12}$] react with these various ligands in the presence of chloroform to form halide adducts of the type $FeCl_2L_2$ or $FeCl_3L$. Solvents such as carbon tetrachloride, tetrachloroethane, and benzylchloride lead to similar oxidation reactions. Dichloroethane and methylene chloride lead to partial decomposition giving rise to impure compounds. The organic ligands which form $FeCl_2L_2$ complexes are formamide, methylformamide and acetamide, whereas those which form the $FeCl_3L$ complex are aniline, benzamide, triphenylphosphine and triphenylarsine. All ligands which have a phenyl group give rise to an iron complex of higher oxidation state.

The formation of complexes of the type $FeCl_2L_2$ or $FeCl_3L$ by the reaction of iron carbonyl with various ligands in the presence of halogenated hydrocarbons may be described by the following reaction mechanism. A substituted carbonyl complex of the type $Fe(CO)_4L$ forms and reacts with halogenated hydrocarbons

to give rise to a halide complex. To verify this, $Fe(CO)_4L$ was
allowed to react with chloroform, and the halide complex was
obtained. In order to establish the reaction mechanism, Singh
and Rivest[28] tried unsuccessfully to isolate the product formed
from the oxidation reaction in chloroform. Therefore, they
replaced the chloroform with diphenyldichloromethane and allowed
this to react with $Fe(CO)_4$ $(AsPh_3)$. They could then isolate
tetraphenylethylene and the halide complex. A possible
mechanism for the reaction, therefore, can be suggested:

$$Fe(CO)_5 + 2L \longrightarrow Fe(CO)_3L_2 + 2CO$$

$$2Fe(CO)_3L_2 + 2(C_6H_5)_2CCl_2 \longrightarrow 2FeCl_2L_2 + (C_6H_5)_2C=C(C_6H_5)_2 + 6CO$$

In view of the foregoing, the decomposition of $Fe(CO)_5$ in
polymers of differing molecular structures (and using different
mutual film casting solvents) may be expected to lead to
significant differences in polymer interactions and
decomposition chemistry. Polymers with different backbone
structures and side chain groups were therefore studied to help
understand the different physicochemical interactions which
might occur with iron pentacarbonyl and other organometallic
complexes.

EXPERIMENTAL

1. Preparation of Composite Films:

a. Polycarbonate/$Fe(CO)_5$/Methylene Chloride

Polycarbonate pellets (10.0 g) and 43.9 ml methylene
chloride (Fisher Reagent) were mixed 24 hours at room
temperature with a Teflon stir bar on a magnetic stir plate in
a 50 ml glass-stoppered erlenmeyer flask. Methylene chloride
used for solvation of the polymer was dried using molecular
seive pellets (Matheson, Coleman and Bell). Iron pentacarbonyl
(Alfa Products, Thiokol/Ventron Division) was filtered through
filter paper circles (Whatman 7.0 cm, qualitative 4) into a foil
covered test tube. Then 1.0 ml was added dropwise to the above
erlenmeyer, now completely covered with aluminum foil. This
solution of iron pentacarbonyl and polymer in methylene chlcride
was mixed for approximately five minutes and then film cast onto
sheets of glass using a 0.005" or 0.020" steel doctor blade.

Resultant films theoretically contained 13.0 wt. % Fe(CO)$_5$.
Films containing 9.1 and 4.8 wt. % Fe(CO)$_5$ in polycarbonate were
also prepared. All films were analyzed for iron content to
determine actual Fe(CO)$_5$ retained.

    b. Polymethylmethacrylate (PMMA)/Fe(CO)$_5$/Benzene

    PMMA pellets (5.0 g) and 32.2 ml benzene (Fisher-Reagent)
were mixed for 48 hours at room temperature in an erlenmeyer
flask. The benzene was dried using molecular seive pellets.
Higher loadings of iron pentacarbonyl were used for the PMMA
studies, primarily because iron analysis of composite PMMA films
indicated that less iron pentacarbonyl was retained in the PMMA
film (in comparison to polycarbonate composites). This may be
due to loss of the carbonyl with the benzene solvent during
drying of the films. Filtered iron pentacarbonyl was added
dropwise to a foil covered flask of PMMA-benzene solution and
mixed for five minutes. This solution was used for film
casting. PMMA composite films with initial loadings of 43, 37,
31, 23 and 11 wt.% Fe(CO)$_5$ were prepared by dropwise addition of
2.5, 2.0, 1.5, 1.0 and 0.4 ml Fe(CO)$_5$, respectively. As before,
all films were analyzed for iron content to determine the actual
Fe(CO)$_5$ concentration retained. The final concentrations were
found to be 0.8, 0.5, 0.4, 0.4 and 0.2 wt. % Fe(CO)$_5$ retained,
respectively.

2. Polyvinylidene Fluoride (PVF$_2$)/Fe(CO)$_5$/Dimethylformamide (DMF):

    Two methods were used to produce PVF$_2$ films for (1) TEM
studies and (2) kinetic studies. Films for TEM were prepared as
follows: PVF$_2$ pellets (10.0 g) and 94.8 ml N,N-dimethylformamide
(Fisher Reagent) were mixed for six days in a flask. Filtered iron
pentacarbonyl (4.1 ml) was added dropwise to a covered flask of
PVF$_2$-DMF. This solution was cast onto hot glass (110°C) and held
at this temperature for 1 hour. This was followed by heating for
24 hours at 140°C. This procedure simultaneously volatilized the
DMF and decomposed the carbonyl. For kinetic studies, solutions of
Fe(CO)$_5$-PVF$_2$-DMF were prepared and film cast at room temperature
and dried for 48 hours. Though poorer quality films resulted, they
were adequate for kinetic studies. Actual Fe(CO)$_5$ content was
determined by pyrolysis of the composite films followed by iron
analysis. Appropriate polymer controls were prepared by
eliminating the addition of iron pentacarbonyl to identical
polymer-solvent solutions. All cast films were foil covered to
prevent photolytic decomposition of the carbonyl during drying and
storage.

3. Quantitative Iron Analysis of Films:

The actual weight percent of iron in the untreated films was determined by Vogel's 1,10-phenanthroline method.[34] This method measures the UV absorption at 396 nm of dissolved samples. Absorption was compared to a standard curve to quantify total iron. Samples for this analysis were prepared by dissolving 0.02-0.10 g of the composite film in 15 ml dimethylformamide in a 25 ml volumetric flask. After solvation of the polymer, the following were added: 2.5 ml of 10% hydroxylammonium chloride (Fisher Reagent), 2.5 ml 0.2M sodium acetate (Fisher Scientific) to adjust pH to 3.5± 1.0, and 2.0 ml 1,10-phenathroline (Fisher Reagent) solutions were vortexed (Vortex-Genie, Scientific Industries) after each addition. This solution was then brought up to 25 ml with dimethylformamide and again mixed. After standing 1 hour (to develop full color), the solution was filtered through a Whatman 934-AH glass fiber filter to remove the precipitated polymer and the UV absorption of the filtrate was measured in quartz cuvettes. The phenanthroline iron analysis was checked using atomic absorption analysis (Perkin-Elmer 460 AA Spectrophotometer with iron hollow cathode lamp and air-C_2H_2 flame).

4. Effect of Iron Pentacarbonyl Concentration Upon Extinction Coefficient in a Polymer:

The IR carbonyl stretching absorption at 1996 cm^{-1} of the same films was measured. Using this absorption, the iron concentration (as determined from the 1,10-phenanthroline method), and Beer's Law (Abs = εdc), the extinction coefficient, ε, for $Fe(CO)_5$ in the various polymers was calculated as a function of polymer and concentration.

5. Decomposition Methods:

Thermal decompositions were carried out in a temperature controlled vacuum oven with controlled atmosphere capability. Decomposition times and temperatures were varied for kinetic measurements.

Photolytic (UV) decompositions were performed using a 100 watt mercury arc lamp (Ealing Optics Corp.) or a 75 watt xenon lamp (Ealing) mounted on an optical bench with a quartz water filter between the sample and source to avoid heating by infrared radiation. UV irradiance and time were varied.

a. Kinetics for Decomposition of $Fe(CO)_5$ in Polymer Matrix

Infrared absorption bands (Perkin-Elmer 283B Infrared Spectrophotometer and Nicolet MX-1 FTIR) at 1996 cm^{-1}, and 645 cm^{-1} (carbonyl stretching and bending, respectively) were observed to change during decomposition of the $Fe(CO)_5$-polymer films. The 1996 cm^{-1} band was used to follow the kinetics of the decompositions. Kinetics for thermal and UV decompositions and activation energies for the thermal decomposition reactions in composite films were determined.

b. Effect of $Fe(CO)_5$ Decomposition on Polymer Molecular Weight

Intrinsic viscosities ([η]) of polymers and composite compositions was determined using a Ubbelohde OB viscometer. Solutions of the polycarbonate and polysulfone films were prepared in dioxane (Fisher Reagent) and then filtered through glass fiber filters just prior to viscosity measurements. Solutions of polystyrene and polymethylmethacrylate were prepared in benzene. Measurements were carried out at 30°C. The [η] was used to calculate viscosity-average molecular weights, M_v, using the Mark-Houwink equation[35] and appropriate constants.

6. Microscopic Characterization of Composite Film Morphology:

Particle size and size distributions were measured by transmission electron microscopy (JEOL Model 200CX STEM and Philips Model 301). Electron diffraction patterns from a TEM field emission gun were used to study the composition and morphology of metallic particles. Samples were prepared by two methods. A 0.1% solution of polymer and $Fe(CO)_5$ mixture in methylene. chloride was applied dropwise to carbon coated (100 Å thick) TEM grids. The films produced on the grids were then treated to decompose the $Fe(CO)_5$. Composite films were also embedded in epoxy resin (Epon 812) and ultramicrotomed (LKB Ultratome) for TEM studies. Composite films were examined further by optical microscopy (Zeiss) and scanning electron microscopy (SEM) (JEOL Model 35C) with electron energy dispersive spectra capability (EEDS).

7. Kinetics of $Fe(CO)_5$ Decomposition in Ethylbenzene Solution:

Fifty ml of ethylbenzene (Matheson Coleman & Bell Reagent) was placed in a foil-covered 100 ml three-neck flask. Filtered iron pentacarbonyl (0.13 ml) was added to the ethylbenzene and stirred with a magnetic stir bar. This reaction vessel was fitted with a

condenser and stoppers (to allow for sampling) and placed in a
heated oil bath. Decompositions were measured at 70°, 90°, 110°,
and 130°C and followed by infrared spectroscopy in sealed
demountable NaCl liquid cells. An ethylbenzene reference was
used. Noack's[25] extinction coefficient of 8000 1 mol^{-1}cm^{-1} was
used for determination of Fe(CO)$_5$ concentration in the ethylbenzene
solution.

RESULTS AND DISCUSSION

1. Sample Preparation:

      One of the major problems in producing homogeneous solid
solutions of a metal carbonyl in a polymer is to insure that they
are comiscible in suitable film casting solvents. It is also
necessary that there be no interaction between the solvent and the
metal carbonyl. These conditions are sometimes difficult to
satisfy. For example, in the system PVF$_2$-Fe(CO)$_5$, PVF$_2$ is soluble
in DMF which is also a good solvent for Fe(CO)$_5$. However, the IR
spectra of Fe(CO)$_5$ in PVF$_2$ (Figure 3), shows the appearance of a
band at 1880 cm^{-1} corresponding to a carbonyl anion. This
indicates that a reaction occurs between Fe(CO)$_5$ and DMF; a
disproportionation reaction[29]. Figure 4 summarizes the possible
reactions that may occur between Fe(CO)$_5$ and DMF. The
disproportionation of Fe(CO)$_5$ in DMF is slow but cannot be
neglected.

      The boiling point of DMF is 136°C. Hence, the drying time for
the film is quite long at room temperature (2-3 days). There is
therefore sufficient time for the carbonyl-solvent reaction. This
is supported by data given in Table 1. Two different film casting
methods are compared. The only difference is the length of time
during which the iron carbonyl is in contact with DMF. In the
"hot" method, the contact time is only 1-2 hours until the film is
dry. For the "cold" method, where the film is dried at room
temperature, the time is several days. The resulting films are
quite different. With the "hot" method, very good, tough and
uniform films are obtained. With the "cold" method, it appears
that the polymer is attacked and degraded to some extent thus
yielding poorly formed films. The iron carbonyl anion which forms
due to the disproportionation reaction, is reactive and probably
attacks the polymer.

      In the PMMA-Fe(CO)$_5$ system, another problem arises. As a
rule, it is preferable for the solvent to have a boiling point well
below that of Fe(CO)$_5$, (101°C). Therefore, in most of the systems
studied, CH$_2$Cl$_2$ (boiling point of 40°C) was chosen. There is no
interaction between Fe(CO)$_5$ and methylene chloride under the
experimental conditions.

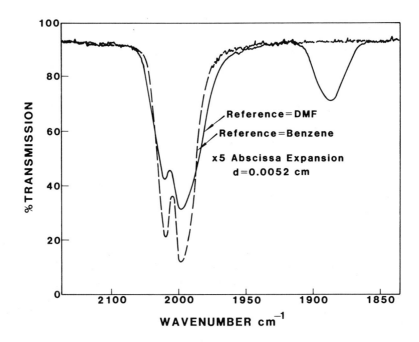

Figure 3.   Infrared Spectra of Fe(CO)$_5$ in DMF and Fe(CO)$_5$ in
Benzene

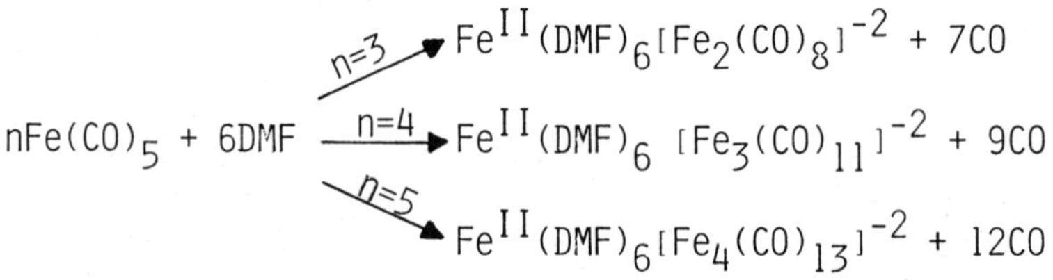

$$nFe(CO)_5 + 6DMF \xrightarrow{\quad n=3 \quad} Fe^{II}(DMF)_6[Fe_2(CO)_8]^{-2} + 7CO$$

$$nFe(CO)_5 + 6DMF \xrightarrow{\quad n=4 \quad} Fe^{II}(DMF)_6 \, [Fe_3(CO)_{11}]^{-2} + 9CO$$

$$nFe(CO)_5 + 6DMF \xrightarrow{\quad n=5 \quad} Fe^{II}(DMF)_6[Fe_4(CO)_{13}]^{-2} + 12CO$$

Figure 4.   The Disproportionation Reaction of Fe(CO)$_5$ in DMF

Table 1. Casting Methods for $Fe(CO)_5/PVF_2/DMF$ films

Solution	Casting Method	Quality of Film	Solubility in DMF	Color
PVF_2/DMF	Hot (114°C)	Good	No	Clear
	Cold (R.T.)	Good	Yes	Clear
$PVF_2/DMF/$	Hot	Good	No	Clear
$Fe(CO)_5$	Cold	Fair Cracked Area	No	Opaque

As stated previously[27,28], interactions between $Fe(CO)_5$ and chlorinated solvents can occur only when the latter have geminal dihalides and are activated by an electron donor group. This is not the case for CH_2Cl_2. For PMMA-$Fe(CO)_5$ however, CH_2Cl_2 could not be used because PMMA does not dissolve. Benzene was therefore used since it dissolves PMMA and because there is no $Fe(CO)_5$-benzene interaction. Since the boiling point of benzene is 80°C, $Fe(CO)_5$ evaporation occurs with benzene evaporation during the drying period. As a result, the initial carbonyl concentration decreased significantly during film forming. For example, an initial $Fe(CO)_5$ loading of 40% wt. was found to correspond to an actual loading of less than 10% wt. in the final cast films.

An additional problem arises because $Fe(CO)_5$ is very sensitive to UV light. The reaction:

$$2Fe(CO)_5 \xrightarrow{UV} Fe_2(CO)_9 + CO$$

occurs readily upon UV irradiation of $Fe(CO)_5$[15,16]. The product of this photolytic process, $Fe_2(CO)_9$, is more susceptible to oxidation than $Fe(CO)_5$, and hence upon exposure of the films to air and light at room temperature, $Fe(CO)_5$ may decompose in a very short time to iron oxide. To prevent this undesired decomposition the films must be shielded at all times from exposure to light and preferably stored in an inert atmosphere. If the first requirement is met, then $Fe(CO)_5$ is not converted to $Fe_2(CO)_9$, and the oxidation of $Fe(CO)_5$ at room temperture is very slow. Thus shielded films can be kept for long periods of time even in air without significant decomposition.

2. Quantitative IR Spectroscopic Measurements:

The identification of metal carbonyl species in solution or in a solid polymer matrix as well as the quantitative determination of $Fe(CO)_5$ concentrations in the polymer was carried out by infrared spectroscopy. IR was used because (a) most solvents and polymers do not have strong absorption bands in the region between 1800 and 2200 cm^{-1} where metal-carbonyls absorb (if a solvent does absorb in this region it can be compensated with a reference) (b) the carbonyl bands of metal-carbonyls, both terminal and bridging, are very sharp and well contoured in the 1800-2200 cm^{-1} region and permit an accurate evaluation of concentration and (c) the infrared spectra of a great number of metal-carbonyls, including $Fe(CO)_5$, are well known.[25,30] Moreover, the influence of various factors on the spectra (such as solvent effects, temperature or substituting ligands) are also well known[25,26,30], and therefore permit evaluation of changes in the spectra of the metal-polymer composite films.

The infrared spectrum of $Fe(CO)_5$ in a hydrocarbon solvent is shown in Figure 5. There are two strong bands at 2019 and 1996 cm^{-1}, and a weak band at 645 cm^{-1}. The two main absorption bands are retained also in the infrared spectra of $Fe(CO)_5$ in the various polymeric matrices as shown in Figure 6. A slight shift to lower wavenumber is observed due to the matrix polymer. One can conclude by comparing these spectra that $Fe(CO)_5$ retains its characteristic absorption in a solid polymer matrix. Hence, the spectra, can be used for quantitative measurements.

In 1960, Noack reported some extinction coefficient values for the main IR bands of $Fe(CO)_5$,[25] both in hydrocarbon and chlorinated solvents. Since we used both types of solvents it was of interest to compare our extinction coefficient values with those of Noack. The extinction coefficient for the 1996 cm^{-1} band calculated by Noack is 11,000 1 mol.$^{-1}$ cm^{-1} in a hydrocarbon solvent and 8000 1 mol^{-1} cm^{-1} in a chlorinated solvent. Table 2 shows the extinction coefficients we calculated for $Fe(CO)_5$ in a solid polymer solution. In all polymeric matrices, we observed that the extinction coefficient for the 1996 cm^{-1} band is lower than in solution, even when taking into consideration the solvent effect on this value. The same phenomenon has also been observed in the case of $Co_2(CO)_8$ in polystyrene.[31,32] However, the concentrations of $Fe(CO)_5$ used in the various polymeric matrices were higher than those used for extinction coefficient measurements in solution and one would expect lower extinction coefficient at higher concentrations in accordance with the Beer-Lambert law.[33] In addition, it is likely that the polymeric matrix may impose restrictions on the vibrational degrees of freedom of the metal-carbonyl resulting in bands of lower intensity.

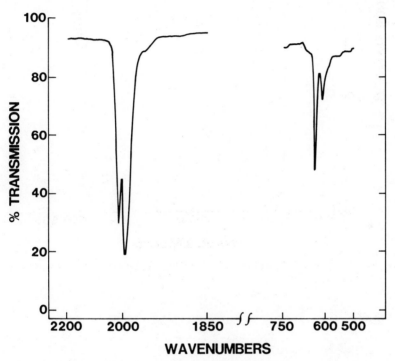

Figure 5. Infrared Spectra of Fe(CO)$_5$ in Benzene

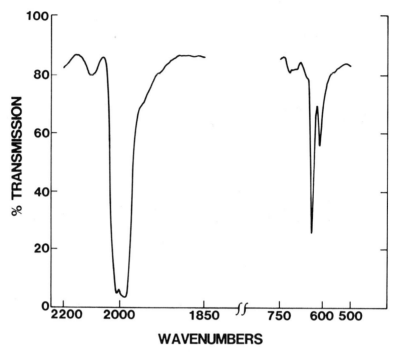

Figure 6.   Infrared Spectra of Fe(CO)$_5$ Polystyrene Composite Before
            Decomposition

3.   Decomposition Studies:

     Fe(CO)$_5$ thermal decompositions in various polymeric matrices
were studied at different temperatures and Fe(CO)$_5$
concentrations. The actual concentrations of Fe(CO)$_5$ in the
polymeric matrices are given in Table 3. Two typical series of
infrared spectra corresponding to two decomposition experiments are
presented. Figure 7 shows the thermal decomposition of Fe(CO)$_5$ in
PSF at 118°C. Figure 8 shows the thermal decomposition of Fe(CO)$_5$
in PVF$_2$ at 140°C. The Fe(CO)$_5$-PVF$_2$ system is presented separately
because the films intended for decomposition studies were initially

Table 2. Extinction Coefficients of $Fe(CO)_5$ at 1996 cm^{-1} in Various Polymeric Matrices.

POLYMER	$\epsilon_{Obs} \left(\dfrac{L \, Film}{cm \cdot Mole \, Fe} \right)$
Polycarbonate	2,524 [a.]
Polysulfone	5,189 [a.]
Polystyrene	7,148 [a.]
Polymethylmethacrylate	7,458 [b.]
Polyvinylidene Fluoride	---- [c.]

a. Films cast from methylene choride solution.

b. Films cast from benzene solution.

c. Data is not comparable with other systems due to disproportionation reaction of Fe $(CO)_5$ with solvent, dimethylformamide.

prepared by "cold" casting and hence there was a disproportionation reaction involving $Fe(CO)_5$ and DMF.

Therefore, all kinetic calculations for the PVF_2 system have been performed using the 1996 cm^{-1} IR carbonyl band. In all cases, the $Fe(CO)_5$ thermal decomposition follows first order reaction kinetics (see Figure 9 and Figure 10).

From the slopes shown above the rate constants for the thermal decompositions were calculated for each system and are summarized in Table 4. This table also includes the decomposition rate constant for a solution of $Fe(CO)_5$ in ethylbenzene. Note that the decomposition in a polymer matrix is _faster_ than in solution. This[31,32] observation has been also made in systems containing cobalt where the decomposition and oxidation of $Co_2(CO)_8$ was found to be two orders of magnitude faster in a solid polymeric matrix than in solution. The thermal decompositions were performed at 118°C, 125°C, 132°C and 142°C. The rate constants were plotted versus temperature as shown in Figure 11 and activation energies were calculated from the slopes (Table 5).

Table 3. Observed Fe(CO)$_5$ Concentration as Determined by
Vogel's 1,10-Phenanthroline Method

Polymer	Fe(CO)$_5$ Loading	Actual Fe(CO)$_5$ Concentration
Polycarbonate	10 wt. %	6.8 wt. %
Polysulfone	10 wt. %	2.0 wt. %
Polystyrene	10 wt. %	1.6 wt. %
Polymethylmethacrylate	43 wt. %	0.8 wt. %
Polyvinylidene fluoride[1]	60 wt. %	4.6 wt. %

[1] Resulting composite PVF$_2$ films were crosslinked and therefore
insoluble. Thus, a pyrolytic technique was necessary prior to
iron analysis.

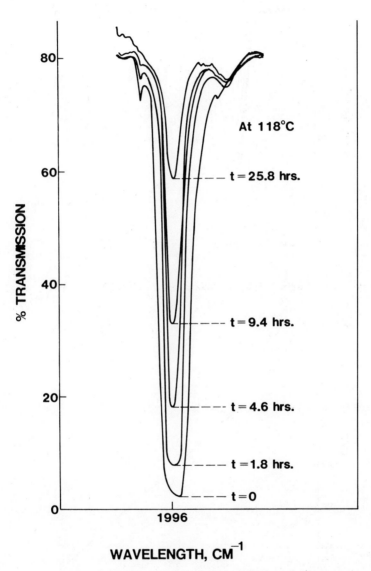

Figure 7. Thermal Decomposition of Fe(CO)$_5$ in PSF at 118°C

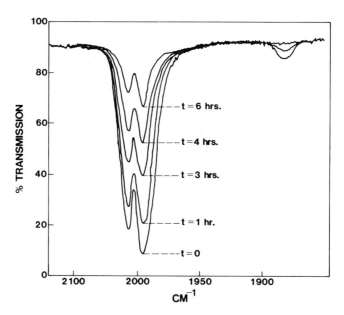

Figure 8. Thermal Decomposition of Fe(CO)$_5$ in PVF$_2$ at 140°C

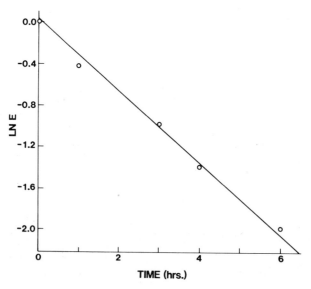

Figure 9. Decomposition kinetics of Fe(CO)$_5$-PVF$_2$ Composites at
           140°C

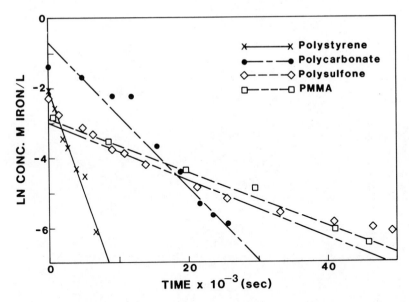

Figure 10.   Decomposition kinetics of Iron Pentacarbonyl-Polymer
Composites at 132°C

Table 4.   Observed Rate Constants for the Thermal Decomposition of
Fe(CO)$_5$-Polymer Composites

Polymer Composite	Rate Constant at 132°C (sec^{-1})
Fe(CO)$_5$-PC	-1.97 x 10^{-4}
Fe(CO)$_5$-PSF	-7.27 x 10^{-5}
Fe(CO)$_5$-PS	-5.59 x 10^{-4}
Fe(CO)$_5$-PMMA	-7.65 x 10^{-5}
Fe(CO)$_5$-ethyl benzene	-8.69 x 10^{-5}

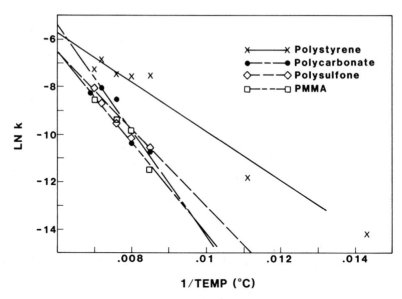

Figure 11.   Arrhenius Plot for Thermal Decomposition of Iron
Pentacarbonyl-Polymer Composites

Table 5.   Activation Energies for the Decomposition of
$Fe(CO)_5$-Polymer Composites

Composite System	Activation Energy ($\frac{Kcal}{mole}$)
$Fe(CO)_5$-PC	45.1
$Fe(CO)_5$-PSF	30.2
$Fe(CO)_5$-PS	29.0
$Fe(CO)_5$-PMMA	35.4

In addition to thermal decomposition, UV and electron beam decompositions were also performed on the $Fe(CO)_5$-polymer composites. The UV decomposition reactions were of particular interest since it is well known that $Fe(CO)_5$ is converted to $Fe_2(CO)_9$ upon irradiation[15, 16] and hence may provide a different mechanistic pathway for the decomposition of $Fe(CO)_5$. These decompositions were also followed by means of infrared spectroscopy, but the characteristic bands for $Fe_2(CO)_9$ at 2066, 2038, 1855 and 1851 cm^{-1}[38] were not observed. Since the conversion of $Fe(CO)_5$ to $Fe_2(CO)_9$ is diffusion dependent and since the experiment was performed in air, it is required that the rate of diffusion of $Fe(CO)_5$ through the polymeric matrix be greater than the oxidative decomposition rate. From preliminary experiments, this did not seem to be the case. At this point, it is not possible to speculate on the UV decomposition mechanism in these systems. Results obtained from electron-beam decompositions also do not permit mechanistic conclusions at this point.

4. Characterization of Metal-Polymer Composites:

The thermal decomposition of $Fe(CO)_5$ in solid polymeric matrices produced heterophase metallic domains. In the $Fe(CO)_5$-PVF_2 system, ferrimagnetic particles are obtained upon thermal decomposition. TEM indicated that the size of most particles was between 100Å and 300Å. Figures 12 and 13 are a typical TEM photo micrographs showing particles formed by thermal decomposition of $Fe(CO)_5$ in a polycarbonate matrix. Electron diffraction patterns for the $Fe(CO)_5$-PC composite product (Figure 14) are indicative of γ-Fe_2O_3. Figures 15-17 show the interesting morphologies observed in the various composites studied. In all cases, γ-Fe_2O_3 was produced in all composites formed by decompositions in air. Other particles (e.g. FeF_2, FeS, β-FeOOH) were also formed from $Fe(CO)_5$ which were dependent upon the polymer used. Table 6 summarizes the composition of particles obtained in the various systems studied.

The electron diffraction analyses for FeF_2 and FeS are presented in Table 7. The formation of FeF_2 particles in the PVF_2 system and FeS particles in the PSF system indicate an interaction between the $Fe(CO)_5$ and polymer with abstraction of F or S. The fluorine atom in PVF_2 (see Figure 2) is not part of the main chain while the sulfur atom in PSF is part of the sulfone group in the polymer chain. Therefore, attack at the sulfur atom would be expected to affect the PSF chain whereas abstraction of the fluorine atom could lead to secondary carbonium ions on PVF_2 which would react further.

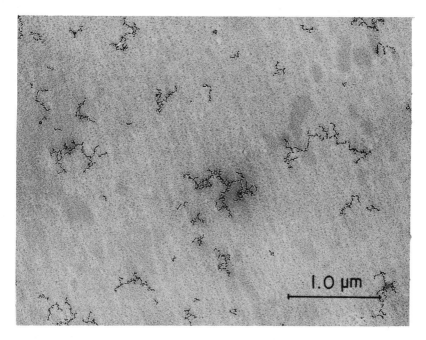

Figure 12.   Particles Formed During the Thermal Decomposition of
             Fe(CO)$_5$ in a Polycarbonate Matrix

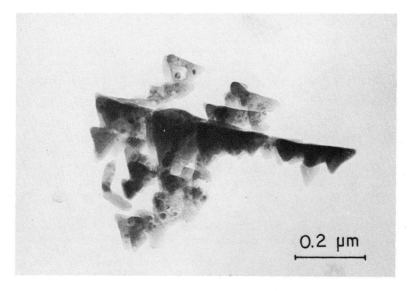

Figure 13.   Particles Formed During the Thermal Decomposition
             of Fe(CO)$_5$ in a Polycarbonate Matrix

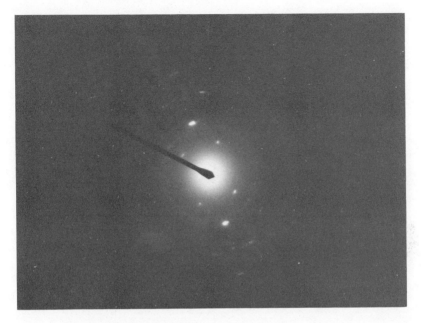

Figure 14.   Electron Diffraction Patterns Indicative of γ-Fe_2O_3
Formed During the Thermal Decomposition of $Fe(CO)_5$ in a
Polycarbonate Matrix

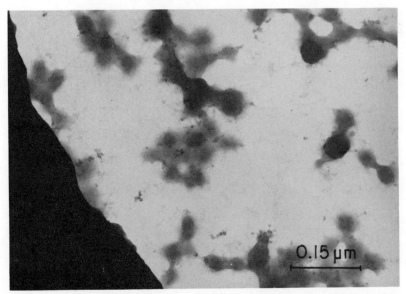

Figure 15.   Particles Formed in $Fe(CO)_5$-PVF_2 Composite, Cast by
"Cold Method" and Thermally Treated

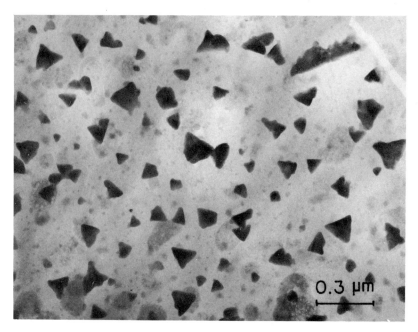

Figure 16.   Particles Formed in $Fe(CO)_5-PVF_2$ Composite Cast by "Hot Method"

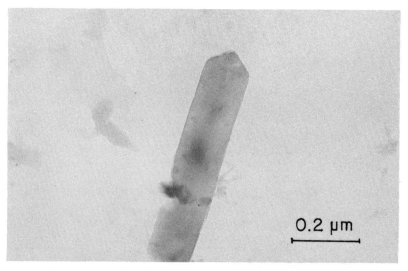

Figure 17.   Single Crystal Formed During UV Decomposition of $Fe(CO)_5$ in PMMA Matrix (indexing is indicative of $\gamma-Fe_2O_3$)

Table 6.   Particles Obtained in Various Systems (as determined from
            electron diffraction patterns)

System	Particles
$Fe(CO)_5$-PVF_2	γ-Fe_2O_3, α-Fe, FeF_2
$Fe(CO)_5$-PC	γ-Fe_2O_3, β-FeOOH
$Fe(CO)_5$-PSF	γ-Fe_2O_3, δ-FeOOH, α-Fe, FeO, FeS
$Fe(CO)_5$-PMMA	γ-Fe_2O_3, δ-FeOOH

Viscosity molecular weight measurements were made for the
various composites before and after thermal treatment. Changes in
molecular weights were observed after thermal treatment for all
polymers. The results are summarized in Table 8. Polycarbonate is
most extensively attacked by the iron carbonyl during thermal
decomposition with a large reduction in molecular weight. This is
accompanied by evolution of CO as well as a considerable increase
in the OH stretching IR band.

The formation of β-FeOOH indicates reaction with water present
during decompositions in air:

$$2FeOOH \longrightarrow Fe_2O_3 + H_2O$$

Reaction of the passive γ-Fe_2O_3 particles with water is well known
and has been discussed by Griffith et. al.[5] It should be noted,
however, that it is the α-polymorph of FeOOH which is normally
formed. The β structure has been observed to form only during the
breakdown of γ-Fe_2O_3 in the presence of chlorinated solvents. Thus,
it is suggested that traces of CH_2Cl_2 present in the PC and PSF
systems (supported by electron energy dispersive spectra) and
athospheric moisture are responsible for this reaction.

The formation of metallic iron particles must also occur even
though the concentration of the α-Fe particles is low due to the
oxidative atmosphere in which the decompositions were conducted.
In an inert atmosphere, these species could become the dominant
particles, as observed for decompositions of $Co_2(CO)_8$ in N_2 which
yield α-Co particles almost exclusively.[31,32]

Table 7. Electron Diffraction Analysis for FeF_2 and FeS in
PVF$_2$ and PSF Composites, Respectively

Electron Diffraction d(Å)	FeF$_2$ d(Å)	Electron Diffraction d(Å)	FeS d(Å)
3.31	3.32	3.59	3.65
2.67	2.70	3.01	3.24
2.33	2.34	2.87	2.84
2.09	2.10	2.53	2.59
1.76	1.73	2.45	2.44
1.64	1.66	2.07	2.11
1.48	1.49	----	1.98
----	1.48	1.81	1.75
1.41	1.42	1.64	1.63
1.36	1.39	1.51	1.50
----	1.21	1.41	1.41
1.19	1.17		
0.99	1.11		
0.92	1.05		

For the PMMA composite, some reduction in molecular weight was
also noted upon decomposition of $Fe(CO)_5$. This was accompanied by
an increase in OH IR band at 2940 cm^{-1}. Although a slight drop in
viscosity was also noted for the PMMA control as well it was not
accompanied by changes in the IR spectra. The glass transition
temperature for PMMA, ~95°C, is substantially lower than that of PC
(147°C) and PSF (170°C). Both PC and PSF controls showed no change
in molecular weight with the 142°C thermal treatment used prior to
viscosity measurements. In general, it appears that $Fe(CO)_5$-
polymer interactions occur in most systems often leading to changes
in molecular weight and branching or cross-linking of the
polymers. Metal binding to polymer is also likely in some cases.

Table 8. Viscosity–Molecular Weight Averages for Iron
Pentacarbonyl-Polymer Composites

Polymer[a]	Treatment	
	None	Thermal[b]
Polycarbonate[c]	32,200	32,200
Fe(CO)$_5$-PC	30,500	4,400
Polysulfone[c]	30,100	31,600
Fe(CO)$_5$-PSF	27,000	21,250
Polystyrene[d]	——	127,100
Fe(CO)$_5$-PS	——	77,800
Polymethyl methacrylate[d]	——	323,400
Fe(CO)$_5$-PMMA	——	211,800

[a]Molecular weight values for composites are based on polymer
weight (i.e. actual iron or iron pentacarbonyl weight excluded).

[b]Samples thermally treated to completely decompose carbonyl
(as indicated by infrared spectra).

[c]Ubbelohde OB viscometer; 1,4-dioxane solvent; 30°C.

[d]Ubbelohde OB viscometer; benzene solvent; 30°C.

5. Mechanistic Aspects:

The results of this study, permit a general mechanistic view
of the thermal decomposition of Fe(CO)$_5$ in polymeric matrices. For
polycarbonate the molecular weight is drastically reduced. The
appearance of an OH stretching IR band at 3510 cm^{-1} indicates the
formation of hydroxyl functions accompanying polycarbonate chain
degradation. A possible mechanism for the decomposition of Fe(CO)$_5$
in PC is given in Figure 18. It seems reasonable to assume that
the first step of the Fe(CO)$_5$ decomposition is an attack by the
iron carbonyl on the ester function of the polymer, followed by
iron oxidation and disruption of the polymer chain. The site of
the Fe(CO)$_5$ attack cannot be assigned specifically, but, taking
into account the relatively neutral character of the iron atom, the
C=O bond is most likely the primary site of an nucleophilic
attack. A molecule of water must also be involved in this
hydrolylic process giving rise to the OH groups. The presence of
H$_2$O was also shown by the formation of β-FeOOH in the presence of
chlorinated solvents.

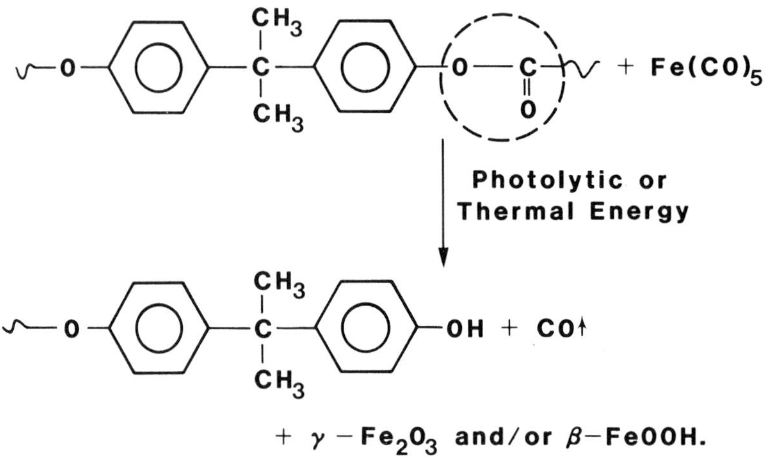

Figure 18.   Proposed Mechanism for Decomposition Reaction of
             $Fe(CO)_5$ in a Polycarbonate Matrix

In the PMMA system, the appearance of a 2940 cm^{-1} IR band
corresponding to a carboxylic OH group indicates formation of
carboxylic groups. The involvment of H_2O is also indicated by the
formation of δ-FeOOH. $Fe(CO)_5$ attack on PMMA must also occur at
the C=O bond. In this case, unlike the PC system, the ester group
is not a part of the polymer backbone, but only a side group.
Therefore, the integrity of the PMMA chain should not be directly
affected.

In the PSF system, a slight change in molecular weight was
observed (see Table 8). Only limited degradation of the polymer
chain occurs in spite of the formation of FeS. At this point, we
cannot suggest any good mechanism. Limited nucleophilic attack by
$Fe(CO)_5$ and/or water (or a more reactive iron species formed as a
first step upon exposure to thermal treatment) on sulfur or oxygen
of the SO_2 group within the polymer backbone must occur and some
recombination of polymeric phenylene species may result in
substantial maintenance of polymer chain length (via formation of
biphenylene or phenyl ether linkages).

In the PVF_2 system, two events may explain the formation of
γ-Fe_2O_3 and FeF_2. The first is that a more reactive species than
$Fe(CO)_5$ such as anionic carbonyls of the type
$Fe^{II}(DMF)_6[Fe_3(CO)_{11}]^{-2}$ are more likely to abstract fluorine.
These anionic complexes are formed during the interaction between
$Fe(CO)_5$ and DMF (see Figure 4). In the $Fe^{II}(DMF)_6$ moiety the iron
atom is cationic and very likely to abstract fluorine in PVF_2.
This mechanism is more probable in films cast by the "cold"

method. In the "hot" method there is less time for the $Fe(CO)_5$ to
react with DMF to produce the anion-carbonyl complexes. Therefore,
in this latter case, we suggest a second mechanism involving a DMF-
carbonyl complex as shown in Figure 19. No infrared evidence was
found for the formation of a $(DMF)Fe(CO)_4$ complex. However, for
films were cast by the "hot" method, the anionic carbonyl pathway
may also occur very rapidly at the elevated temperature as well as
the reaction in Figure 19. It isn't possible from the available
data to draw strong mechanistic conclusions at this time.

6. Related Metal-Polymer Composite Systems:

We have studied other metal-polymer composites which should be
briefly noted here. Results for dicobaltoctacarbonyl in PVF_2 were
similar to the $Fe(CO)_5-PVF_2$ system. $Co_2(CO)_8$ undergoes a vigorous
disproportionation reaction with DMF:[36]

$$3Co_2(CO)_8 + 12DMF \longrightarrow 2Co^{II} (DMF)_6 [Co(CO)_4]_2^- + 8CO$$

The reaction is several orders of magnitude faster than with
$Fe(CO)_5$ and the resulting $Co(CO)_4$ anion is a very reactive
species.[37] The films obtained with this system were brittle and
insoluble indicating substantial polymers-carbonyl interaction.
Electron diffraction, showed the particles formed after thermal
treatment to be were cobalt oxides (CoO and Co_2O_3) and cobalt
fluoride, CoF_2.[37] Figure 20 shows a TEM photo micrograph of the
particles formed in thermally treated $Co_2(CO)_8-PVF_2$ composites.
The formation of CoF_2 in this system parallels the formation of
FeF_2 in the $Fe(CO)_5-PVF_2$ system, and points to the fact that in
both cases the polymer matrix was attacked by a reactive metal-
carbonyl species (perhaps the anion formed during the
disproportionation reaction). This resulted in abstraction of a
fluorine atom, and in the case of cobalt, also resulted in severe
degradation of the polymeric matrix. The faster reaction rates for
cobalt systems may be expected since cobalt carbonyls are known to
be more sensitive to oxidative environments than iron carbonyls.

Another system studied was $Co_2(CO)_8-PS$. Here the rate
constant of the oxidative decomposition of $Co_2(CO)_8$ was two orders
of magnitude faster in the solid polymeric matrix than in
hydrocarbon solution.[31,32] This was similar to $Fe(CO)_5$ (Table 4)
although the iron carbonyl reaction showed a difference of only one
order of magnitude between solid state and solution kinetics. One
possible explanation for this phenomenon is that in solid polymeric
matrices, the metal carbonyl is dispersed in a very thin film with
greater exposure to the oxidative atmosphere than in solution.

$$Fe(CO)_5 + DMF \longrightarrow (DMF)\,Fe(CO)_4 + CO$$

$$(DMF)Fe(CO)_4 + (R - C\overset{F}{\underset{F}{\diagdown}}) \longrightarrow FeF_2 + R^1$$

Figure 19.   Proposed Mechanism for Decomposition Reaction of
Fe(CO)$_5$ in a Polyvinylidene Fluoride Matrix

Therefore the oxidative decomposition of the metal carbonyl
proceeds faster in the polymeric matrix. The differences between
the decomposition rates in solution and in the solid polymeric
matrix are larger for Co$_2$(CO)$_8$ than for Fe(CO)$_5$ because the cobalt
carbonyl is more sensitive to O$_2$ than Fe(CO)$_5$. In an invert O$_2$
free environment, we have also demonstrated the predominant
formation metal rather than oxide microparticles. Other metal
carbonyls and organometallics are under investigation to
demonstrate the generality of polymer matrix decompositions.

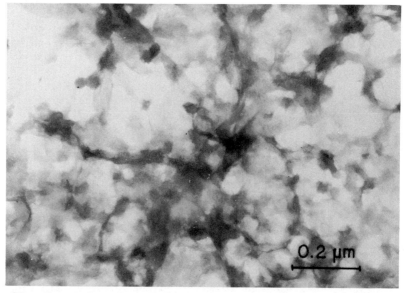

Figure 20.   Particles Formed During the Thermal Decomposition of
Co$_2$(CO)$_8$ in PVF$_2$ Matrix

SUMMARY

The systems discussed in this chapter indicate the complexity of the decompositions of organometallic complexes in solid polymer matrices and the potential for preparing new and interesting metal-polymer composites. The following conclusions may be drawn:

The formation of uniform dispersions of metals or oxides by the decomposition of metal carbonyls or other organometallic complexes in a solid polymeric matrix may give rise to a new class of composite materials with very interesting physical or electronic properties. These materials may also be designed to exhibit novel catalytic activity. The technique described in this study may also be of value for development of novel solid matrix isolation methods to substitute for conventional gas matrices for infrared matrix isolation studies of excited atomic or molecular species.[39, 40] By understanding the variables which govern decomposition chemistry, it is hoped to improve this phase separation technique and thereby better control the decompositions and the composite compositions formed.

ACKNOWLEDGEMENTS

This work was funded in part by a State of Florida Materials Center of Excellence. The authors are indebted to Dr. Shimon Reich who pioneered the work on $\gamma Fe_2O_3-PVF_2$ during 1980 when he was a Visiting Professor at the University of Florida. The authors also wish to thank Dr. M. Levy for contributions to early studies; Drs. J. Hren, J. Newkirk and E. J. Jenkins for their advise and help with electron microscopy; and M. Smith for her diligence and patience in the preparation of this chapter.

REFERENCES

1. K. J. Klaubunde, Y. Tanaka, Proceedings of the 57th Colloid and Surface Science Symp. (1983).
2. Y. Inizu and K. T. Klaubunde, Proceedings of the XI International Conference on Organometallic Chemistry, (1983), pg. 77.
3. G. F. Emerson, K. Ehrlich, W. P. Giering, D. Ehntholt, Trans. N. Y. Acad. Sciences, 30(7), 1001-1010 (1968).
4. G. F. Emerson, J. E. Mahler, R. Kochlar, R. Pettit, J. Org. Chem. 29, 3620-3624 (1964).
5. C. H. Griffiths, M. P. O'Horo, T. W. Smith, J. Appl. Phys., 50, (11) 7108-7115 (1979).
6. T. W. Smith, D. Wychick, J. Phys. Chem. 84, 1621-1629 (1980).

7. P. H. Hess, P. H. Parker, Jr., J. Appl. Polym. Sci. <u>10</u>, 1975-1927 (1966).

8. C. L. Flenniken, S. Reich, M. Levy and E. P. Goldberg, Proceedings of the 57th Colloid and Surface Science Symp. (1983)

9. C. L. Flenniken, M. Levy, S. Reich and E. P. Goldberg, in preparation.

10. S. Reich, E. P. Goldberg, J. Poly, Sci. Phys. Ed. <u>21</u>, (6) 869-879 (1983).

11. F. A. Cotton and G. Wilkinson, "Advanced Inorganic Chemistry" 3rd Ed., John Wiley and Sons, 1972, p. 730.

12. J. W. Moore, Acta. Chem. Scan <u>20</u>, 1154 (1966).

13. J. P. Ysinowski, T. L. Brown, Inorg. Chem. <u>70</u>, 1097 (1971).

14. S. F. A. Kettle and L. E. Orgel, Chem. Ind. (London) 49 (1960).

15. G. O. Schenck, E. Koerner van Gustorf and Mon-Jon Tun, Tetrahedron Letters, 1059 (1962).

16. E. Speyer, H. Wolf, Chem. Ber. <u>60</u>, 1424 (1927).

17. A. R. Luxmoore and M. R. Truter, Proc. Soc. Chem. p. 466 (1960).

18. H. D. Murdock and E. Weiss, Helv. Chim. Acta. <u>45</u>, 1156 (1962).

19. H. Reihlen, A. Gruhl, G. von Hessling and O. Pfrengle, Ann. Chem. <u>482</u> 161 (1930).

20. W. Reppe and H. Vetter, Ann. Chem. <u>582</u>, 133 (1953).

21. B. F. Hallam and P. L. Pauson, J. Chem. Soc. 642 (1958).

22. F. A. Cotton and G. Wilkinson, "Advanced Inorganic Chemistry" 3rd Ed. John Wiley and Sons, 1972, p. 741.

23. H. J. Dauben and D. J. Bertelli, J. Am. Chem. Soc. <u>83</u> 497 (1961).

24. A. Davison, M. L. H. Green and G. Wilkinson, J. Chem. Soc. 3172 (1961).

25. K. Noack, Helv. Chim. Acta <u>45</u> 1987 (1962).

26. F. A. Cotton and G. Wilkinson, "Advanced Inorganic Chemistry" 4th Ed. John Wiley and Sons, 1980, p. 85.

27. L. Eugene Coffey, J. Am. Chem. Soc. <u>83</u>, 1623-1626 (1961).

28. P. P. Singh and R. Rivest, Can. J. Chem. <u>46</u>, 1773-1779 (1968).

29. W. Hieber and N. Kahlen, Chem. Ber. <u>91</u> 2223 (1958).

30. P. S. Braterman, "Metal Carbonyl Spectra" Academic Press, 1975.

31. R. Tannenbaum, C. L. Flenniken, E. P. Goldberg, XI International Conference on Organometallic Chemistry, 1983, p. 77.

32. R. Tannenbaum, C. L. Flenniken, E. P. Goldberg, paper in preparation.

33. P. W. Atkins, "Physical Chemistry" 2nd Ed. W. H. Freeman and Co., 1982, p. 605.

34. J. Bassett, R. C. Denney, G. H. Jeffery and J. Mendham, Vogel's Textbook of Quantitative Inorganic Analysis, 4th edition, Longman, London and New York 1978, p. 742.

35. F. W. Billmeyer, Introduction to Polymer Science and Technology, H. S. Kaufman and T. T. Falcetta editors. Wiley-Interscience, 1977, p. 186.

36. I. Wender, H. W. Sternberg and M. Orchin, J. Am. Chem. Soc. 74. 1216 (1952).

37. S. Reich, R. Tannenbaum, C. L. Flenniken and E. P. Goldberg, Proceedings of the 57th Colloid and Surface Science Symp. (1983).

38. M. Poliakoff and J. J. Turner, J. Chem. Soc. (A) 2403 (1971).

39. R. H. Hooker and A. J. Rest, J. Organometal. Chem. 249 137-147 (1983).

40. R. H. Hooker, K. A. Mahmoud and A. J. Rest, J. Organometal. Chem. 254, C25-C78 (1983).

MICROEMULSION; SURFACTANT VESICLE AND POLYMERIZED
SURFACTANT VESICLE ENTRAPPED COLLOIDAL CATALYSTS AND
SEMICONDUCTORS: PREPARATION, CHARACTERIZATION, AND
UTILIZATION

Janos H. Fendler and Kazue Kurihara

Department of Chemistry and
Institute of Colloid and Surface Science
Clarkson College of Technology
Potsdam, New York 13676

INTRODUCTION

Noble metal catalysts have been extensively utilized in many important
industrial processes for some time. Typically, they are deposited on
solid surfaces in reactors. Physical chemical characterizations of catalytic
surfaces have only recently become available.[1] In general, catalytic
activity is related to surface area. Smaller particles have larger surface
areas and are, therefore, better catalysts than larger particles.

Semiconductors have found applications in electrical and photoelectrical
processes and in solar energy conversion.[2-5] They are deposited on
a variety of solid surfaces or on electrodes. Particularly noteworthy
is the recent utilization of dispersed colloidal doped semiconductors
in artificial photosynthesis.[2-5]

The reproducible preparation of stable, uniform, catalitically active
and selective colloidal catalysts or semiconductors are of obvious importance.
Utilization of membrane mimetic systems[6] hold the key for such preparations
since they provide well defined compartments for the in situ generation
and stabilization of colloidal species. Water-in-oil, w/o, microemulsions,
surfactant vesicles and polymerized surfactant vesicles have provided
the best systems. The size of the catalysts or semiconductors is quite
simply restricted by the volume available inside the microemulsions and
vesicles.

Concentration dependent self-association of suitable surfactants,
such as sodium di-(2-ethylhexyl)sulfosuccinate, AOT, in apolar solvents
result in the formation of reversed micelles or w/o microemulsions.
Definitions of reversed micelles and microemulsions are not agreed upon.[6]

341

In general, aggregates containing fewer than 20 - 50 surfactants in a
hydrocarbon solvent are referred to as reversed micelles whereas larger
ones, containing excessive water, are referred to as w/o microemulsions.
Importantly, AOT can solubilize up to 60 molecules of water/molecule
of surfactants in the apolar solvent. The size of the aggregates depend
on the water to surfactant ratios, w-values. The hydrodynamic diameters,
at constant w-values, are independent of the temperature for reversed
micelles, but they are dependent on it for microemulsions. For example,
the diameter of aggregates in isooctane at w = 15, is 80 Å (reversed
micelles) while at w = 60, and at 20° C, it is 300 Å (microemulsions).
w/o microemulsions are dynamic. Their water pools exchange on the
millisecond timescale.[7]

Swelling of dried phospholipids or surfactants in water results in
the formation of onion-like multicompartment vesicles.[6] Sonication above
the phase transition temperature yields single compartment bilayer vesicles
whose diameters are in the 300-2000 Å range. The kinetic stability of
vesicles is considerably greater than that of microemulsions. Even greater
stabilities, controllable permeabilities and sizes have been obtained by
developing polymerized surfactant vesicles.[9] Vesicle forming surfactants,
functionalized with vinyl, methacrylate, diacetylene, isocyano, and styrene
groups have been synthesized. Polymerizable double bonds in these surfactant
are located either at the end of the hydrocarbon tail of the surfactant
or at their headgroups. Subsequent to vesicle formation, irradiation by
ultriviolet light, or exposure to an initiator (azobisisobutyronitrile, AIBN,
potassium persulfate, for example) resulted in the loss of the polymerizable
double bonds. Depending on the position of the double bonds, vesicles
could be polymerized either across their bilayers or headgroups.[9]

Vesicles and polymerized vesicles are capable of solubilizing a large
number of substrates per aggregate. They also have the largest number
of solubilization sites. Hydrophobic molecules can be distributed among
the hydrocarbon bilayers of the vesicles. Polar molecules may move about
relatively freely in vesicle-entrapped water pools, particularly if they
are electrostatically repelled from the inner surface. Small charged ions
can be electrostatically attached to the oppositely charged outer or inner
surfaces of vesicles. Species having charges identical to those of the
vesicles can be anchored onto the vesicle surface by a long hydrocarbon
tail.

Properties of the media need to be kept in mind in preparing com-
partmentalized colloidal particles. Thus, in microemulsions it is not possible
to restrict the growth of the colloid by controlling the concentration
of the precursor. Even though initially there may be only 3 - 5 molecules
of precursors per microemulsion interchanging of the water pools will
lead to the growth of colloidal particles.[10] The ultimate stability of the
colloid bearing microemulsion is the only factor which determines the
sizes of particles. Conversely, vesicles and polymerized vesicles allow

fine controls since, once entrapped, the precursors and the colloids formed from them remain locked inside the compartments. Sizes of colloidal particles can be controlled, therefore, by the concentration of precursors entrapped in each vesicle and by the available internal volume.

Microemulsions, w/o, have been utilized for the preparation of monodispersed 80 A , Pt, Pd, Rd, Ir, Au, Ag and Ir, particles which were subsequently transferred to solid support or used catalysts.[10-14]

Emphasis in the present article will be placed on recent work pertaining to the mechanism of colloidal catalyst and semiconductor formation in microemulsions, surfactant vesicles and polymerized vesicles, their characterization and utilization.

EXPERIMENTAL

Both neutral pentaethylene glycol dodecyl ether, PEDGE, and ionic AOT were used for forming w/o microemulsions. PEDGE was prepared by molecular distillation of the commercial surfactant Berol 050, which according to the manufacturer (Berol, Stenungsund, Sweden) has a distribution of chain lengths with a mean value of 5 units in the polyoxyethylene chain. The main impurities are dodecanol and sodium chloride. Low-boiling impurities were removed by six consecutive distillations at 10^{-3} mm Hg and 50, 70, 80, 90, 100, and 115°C, respectively. The residue of the last distillation was finally distilled at 10^{-3} mm Hg and 125°C, and the distillate was used as the final product.

AOT (Fisher), was purified by two different methods. In the first method, due to Eicke,[7] 100 g AOT were mixed with 30 g activated carbon and 1 liter MeOH. This mixture was stirred during 24h and separated from the activated carbon by filtration with sintered glass filters no. 4 and no. 5. The MeOH was evaporated by means of a rotary evaporator (40°C) and the AOT was dissolved in 750 ml petroleum ether. This solution was washed with 2 x 200 ml H_2O. After phase separation the organic phase was evaporated to a gel which was dissolved in 500 ml MeOH. This solution was washed three times with portions of 300 ml/100 ml/ 100 ml petroleum ether. The MeOH phase was evaporated to dryness. The residue was dried with a vacuum pump. Afterwards the AOT was dissolved in ethylether, the solution was evaporated to dryness and dried again with the vacuum pump at 0.02 Torr. In the second method, 100g AOT was dissolved in 1000 ml methanol and 10g activated charcoal was added. The solution was filtered, the solvent rotary evaporated (below 40°C). In the limits of our experimentation no differences were found between these two methods of AOT preparations.

Dihexadecylphosphate, DHP, and dipalmitoylphosphatidylcholine, DPPC, (Sigma) were used as received.

Preparation, purification and characterization of a styrene containing surfactant $[H_2C=CHC_6H_4NHCO(CH_2)_{10}][C_{16}H_{33}]N^+[CH_3]_2$, Br^-, **1**, have been described. $HAuCl_4$, K_2PtCl_4, $FeCl_3$, $CdCl_2$, H_2S, $RhCl_3$, PhSH, EDTA were the best commercial available compounds.

Microemulsions were prepared by dissolving weighted amounts of dried surfactant in hydrocarbon and adjusting the water content to the desired value. Care was taken to keep the w–values constant throughout given sets of experiments.

Vesicles were prepared by sonication using a Bransonic sonicator. Typically, sonications were carried out at 60°C for 20–30 minutes.

Colloidal Pt coated CdS semiconductors were prepared _in situ_ in AOT microemulsions. Typically, 3×10^{-4} M $CdCl_2$ were dissolved in 0.1 M AOT in isooctane at w–values of 20. Gentle exposure to H_2S resulted in the formation of CdS inside the AOT reversed micelles. Concentration of CdS were determined to be 10^{-4} M by ultraviolet spectroscopy. Platinization was carried out by adding the appropriate concentrations of $K_2Pt_2Cl_4$ (maintaining the same w–values) to the CdS carrying reversed micelles and irradiating with a 450 Xenon lamp with Argon bubbling for 30 minutes.

Preparation of vesicle and polymerized vesicle entrapped catalyst and semiconductors were somewhat more complex. The appropriate precursors were entrapped by cosonication with the surfactants. Those remaining on the outside of the vesicles were removed by gel filtration or by passages through ionic exchange resins. Absence of extravesicular precursors or colloids were verified by suitable analytical methods. In case of using vesicles which were prepared from polymerized or mixtures of polymerized or non–polymerized surfactants, polymerization had been carried out by irradiating with ultraviolet light. Irradiation by ultraviolet light in the presence of $K_2Pt_2Cl_4$ or $RhCl_3$, resulted in the concurrent platinization or rhodination of the vesicle entrapped colloidal semiconductor and in surfactant photopolymerization.

Colloid formation, polymerization and substrate reductions were monitored absorption spectrophotometrically using a CARY 118C spectrophotometer.

Hydrodynamic diameters of the particle containing vesicles were determined by a Malvern 2000 light scattering system using a Spectra Physics 171 Ar^+ ion laser as the excitation source. Data were collected at 90° and 23° C. Typical sampling times were 3–5 μsec. Each measurement was carried out at least in triplicate. Q–values (polydispersity indices) were less than 0.3 ($\pm$ 3%).

Electron micrographs were taken on a Philips model 100C instrument. Vesicles were deposited on an Ernest F. Fullans Inc. 11270 (collodion coat) grid by ultracentrifugation at 43,000 g for 30 minutes. Flash photolysis measurements were carried out by a Nd:YAG laser (Quanta Ray DCR–1A)

laser system. Fluorescence lifetime measurements were performed on
a single photon counting instrument using a Spectra Physics Sync-pump
laser as the excitation source.

COLLOIDAL PARTICLES IN MICROEMULSIONS

The method of colloidal metal particle formation in microemulsions
depends on the nature of metal, the precursors and the system used.
Colloidal platinum particles are the easiest to prepare. Reduction of
microemulsion entrapped H_2PtCl_6 by hydrogen takes only a few minutes.
Pd particles are obtained by bubbling hydrogen through the microemulsions
containing $PdCl_2$. $RhCl_3$ cannot be produced by hydrazine reduction
of the precursors due to the formation of stable hydrazine-rhodium complex.
Colloidal rhodium can be prepared by extensive bubbling of $RhCl_3$ with
hydrogen. Ir particles are the most difficult to prepare. Reduction requires
the use of Al_2O_3 with 2% metallic platinum and extensive hydrogen bubbling.
Iron and nickel boride particles are prepared by $NaBH_4$ reduction of Fe^{3+}
and Ni^{2+} ions entrapped in the water pools of cationic n-hexanol-hexadecyl-
trimethylammonium bromide, CTAB, reversed micelles. Colloidal gold
and silver are formed by pulse radiolysis or photolysis of the metal chlorides
in AOT reversed micelles.

Generally, colloidal metal particles retain their stability in microemulsions
for several months. They have been profitably used as hydrogenation
catalysts subsequent to depositing them on a carrier and washing away
the surfactant by ethanol.[11] These depositions could be performed without
extensive agglomeration of the particles. Colloidal microemulsion entrapped
catalysts were also used directly for alkene reductions.[13,14]

Formation of the colloidal particles can be monitored by following
the development of the characteristic absorption bands. Hydrodynamic
diameters of colloidal particles, prepared in microemulsions, are in the
order of 30 - 150 Å.

The advantage of w/o microemulsions is that high concentrations
of precursors can be dissolved in the surfactant solubilized water pools.

Stabilization of colloids in microemulsions is likely to be due to the
adsorption of surfactants on the metal particles. The particle growth
in the surfactant entrapped water pools is rather complex. It depends
on the composition of the microemulsions, the type of colloidal particles
and the method of preparation. Colloidal gold particles, formed in the
photoreduction of $HACl_4$ in the PEDGE entrapped water pools, have been
shown to undergo growth through exchange of the contents of neighboring
microemulsions. Under the experimental conditions, initially there had
been only 6 gold particles present in each microemulsion. Formation of
much larger colloidal particles are clearly the consequence of inadequate
binding of the small nucleating gold colloids to the surfactant covering
the inner pools of microemulsions.[10]

Quite a different situation had been observed for the preparation of suspended AgCl microcyrstals in w/o microemulsions.[15] The AgCl microcrystals were prepared by mixing 2 populations of AOT microemulsions (with identical w-values) containing separately AgCl and AgNO$_3$. Measurements of turbidity and small angle scattering showed 60 Å AgCl particles whose size remained constant regardless of the sizes of the water pools used. Formation of stable microcyrstalline silver particles are important, of course, in photography.

It is essential to develop microemulsions which preclude the growth of in situ formed colloidal particles. There are, at least, two possibilities for this. The first is to add appropriate cosurfactants which enhance the binding of the colloid to the aggregate. The alternative is to use polymerized or polymer coated microemulsions. All these approaches are actively pursued in different laboratories around the world.

Empty AOT microemulsion (w = 20) had grown from 102 Å to 150 Å upon the in situ formation of platinized CdS.[16] Electron microscopy established the diameters of the CdS colloids to be smaller than 100 Å Platinization was monitored by following the aapearance of band absorbances due to colloidal Pt, possibly formed by equations

$$K_2PtCl_4 \longrightarrow PtCl_3 + Cl^-$$

$$2PtCl_3 \longrightarrow Pt + PtCl_4{}^{2-} + Cl_2$$

Unplatinized air saturated reversed micelle entrapped colloidal CdS showed the characteristic weak fluorescence emission, due primarily to the recombination of electrons with positive holes, previously observed in homogeneous solutions. Fluorescence intensities of these solutions decayed biexponentially with lifetimes of 2.4 (25 + 5%) and 28.5 (75 + 5%) ns. The fluorescence lifetimes of reversed micelle entrapped colloidal CdS are considerably longer than those observed in water (0.3 ns) indicating dramatic stabilization by decreasing the probability of electron–hole recombination. Addition of 2.3 x 10^{-4} M methylviologen, MV^{2+}, decreased the fluorescence lifetime of the longer lived component to 18.7 ns. Similarly, addition of a surface active viologen, CH$_2$=C(Me) CO$_2$[CH$_2$]$_{11}$(C$_6$H$_4$N$^+$)$_2$Me, RMV^{2+}, and PhSH quenched the emission intensity of reversed micelle entrapped CdS presumably by removing electrons (MV^{2+}, RMV^{2+}) or holes (PhSH). Apparent Stern–Volmer constants for quenching the fluorescence of colloidal CdS, entrapped in AOT reversed micelles by MV^{2+}, RMV^{2+}, and PhSH, were calculated to be 2.6 x 10^3, 4.6 x 10^3, and 11.6 M^{-1}, respectively.

Irradiation of degassed micelle entrapped platinized CdS by visible light (450 W Xenon lamp, χ x 350 nm) resulted in hydrogen formation upon addition of 1.0 x 10^{-3} M PhSH and it could be sustained for 12 hours (Figure 1).

This is the consequence of electron transfer from PhSH to the positive holes in the colloidal CdS, which diminish electron–hole recombinations (Figure 2).

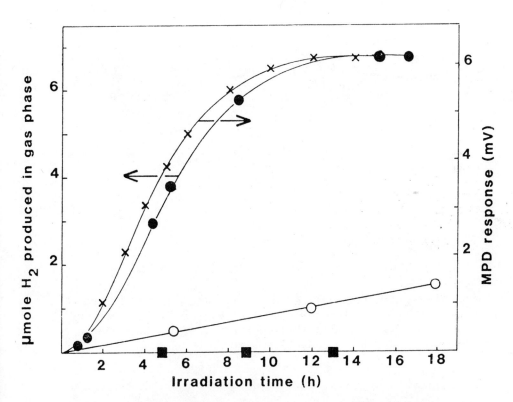

Figure 1. Hydrogen production in 1.9×10^{-4} M colloidal platinized CdS in 0.1 M AOT reversed micelle in isooctane in the presence of 1.0×10^{-3} M PhSH in argon bubbled solution at 30^0 as a function of irradiation time, using 350 nm cut-off and water filters. Plotted are the amounts of H_2 formed in a 25 ml solutions, measured in the gas phase (16 ml) by glpc ($\bullet$). H_2 formation, determined by glpc, in the absence of PhSH ($\square$) and in the absence of colloidal platinized CdS ($\circ$) are also given.

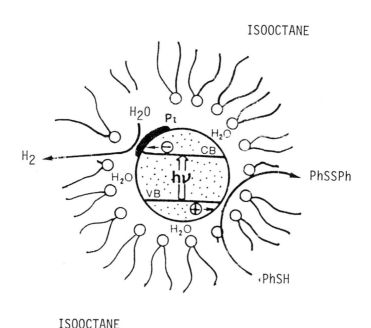

Figure 2.   An idealized model for the CdS sensitized
water photoreduction by PhS in AOT reversed
micelles in isooctane. VB = valence band,
CB = conduction band.

COLLOIDAL PARTICLES IN VESICLES AND IN POLYMERIZED VESICLES

In principle there are two alternative ways of making colloidal particles
in surfactant vesicles. Particles with appropriate isoelectric points and
sizes (smaller than the available volume in the vesicle interiors) can be
incorporated by co-sonication. The advantage of this method is that it
affords means for entrapping large variety of structually different colloids,
semiconductors and their doped counterparts. The disadvantage is that
extremely small monodispersed stable particles need to be prepared.

Alternatively, particles can be generated in situ. The requirement
for the in situ preparation of the colloidal particles are rather stringent.
Temperatures cannot be raised appreciably above the phase transition
temperature since this would cause leakage of the precursors out of the
vesicles. Chemical reactions can only be performed with reagents which
are permeable across and yet impervious to the vesicles. This limits the

reagents to gases (H_2 and H_2S for example), acids, bases and small hydrophobic molecules.

Taking advantage of suitable electron carriers incorporated in vesicle bilayers allowed the Pt catalysed hydrogen gas mediated reduction of extravesicular compounds (Figure 3).[17]

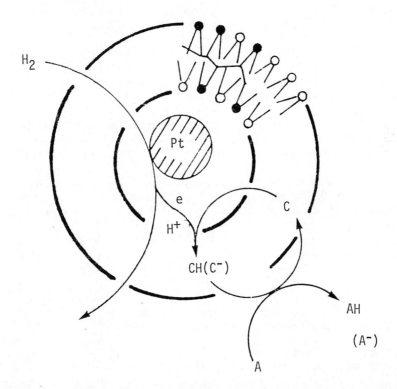

Figure 3. Use of polymerized vesicle entrapped colloidal platinum in catalysis. Electron and/or hydrogen carriers distributed in vesicle bilayers mediate the colloidal platinum-catalyzed reduction of extravesicular molecules by hydrogen bubbling. C and CH (or C⁻) are the oxidized and reduced forms of the electron and/or hydrogen carrier, A and AH (or A⁻) are the oxidized and reduced electron and/or hydrogen acceptor, and Pt is the polymerized vesicle entrapped colloidal platinum catalyst.

Surfactant vesicles were prepared by cosonicating DPPC, either with a styrene-containing surfactant 1 or with $[CH_2-CH(CH_2(CH_2)_2CO)]$ NPO(OH)$_2$ (2) for 10 minutes at 60° C and 150 W. Platinum ions were entrapped in the mixed DPPC/1 or DPPC/2 vesicles by the addition of solid K_2PtCl_4 and further sonication. Vesicle entrapped ions were separated from those in the bulk and/or attached to the outer surface by gel filtration and passages through an anion exchange resin. Irradiation of Ar-bubbled K_2PtCl_4-containing vesicles by a 450-W Xenon lamp at room temperature for 30 minutes resulted in the formation of colloidal platinum and the concomittant polymerization of 1 or 2 in the matrices of DPPC/1 or DPPC/2 vesicles. Colloid formation and polymerization were monitored absorption spectrophotometrically. Importantly, no appreciable K_2PtCl_4 could be entrapped in vesicles prepared exclusively from DPPC.

Hydrodynamic diameters of K_2PtCl_4 containing DPPC/1 vesicles prior and subsequent to colloidal platinum formation and vesicle polymerization were determined to be 1430 and 1433 A, respectively. DPPC/2 vesicles were of comparable morphology.

The catalytic efficiency of vesicle entrapped colloidal platinum was demonstrated by the reduction of methylene blue, 3, and 10-methyl-5-deazaisoalloxazine-3-propanesulfonic acid, 4. Bubbling of hydrogen through a solution that contained polymerized DPPC/1 vesicle entrapped colloidal platinum resulted in the prompt reduction of 3 or 4, localized in the vesicle bilayers (Figure 3). No reduction of 3 and 4 occurred in the absence of colloidal platinum either in vesicles or in homogeneous solutions.

Reduced polymerized DPPC/1 vesicle entrapped 3 or 4 could be reoxidized by the addition of $FeCl_3$. The process could be recycled. Bubbling of hydrogen for a few minutes caused the reduction of reoxidized 3, entrapped in the bilayers of polymerized colloidal-platinum-containing DPPC/1 vesicles (Figure 4).

Stopping the hydrogen bubbling and adding more $FeCl_3$ reoxidized 3, and so on. Hydrodynamic diameters of DPPC/1 vesicles containing colloidal platinum and 3 remained 1420 A after several reduction and oxidation cycles. These observations demonstrate the feasibility of using polymeric surfactant vesicle stabilized colloidal catalysts for reducing extravesicular molecules by hydrogen bubbling via appropriate electron and/or hydrogen carriers.

Crystalline, CoS and Fe_3O_4 have also been prepared in compartments provided by liposomes. Liposome entrapped magnetite found can be used in nmr spectroscopy and electron microscopy. These systems have the potential to act as magnetic drug carriers.

Colloidal rhodium coated CdS semiconductors, prepared in situ in dihexylphosphate, have been utilized for hydrogen generation (Figure 5).[20]

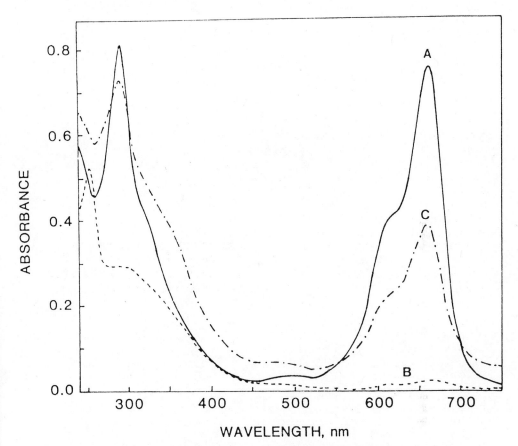

Figure 4. Absorption spectra of polymerized DPPC/1 vesicle
entrapped colloidal platinum containing 1.07 x 10^{-5}
M $\underline{3}$ in its bilayers prior to hydrogen bubbling (A),
subsequent to 3-min hydrogen bubbling (B), and
followed by an additions of $FeCl_3$ to give an overall stoichiometric iron concentration of 2 x 10^{-4} M
(C). The blank contained polymerized DPPC/1
vesicles of the same concentration, and it had been
bubbled by hydrogen for the same time as the sample.

Band gap excitation by visible light resulted in sustained hydrogen formation
in the presence of PhSH as the sacrificial electron donor (D). Our ultimate
goal is to develop separate surfactant vesicle entrapped colloidal semiconductors
based reduction and oxidation half cells and to connect them through
a real membrane (Figure 6).

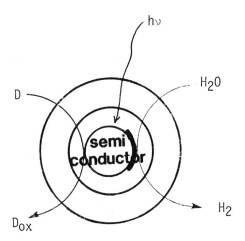

Figure 5.   Schematics of photosensitized water reduction
via band gap excitation of colloidal Rd coated
CdS, localized in the interior of DHP vesicles.
PhSH is the sacrificial electron donor. (D).

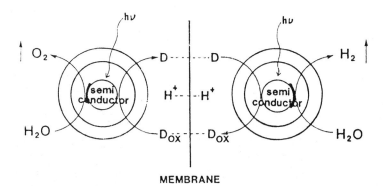

Figure 6.   A idealized scheme for the proposed cyclic
catalytic water splitting apparatus. Oxidation
and reduction half cells are coupled through
a charge carrier, D, capable of permeating
polymer membrane which separates the hydrogen
and oxygen producing vesicle stabilized catalytic semiconductors.

ACKNOWLEDGEMENTS

Support of this work by the National Science Foundation and by the U. S. Army Research office is gratefully acknowledged.

REFERENCES

1. G. A. Somorjai, "Chemistry in Two Dimensions : Surfaces", Cornell University Press, New York, 1981.
2. M. Gratzel, Acc. Chem. Res., 14, 376 (1981).
3. J. R. Darwent and G. Porter, J. Chem. Soc., Chem. Commun., 145 (1981).
4. J. R. Darwent, J. Chem. Soc. Faraday Trans. 2, 77, 1703 (1981).
5. K. Kalyanasundaram, E. Borgarello, and M. Gratzel, Helv. Chim. Acta, 64, 362 (1981).
6. J. H. Fendler "Membrane Mimetic Chemistry", Wiley-Interscience, New York, 1982.
7. H. F. Eicke, Topics Curr. Chem., 87, 85 (1982).
8. D. J. Kippenberger, K. Rosenquist, L. Odberg, P. Tundo, and J. H. Fendler, J. Am. Chem. Soc., 105, 1129 (1983).
9. J. H. Fendler, P. Tundo, Acc. Chem. Res., 16, 000 (1983).
10. K. Kurihara, J. Kizling, P. Stenius, and J. H. Fendler, J. Am. Chem. Soc., 105, 2574 (1983).
11. M. Boutonnet, C. Andersson, and R. Larsson, Acta Chem. Scand., A 34, 639 (1980).
12. M. Boutonnet, J. Kizling, P. Stenius, G. Maire, Colloids and Surfaces, 5, 209 (1982).
13. J. B. Nagy, A. Gourgue, and E. G. Deroune, Stud. Surf. Sci. Catal. 1982. Third International Symposium on Scientific Bases for the Preparation of Heterogeneous Catalysts, Louvan-la-Neuve, Sept. 1982, Elseuier, Amsterdam, 1982.
14. N. Lufimpadio, J. B. Nagy, E. G. Deroune, Proceedings of the Symposium on Surfactants in Solution, Lund, Sweden, June, 1982.
15. M. Dvolaitzky, R. Ober, C. Taupin, R. Anthore, X. Auvray, C. Petipas, and C. Williams, J. Dispersion Sci. and Technol., 4, 29 (1983).
16. M. Meyer, C. Wallberg, K. Kurihara and J. H. Fendler, J. Chem. Soc. Chem. Commun., 000 (1983).
17. K. Kurihara and J. H. Fendler, J. Am. Chem. Soc., 105, 6152 (1983).
18. A. J. Skarnulis, P.J. Strong, and R. J. P. Williams, J. Chem. Soc. chem. Commun. 1030 (1978)
19. J. Mann, A. J. Skarnulis and R. J. P. Williams, J. Chem. Soc. Chem. Commun. 1067 (1979).
20. Y.-M. Tricot and J. H. Fendler, unpublished results.

METAL-CHELATE POLYMERS: STRUCTURAL/PROPERTY

RELATIONSHIPS AS A FUNCTION OF THE METAL ION

Ronald D. Archer, Christopher J. Hardiman, Kong S. Kim,*
Edward R. Grandbois and Madeline Goldstein

Department of Chemistry and
Materials Research Laboratory
University of Massachusetts
Amherst, MA 01003

INTRODUCTION

Metal ions chelated to preformed polymers can produce marked property changes, especially when the metal ion is bonded directly to the polymer backbone. Such metal-chelate polymers can show superior properties over their organic counterparts in certain applications,[1] although properties such as thermal stability are quite polymer and metal ion dependent.[2] A good example of improved thermal stability is provided by the zinc chelate of PTO; i.e., poly(terephthaloyloxalic-bis-amidrazone)[3,4]. This metal-chelate polymer and many others are also good fire retardants. The existence of polymer-anchored metal catalysts is well documented, too.[1,5-11] However, superiority is not always the case. Brittleness and intractability are two properties often associated with metal-containing polymers.[12] We are attempting to modify polymers through metal ion chelation and to vary the metal ions therein, in order to learn more about how such chelation modifies the properties of the polymers. The polymeric hydrazones under investigation in our laboratory provide examples of the changes which are possible.

When the metal ion is a part of the backbone itself, even more drastic property changes are possible. However, intractibility is often very severe in such polymers. In order to minimize intracti-

---

*Permanent address: Graduate School, Chungbuk University, Cheong-ju City, Korea. All work done at the University of Massachusetts.

bility problems, we are synthesizing ligands and polymers with
    1) oxo metal ions,
    2) bulky ligands, and/or
    3) non-rigid coordination centers.
Our initial studies using the latter have already been published.[13]
This paper notes our progress in obtaining oxo-metal ion polymers
for film-forming applications. To date these studies have concentrated on uranyl polymers. Ligand syntheses related to the other two
methods are also noted.

EXPERIMENTAL

Synthesis

    Poly(terephthaloyloxalic-bis-amidrazone), PTO, was obtained
commercially and poly(terephthaloylbutane-2,3-dihydrazone), PTBH,
was synthesized from butane-2,3-dihydrazone and terephthaloyl
chloride in dimethylacetamide with a triethylamine catalyst. <u>Anal.</u>
Calcd for $C_{12}H_{12}N_4O_2$: C, 59.0; H, 4.9; N, 22.9. Calcd for $C_{12}H_{12}N_4O_2$
H_2O: C, 55.0; H, 5.3; N, 21.4. Found*: C, 56.5; H, 5.1; N, 21.5. For
GPC of product see Fig. 1. The butanedihydrazone was synthesized by
slowly adding a methanolic solution of 2,3-butanedione (diacetyl) to
64% aqueous hydrazine. The model ligand dibenzoylbutane-2,3-dihydra-
zone was prepared from benzoyl chloride and butane-2,3-dihydrazone
in an analogous manner.

    The metallated chelate polymers of PTO and PTBH were synthe-
sized for several metals from aqueous ammonia solutions of the metal
salts and the respective polymer. For less labile metal ions, e.g.,
nickel(II), elevated temperature reactions under pressure were used
to avoid the loss of ammonia and to obtain stoichiometric metal
content. Although a few hours at room temperature suffices for metal
ions as labile as zinc(II), longer time periods at 90° are necessary
for nickel(II). Chelation with PTBH requires longer times as the
polymer is not swelled by concentrated aqueous ammonia like PTO. See
Table 1.

    Bis(8-quinolinol) was prepared by modifications of a published
procedure of Bratz and von Niementowski.[14] A warm (40°) solution of
ferric sulfate (210 g [0.5 mol] in 1400 mL H_2O) was added rapidly to
a 40° solution of 8-quinolinol (70 g [0.5 mol] in 1400 mL H_2O). After
15 min of stirring, a NaOH solution (75.6 g in 700 mL H_2O was added
over a period of an hour. After standing overnight, the black pasty
material was filtered and dried. The paste was dissolved in 240 mL

---

*Elemental analyses by the University of Massachusetts Microanalysis
Laboratory.

Table 1. PTO and PTBH Metal Chelate Polymers[a]

Polymer	Metal Ion[b]	Temp.	Time	$[NH_3]$	Satn[c]
PTO	Zn^{++}	25°	1 da	6 M	100 %
		25°	1 da	15 M	100 %
		90°	2 hr	15 M	100 %
PTO	Ni^{++}	25°	2 da	6 M	80 %
		90°	2 hr	15 M	80 %
		90°	2 da	15 M	100 %
PTO	Cu^{++}	25°	4 da	15 M	100 %
PTO	Cd^{++}	25°	3 da	15 M	100 %
PTO	Pb^{++}	25°	3 da	15 M	75 %
PTBH	Zn^{++}	25°	1 da	15 M	70 %
		90°	2 da	15 M	100 %
PTBH	Ni^{++}	90°	3 da	15 M	95 %
PTBH	Cu^{++}	90°	3 da	15 M	55 %
PTBH	Cd^{++}	90°	3 da	15 M	75 %
PTBH	Pb^{++}	90°	3 da	15 M	100 %

[a] PTO = poly(terephthaloyloxalic-bis-amidrazone);
   PTBH = poly(terephthaloylbutane-2,3-dihydrazone).
[b] Metal ions in 5 to 10 fold excess during reactions.
[c] Satn = % saturation for 1 metal ion/polymer unit.

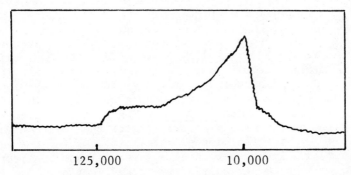

125,000   10,000

Fig. 1. GPC of PTBH vs polystyrene standards in NMP on Ultrastyragel

conc HCl and 320 mL glacial acetic acid via vigorous stirring and
heating. Acetone (6.4 L) was used to precipitate the bright yellow
dihydrochloride salt; yield, 22%. Addition of conc. aqueous ammonia
to pH = 8 followed by 1 hour stirring, filtering, washing with
water, and drying in vacuo, two recrystallizations from dimethylformamide, dissolved in HCl/H_2O and reprecipitated with conc.
aqueous ammonia as before; yield, 12%. Anal. Calcd for $C_{18}H_{12}N_2O_2$:
C, 75.0; H, 4.2; N, 9.7. Found: C, 74.7; H, 4.2; N, 9.8.

Bis(7-propyl-8-quinolinol) was prepared by treating 7-allyl-8-
quinolinol[15] with two equivalents of p-toluenesulfonhydrazide[16] in
diglyme under reflux conditions, precipitated with water, dissolved
in CH_2Cl_2, concentrated, ether added and HCl salt precipitated with
HCl gas. The salt was dissolved in water, neutralized to pH = 7.6,
extracted with ether, dried with Na_2SO_4, concentrated to dryness at
room temperature, and distilled ($126°/5$ mm Hg). The alkyl quinolinol
was obtained in a 73% yield and was oxidatively coupled by adding an
ethanolic solution to a basic aqueous solution of $K_3Fe(CN)_6$. The
coupled product, alternatively designated as 8,8'-dihydroxy-7,7'-
dipropyl- 5,5'-biquinolyl, was obtained in a 21% yield after recrystallization from dimethylformamide. Anal. Calcd for $C_{24}H_{24}N_2O_2$:
C, 77.4; H, 6.4; N, 7.5. Found: C, 76.8; H, 5.8; N, 7.5.

1,5-Diazaanthraquinone has been prepared through a simplified
synthesis of a 1,5-diazaanthracene derivative. First, 2,6-dimethyl-
1,2,3,4-tetrahydro-1,5-diazaanthracene-2,4,8-tricarboxylic acid was
prepared from p-phenylenediamine and pyruvic acid according to the
procedure of Giuliano and Stein;[17] yield, 39%. This compound was
simultaneously decarboxylated and dehydrogenated with palladium on
charcoal in 1,2,4-trichlorobenzene under reflux for three hours. The
crude 2,6- dimethyl-1,5-diazaanthracene was heated with concentrated
nitric acid for seven hours. Upon cooling, 1,5-diazaanthraquinone-
2,6-di-carboxylic acid crystallized as the dihydrate, yield, 14%.
After removing the water of hydration, this diacid was decarboxylated by mixing it with twice its weight of calcium oxide and then
heating the mixture rapidly in an evacuated sublimator. After one
crystallization of the sublimate, 1,5-diazaanthraquinone was
obtained as yellow/brown flakes; yield 15%, net yield overall, 1%.
Anal. Calcd. for $C_{12}H_6N_2O_2$: C, 68.6; H, 2.9; N, 13.3. Found: C,
68.2; H, 3.0; N, 13.2.

Uranyl polymers have been synthesized through solution reactions as follows: A donor solvent such as dimethyl sulfoxide (DMSO)
was used to dissolve stoichiometric quantities of a bridging ligand
and uranyl acetate dihydrate. Typically, 100 mL of DMSO is sufficient for a one gram reaction. The DMSO/acetic acid azeotrope was
distilled from the solution in vacuo using a minimum amount of heat
and then taken to dryness. The progress of the reaction was monitored via NMR examination of the distillate. Drying the product
overnight at $100°$ in vacuo typically gave analyses consistent with
one to two moles DMSO per mole of the uranyl cation. See Table 2.

Table 2. Uranyl Polymers of Dicarboxylic Acids

Ligand	$\bar{M}_n{}^a$	Empirical Formula/Analyses
2,2-Dimethylsuccinate[b]		$UO_2[O_2CC(CH_3)_2CH_2CO_2](C_2H_6SO)$
		Calcd: C, 19.5; H, 2.9; S, 6.5; U, 48.4
	7,000	Found: C, 19.6; H, 2.9; S, 6.6; U, 48.3
2,2-Dimethylglutarate		$UO_2[O_2CC(CH_3)_2CH_2CH_2CO_2](C_2H_6SO)$
		Calcd: C, 21.4; H, 3.2; S, 6.3; U, 47.0
	30,000	Found: C, 21.4; H, 3.2; S, 6.9; U, 46.1
3,3-Dimethylglutarate		$UO_2[O_2CCH_2C(CH_3)_2CH_2CO_2](C_2H_6SO)$
		Calcd: C, 21.4; H, 3.2; S, 6.3; U, 47.0
	10,000	Found: C, 21.1; H, 3.2; S, 6.7; U, 47.1
2,2,6,6-Tetramethylpimelate[c]		$UO_2[O_2C(CH_3)_2(CH_2)_3C(CH_3)_2CO_2](C_2H_6SO)$
		Calcd: C, 27.8; H, 4.3; S, 5.7; U, 42.3
	6,000	Found: C, 27.8; H, 4.3; S, 5.6; U, 42.5
Thiodiglycolate		$UO_2(O_2CCH_2SCH_2CO_2)(C_2H_6SO)_{1.5}$
		Calcd: C, 15.7; H, 2.5; S, 15.0
	4,000	Found: C, 16.0; H, 2.6; S, 15.1
Maleate		$UO_2(\underline{cis}\text{-}O_2CCH=CHCO_2)(C_2H_6SO)_{1.75}$
		Calcd: C, 17.3; H, 2.3; S, 10.8; U, 45.8
	8,000	Found: C, 17.5; H, 2.5; S, 10.5; U, 45.4
Fumarate		$UO_2(\underline{trans}\text{-}O_2CCH=CHCO_2)(C_2H_6SO)_2$
		Calcd: C, 17.8; H, 2.6; S, 11.9; U, 44.1
	10,000	Found: C, 17.8; H, 2.6; S, 11.5; U, 43.8
Phthalate		$UO_2(\underline{o}\text{-}O_2CC_6H_4CO_2)(C_2H_6SO)_2$
		Calcd: C, 24.4; H, 2.7; S, 10.9; U, 40.3
	10,000	Found: C, 24.3; H, 2.8; S, 10.6; U, 40.5
Isophthalate		$UO_2(\underline{m}\text{-}O_2CC_6H_4CO_2)(C_2H_6SO)_{1.25}$
		Calcd: C, 23.7; H, 2.2; S, 7.5; U, 44.8
	lim sol[b]	Found: C, 23.8; H, 2.0; S, 7.6; U, 44.9
Terephthalate		$UO_2(\underline{p}\text{-}O_2CC_6H_4CO_2)(C_2H_6SO)_2$
		Calcd: C, 24.4; H, 2.7; S, 10.9; U, 40.3
	insol[b]	Found: C, 24.2; H, 2.7; S, 10.9; U, 40.2
Acetylenedicarboxylate		$UO_2(O_2CC{\equiv}CCO_2)(C_5H_9NO)_{1.75}$
		Calcd: C, 27.6; H, 2.9; N, 4.4; U, 42.8
	6,000	Found: C, 26.9; H, 3.5; N, 4.6; U, 44.4

(continued)

Table 2. (continued)

[a] $\bar{M}_n$ values based on GPC & viscosity in NMP relative to polystyrene.
[b] The uranyl polymer of dimethylsuccinate was also synthesized in
   pyridine (the monopyridine product requires N, 2.8; found: N, 2.5)
   and interfacially (the diaqua product requires C, 16.0; H, 2.7;
   found: C, 15.8; H, 2.1).
[c] The uranyl polymer of tetramethylpimelate was also synthesized in
   pyridine (the monopyridine product requires N, 2.5; found: N, 2.4)
   (the diaqua product requires C, 25.4; H, 4.3; found: C, 25.8; H,
   3.9).

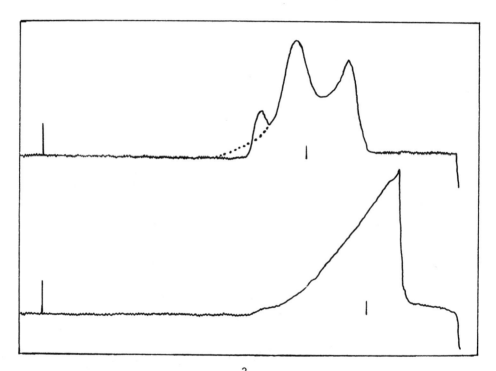

Fig. 2. GPC spectra (in NMP on 10^3 A Ultrastyragel) of UO_2(2,2-
       dimethylsuccinate)·$2H_2O$ polymer isolated during an inter-
       facial polymerization. The Upper curve is the precipitate
       formed at the interface during the polymerization (M_n ca.
       25,000 vs. polystyrene). The dotted line represents an esti-
       mate for the polymer which exceeds the exclusion limit of
       the column. The Lower curve represents the precipitate which
       occurs from the aqueous layer upon cooling (M_n ca. 7,000 vs.
       polystyrene).

Uranyl polymers have also been synthesized interfacially based on the work of Carraher and Schroeder.[18]   Initially, 0.02 mol of an appropriate bridging ligand, such as a dicarboxylic acid, was added to 100 mL of water, neutralized with a quantitative amount of NaOH, and placed in a high-speed blender. A solution of 0.02 mol uranyl nitrate hexa- hydrate in diethyl ether was also placed in the blender and the two phase system was mixed at high velocity for 30 seconds and the product   was filtered from the solution. Drying overnight at 100° in vacuo typically provided analyses consistent with diaquo species.

Properties

Standard methods were used for all studies. Of particular note are the gel permeation studies of the uranyl polymers. Since they are generally insoluble in tetrahydrofuran, the usual solvent of choice, the polymers could not be easily studied until the recent availability of columns compatible with N-methyl-2-pyrrolidone (NMP) (Ultrastyragel/Waters).

RESULTS AND DISCUSSION

PTO and PTBH Polymers

Metallated PTO polymers show that extreme differences in behavior which can result from the direct attachment of metal ions directly onto a polymeric backbone. Whereas the zinc derivative is appreciably more thermally stable than the parent polymer,[2,3] we find that the nickel(II) and copper(II) derivatives are less thermally stable than the parent polymer. See Table 3. This modification thermal stability is dependent on saturation of the polymer by the metal ion; e.g., Frank and coworkers[19] have claimed no enhanced stability for the zinc PTO derivative, in direct opposition to the results of van Krevelen[2] and ourselves. Similarly, the mechanical properties of the zinc PTO derivative are excellent, but the nickel-

Table 3. Decomposition Temperatures for PTO and PTBH Metal Polymers

Polymer	Metal-free	Zn(II)	Ni(II)	Cu(II)	Cd(II)	Pb(II)
PTO	325°C	500°	300°	300°	450°	unsatd
PTBH	290°C	350°	330°	unsatd	unsatd	340°

[a]unsatd = unsaturated.

(II) derivative is very brittle after annealing at 100° for the normal drying period of several hours. One potential bonding mode for the zinc PTO derivative is shown in structure 1.

1

Note that the coordination of zinc to four of the donor atoms leaves four other nitrogen donors free as potential crosslinking sites. Our choice of 1 as the mode of coordination is based on the loss of carbonyl stretching transitions in the infrared upon coordination of zinc, and the other metal ions, as well.[2,3,19,20] The mode of coordination had been in dispute previously,[2,3,19,20] although Frank et al.[19] had come to a similar conclusion. The electronic spectra of the metallate PTO species show large bathochromic (red) shifts as expected for a more conjugated system. The parent polymer backbone of PTO ,2, lacks total conjugation.

2

The brittleness of nickel(II) derivative coupled with the knowledge that nickel(II) is often six-coordinate led us to synthesize the PTBH polymer, in which the amine groups in 1 are replaced with methyl groups. We felt that this substitution should minimize the crosslinking possibilities for the nickel ions. To our surprise, the nickel(II) derivative is once again appreciably less stable than the analogous zinc derivative. The stability of the nickel derivative relative to the parent polymer appears to be dependent on the stoichiometry of the product. Again the copper(II) derivative is similar in stability to the nickel species. Cadmium and lead derivatives, even when not fully stoichiometric, are better than the transition elements. Together these results suggest the rigid stereochemistry of the partially filled d level ions (tetragonal planar for four coordination for these types of ligands) causes the rigidity. The d^{10} ions have no favored angles (although ligand donor atom repulsions favor movement toward a tetrahedron). The lower thermal stability was unexpected, but as noted by Carraher,[18] each system of chelated polymeric materials appears different.

Bridging Ligands

Conjugated bridging ligands have been synthesized in order to prepare thermally stable metal chelate polymers which might show electron transport properties. 5,5'-Bis(8-quinolinol) is difficult to obtain pure. The coupling reaction involves iron(III), which stays coordinated except in strongly acidic or basic media. Unfortunately, after being purified by sequential treatment with acid and then base, the product exhibited low solubility and was difficult to use in polymer synthesis. However, it is known that alkyl substitution often increases the solubility of 8-quinolinol derivatives quite markedly, and 5,5'-bis(7-propyl-8-quinolinol) is no exception. Its solubility is much greater than the non-alkyl substituted species. Although we had hoped to oxidize these to quinone forms so we could oxidize and polymerize tungsten(II) precursors as we had done with quinoxaline-5,8-dione,[13] such oxidations have been unsuccessful to date. An alternative ligand, 1,5-diaza-anthra-9,10-quinone, has been synthesized for analogous purposes. The synthesis is much improved over the literature procedure for its precursor, 1,5-diazaanthracene.[21] The earlier procedure has several more steps and an even poorer yield. The reaction of the quinone with $W(CO)_3$-$(PPh_3)_2Cl_2$ is negligible thermally, but photochemical activation allows some product formation, although the details have not been worked out as yet. Thus, although the oxidizing ability of diaza-anthraquinone is appreciably less than that of quinoxalinedione, photochemical activation may allow its use in successful polymer synthesis.

Uranyl Polymers

Polymeric species with the metal ions directly in the backbone can alter the polymer properties significantly, even when cross-linking is not taking place. To mimimize chain stacking forces, we are using the dioxouranate(VI) ion. We have synthesized a wide variety of amorphous, film-forming, uranyl dicarboxylate polymers. On the other hand, Beilstein[22] lists a large number of metal derivatives of 2,2-dimethylsuccinic acid, maleic acid, and fumaric acid as crystalline. Although fewer derivatives of the other acids have been listed,[23] in general, the metal ion derivatives of the dicarboylates have been isolated as micro-crystalline species. Our success in producing non-crystalline polymers may be related to the careful control of stoichiometry in the solution polymerizations of the diacids with uranyl acetate (equation 1) in polar solvents.

$$UO_2(C_2H_3O_2)_2 + HOOC-R-COOH \longrightarrow [UO_2(OOC-R-COO)] + 2 HC_2H_3O_2 \quad (1)$$

The acetic acid is removed in vacuo to drive reaction 1 to the right. Although we have had some success in pyridine, DMSO appears to be the solvent of choice in most cases. The sulfur analysis provides unambiguous evidence for the degree of solvation in the products. Pyridine can be similarly analyzed for nitrogen, but the lower solubilities of the polymers in pyridine causes precipitation before adequate chain lengths had been obtained. The amorphous nature of the products is probably aided by the dioxouranate(VI) ion, which interacts strongly with the carboxylates, and even cross-links in the absence of coordinating solvents, but has enough bulk to minimize chain interactions. The number-average molecular weights of the uranyl polymers relative to polystyrene range from 4000 to 30,000, consistent with expectations for step-growth condensation polymerizations which are stoichiometrically limited. Ligand purity varied from acid to acid, and the thiodiglycolic was the poorest in quality.

Interfacial polymerizations, as in equation 2, have also been used, where the uranyl nitrate, dissolved in diethyl ether, and the disodium salt of the diacid, dissolved in water, were stirred together at high speeds for up to 30 seconds.

$$UO_2(NO_3)_2 + {}^-OOC-R-COO^- \longrightarrow [UO_2(OOC-R-COO)]_n + 2\ NO_3^- \qquad (2)$$

The procedure is analogous to that used by Carraher and Schroeder.[18] Molecular weight improvement occurred by interfacial synthetic procedures for 2,2-dimethylsuccinate. However, the lower solubility of the product and the necessity of finding the proper concentration conditions for enhanced molecular weights, not necessarily stoichiometric, caused us to continue with solution polymerizations for most of our studies.

Infrared spectra are consistent with coordinated carboxylate with very weak bands due to uncoordinated carbonate occasionally observed at intensities low enough to suggest that the molecular weight estimates are real. In other samples weak methyl NMR signals suggest acetate end groups.

These yellow polymers display good glass adhesion and are being evaluated for radiation sensitivity for lithography. Preliminary results are very encouraging.

Conclusion

Overall, the prospects for significant improvements in polymers through the appropriate incorporation of metal ions is evident. The properties of such polymers are very metal-ion dependent. Minor changes in metal ion content can also make significant differences in the resulting properties of the polymeric materials produced. Modification of the coordination sphere of a given metal provides another pathway to property modification that remains to be explored.

Acknowledgment

The authors gratefully acknowledge the support of the Office of Naval Research and the National Science Foundation Materials Research Laboratory program for support of this research.

REFERENCES

1.   J. E. Sheats, Kirk-Othmer Ency. Chem. Tech., 3rd ed., 15, 184 (1981).

2.   C. E. Carraher, Jr., J. Macromol. Sci.-Chem. A17, 1293 (1982).

3.   D. W. van Krevelen, Chem. and Ind., 49, 1396 (1971); Angew. makromol. Chem., 22, 133 (1972).

4.   Anon., Textile Progr., 8, 124, 156 (1976).

5.   C. U. Pittman, Jr., "Polymer-Supported Reactions in Organic Synthesis (P. Hodge and D. C. Sherrington, Eds.), Wiley, New York, 1980, Chpt.5.

6.   C. Carlini and G. Sbrana, J. Macromol. Sci.-Chem., A16, 323 (1981).
7.   D. W. Slocum, B. Conway, M. Hodgman, K. Kuchel, M. Moronski, R. Noble, K. Webber, S. Duraj, A. Siegel, and D. A. Owen, J. Macromol. Sci.-Chem., A16, 357, (1981).

8.   Y. Chauvin, D. Commereuc and F. Dawans, Progr. Polym. Sci., 5, 95 (1977).

9.   J. P. Collman and L. S. Hegedus, "Principles and Applications of Organotransition Metal Chemistry," University Science Books, Mill Valley, CA, 1980, p. 370.

10.   F. R. Hartley and P. N. Vezey, Adv. Organomet. Chem., 15, 189 (1977).

11.   J. C. Bailar, Jr., Catal. Rev.-Sci. Eng. 10, 17, (1974).

12.   J. C. Bailar, Jr., "Organometallic Polymers" (C. E. Carraher, Jr., J. E. Sheats, and C. U. Pittman, Jr., Eds.) Academic Press, N. Y., 1978, p. 313.

13.   R. D. Archer, W. H. Batschelet and M. L. Illingsworth, J. Macromol. Sci.-Chem., A16, 261 (1981).

14.   L. T. Bratz and S. von Niementowski, Ber., 52B, 189 (1919).

15.   B. Mander-Jones and V. M. Trikojus, J. Proc. Roy. Soc. N. S. Wales, 66, 300 (1932).

16.   R. S. Dewey and E. E. vanTamelin, J. Am. Chem. Soc., 83, 3729 (1961).

17.   R. Giuliano and M. L. Stein, Gazz. Chim. Ital., 84, 284 (1954).

18.   C. E. Carraher, Jr. and J. A. Schroeder, J. Polym. Sic. Polym. Lett. Ed., 13, 215 (1975).

19.   V. D. Frank, W. Dietrich, J. Behnke, A. Kobock, G. Schuck and M. Wallrabenstein, Angew. Makromol. Chem., 40, 445 (1974).

20.   D. W. van Krevelen, Chem. Eng. Tech., 47, 793 (1975).

21.   P. Ruggli and E. Preiswerk, Helv. Chem. Acta, 22, 478 (1939).

22.   Beilstein, II, 662, 740, 751 (and analogous pages in suppl. 1 - 4).

23.   Beilstein, II, 677, 684, 728, 802 (and analogous pages in suppl. 1 - 4).

A STUDY OF COBALT CONTAINING POLYIMIDE FILMS

Eugene Khor and Larry T. Taylor[*]

Department of Chemistry Virginia Polytechnic Institute

and State University, Blacksburg, VA 24061

INTRODUCTION

The need to replace current technological materials by lightweight, easily processible and cheaper substances has been reflected in active research toward the modification of polymer characteristics. One of the more prominent demonstrations has been the enhancement of electrical conduction in polyacetylenes and related polymers by addition of either donor or acceptor species.[1] These doped materials suffer, however in stability for most routine applications; consequently their use will probably be limited to "hermetically sealed" systems.[2] Our laboratory has also been interested in lowering polymer electrical resistance. Limiting ourselves to polyimides, we have achieved to varying degrees air stable electrical conduction.[3] This achievement, however, has been restricted to films with additives that showed migration of metal to the surfaces of these doped films during thermal imidization. The formation of a metal(0) layer during thermal imidization in air or the interaction of a surface metal species with moisture have been rationalized to account for the lowered resistivity. The quest for additives that will enhance conduction through the bulk of the polymer rather than on the polymer surface remains challenging. As one way of obtaining some form of regularity in metal disposition, we have employed polyimides which contain potential coordinating groups in the polymer backbone. We have also sought to use a wide range of additives from each metal system, since the potential for thermal decomposition, polyimide coordination and the like are dependent on the oxidation state liability, etc. of the metal additive. In this report, we describe the preparation and characteristics of polyimide films doped with cobalt(II) and cobalt(III) additives.

EXPERIMENTAL

Materials

The monomers used are shown in Figure 1. 3,3',4,4'-Benzophe-
nonetetracarboxylic acid dianhydride (BTDA) was obtained from Gulf
Oil Chemicals Co. and dried before use. 4,4'-Oxydianiline (ODA)
was secured as a free sample from Mallinkrodt and purified by
recrystallization and sublimation. 4,4'-Dianiline sulfide (DAS)
and 3,3'-bis(aminopyridine)sulfide (BAPS) were supplied by NASA,
Langley Research Center, Hampton, VA.
2,2'-Dioxy-phenylene-5,5'-bis(aminopyridine) (DOPBAP) was
synthesized as per the original reference.[4] N,N-dimethylacetamide
(DMAC) was obtained from Burdick and Jackson and used as received.
Anhydrous cobalt(II) chloride was synthesized by drying the
hexahydrate in a vacuum oven at 120°C for 2 days and used
directly. Bis-(trifluoroacetylacetonato)-cobalt(II) was prepared
by a reported procedure.[5] The remaining dopants that were
screened as additives were from "in-house" stocks which had been
previously synthesized. They were:
Dichlorobis(ethylenediamine)cobalt(III) chloride $[Coen_2Cl_2]Cl$[6],
Carbonatotetramminecobalt(III) nitrate $[Co(NH_3)_4(CO_3)]NO_3$[7],
Hexaamminecobalt(III) chloride $[Co(NH_3)_6]Cl_3$[8], (N,N'-ethylene
bisacetylacetoiminato) cobalt(II) $[Co(BAE)]$[9],
Bis(1,1,1,5,5,5-hexafluoro-2,4-pentanedionato)cobalt(II)
$Co(HFA)_2$[10], Dithiocyanato-bis(ethylenediamine)cobalt(III) chloride
$[Coen_2(SCN)_2]Cl$[11], methylcobaloxime $[Co(DMG)2(CH_3)]$[12],
(N,N',-diethylenetriamine bis(2-pyridene-iminato)cobalt(II)
hexaflurophosphate $[Co(pyDIEN)(PF_6)_2]$[13].

Polymerization

Polymerization was achieved by first dissolving diamine in
DMAC (15-20% w/w final solution) followed by stoichiometric
dianhydride. The resulting solution was stirred at room
temperature for 6h under nitrogen. The cobalt additive was then
introduced and stirring continued for 1h under nitrogen.
Solutions were kept in a refrigerator in the polyamic acid-metal
"adduct" state until cast (diamine:dianhydride:metal ion; 4:4:1;
mole ratio).

Film Preparation

Polyamic acid-metal "adduct" solutions were centrifuged at
~2000 rpm for 15 min and then poured onto glass plates. Solutions
were spread using a doctor blade with a 12-18 mil gap, dependent
on solution viscosity. Films were preliminarily dried in static
air at 333K for two hours. For ODA films, imidization was
thermally achieved by heating in a forced air oven 1 h each at
373, 473 and 573K. All other diamine derived polymers were

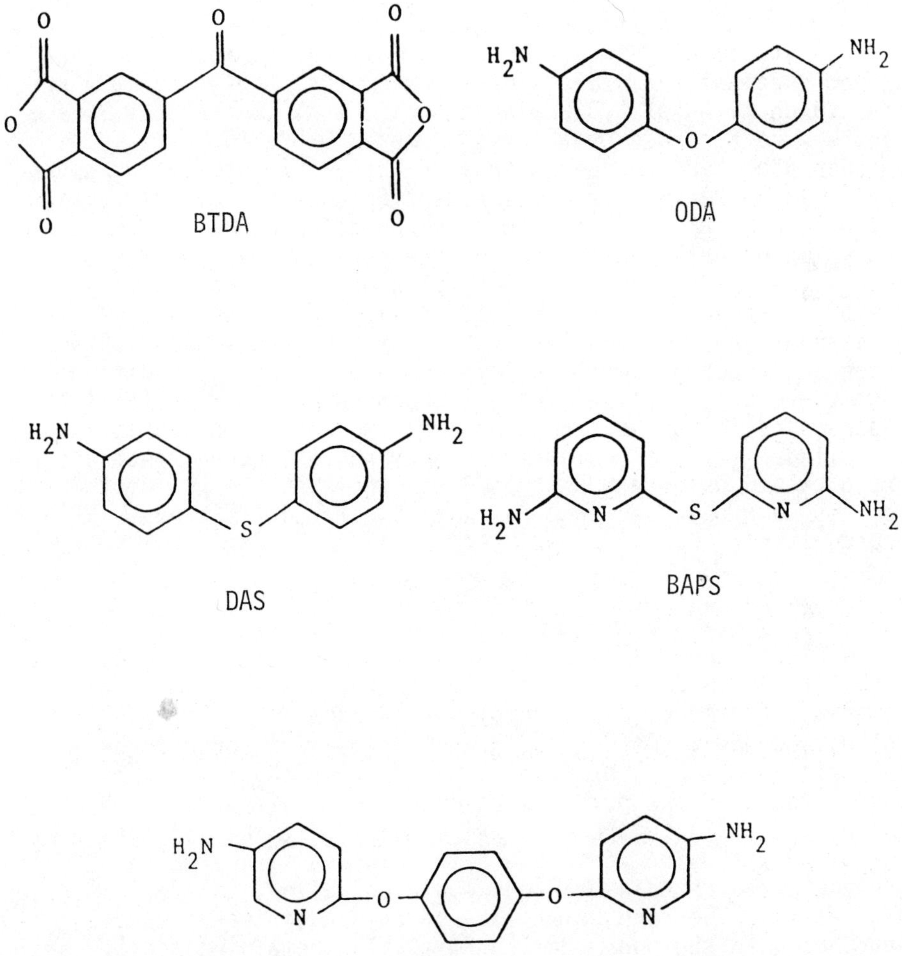

Figure 1: Structures and Acronyms of Monomers

imidized in a vacuum oven for 1 h each at 373 and 473K. Films
were removed by soaking the glass plates in distilled water at
room temperature. All films were air dried and cleaned with a
methanol/ether mixture (1:1).

Characterization

Samples were sent to Galbraith Labs Inc., Knoxville, TN for
metal analyses. Thermomechanical analyses (TMA) were performed on
films in static air at 5°C/min on an E.I. DuPont model 990
thermomechanical analyzer. Thermogravimetric analyses (TGA) were
obtained on films at 2.5°C/min in static air. Surface UV-VIS
spectra were recorded on a Perkin-Elmer 330 UV-VIS
spectrophotometer equipped with a PE H210-2101 integrating sphere
(performed by Perkin-Elmer Instrument Sales and Service Division,
Newton Square, PA). Samples were scanned from 1600-185 nm.
Spectra were subsequently checked in-house in the 750-185 nm
range. Transmittance spectra of films were recorded in the
750-400 nm range. Electrical resistivity was measured following
the standard ASTM method (D257-66) employing a Keithley voltage
source and electrometer. The electrode assembly was modified to
accommodate cartridge heaters and thermocouples and to facilitate
measurements in vacuum. Heating and temperature monitoring were
controlled by a HP-85 computer. X-ray photoelectron spectra (XPS)
were obtained using a DuPont 650B spectrometer. Auger depth
profiles were obtained on a Physical Electronics 550 Auger/ESCA
spectrometer.

RESULTS AND DISCUSSION

Candidate cobalt compounds were first screened for their
solubility in DMAC, i.e.the solvent in which the polymers were
prepared. A number (5) of complexes [Co(acac)$_3$], [Co(en)$_2$Cl$_2$]Cl,,
[Co(NH$_3$)$_4$CO$_3$]NO$_3$, [Co(NH$_3$)$_6$]Cl$_3$, [Co(BAE)] were found to be
insoluble in DMAC. [Co(II)(pyDIEN)](PF$_6$)2 although soluble in
DMAC, coagulated the polyamic acid probably by coordination of the
labile cobalt(II) to the carboxylic acid and amide linkages of the
pre-polymer, thus preventing a film being cast. Co(HFA)$_2$ was also
found to be soluble in DMAC but would not give a smooth film after
imidization. The high volatility of this additive no doubt
contributed to the poor quality film. The remaining additives
[Co(DMG)$_2$CH$_3$], CoCl$_2$, Co(TFA)$_2$ and, [Coen$_2$(SCN)$_2$]Cl yielded with
the monomer combination BTDA-ODA, films that were of reasonable
quality. Except for CoCl$_2$ which gave a green-colored film, all
other dopants gave light to dark brown colored films. CoCl$_2$ was
used as the primary dopant since it was the simplest compound used
and had the best potential for coordination with the pyridine
containing functionality in the polyimide backbone.

The results of thermal and elemental analyses for cobalt
doped BTDA-ODA derived polyimide films are summarized in Table I.

The apparent softening temperatures (AST) were found to be higher than their undoped counterpart with the reverse being true for the polymer decomposition temperature (PDT). A survey of the AST of these doped films reveals an interesting feature. There is apparently a bigger increase in AST by dopants that contain bulky ligands. This is exemplified with $CoCl_2$ and $[Coen_2(SCN)_2]Cl$. The AST for the $CoCl_2$ film is 335°C when the amount of cobalt present is 1.51%. The AST of the $[Coen_2(SCN)_2]Cl$ doped film is 10° higher (345°C) but is achieved with only a 0.9% cobalt present. This suggests that the ligand plays a role in increasing the AST. The organic ligand in the example cited, might perhaps interact more with the polymer backbone than a simple chloride ligand causing decreased polymer mobility, thus raising the AST.

Table I
Thermal Characteristics of Cobalt
Containing BTDA-ODA Polyimide Films

Additive	AST (°C)	PDT (°C)	Wt% Metal Found	Calcd.
undoped	286	540	-	-
$CoCl_2$	335	478	1.51	2.85
			(2.31)	(2.85)[a]
$Co(DMG)_2CH_3$	340	450	1.03	2.51
$[Coen_2(SCN)_2]Cl$	345	490	0.90	1.06
$Co(TFA)_2$	347	450	1.84	2.55
	(322)	(450)	(2.54)	(2.55)

There appears to be a wide scatter in the PDT of these doped films. Above the softening temperature, the additive may facilitate greater polymer mobility (which would aid in the decomposition process) by lessening close range effects in polymer chains. This would be more pronounced with additives containing bulky ligands since there is a relatively larger volume for polymer motion to occur (above the AST) when compared to additives possessing simple ligands. This reasoning is supported by the PDT data which shows lower decomposition temperatures for the majority of additives containing bulky ligands.

The other polyimides used in this study were derived from the following monomer pairs: BTDA-DAS, BTDA-DOPBAP, BTDA-BAPS and PMDA-ODA. They were chosen in order to investigate their potential for coordination with $CoCl_2$ (BAPS, DAS and DOPBAP) as well as to determine the general effects, that dianhydride variation (PMDA) has on film properties. It should be pointed out that film imidization with these monomer pairs was achieved at 200°C under vacuum except for PMDA-ODA which was cured at 300°C in a nitrogen environment. Air curing gave brittle films with these diamines. In the case of BAPS, this may have been caused by oxidation of the thioether bridge. Since previous curing of such pyridine-derived polyimides was achieved under vacuum[4], the same procedure was adopted here, which ultimately allowed good quality films to be obtained. These undoped films were typically yellow in color, smooth and flexible. The $CoCl_2$ doped films were green in color.

AST and PDT for undoped polyimide films derived from the four previously noted monomer combinations are reported in Table II. A survey of the monomer pairs (Table II) shows a range of softening temperatures. AST was found to follow the trend BTDA-BAPS < BTDA-DAS < BTDA-DOBPAP < PMDA-ODA. The lowest AST value was observed for the polyimide film derived from BAPS, which could be attributed to flexibility[14] imparted by the m,m' positioning of the amines. Another reason would be the better packing afforded by the para over the meta isomer which should lower the number of "voids". This is exemplified with the undoped DAS film which has essentially the same softening point as the undoped ODA derived film. This suggests that the sulfide linkage has little effect on the apparent positioning most likely accounts for the DOPBAP film having the highest softening temperature in the BTDA derived films. The PMDA-ODA monomer combination probably offers the best fit in packing, resulting in fewer voids and greater intermolecular interaction. Films derived from the two pyridine based diamines exhibited the lowest PDT in the series. A likely reason for this could probably be attributed to the pyridine functionality which was previously noted by Kurita and Williams.[4]

Table II
Thermal and Elemental Analyses for
Polyimide Films

Monomer Pair	AST(°C)	PDT(°C)	Position of Diamines
BTDA-DAS	282	578	p'p'
BTDA-BAPS	257	512	m,m'
BTDA-DOPBAP	344	530	p'p'
PMDA-ODA	405	547	P,P'

AST and PDT for films doped with $CoCl_2$ from the same four monomer combinations are reported in Table III. Again the AST follows in much the same manner as the undoped films i.e. BAPS < DAS < DOPBAP. The PMDA-ODA doped film appears to be the only exception. Figure 2 shows thermomechanical analyses plots for the doped and undoped PMDA-ODA films. The undoped film followed normal behavior, displaying a softening point around 400°C. For the doped film no apparent softening was observed, but around 450°C, contraction of the sample appears to have occurred. Based on previous work which has noted higher AST values for doped polyimides, one might estimate for this cobalt film an AST of approximately 450°C. It is possible to envision that since the

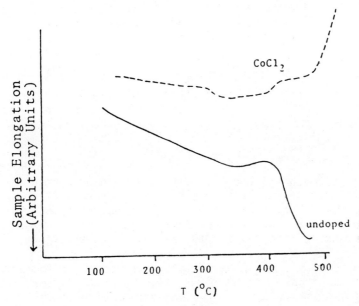

Figure 2: Thermomechanical Analyses of Undoped and CoCl$_2$ Doped PMDA-ODA Films

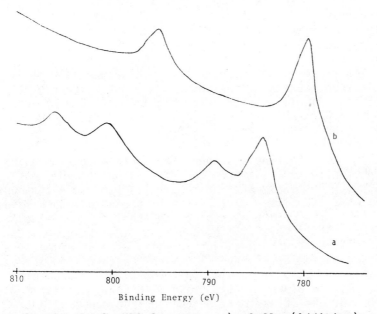

Figure 3: Cobalt 2p XPS Spectra: a) CoCl$_2$ (Additive)
b) BTDA-ODA/CoCl$_2$ Air (Film)

TABLE III
Thermal and Elemental Analyses for
CoCl$_2$ Doped Polyimide Films

Polymer Constituents	AST(°C)	PDT(°C)	Wt.% Found	Cobalt Calcd.	Curing Atm.
BTDA-DAS	287	504	1.99	2.76	vacuum
BTDA-BAPS	270	505	3.21	2.75	vacuum
BTDA-DOPBAP	368	460	2.28	2.41	vacuum
PMDA-ODA	-	440	2.84	3.56	N$_2$

difference between AST and PDT with this monomer combination would be only 142°C (the smallest observed in the series of monomer pairs studied) that perhaps the estimated AST of 450°C coincides more with the onset of decomposition. This would explain the non-observance of the AST and the resulting contraction.

PDT for CoCl$_2$ doped films in all instances were found to be lower than their undoped counterpart. The observed lower PDT suggests that the additives play a detrimental role in thermal-oxidative stability. A typical mechanism proposed for the thermooxidative degradation of polyimides would consider the formation of radical species as intermediates which eventually yield carbon dioxide, carbon monoxide, water and char.[15] The additives might be expected to catalyze this process by stabilizing such radical intermediates. The lowered PDT apparently is more pronounced with films that do not contain the pyridine functionality. With the DAS film, the earlier degradation suggests not only that the additives hasten the decomposition process, but also that the thioether linkage is more susceptible than the ether linkage in ODA. A smaller repeat unit appears to be a likely cause for the significant (100°C) decrease in PDT for the PMDA-ODA polyimide (i.e. it is easier to degrade one benzene ring (PMDA) than two (BTDA)).

All films were routinely scanned for surface metal via XPS. Only the photoelectron region of the 2p$_{1/2}$, 2p$_{3/2}$ levels was investigated. The XPS spectra of cobalt(II) high spin compounds display intense satellite peaks on the high binding energy side of the main peaks with main peak separation equal to approximately 16 eV.[16,17] Low spin cobalt(III) compounds on the other hand do not display satellite structure and are usually characterized by a main peak separation of 15 eV.[16,17] With the cobalt additives employed in this study (i.e. high spin Co(II) and low spin Co(III)), only the BTDA-ODA film doped with CoCl$_2$ displayed surface cobalt, and it was only on the surface exposed to air during imidization. The glass surface of the BTDA-ODA/CoCl$_2$ film and all other cobalt doped BTDA-ODA films (regardless of surface),

displayed no signal in the $2p_{1/2}$, $2p_{3/2}$ photoelectron region even
after long-term scanning. The XPS spectrum of $CoCl_2$ (Figure 3a)
shows the satellite structure common to cobalt(II) high spin
compounds with a main peak separation of 16.1 eV. The cobalt in
the film however, displays no satellite structure (Figure 3b) and
has a main peak separation of 15.2 eV. The binding energy of the
$2p_{1/2}$, $2p_{3/2}$ main peaks at 794.8 eV and 779.6 eV in the film are
shifted to lower binding energy relative to $CoCl_2$. This suggests
that the $CoCl_2$ in the film has been modified during thermal
imidization to probably a cobalt(III) species. Some of the more
likely cobalt(III) species that could have been generated during
thermal imidization would be $CoCl_3$ (yellow) Co_2O_3 or Co_3O_4 (black).
These materials cannot explain the green color of this film. If
however, only the surface cobalt is modified, while the bulk
cobalt remains as $CoCl_2$, the combination of the blue color of
$CoCl_2$ with the yellow color of the polymer could give rise to a
green color, thus accounting for the observation. An obvious
approach to proving that $CoCl_2$ is present in the bulk would be to
argon ion etch away the surface "cobalt(III)" layer of the
BTDA-ODA/$CoCl_2$ film and subsequently XPS scan the new exposed
layer for cobalt(II). However, the validity of this new data
would be in question since it is known that reduction of surface
metal species can occur under such conditions.[18] Thus, even if
cobalt(II) were present in the bulk, the uncertainty arising from
possible photoreduction of cobalt(III) would make suspect the
interpretation.

    Auger depth profiling with argon ion etching of the
BTDA-ODA/$CoCl_2$ film verifies that the majority of cobalt is
present only on the air surface. The data are presented as
relative atomic concentration (i.e. only the elements scanned were
normalized to constitute a hundred percent). The estimated
sputter rate is 500Å per minute. Figure 4a shows the initial rise
in cobalt concentration as the surface of the film is reached
(after sputtering through the conducting Au surface coat). This
high cobalt level subsequently drops off to a low constant
concentration into the bulk of the film. Sputtering the glass
surface of the film reveals a low constant amount of cobalt from
the surface into the bulk (Figure 4b) which explains the
non-observance of cobalt XPS peaks on the glass side. Similar low
cobalt levels were obtained on extending the auger profiling to
other cobalt doped BTDA-ODA films. This suggests that, except for
$CoCl_2$, the other cobalt dopants distribute themselves well
throughout the film. A possible explanation for this observation
lies in the environment of the cobalt additives. With $CoCl_2$,
formation of $Co(DMAC)xCl_2$ most likely occurs on dissolution. This
would tend to make $CoCl_2$ very soluble in DMAC and during thermal
imidization, migration with the solvent to the atmosphere surface
may occur. Subsequent volatilization of solvent would concentrate
cobalt on this surface. The other additives are less soluble in

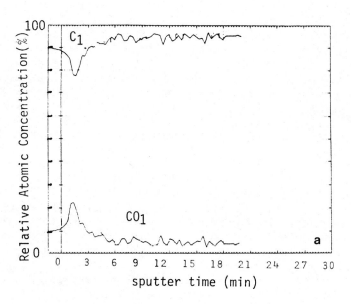

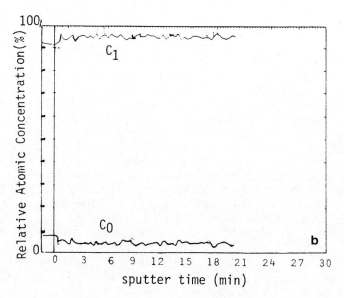

Figure 4: Auger Depth Profile of BTDA/ODA/CoCl$_2$ Film
           a) Air Surface   b) Glass Surface

DMAC (relative to $CoCl_2$) which affords the dopant less mobility within the confines of the pre-cured film. Furthermore, the other ligands (DMG, en and TFA) are better donors and may keep the additive intact thereby leading to some intermolecular interaction of the ligand with the polymer, which would give rise to decreasing metal ion mobility. The non-observance of XPS features for non $CoCl_2$ doped BTDA-ODA films was also extended to $CoCl_2$ doped polyimide films derived from DAS, BAPS and DOPBAP diamines with BTDA and PMDA-ODA. In an attempt to obtain an idea of the nature of cobalt in these films, UV-VIS reflectance spectra were obtained. The UV-VIS spectrum of typical low spin cobalt(III) compounds displays two spin allowed absorbances at about 350 nm and 500 nm corresponding to transitions $^1T_{1g} \leftarrow {}^1A_{1g}$ and $^1T_{2g} \leftarrow {}^1A_{1g}$.[19] These two transitions would tend to be obscured by absorbances due to the polymer. High spin cobalt(II) compounds are usually octahedral or tetrahedral in geometry. Three allowed transitions are possible with octahedral geometry namely $^4T_{2g} \leftarrow {}^4T_{1g}$, $^4A_{2g} \leftarrow {}^4T_{1g}$ and $^4T_{1g}(P) \leftarrow {}^4T_{1g}$.[20] An example is anhydrous $CoCl_2$ which exhibits absorbances at around 1500, 750 and 580 nm. Tetrahedral cobalt(II) compounds have been extensively studied and display two transitions $^4T_1(F) \leftarrow {}^4A_2$ (absorbance in the near infrared) and $^4T_1(P) \leftarrow {}^4A_2$ (abosrbance in the visible). The usual distinction between octahedral and tetrahedral geometries is normally achieved by comparing the intensities of the absorbances arising from each, with tetrahedral being much more intense than octahedral.

Figure 5 shows the air surface UV-VIS reflectance spectra of the BTDA-ODA/$CoCl_2$, BTDA-DOPBAP/$CoCl_2$ and BTDA-DAS/$CoCl_2$ films. The absorbance in the near UV from about 250 to 500 nm is due to the polymer.[21] The presence of cobalt(III) (suggested from XPS results) would be obscured by the absorbances from the polymer. However, an additional shoulder at about 550-650 nm was observed and it is reasonable to attribute this to cobalt(II). The shoulder could either be assigned $^4T_{1g}(P) \leftarrow {}^4T_{1g}$ transition for an octahedral environment or $^4T_1(P) \leftarrow {}^4A_2$ in the case of a tetrahedral field. It is unfortunate that because these are surface spectra, relative absorbance intensities could not be easily obtained. The greenish color of the film does suggest that tetrahedral coordination is more likely because the addition of blue (tetrahedral environment) with yellow (polymer) would give a green color; whereas, octahedral fields usually give a pink color and addition to yellow would probably not result in a green color.

In an attempt to resolve the coordination question, a transmission absorbance spectrum was obtained in the visible region for the BTDA-ODA/$CoCl_2$ film (Figure 6a). A reasonably well-resolved shoulder is observed between 600-700 nm (0.25 absorbance unit). An approximate calculation was performed to obtain an idea of ε, the molar extinction coefficient, using

Beer's law (A = $\varepsilon c l$) where A is the absorbance, l, is the path
length (typically films are 0.003 cm thick) and c is the
concentration of cobalt in the film. From elemental analysis,
1.51% cobalt in the BTDA-ODA/CoCl$_2$ film is 0.026 moles of cobalt
for a 100 gm film sample. The density of typical polyimides is
about 1.4 g/c.c.,[22] which would give a volume of about 70 mL. The
resultant molar concentration would be 0.37M, and ε is then
estimated to be 2.25x10^2 L-cm^{-1}-mole^{-1}. This falls in the range
for tetrahedral cobalt(II) and thus suggests that tetrahedral
coordination is present. The lack of an absorbance in the near IR
region for the BTDA-ODA/CoCl$_2$ film is probably due to the relative
low concentration of cobalt(II) in the polymer film and
cobalt(III) surface species. It is therefore concluded that the
UV-VIS spectrum probably consists of all three components
(polymer, cobalt(III) (since the surface is cobalt(III) rich) and
cobalt(II)) resulting in the broad absorbance bands observed.

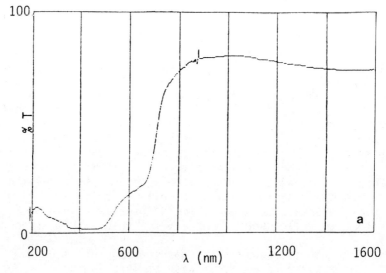

Figure 5:   UV-VIS Reflectance Spectra of CoCl$_2$ Doped Polyimide
            Film: a)   BTDA-ODA
                    b)   BTDA-DAS
                    c)   BTDA-DOPBAP

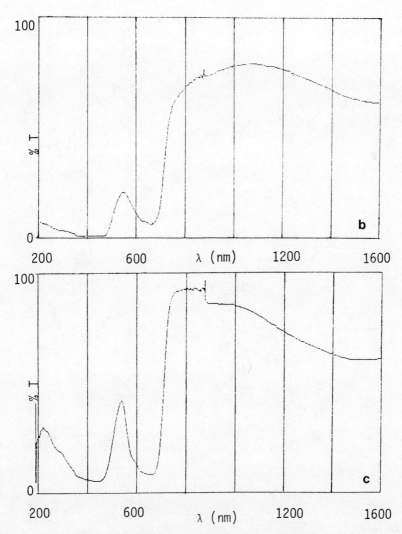

Figure 5 (continued) : UV-VIS Reflectance Spectra of $CoCl_2$ Doped
Polyimide Film: a) BTDA-ODA
b) BTDA-DAS
c) BTDA-DOPBAP

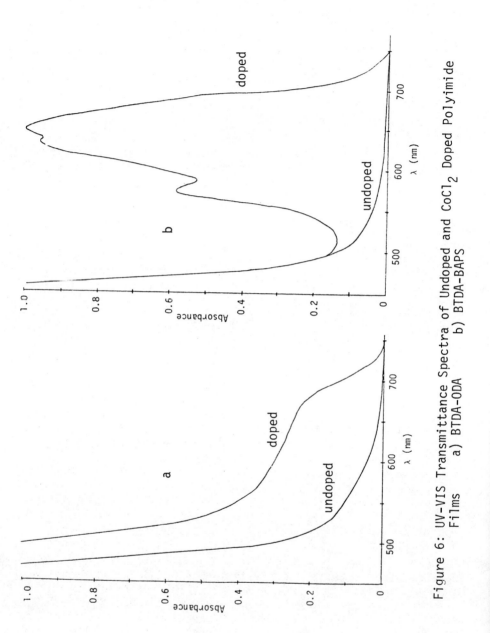

Figure 6: UV-VIS Transmittance Spectra of Undoped and CoCl$_2$ Doped Polyimide
Films   a) BTDA-ODA   b) BTDA-BAPS

Two absorbances are observed the surface UV-VIS spectra of $CoCl_2$ doped DAS and DOPBAP derived films. A more intense band ca 650 nm is seen; while, a weaker absorbance appears at 1600 nm. The clear observance of two absorbances is indicative of the presence of cobalt(II). The reflectance UV-VIS spectrum of BTDA-BAPS/$CoCl_2$ is markedly different from the ODA derived film. There are three distinct absorbances at 650 nm (strong), 959 nm (weak) and 1308 nm (midrange) which is rather typical of high spin cobalt(II) compounds. The color of the film would again suggest a tetrahedral environment although the three absorbances would tend to indicate octahedral coordination.

The $^4A_{2g}$ transition is usually not observed in octahedral cobalt complexes. In order for the 959 nm absorbance to be assigned as such, theory requires that the ratio of the energy of the transitions (i.e. $^4A_{2g}/^4T_{2g}$) be in the range of 2.1-2.2. In this instance, the ratio was calculated as 1.36; therefore it is unlikely that the $^4A_{2g}$ transition is responsible for the 959 nm absorbance. The 959 nm absorbance may arise from spin forbidden transition. The transmission spectrum for the highest intensity peak was obtained (Figure 6b) and ε was calculated to be 440 L cm^{-1} $mole^{-1}$ which is within the range for tetrahedral cobalt(II). This suggests that in this instance, modification of cobalt(II) to cobalt (III) under thermal imidization did not occur. The UV-VIS spectrum of the PMDA-ODA/$CoCl_2$ film is very similar to the BTDA-BAPS/$CoCl_2$ film with absorbances at 686 nm, 940 nm and 1600 nm. The calculated molar extinction coefficients are 380, 250 and 230 $L-cm^{-1}-mole^{-1}$ respectively. This again fits within the range observed for tetrahedral cobalt(II). In summary, PMDA-ODA and BTDA combined with coordinating diamines gave films that showed no change in cobalt character upon imidization. Only the air surface of the BTDA-ODA film as demonstrated by XPS showed modification in cobalt character. This is not surprising as no oxygen was present during thermal imidization and most importantly, no cobalt migration transpired.

Electrical resistivity measurements for the cobalt doped films were performed. With one exception, all films do not display changes in surface or volume resistivities to any significant degree over their undoped counterpart which were > 10^{14} ohm and $\approx 10^{16}$ ohm-cm, respectively. Only the BTDA-ODA/$CoCl_2$ film hinted of a slight lowering in surface and volume resistivity. It is interesting to note that this film alone displayed metal migration to the air surface of the film. Interaction of moisture with metal specie is known to lower resistivity[23]; therefore, this phenomenon has been explored.

The $CoCl_2$ doped BTDA-ODA film adsorbs about 0.027 mg H_2O/mg polymer, about the same amount of water as the undoped film (0.023). Upon adjusting for adsorption of water by the polymer,

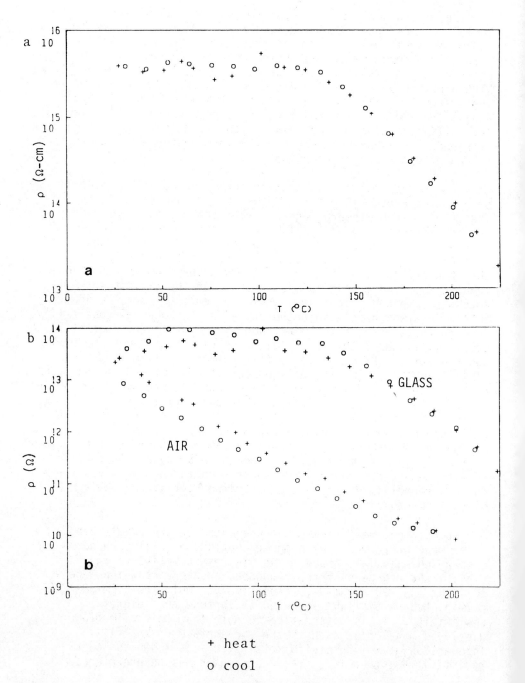

+ heat

o cool

Figure 7: Electrical Resistivity Profile of BTDA-ODA/CoCl$_2$ Film
a) Volume Resistivity   b) Surface Resistivity

it appears that the cobalt is responsible for very little
interaction with moisture since only one mole of water actually
interacts with the metal. If moisture contributed to lowered
electrical resistivity, an initial increase in resistivity might
be expected as the temperature is raised, which could be
attributed to loss of water. This observation would be absent if
water did not play a role. The electrical resistivity-temperature
profile (20-225°C) of this cobalt film was performed to obtain an
idea of its moisture interaction. As can be seen (Figure 7a) the
volume resistivity remains constant until above 100°C, whereupon
resistivity decreases with an increase in temperature.
Furthermore, the cooling curve approximately matches the heating
curve at the lower temperature end of the profile. This would not
be expected if water played an active role in lowering volume
resistivity at lower temperatures. The air surface (Figure 7b)
surprisingly exhibited a steady reduction of resistivity with an
increase in the temperature. The most probable reason for this
linear-like behavior is that a semi-conductive surface was
generated during thermal imidization on the air surface. This is
not unreasonable since this surface has been shown to have a
modified cobalt(III) species from XPS. The glass surface exhibits
a resistivity profile much like the volume resistivity profile,
which is expected since it contains little cobalt.

CONCLUSION

The problem of metal migration has been found to be
controllable with cobalt compounds containing bulkier ligands or
by using monomers that might coordinate the metal ion species.
The integrity of the additive was suggested by UV-VIS spectra.
XPS has indicated that when metal migration occurs, the surface
and bulk metal species might be different. This is apparently the
case for the BTDA-ODA/$CoCl_2$ film which produced a modified metal
surface that behaved semiconductively. Although bulk conduction
has not been realized in this work, it has laid the groundwork for
the search for additives that might coordinate (or which possess
bulky ligands) which prevents migration while at the same time be
more susceptible to thermal reduction. If this could be achieved,
bulk conduction might be realized.

Acknowledgements:

We thank the National Aeronautics and Space Administration
(Grant NSG 1428) for sponsoring this research and Lee Houck and
David A. DeRafelo (Perkin Elmer Corporation) for the UV-VIS
reflectance spectra.

REFERENCES

1. Y. W. Park, M. D. Dray, C. K. Chiang A. G. MacDiarmid, A. J.
    Heeger, H. Shirakawa, S. Ikeda, J. Polym. Sci., Polym. Lett.
    Ed., 17, 195 (1979).
2. H. Garrett De Young, High Technology, 65 (1983).
3. V. C. Carver, L. T. Taylor, T. A. Furtsch, A. K. St. Clair,
    J. Amer. Chem. Soc., 102, 876 (1980).
4. K. Kurita, R. L. Williams, J. Polym. Sci., Polym. Chem. Ed.,
    12, 1809 (1974).
5. A. Chopin, J. Mol. Struct., 13, 295 (1972).
6. J. C. Bailar, "Inorg. Synth." (W. C. Fernelius, Ed.), 2, 222
    (1946).
7. G. Sclessinger, "Inorg. Synth." (E. G. Rochow, Ed.), 6, 173
    (1960).
8. J. Bjerrum, J. P. McReynolds, "Inorg. Synth." (W. C.
    Fernelius, Ed.) 2, 217 (1946).
9. G. T. Morgan, J. D. Smith, J. Chem. Soc., 127, 2030 (1924).
10. M. L. Morris, R. W. Moshier, R. E. Sievers, Inorg. Chem.,
    2, 411 (1963).
11. J. A. Mclean, Jr., A. F. Schreiner, A. F. Laethem, J.
    Inorg. Nucl. Chem., 26. 1245 (1964).
12. G. N. Schrauzer, "Inorg. Synth." (W. L. Jolly, Ed.), 11, 68
    (1968).
13. S. R. Edmonson, W. M. Coleman, L. T. Taylor, 23rd Southeast
    Regional Meeting of the American Chemical Society,
    Nashville, TN, Nov., 1979.
14. V. L. Bell, B. L. Stump, H. Gager, J. Polym. Sci., Polym.
    Chem. Ed., 14, 2275 (1976).
15. R. A. Gaudiana, R. T. Conley, J. Macromol. Sci. Chem., A4,
    441 (1970).
16. D. C. Frost, C. A. McDowell, L. S. Woolsey, Chem. Phys.
    Lett., 17, 320 (1972).
17. D. C. Frost, C. A. McDowell, L. S. Woolsey, Mol. Phys., 27,
    1473 (1974).
18. K. S. Kim, W. E. Baitinger, J. W. Amy, N. Winograd, J.
    Electron. Spectros. Relat. Phenom., 5, 351 (1974).
19. A.B.P. Lever, "Inorganic Electron Spectroscopy", Elsevier
    Publishing Co., Amsterdam, 1968, p. 306.
20. ibid., p. 318-333.
21. B. V. Kotov, T. A. Gordina, V. S. Voishchev, O. V.
    Kolninkov, A. N. Pravednikov, Vysokomol. Soyed., 419, 617
    (1977).
22. C. E. Sroog, A. L. Endrey, S. V. Abramo, C. E. Berr, W. M.
    Edwards, K. L. Olivier, J. Polym. Sci., Part A, 3, 1373
    (1965).
23. E. Khor, L. T. Taylor, Macromolecules, 15, 379 (1982).

POLYMERIC BIPYRIDINES AS CHELATING AGENTS AND CATALYSTS

Douglas C. Neckers

Department of Chemistry
Bowling Green State University
Bowling Green, Ohio   43403

INTRODUCTION

Anchoring reagents to insoluble supports has come to be known as solid phase synthesis. Based on the pioneering of Merrifield in polypeptide synthesis[1], rapid developments have occurred in this field such that many of these applications are now considered routine.

The immobilized transition metal catalyst offers potentially a plethora of practical advantages. Many examples employing phosphines were in the literature at the time of our original publication of the synthesis and basic applicability of (P)-bipy[2]. These many phosphine-based transition metal catalysts have been reviewed[3].

To increase the general availability of immobilized transition metal catalysts, we first reported the synthesis of bipyridine attached to 2% crosslinked polystyrene (Equation 1). Bipyridine had been known for years to be an excellent chelating agent for low valent metals and it was this characteristic plus the characteristic of significant potential versatility that attracted us to this ligand at the outset[4].

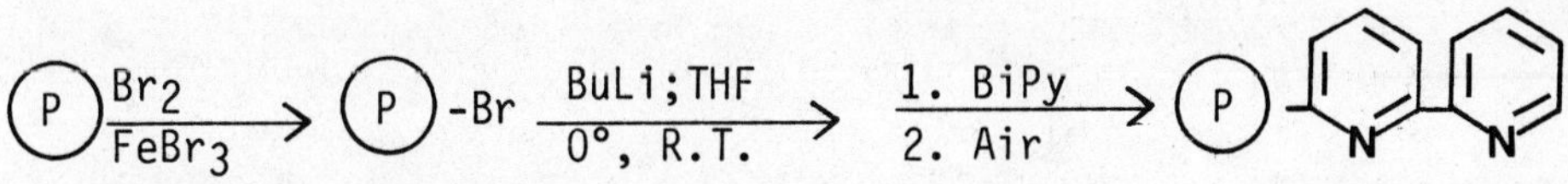

385

Table 1

The Amounts of Various Metal Salts Which Become Bound
to Polymer 1 in Tetrahydrofuran

Metal Salt	Quantity Bound (meq metal/g polymer)
$Cr(NO_3)_3$	0.22
MnI_2	0.14
$FeCl_2$	0.36
$FeCl_3$	0.83
$RuCl_3$	not determined
$CoCl_2$	0.35
$Ni(NO_3)_2$	0.70
$Pd(O_2CCH_3)_2$	0.73[a]
$CuBr_2$	0.34 and 0.37[b]
$AgNO_3$	not determined

[a] Determined by elemental analysis for Pd(Spang
    Microanalytical Lab).
[b] Two different batches of polymer were used.

The synthesis of $\textcircled{P}$-bipy was patterned after the synthesis of 2-phenylbipyridine and the yellow gold polymer derives from the addition of bipyridine in THF to polystyryllithium

Complexation of metal salts with 1 is strikingly facile. In a typical procedure 1 is added to an approximately 10^{-3}M THF solution of ferric chloride or other metal salt. The resulting solution is shaken for several minutes and the polymer filtered off. The amount of metal bound to the polymer surface can be determined either from quantitative analysis of the visible spectrum of the metal ion solutions before and after exposure to the polymer or by atomic absorption techniques. Some typical examples of metal ion complexes prepared by this technique are given in Table 1.

One of the potential advantages of using a polymer-based chelating ligand involves the expected decrease of metal complexibility due to the chelation effect[5]. This prediction was taken advantage of for the synthesis of immobilized metal ions which could be either reduced or oxidized for catalytic purposes. The metal ion to which most of our attention has been directed is palladium and in that application we discovered that the polystyrene sheath onto which the chelate and the metal ion were immobilized provided several advantages in application. A major advantage derived from the protection afforded the zero valent metal when palladium (II) was reduced as its chelate bound to $\textcircled{P}$-bipy. The reduction could be affected either by catalytic hydrogenation procedures or with LiAlH$_4$ in THF. Said reduction produced palladium metal clusters[6], a fact shown several years later by Drago's group by electron microscopy. Thus, upon reduction of palladium (II) as its $\textcircled{P}$-bipy complex an active metal cluster catalyst was produced. Of course, the polystyrene sheath provided other advantages in catalytic hydrogenation applications; these advantages being those of differential selectivity based on pore size, a phenomenon reported by Grubbs and coworkers some years earlier[7]. Reaction of (poly-(styryl)bipyridine) palladium acetate with LAH results in formation of (poly)-styryl)bipyridine) palladium (0) (Equation 2). This reaction is accompanied by a color change from brown to black and the loss of acetate bands in the IR spectrum. The palladium (0) complex can be reoxidized to the +2 oxidation state by treatment with dilute nitric acid or ceric ammonium nitrate.

$$\textcircled{P}\text{-Bipy} + \text{Pd(OAc)}_2 \longrightarrow \textcircled{P}\text{-Bipy} \cdot \text{Pd(OAc)}_2 \xrightarrow{\text{LAH}} \textcircled{P} \cdot \text{Bipy} \cdot \text{Pd(0)}$$

A large number of catalytic processes are known for various palladium species. The cost of palladium, however, makes easy recovery and reuse of the catalysts financially attractive. Therefore, low lability of a polymer-bound palladium catalyst in contrast to other palladium systems should be an important asset.

(P)-bipy-Pd^{2+} is an active hydrogenation catalyst which has properties different from both those of palladium acetate and palladium on carbon. In the first place, it appears to be selective for the reduction of carbon carbon multiple bonds. We have no evidence that it reduces nitro groups to amino groups nor do we have evidence that it reduces carbonyl groups in any of their various configurations. In our hands, it only is catalytically active for the olefins and the alkynes. Its selectivity is substantially different from palladium on carbon as is its reactivity and it functions as a catalyst for reduction processes under conditions where palladium acetate itself is not applicable. The hydrogenation of simple olefins occurs readily at ambient temperature and 1 atm of hydrogen.

In general, the following characteristics pertain: It appears to reduce hindered olefins less rapidly than less hindered olefins and varying hindered species such as pinene are not reduced at all with the catalyst. Second, the hydrogenation of olefins containing hydrogen on a carbon alpha to the double bond is accompanied by isomerization. Third, turnover numbers are in the neighborhood of 1,000 to 2,000 with these catalysts. Fourth, the amount of leaching of metal during the hydrogenation process is not significant at room temperature and under one atmosphere of hydrogen, though higher pressure and higher temperatures do cause palladium metal leaching. Relative to other catalysts developed in our laboratories later[8], these (P)-bipy-Pd catalysts are less satisfactory and leach much more.

In the years since its original discovery, many other applications have been reported for (P)-bipy. Among those is one reported from our laboratories on the (P)-bipy-Pd(0) catalyzed isomerization of quadricyclene[9]. In that application intended as a model system for solar energy storage and utilization, (P)-bipy-Pd(0) was intended to reisomerize quadricyclene to norbornadiene. Quadricyclene reisomerizes with the release of something over 80k/mol and can be prepared from norbornadiene by a photosensitized process[10].

BIPYRIDINE POLYMERS FROM SPIRO'S MONOMER[11]

As useful as (P)-bipy and its metal ion complexes are for

various catalytic applications, it does suffer from some disadvantages. The most important of these is that metal ion retention by the polymeric catalyst is not complete. This has two effects. Over time the metal ion concentration in the catalyst decreases and the catalytic activity of the complex decreases over time. A second disadvantage of systems prepared by immobilization on polystyrene is that one cannot control the properties of the polymer to meet specific catalytic needs. Though crosslinked polystyrenes serve well in immobilization applications, designing synthetic polymers offers many other advantages and much more versatility.

Spiro's monomer is 4-methyl 4'-vinylbipyridine[11]. It was the first polymerizable monomer containing bipyridine units to appear in the literature. Actually, many other vinyl monomers containing bipyridine would serve the same purpose as does Spiro's monomer; for example, 2-vinylbipyridine or 4-vinylbipyridine. Both of these monomers are very easily polymerized and at least in our hands[12] were completely polymerized during synthetic reactions required for their preparation.

Accordingly, in investigating template effects in polymer synthesis for possible metal ion selectivity based on bipyridine polymer complexes, Gupta chose to use Spiro's monomer as the principal complexing agent to test the concept.

The template effect is a phenomenon originally attributed to Pauling[13]. What was reported from Pauling's laboratories was that one could leave the imprint of methyl orange in silica gel synthesized from sodium silicate and dilute acid in the presence of the dye. After the silica gel was synthesized, the dye was eluted and comparative absorption experiments performed using the silica gel prepared in this manner to absorb methyl and ethyl orange in competition. It was observed that methyl orange was absorbed slightly more strongly than ethyl orange under identical conditions.

Over the years, there were a number of reports of similar phenomena of some involving other vinyl polymers[14]. Spiro's monomer provides an opportunity to test the concept of a metal ion template effect. The question is "does a metal ion complex of Spiro's monomer when polymerized and the metal ion removed produce a polymer which is selective for the metal ion of the original complex?" This is demonstrated in Figure 1.

Thus, Gupta[15] prepared a series of bipyridine containing polymers (Table 2) in which the bipyridine-function was incorporated in the form of Spiro's monomer.

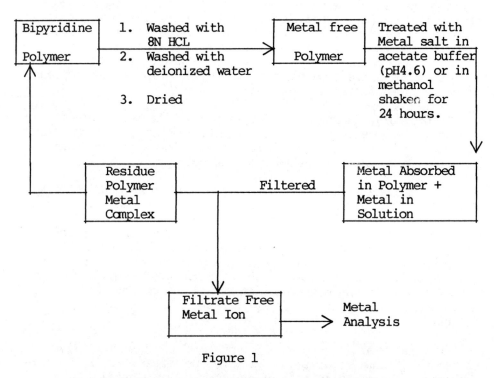

Figure 1

Schematic Diagram of Metal Absorption Experiments

Table 2. Synthesis of Bipyridine Polymers[a]

Polymer Number	VBP M	DVB M	Salt Type	Concentration (M)	Polymer Name	Bipyridine in polymer (mol %)[b]
A1	0.34	0.49	Ni	0.28	Ni polymer	36.14
A2	0.34	0.49	Co	0.29	Co polymer	43.08
A3	0.34	0.49	X	X	Control	35.78
B1	0.21	0.58	Ni	0.16		14.84
B2	0.21	0.58	Co	0.16		20.40
B3	0.21	0.58	X	X		15.66
C1	0.20	1.38	Ni	0.20		13.91
C2	0.09	0.58	Co	0.09		8.01
C3	0.09	0.58	X	X		8.45
D1	0.09	0.87	Ni	0.09		5.56
D2	0.09	0.87	Co	0.09		5.82
D3	0.09	0.87	X	X		7.09
E	0.34	X	X	X	Poly(vinyl bipyridine)	100

[a]Solvent methanol; $Ni = Ni(NO_3)_2 \cdot 6H_2O$; $Co = Co(NO_3)_2 \cdot 6H_2O$; temperature 60°C; polymerization time 18h; [AIBN] = 5 X 10^{-3} M.

[b]From nitrogen analysis of metal-free polymer.

The polymers (Table 2) were studied for their capacity to absorb Cu^{2+}, Ni^{2+}, and Co^{2+} from water and methanol. In acetate buffer the absorption of these metals by highly divinylbenzene-crosslinked, low-bipyridine-content polymers was low. Polymer A3, polymerized in the absence of metal chelate, contains 35.78% Spiro's monomer and 64.22% DVB. It absorbs little or none of the three metal ions and the ratio of the bipyridine units in the polymer to [M^{2+}] ion absorbed by it is large—100 or more.

It is clear that the polymers made with a metal ion chelated in the monomer as a template absorb much more metal per resident bipyridine when compared to the polymer template made in the absence of any chelated metal. The homopolymer of 4-vinyl-4'methylbipyridine (PVBP), of course, absorbs the largest quantity of metal and the ratio of bipyridine to metal absorbed by it approaches unity in all cases.

A significant variation is observed when rigid crosslinked polymers are considered. In the case of the homopolymer PVBP, when no crosslinking agent is used, the mobilities of the polymeric chains are more free than they are in the crosslinked polymer. Although the homopolymer is not soluble in acetate buffer, acid, or organic solvents, it does swell somewhat, so the complexes it forms with the metals used approach the maximum, i.e., 1:1 mole ratio of resident bipyridine to metal ion.

The quantity of chelate formed from each crosslinked polymer follows the normal trend of complexation Cu>Ni>Co, which is in turn based on the metal ion complex stability constants. In the case of rigid crosslinked polymers, a huge difference in the metal absorbed by the template polymer is observed when these polymers are compared to control polymers made such that no metal was used during polymerization process.

Metal absorption by the copolymers, in general, is increased by metal template polymerization. When a template metal chelate monomer is polymerized, the polymer derived absorbs more metal ion. The question is how specific are these template polymers in absorbing their template metal? In the absence of any template metal ion effect, the ratio of bipyridines/metal absorbed ratio should increase with increasing crosslinking since the resident bipyridines would be less accessible to the metal ions. This is not observed.

In Figure 2, for example, a comparison is made between Cu^{2+} and Ni^{2+} absorption by the simple homopolymer (PVBP) and by the Ni template polymer. Whereas in PVBP absorption of Cu^{2+} is always higher than Ni^{2+} (expected because of the higher stability constant for Cu^{2+}), in the Ni template polymer the Ni^{2+} absorption is either equal to or actually higher than Cu^{2+} absorption. This demonstrates that there is a degree of specificity to the template effect. Even though the $BiPy \cdot Cu^{2+}$ stability constant exceeds that of $BiPy \cdot Ni^{2+}$, VBP copolymers actually absorb more Ni^{2+} than Cu^{2+}. When Ni^{2+} absorption is compared with the absorption of Co^{2+} (Fig. 3), the general trend of higher Ni^{2+} absorption is actually reversed with the Co^{2+} template polymer. In other words the amount of Co^{2+} absorbed by the Co^{2+} template polymer is greater than the quantity of Ni^{2+} absorbed by the same Co^{2+} template polymer, exactly the opposite of what is found with either PVBP or the Ni^{2+} template polymer and the homopolymer than Co^{2+}. This is a clear demonstration that the Co^{2+} template polymer also has a degree of specificity for Co^{2+}.

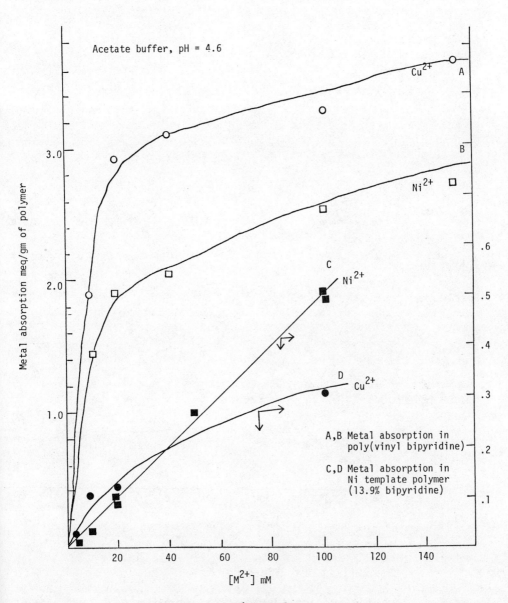

Figure 2

Absorption of Ni^{2+} and Cu^{2+} by Ni^{2+} Template Polymers (C,D)
(13.9% Bipyridine) Compared with Nontemplate
Polymer (A,B) [Poly(vinyl bipyridine)].
Acetate Buffer, pH 4.6.

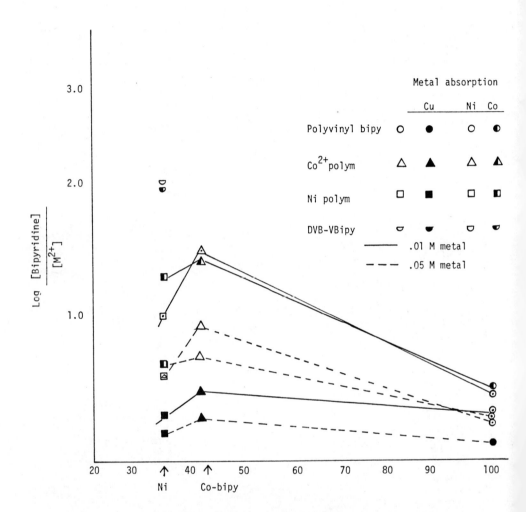

Figure 3

Metal Ion Absorption by Various
Template Copolymers: (——) 0.01M Metal,
(— — —) 0.05M Metal.

Since the wettability of the polymers in water decreases with increasing DVB content the absorption of metal ions in solvent containing 50% (by vol) methanol was also measured. As shown in Figure 4, the absorption of Cu^{2+}, Ni^{2+}, and Co^{2+} is substantially higher with templated polymers than with the controls, but no specificity with respect to absorption of a particular metal ion by a particular template polymer was observed. Although the overall extent of metal ion absorption increased in the template polymer over the control,(possibly due to more swelling or more mobility of the polymeric chains, the template specificity observed with the same polymers when metal ions were absorbed from a water solution was lost. We suggest that a molecular compromise between rigidity and wettability is necessary to maximize specific metal ion separation. Perhaps divinylbipyridine should be used as crosslinking agent rather than divinylbenzene.

DIAMINOBIPYRIDINE - TDI POLYUREAS: SYNTHESIS, METAL COMPLEXES AND CATALYTIC ACTIVITY

Polymers based on 4-methyl 4'-vinylbipyridine have provided important new information about the three-dimensional structure of chelate complex polymers containing bipyridine residue. Polyureas prepared from 4-4'diaminobipyridine have greatly improved the metal ion stability of metal ion bipyridine complex polymers. Thus a series of polyureas deriving from diisocyanates and the bipyridine monomer 4-4'diaminobipyridine were prepared by Keda Zhang and the palladium complexes of these polymers were used as hydrogenation catalysts in the manner of Card[2]. Table 3 outlines the relative rates of olefin hydrogenation using the palladium (0) complex of TDI 4-4'diaminobipyridine copolymer.

The more hindered carbon carbon double bonds are reduced less rapidly than are less hindered ones. Linear olefins are reduced more rapidly than cyclical ones. The greater the number of alpha-substituents, the slower the reduction. Thus, in the reduction of dienes the least hindered double bonds will be reduced preferentially. Second, in cases where the steric hindrance of a pair of double bonds in a diene is essentially the same, the more strain to the pair will be reduced first. An example is given by 5-ethylidene-2-norbornene. In this compound, the ring double bond reduces much before the exocyclic double bond. Whereas in 5-vinyl-2-norbornene the exocyclic double bond is reduced at about the same rate as the double bond of the ring even though the former is much less hindered than the latter. Third, the attachment of polar functions to the double bond has a substantial influence on the rate of hydrogenation. Olefins containing -OH groups, for example, are hydrogenated much more rapidly than our identical olefins without the -OH function.

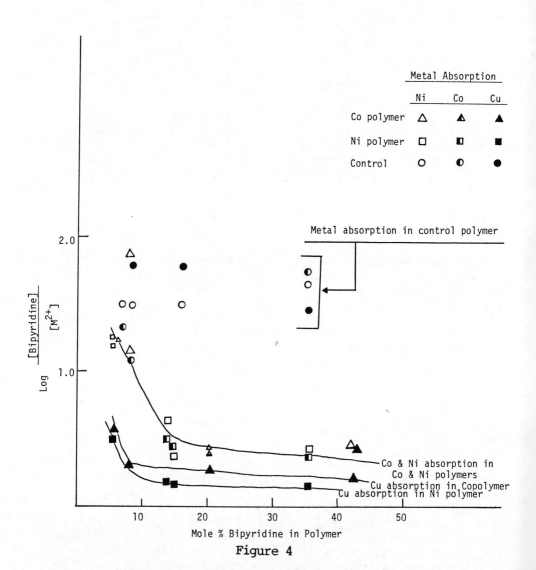

Figure 4

Absorption Capacity of Different Metal Ions
in Different Polymers (Solvent: Methanol).

Table 3

Relative Rates of Olefin Hydrogenation[a]

Olefin	Structure[b]	Rel Rate
(+) - Limonene		0.41
1.5-Cyclooctadiene		0.59
Cyclohexene		1.00
Acrylic acid	$CH_2=CH-COOH$	1.01
Methylmethacrylate	$CH_2=C\overset{CH_3}{\underset{COOCH_3}{}}$	1.25
Vinyl acetate	$CH_3-\overset{O}{\overset{\|}{C}}-OCH=CH_2$	1.34
5-Ethylidene-2-norbonene		1.46
5-Vinyl-2-norbonene		1.57
1-Octene		1.59
1-Hexene		1.60
4-Vinyl-1-cyclohexene		1.62
β-Hydroxyethyl methacrylate	$CH_2=C\overset{CH_3}{\underset{COOCH_2CH_2OH}{}}$	1.62
2-Methyl-3-buten-2-01		1.81
Allyl alcohol	$CH_2=CH-CH_2-OH$	1.82
Allyl acetate	$CH_3CO_2CH_2CH=CH_2$	2.08

[a]10 mmol olefin, 100 mg polymer pd(O) complex, in 10mL dry tetrahydrofuran, 1 atm of hydrogen at room temperature.

[b]The rate is first double bond if the reactant is diene.

The rate of reduction using DABP-Pd(0) polyureas as hydrogenation catalysts depends upon the polarity and the molecular
size of the solvent used for the hydrogenation reaction. The
more polar the solvent, the more rapidly the hydrogenation of the
same olefin occurs. Solvent effects on the hydrogenation of
cyclohexene are shown in Figure 5. The slope of the line indicates the hydrogenation rate. For discussion, this solvent
effect problem can be divided into four limiting cases.

A. Non-hydroxylic solvent - non-hydroxylic reactant

An example which illustrates this case is the reduction of
cyclohexene in the solvent, THF. The rate of reduction
reaches a maximum at 15% cyclohexene and remains constant to
100% olefin.

B. Hydroxylic solvent - non-hydroxylic reactant

An example illustrating this case is the reduction of cyclohexene in methanol. In this case, the hydrogenation rate
goes through a maximum at 15% and then decreases.

C. Non-hydroxylic solvent - hydroxylic reactant

This case is illustrated by the reduction of allyl alcohol in
THF. The hydrogenation rate again goes through a maximum,
this time at 13% substrate.

D. Hydroxylic reactant - hydroxylic solvent

The fourth case is illustrated by the reduction of allyl
alcohol in methanol. In this hydrogenation, the reaction
reaches a maximum rate at about 10% and levels off.

Activity and Lifetime of the Pd(0) Complex in Hydrogenation
Reactions.

We compared the hydrogenation activity of the polymer Pd(0)
complex polymer catalyst with Pd(0)/C. In general the activity
of Pd(0)/C is somewhat greater than is our catalyst, though the
Pd content of the former is also 40% higher. Under some conditions, the activity of our catalyst, even with less palladium,
is larger than palladium on charcoal, Table 4.

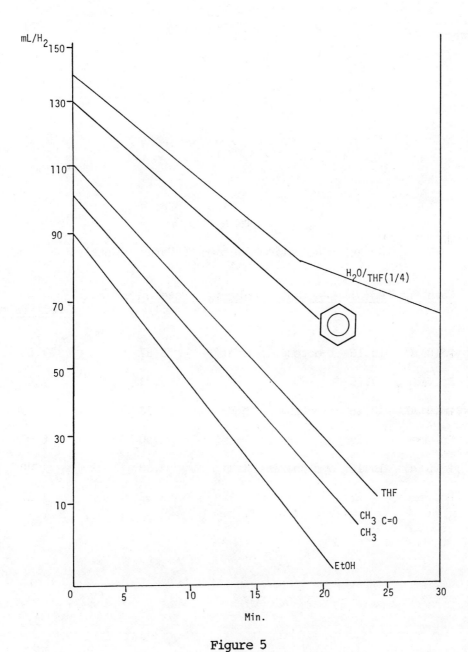

Figure 5

The Rate of Hydrogenation of
Cyclohexene in Different Solvents

Table 4

Relative Reduction Rates with Various Catalysts

Catalyst	Pd(O)	Reactant	Solvent	Rate of Hydro. ml/min	Rel. Rate
Pd(0)/C	10.16	1-octene	EtOH	3.67	1.00
Polymer	7.25	--	--	4.33	1.65
Pd(0)/C	10.16	1-octene	THF	3.58	1.00
Polymer	7.25	--	--	2.50	0.97
Pd(0)/C	10.16	cyclohexene	MeOH	3.50	1.00
Polymer	7.25	--	--	3.02	1.21

The activity of Pd(0)/C is not as sensitive as is the polymer catalyst to a change in solvent. Thus the polarity of the solvent has a much greater effect on the polymer which binds the catalyst to a polar segment than it does on the non-polar carbon which serves the same purpose.

The catalysts which are polymer based have very good stability, and the polymer based palladium function can be used over and over again with no change in catalytic activity. We have actually used the same polymer catalyst for 42 successive catalytic hydrogenation reactions. The decline in its catalytic activity was nil and its loss of palladium metal essentially non-existent.

CONCLUSIONS

A palladium(0) polymeric polyurea containing a bipyridine has been synthesized and used for the hydrogenation of olefins. To the best of our knowledge, this is the first example of a polar solvent compatible polymer based Pd complexing agent and the system developed opens the door for the study of catalytic processes involving polar reactants.

The polymeric palladium catalyst is somewhat selective as a hydrogenation catalyst. This selectivity derives from both thermodynamic and kinetic effects.

The polarity of the reactant and the solvent, and their relative concentrations, affect the rate of olefin hydrogenation a great deal. The rate of hydrogenation is determined by the microenvironment of the Pd(0). On the polymeric reagent deriving from a polyurea and that microenvironment is most polar.

The catalyst has good activity and excellent long-term stability.

SUMMARY

In summary, we have synthesized three general types of polymers containing the bipyridine chelating ligand.

1. (P) -bipy bipyridine immobilized to 2% divinylbenzene styrene copolymer beads. The advantage of this polymer system is its simplicity of preparation and its versatility of application. Disadvantages in catalytic applications the metal ion is not rigidly held to the polymer backbone and tends to leach with extensive use.

2.   Vinyl polymers based on 4-methyl 4'vinyl bipyridine.
     The advantages of polymers prepared from this monomer
     are that they can be highly crosslinked and formed and
     relatively rigid structures. Disadvantages are that
     various inopportune crosslinkings lead to highly inso-
     luble products in this particular system.

3.   Diaminobipyridine-TDI polyureas. These polymers are the
     first examples of polar backbone polymers prepared with
     bipyridine residues contained therein. Metal ion
     complexes of these polymers are stable to extended use
     and polar solvents are much more compatible with the
     polymer backbone. Therefore, reactions under more polar
     conditions can be carried out.

ACKNOWLEDGMENT

    This work was supported by the Division of Materials
Research, National Science Foundation. The authors acknowledge
with gratitude their support. The work has been carried out by
the following students and post-doctorals: Dr. Roger Card, Dr.
S. N. Gupta, Mr. Keda Zhang. The senior author is very grateful
for their fine contributions to this program.

REFERENCES

1. R. B. Merrifield, *Science,* 150, 178 (1965).

2. R. J. Card and D. C. Neckers, *J. Amer. Chem. Soc.,* 99, 7733 (1977).

3. B. M. Trost and E. Keinan, *J. Amer. Chem. Soc.,* 100, 7779 (1978).

4. F. A. Cotton and G. Wilkenson, Basic Inorganic Chemistry, J. Wiley & Sons, New York (1976); p. 491.

5. A. J. Moffat, *J. Catl.,* 18, 193 (1970).

6. R. S. Drago and J. H. Gaul, *Inorg. Chem.,* 18, 2019 (1979).

7. R. Grubbs, C. P. Lau, R. P. Cukier, and C. Brubaker, Jr., *J. Amer. Chem. Soc.,* 99, 4517 (1977).

8. K. Zhang and D. C. Neckers, *J. Poly. Sci.,* 21, 3115 (1983).

9. R. J. Card and D. C. Neckers, *J. Org. Chem.,* 43, 2958 (1978).

10. R. R. Hautala, J. Little, and E. Sweet, *Sol. Energy,* 19, 503 (1977).

11. P. K. Ghosh and T. G. Spiro, *J. Amer. Chem. Soc.,* 102, 5543 (1980).

12. S. N. Gupta and D. C. Neckers, unpublished.

13. F. H. Dickey, *Proc. Natl., Acad., Sci,,* U.S., 227 (1949).

14. G. Wulff, A. Sarhan, and K. Zabrocki, *Tetrahedron Letters,* 4329 (1973).

15. S. N. Gupta and D. C. Neckers, *J. Poly. Sci., Poly. Chem.,* 20, 1609 (1982).

ALKALI METAL ORGANOMETALLIC POLYMERS: PREPARATION AND APPLICATIONS

IN POLYMER AND REAGENT SYNTHESES

David E. Bergbreiter

Department of Chemistry
Texas A&M University
College Station, Texas   77843

Organometallic polymers include a variety of materials with diverse properties ranging from stable refractory polymers to highly reactive transition metal-containing polymeric catalysts. Alkali metal organometallic polymers, polymers containing alkali metal-carbon bonds or which incorporate alkali metal organometallic reagents, are expected to be and are reactive materials in analogy to their well known homogeneous counterparts. In this chapter, we summarize some of the chemistry leading to these useful organometallic polymers and selectively describe some of their applications. Illustrative examples from our work and the work of others are included in these discussions. Although these discussions mainly concern organometallic derivatives of organic polymers, we have also included a brief discussion of some similar chemistry of alkali metal-graphite intercalates.

Alkali metal-graphite intercalation compounds are perhaps the earliest examples of reactive organometallic polymers. These materials, which readily form from the direct reaction of sodium, potassium, rubidium or cesium and graphite (equation 1), were

$$\text{Graphite} + \text{Metal} \longrightarrow C_n\text{Metal} \qquad (1)$$

$$\text{Metal} = \text{Na, K, Rb, or Cs}$$
$$n = 8, 24, 36, \text{etc.}$$

of interest initially mainly to inorganic and physical chemists.[1] However, renewed broader interest in these unusual materials has developed since their activity as catalysts and as reagents has

been recognized.[2] For example, potassium graphite in the form of
C_8K has been shown to be a catalyst for Fischer Tropsch synthesis
of hydrocarbons from carbon monoxide and hydrogen.[3] Catalytic
activity for this stage of potassium-graphite and for other stages
(a "stage" refers to the relative amounts of potassium/carbon, e.g.
C_nK where n = 24, 36, 48, etc.) of potassium-graphite in other
reactions such as hydrogenation and alkene polymerization have also
been noted.[4,5] Alkali metal-graphites such as C_8K also have been
shown to possess useful synthetic reactivity. For example, several
groups have reported that these alkali metal-graphite intercalates
can be used as bases to deprotonate weakly acidic carbonyl compounds and derivatives of carbonyl compounds to generate potassium
salts of stabilized carbanions (equation 2).[6-8] In this particular
reaction, the potassium-graphite apparently either acts like a

$$C_8K \ + \ RCH_2C(=X)R' \ \longrightarrow \ \underset{R}{\overset{H}{>}}C=C\underset{R'}{\overset{X^-}{<}} \qquad (2)$$

$$X = O, \ NR$$

Lewis base or else generates some base in situ which then actually
deprotonates the weakly acidic carbon acid.[9,10] Carbon acids which
have been deprotonated (and then subjected to subsequent electrophilic substitution reactions) include oxazines, ketones, imines,
esters and nitriles. One advantage cited for the use of C_8K as a
base is the minimization of polyalkylation in reactions like allyl-
ation of 1-tetralone using allyl bromide. In this particular case,
use of C_8K and hexane as a solvent with this ketone produced 98%
monoalkylated product. The high yield of monoalkylated product has
been ascribed to the heterogeneity of this reaction and the hydrocarbon solvent used.

The ability of potassium-graphite to act as a reducing agent
in reactions with substrates such as ketones, alkyl halides (equation 3) or α,β-unsaturated sulfones (equation 4) has also been used

$$C_8K \ + \ R-X \ \longrightarrow \ R-H \qquad (3)$$

$$X = -Cl, \ -Br, \ or \ -I$$

$$C_6H_5SO_2(R)C=CHCH_3 \ \xrightarrow[\text{THF}]{C_8K} \ RCH=CHCH_3 \qquad (4)$$

in several synthetic applications leading to alcohols, alkanes or alkenes respectively.[9,11,12] Mechanistic studies carried out in the case of alkyl halide reductions have indicated that this reduction proceeds by a mechanism similar to that reported for reduction of alkyl halides by soluble alkali metal aromatic radical anions.[9] Radical intermediates have been implicated by the observed cyclization to form methylcyclopentane during reduction of 1-bromo-5-hexene by C_8K in THF. In some cases, these reducing agents have been reported to be very useful as reagents. For example, Cristol has shown that potassium-graphite was more effective than sodium in liquid ammonia as a reagent for reduction of chloro alcohols to alcohols.[13] Potassium-graphite has also recently been used as a heterogeneous analog of soluble alkali metal aromatic radical anions in reductions of sulfonium salts.[14]

Alkali metal-graphite compounds such as potassium-graphite have also been used as precusors of novel supported catalysts. In this last example, the reducing properties of potassium-graphite were used to directly produce highly dispersed metal catalysts from solutions of palladium(II),[14] nickel(II)[15] and iron(III) halides.[16] The resulting zero valent metal catalysts dispersed

$$C_8K + MX_n \longrightarrow C_mM + n \ KCl \qquad (5)$$

$$M = Pd, \ X = Cl, \ n = 2, \ m = 16$$
$$M = Ni, \ X = Cl, \ n = 2, \ m = 16$$
$$M = Fe, \ X = Cl, \ n = 3, \ m = 24$$

on graphite are reported to have high reactivity. For example, palladium-graphite prepared in this manner has reactivity characteristic of a homogeneous palladium(0) complex such as tetrakis(triphenylphosphine)palladium(0) in reactions like those shown in equation 6. Palladium-graphite prepared according to equation 5

$$RI + \underline{trans\text{-}R'HC{=}CHR''} + \underline{n\text{-}Bu_3N} \xrightarrow{\ Gr\text{-}Pd\ }$$
$$RR'C{=}CHR'' + \underline{n\text{-}Bu_3NH^+ \ I^-} \qquad (6)$$

has also been shown to be useful as a hydrogenation catalyst for nitroarenes, alkenes and alkynes with a reactivity comparable to 10% Pd/C but with greater product selectivity and greater stereoselectivity. Studies of nickel-graphite by these same authors have shown that this nickel catalyst is useful for stereospecific semihydrogenations of alkynes. This nickel(0)-graphite catalyst is also reported to be similar to nickel catalysts derived from direct reaction of alkali metals and nickel salts or to nickel catalysts

derived from metal atom vaporization techniques (e.g. nickel(0)-
graphite catalyzes the allylic coupling of allyl halides as is
shown in equation 7). Air oxidation of this nickel-graphite

$$BrCH_2CH=CH_2 \xrightarrow{\text{Ni-graphite}} CH_2=CHCH_2CH_2CH=CH_2 \qquad (7)$$

produces an oxidized nickel-graphite which is reported to have
useful selectivity in partial hydrogenation of polyfunctional molecules such as unsaturated carbonyl compounds and diketones (equation 8). Iron-graphite, accessible by equation 5, has also been
reported to be useful as a dehalogenating agent. Recently, procedures similar to equation 5 were reported for reaction of potassium-graphite with zinc chloride leading to zinc-graphite useful
for synthesis of β-hydroxy esters, homoallylic alcohols and
methylene-γ-lactones in reactions similar to Reformatsky reactions.[18]

Although alkali metal-graphites like those described above do
indeed have interesting chemistry and utility in certain applications, alkali metal derivatives of organic polymers have potentially more versatility because of the variations in chemical and
physical structure possible with various types of organic polymers.
The balance of this paper deals with such sorts of organometallic
materials. For the purposes of this paper, the discussion is
limited to examples of reactive alkali metal derivatives of insoluble polymers. Thus, the important chemistry involved with living
organometallic polymers like those involved in anionic polymerization reactions[19] is not discussed below.

Lithiated polystyrene prepared from divinylbenzene(DVB)-cross-
linked polystyrene is perhaps the most common example of an alkali
metal derivative of a conventional insoluble organic polymer. Several synthetic routes have been developed leading to this lithiated
polymer. First, direct metalation of DVB-crosslinked polystyrene
by an alkyllithium reagent complexed with tetramethylethylenedi-
amine in a hydrocarbon solvent produces a mixture of meta and para
lithiated material (equation 9).[20-24] Apparently the meta and para

protons have equivalent kinetic acidities since a 2:1 statistical
mixture of meta:para products is produced as determined by IR
analysis of a phosphonated product of the intermediate lithiated
polymer. Both n-butyllithium and sec-butyllithium have been used
as the alkyllithium reagent in this procedure. However, using

$$\text{(9)}$$

linear polystyrene and excess alkylllithium reagent-tetramethyl-
ethylenediamine in cyclohexane it has been determined that 40%
lithiation occurs within 4 hours at 35 °C using sec-butyllithium
while only 20 - 25% lithiation occurs in 65 hours at 60 °C using n-
butyllithium as the alkyllithium reagent.[24] In the case of highly
crosslinked macroreticular polystyrene, this method reportedly
produces "unevenly" substituted lithiated polymers. A more even
distribution of lithium throughout the resin bead however can be
achieved using a temperature regime for the reaction in which the
initial addition of n-BuLi-TMEDA is at -78 °C after which the
reaction mixture is slowly allowed to warm to room temperature.[22]
The lithiation occurs only on the pendent aryl ring of polystyrene
because the increased substitution at the benzylic position makes
the tertiary benzylic C-H less acidic than the aryl C-H's.[24,25] In
DVB-crosslinked polystyrene, no ortho metalation has been reported
possibly due to steric effects of the polymer backbone. Some ortho
lithiation was however seen in lithiation of a model linear polystyrene, 1,3,5,7-tetraphenyloctane.[24]

     A second procedure for the preparation of lithiated DVB-cross-
linked polystyrene is to first halogenate the polystyrene and to
then form the lithiated polymer using a halogen-metal exchange
reaction (equation 11).[20] The initial halogenated polystyrene is

$$\text{PS-Br} \xrightarrow[\substack{\text{THF} \\ 25\ ^\circ\text{C}}]{\text{n-BuLi}} \text{PS-Li} \qquad (11)$$

prepared in several ways. Copolymerization of p-bromostyrene with
styrene and divinylbenzene is one route to a brominated DVB-cross-
linked polystyrene. This method is especially useful if polymers
with structural properties (e.g. various percentages of DVB-cross-
linking) are not available commercially.[25] Alternatively, unsubstituted DVB-cross-linked polystyrene can be functionalized using

unexceptional electrophilic substitution reactions to produce predominantly para brominated or iodinated polystyrene. Bromination
can be accomplished either directly using bromine and a catalyst
like ferric bromide or indirectly by initial electrophilic thalla-
tion followed by electrophilic bromination of the intermediate
polystyrene-thallium bond.[20,26,27] Ferric bromide catalyzed bromination has been reported to be capable of introducing up to 3.9
mequiv of bromine per gram of polymer corresponding roughly to
halogenation of nearly 50% of the pendant aryl rings of polystyrene. However, this procedure suffers from the fact that it is
difficult to remove the catalysts for the substitution reaction
from the polystyrene beads and from the low quality and uneven
substitution of the resulting brominated polystyrene. These problems are circumvented in part using the thallium procedure. Thallium trifluoroacetate or other thallium(III) salts can be used
stoichiometrically or catalytically (at a 1 - 2% level) with subsequent or concurrent addition of a carbon tetrachloride solution of
bromine. More polar solvents like acetic acid are unsatisfactory
because of depolymerization of the polystyrene which occurs when
they are used. Iodination of polystyrene via a thallium intermediate requires stoichiometric use of thallium trifluoroacetate.
An alternative to the use of expensive thallium(III) salts is
mercuration of DVB-cross-linked polystyrene. Taylor has recently
described procedures by which high loadings of functionalized polystyrenes may be prepared by means of an intermediate mercurated
polymer.[28]

   Lithiation of a halogenated polystyrene as is depicted in
equation 11 is unexceptional. Typical procedures employ n-butyllithium at room temperature in an ethereal solvent using standard
inert atmosphere experimental techniques.

   A third preparation of lithiated polystyrene shown in equation
12 involves transmetalation of a mercurated DVB-cross-linked polystyrene with an excess of an organolithium reagent like n-butyllithium. This procedure was used by Burlitch as a method of achie-
ving 81% conversion of a mercurated polystyrene to a lightly loaded
but evenly substituted lithiated polymer.[22]

$$PS\text{-}HgCl \xrightarrow{\text{n-BuLi}} PS\text{-}Li \ + \ Bu_2Hg \ + \ LiCl \qquad (12)$$

   One serious and as yet largely unresolved problem facing the
use of reactive alkali metal organometallic polymers is the difficulty with which these reactive materials can be adequately characterized. For example, in the case of the lithiated polystyrenes
prepared according to equations 10-12 even simple questions like
accurate analysis of the amount of lithiated material bonded to an

insoluble polystyrene are difficult to measure conveniently. More
subtle problems such as the amount, nature and distribution of
impurities in these heterogeneous polymeric lithium reagents present further difficulties. Questions concerning the degree of
aggregation of these polymeric lithium reagents and their structures which are difficult to answer for homogeneous systems are
typically beyond the scope of current analytical procedures for
heterogeneous polymeric lithium reagents or have not yet been
adequately studied.

    Quantitative analysis of lithiated polystyrene is a more dif-
ficult problem than it might seem for several reasons. First,
several types of organolithium species can be present on the polymer. DVB-cross-linked polystyrene normally has unreacted terminal
vinyl benzene groups. These unreacted vinyl arene species which
are present at higher concentration in more heavily cross-linked
polymers react with alkyllithium reagents to form benzyllithium
species. If a lithiated polystyrene is desired which contains only
aryllithium species, it is necessary to first pre-treat the DVB-
cross-linked polystyrene that is to be used with n-butyllithium to
consume these unreacted vinyl groups. Even when these benzyllithium species are not present, the degree of lithiation of polystyrene lithium is still difficult to measure because alkyllithium
reagents that are needed for the preparation of polystyryllithium
and that are expected to normally be soluble can aggregate with the
polymeric lithium reagent. In the case of n-butyllithium, the
absorbed n-butyllithium is quite difficult to remove completely.
Thus, a simple acid base titration probably overestimates the
concentration of polymeric organolithium reagent in a sample of
polystyryllithium. A procedure we have developed avoids this problem although it is a batch analysis and is not all that convenient.[29] In this analytical procedure, a solution of a 1,2-
dibromoalkane such as 1,2-dibromodecane which can react with an
aryl- or alkyllithium reagent to form an aryl bromide or alkyl
bromide and decene is added to the polymeric lithium reagent. The
concentration of decene and alkyl bromide and alkyl bromide present
in solution after addition of this dibromide to a polymeric organolithium reagent can be determined relative to an internal standard
by gas chromatography and from these two numbers, the amount of
polystyryllithium reagent can be deduced based on the amount of
decene minus the amount of alkyl bromide formed. While both this
procedure and the simpler expedient of a hydrolysis and acid-base
titration can both determine the total amount of alkyllithium and
polystyryllithium reagent present, the procedure employing a dibromide (which is modeled after a known procedure used for analysis
of homogeneous organolithium reagents)[30] has some advantages.
Specifically, the dihalide procedure also permits the determination
of the concentration of other bases such as lithium alkoxides which
might be present. Alkyllithium reagents in solution typically
contain some alkoxide, the amount of alkoxide impurity depending on

the care in preparation and handling of the alkyllithium reagent. Other lithium salts such as lithium alkoxides or lithium halides are known to profoundly affect the chemistry of homogeneous lithium reagents and can reasonably be expected to be equally influential in the chemistry of a polymeric organolithium reagent. Thus, it is worthwhile to occasionally determine the concentration of species such as lithium alkoxides in polymeric organolithium reagents to insure that the observed chemistry is due to a relatively pure organolithium reagent.

An analytical procedure we have also used with some success in analysis of polymeric lithium reagents is direct titration of lithium reagents using a soluble indicating proton transfer agent. In this procedure, which is modeled after an analytical procedure we developed separately for soluble organolithium and -magnesium reagents,[31] a small amount of N-phenylnaphthylamine is added to a suspension of lithiated polystyrene. Because of the acidity of this diarylamine, lithium N-phenyl-1-naphthylamide readily forms by reaction with an polymeric organolithium reagent. The resulting orangish lithium diarylamide is readily soluble and is present in solution. As small amounts of a standardized xylene solution of butanol are added to this suspension of polymeric lithium reagent and soluble lithium diarylamide, both the polystyryllithium and lithium diarylamide are protonated. As the diarylamide is protonated, the solution's color fades. However, any diarylamine formed is readily deprotonated by unreacted polystyryllithium to regenerate a colored solution. Thus, slow addition of the titrant solution allows one to directly carry out a colorimetric titration of the amount of polymeric lithium reagent.

Direct spectroscopic analysis of lithiated polystyrene using solid state NMR spectroscopy should be a useful non-destructive way of analyzing these alkali metal polymers. While there will be some experimental difficulties in the handling procedures in this type of experiment because of the reactivity of these materials toward air and water, the results of such an experiment would be very interesting in that the resulting spectra can be directly compared to spectra of homogeneous solutions of phenyllithium in solvents such as benzene or toluene.

Several reports in the literature describe alternative types of alkali metal organometallic derivatives of polystyrene. These alkali metal polymers are derived from functionalized polystyrenes containing more acidic carbon acids. For example, reaction of DVB-crosslinked polystyrene with diphenylmethyl chloride in nitrobenzene in the presence of aluminum chloride as a catalyst yields a polymer-bound triarylmethyl group (**2**).[32] This polymer-bound triarylmethyl group is readily deprotonated on addition of an organolithium reagent (equation 13). Another example of a polymer-bound weak carbon acid is the polymer-bound diarylmethyl group **4**.

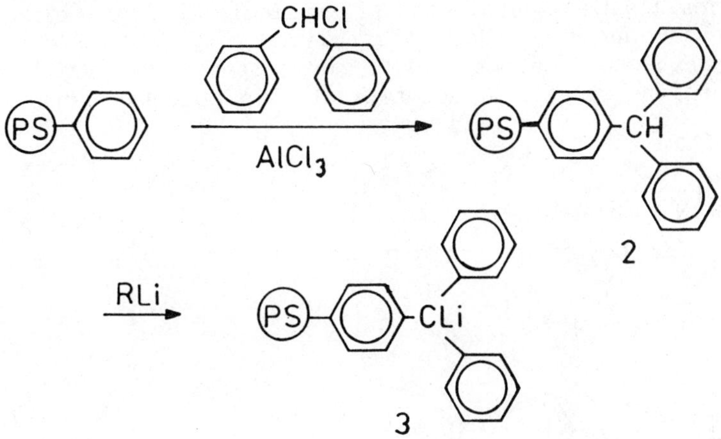

CHCl

$$\text{(PS)} \longrightarrow \xrightarrow{\quad \text{AlCl}_3 \quad} \text{(PS)}-\text{CH}$$

(13)

2

$$\xrightarrow{\quad \text{RLi} \quad} \text{(PS)}-\text{CLi}$$

3

In this case, chloromethylation of DVB-crosslinked polystyrene is first used to prepare a chloromethylated polystyrene. This polymer is also commercially available in the form of 1-2% DVB or 20% DVB crosslinked beads. Subsequently addition of phenyllithium to this polymeric benzyl chloride leads to nucleophilic substitution by the aryllithium reagent to form the polymeric carbon acid. If excess phenyllithium reagent is used in this step, deprotonation occurs *in situ* to form a polymeric lithium reagent **5** (equation 14).[33] In the case of diarylmethyl containing polymers, alkali metal aromatic radical anions such as sodium, potassium, rubidium or cesium naphthalene have also been shown to be useful as bases to deprotonate

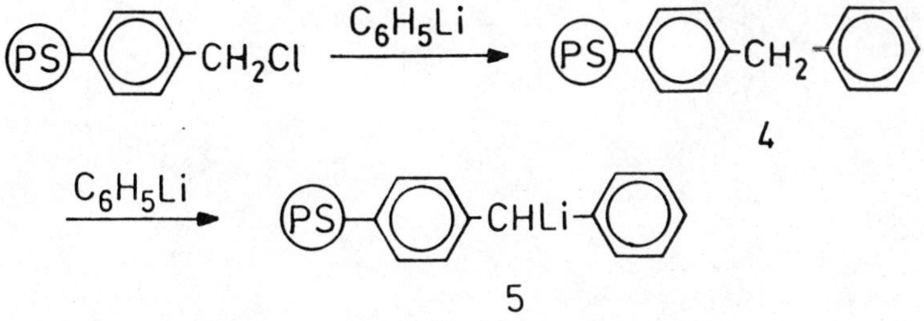

$$\text{(PS)}-\text{CH}_2\text{Cl} \xrightarrow{\quad \text{C}_6\text{H}_5\text{Li} \quad} \text{(PS)}-\text{CH}_2-$$

4

(14)

$$\xrightarrow{\quad \text{C}_6\text{H}_5\text{Li} \quad} \text{(PS)}-\text{CHLi}-$$

5

4 to form the whole family of alkali metal derivatives of diarylmethyl-functionalized polystyrene (equation 15).

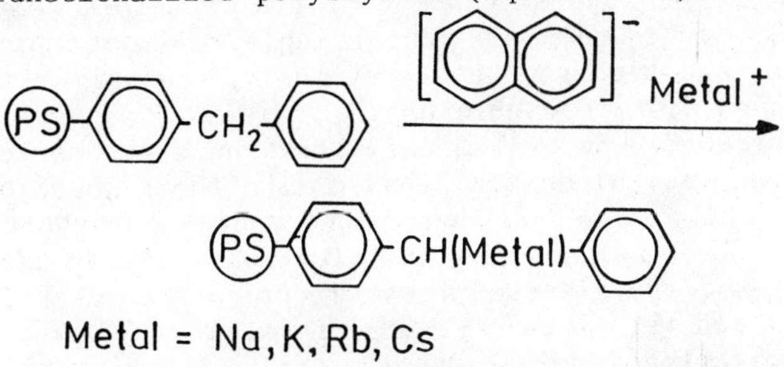

$$\text{(PS)}-\text{CH}_2- \xrightarrow{\quad [\text{naphthalene}]^{-} \text{Metal}^{+} \quad}$$

(15)

$$\text{(PS)}-\text{CH(Metal)}-$$

Metal = Na, K, Rb, Cs

Polymeric alkali metal aromatic radical anions can also be prepared from appropriately functionalized polymers. Using the synthetic sequence of equation 16,[34] we prepared a polystyrene bound anthracene species. Reduction of this polystyrylanthracene

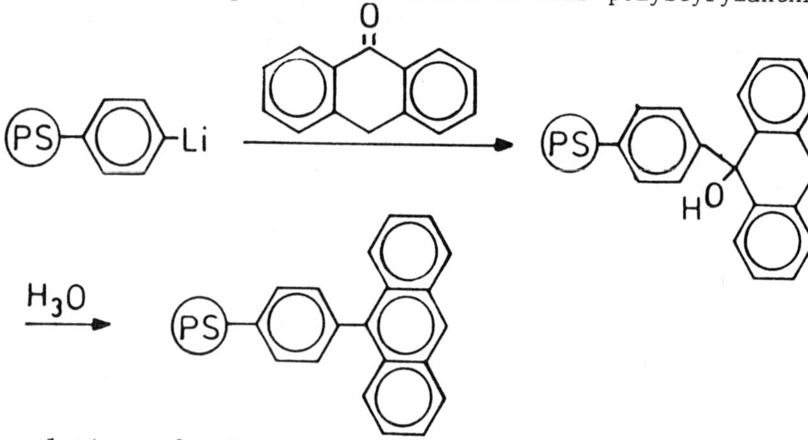

(16)

by solutions of reducing agents such as sodium naphthalene or by solutions of other alkali metal aromatic radical anions according to equation 17 was a very facile process and readily formed a polystyrene bound alkali metal anthracene radical anion. Typically

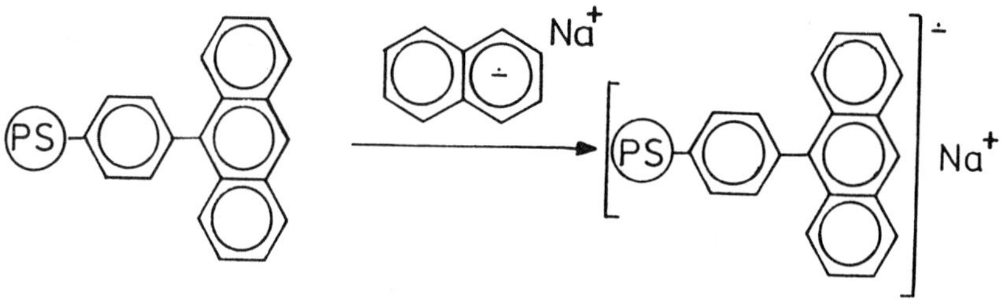

(17)

in these procedures, the color characteristic of the soluble alkali metal aromatic radical anion disappeared on addition of the solution of this reagent to a suspension of the reducible polymer. While non-cross-linked polymers containing alkali metal aromatic radical anions have been reported previously,[35] alkali metal radical anions bound to an insoluble organic polymer had not been previously described. In preliminary work these polymeric alkali metal anthracene radical anions behaved like their solution counterparts. For example, polystyrene bound sodium anthracene was used to reduce alkyl halides and was also successfully used to reduce cyclopentadienyldicarbonyliron dimer to form a nucleophilic iron complex which was in turn trapped in situ by 2% DVB-cross-linked chloromethylated polystyrene.

Although solutions of alkali metal aromatic radical anions
have been used to transfer electrons to reducible polycyclic
aromatic radical anions on polymers and have been used to deprotonate weak carbon acids bound to polymers, similar chemistry has
yet to be used to prepare polymeric lithium reagents from polymeric
halides. This chemistry now has firm precedent in homogeneous
systems.[36] It seems likely that application of equation 18 to systems where "R" is a polymer would be fruitful. A possible advantage of such a procedure would be that absorption of soluble lithium or alkali metal reagents which is a problem with procedures
described in equations 10-12 employing alkyllithium reagents would
be avoided.

$$2 \; Ar^{\overline{\cdot}} \; Metal^{+} \; + \; R\text{-}X \; \longrightarrow \; R\text{-}Metal \; + \; Metal\text{-}X \; + \; 2 \; Ar \qquad (18)$$

$$Metal = Li, \; Na, \; K$$

Carbonyl compounds are perhaps the most ubiquitous examples of
weak carbon acids. Thus, it would not be surprising to find many
examples of deprotonation of polymer-bound carbonyl groups as a
synthetic route to another type of alkali metal organometallic
polymer. There however have been only a limited number of examples
of deprotonation of carbonyl substrates bound to insoluble polymers
leading to organometallic polymers. The principal impetus in ex-
ploring this chemistry to date has been to develop new synthetic
reactions which in some way take advantage of the heterogeneity of
an insoluble polymer. For example, Patchornik and Ford have both
described deprotonation of esters bound to DVB-cross-linked polystyrene.[37-39] The polymer-bound ester enolates formed in this way
have considerably greater thermal stability than their homogeneous
counterparts and permit the use of ester enolates in electrophilic
substitution reactions at 0 °C instead of -78 °C (equations 19 -
21).

$$PS\text{-}CH_2O_2CCH_2CH_2C_6H_5 \; \xrightarrow[\text{THF}]{(C_6H_5)_3CLi} \; PS\text{-}CH_2O_2CCHLiCH_2C_6H_5 \qquad (19)$$

$$6$$

Ford has extensively investigated the effect of changing parameters
of polymer structure on the stability and utility of these alkali
metal ester enolates.[38] In these studies, he has shown that macroreticular polystyrene or gel-type polystyrene crosslinked with high
percentages of divinylbenzene is most useful. Using such types
of polymers, relatively high loadings of ester enolates of 0.67

$$\textbf{6} \xrightarrow{\text{ClCOCH}_3} \text{PS-CH}_2\text{O}_2\text{CCH}(\text{COCH}_3)\text{CH}_2\text{C}_6\text{H}_5 \xrightarrow[\substack{(\underline{n}\text{-C}_4\text{H}_9)_4\text{NOH} \\ \text{THF, 75 °C}}]{\text{KOH}}$$

$$\xrightarrow[\text{heat}]{\text{H}_3\text{O}^+} \text{CH}_3\text{COCH}_2\text{CH}_2\text{C}_6\text{H}_5 \; + \; \text{CO}_2 \; + \; \text{PS-CH}_2\text{OH} \tag{20}$$

$$\textbf{6} \xrightarrow{\text{RBr}} \text{PS-CH}_2\text{O}_2\text{CCH}(\text{R})\text{CH}_2\text{C}_6\text{H}_5 \xrightarrow[\substack{(\underline{n}\text{-C}_4\text{H}_9)_4\text{NOH} \\ \text{THF, 75 °C}}]{\text{KOH}} \xrightarrow{\text{HCl}}$$

$$\xrightarrow{\text{CH}_2\text{N}_2} \text{PS-CH}_2\text{OH} \; + \; \text{CH}_3\text{O}_2\text{CCH}(\text{R})\text{CH}_2\text{C}_6\text{H}_5 \tag{21}$$

mmol of ester per gram of polymer are possible. Using a 10% DVB-cross-linked gel copolymer alkylation reactions proceeded in 73 - 87% isolated yield and acylation reactions can be carried out on 0.1 mol scales with isolated yields of acylation products of 77%. Based on both Ford's studies and on earlier work by Rapport[40] and by Patchornik,[39] more lightly crosslinked polymers (e.g. 2% DVB-crosslinked polystyrene) are less suitable because the ester enolates attached to the polymer can undergo self-condensation reactions. Other factors affecting the yield of acylated product and the stability of the intermediate polymeric ester enolate were the nature of the base used to form the ester enolate and the loading of functionality on the polymer in addition to the percentage of divinylbenzene cross-linking agent in the copolymer.

Polymer-bound chiral reagents are other recent examples of stabilized alkali metal carbanions attached to polymers via heteroatom links.[41,42] Leznoff's group has used chiral amino alcohols derived from aminoacids such as phenylalanine bound to chloromethylated polystyrene via benzyl ether bonds to prepare ketimines.[42] Deprotonation of such a cyclohexanone ketimine followed by electrophilic substitution with methyl iodide produced alkylated products with up to 94% e.e. at 20 °C in 87% chemical yield (equation 22). Similar high stereoselectivities were also reported to result from an asymmetric protonation of an anion derived from 2-methylcyclohexanone ketimine attached by a similar chiral amino ether to polystyrene.[42]

$$\text{[(CH}_3)_2\text{CH]}_2\text{NLi} \xrightarrow{\text{THF}} \text{RI} \quad \text{H}_3\text{O} \quad (22)$$

$$R = CH_3, \ 80\% \ (95\% \ e.e.)$$
$$R = CH_2CH_2CH_3, \ 80\% \ (60\% \ e.e.)$$

Polymeric Wittig reagents are a third example of alkali metal organometallics attached to polymers, primarily for the purpose of increasing their synthetic utility.[43-45] In this case, the primary advantage of using a polymeric system is the ease of separation of alkene products from the by-product triphenylphosphine oxide and the potential for recycling triphenylphosphine oxide by trichlorosilane reduction. While various authors have described versions of such reactions using many different types of DVB-cross-linked polystyrene and many different reaction schemes, the recent synthesis of methyl retinoate with DVB-crosslinked polystyrene-bound Wittig reagents (equation 23)[45] exemplifies the possible advantages to be gained from application of these polymeric alkali metal organometallic reagents.

$$BrCH_2C(CH_3)=CHCO_2Et + \text{(PS)}-PPh_2 \longrightarrow$$

$$\text{(PS)}-\overset{+}{P}(Ph)_2CH_2C(CH_3)=CHCO_2Et$$

$$\xrightarrow{\text{EtO}^-, \text{ EtOH}} \quad (23)$$

While heteroatom stabilized alkali metal organometallic polymeric reagents have proven to be of increasing utility in synthetic procedures, synthetic applications of alkali metal organometallic polymers which are more classical examples of "organometallic" reagents are fewer. One of the most interesting of these examples is the use of polystyrene bound triphenylmethyl anions by Cohen as bases in so-called "Wolf and Lamb" reactions - reactions in which

two mutually incompatible reagents are present in the same reaction
mixture but which are separated by virtue of their being bound to
different insoluble polymers.[32,46] Equation 24 illustrates this
sort of application of an alkali metal organometallic polymer.

$$(24)$$

Recently we have developed an alternative method for synthesis
of aryl–stabilized alkali metal carbanions attached to polymers.
This method produces the sorts of alkali metal organometallics
described by equation 10 above and also has some other applications.[33] First, these reactive polymeric organometallic reagents
are useful as precursors to other organometallic polymers (equation
25). Second, these reagents which are polymeric analogs of alkali
metal salts of diphenylmethane can be useful for preparation of

$$(25)$$

solutions of organometallic reagents by deprotonation of weak carbon acids (equation 26).

$$(26)$$

Metal = Na, K, Rb, Cs

Perhaps the most common application of reactive alkali metal
organometallic polymers is their use to prepare other derivatives
of polymers. In this sense, polystyryllithium's utility mirrors
that of a soluble lithium reagent which is of principal importance
as an intermediate en route to other functional groups. As

expected, an alkali metal polymer such as polystyryllithium undergoes all the reactions of its homogeneous counterpart. A few examples of such chemistry are listed in the equations below (equations 27 - 29).[20,47]

A final example of an application of reactive alkali metal organometallic polymers derived is our recent work in which these polymers were used to synthesize polymeric catalysts.[48] The objective behind this chemistry was the desire to prepare highly dispersed metal catalysts on a support whose chemistry was easily altered. In our studies, we have found that reaction of reactive alkali metal organometallic derivatives of divinylbenzene–crosslinked polystyrene with solutions of dichloro(1,5-cyclooctadiene)palladium(II) formed useful palladium/polystyrene catalysts. Alkali metal polymers used in these reactions included polystyryllithium, polystyrylbenzyllithium (5), polystyrylphenyl(trimethylsilyl)methyllithium (7) and polystyrene–bound sodium anthracene. The resulting palladium/polystyrene catalysts formed in such reactions (equations 30 - 33) were first

$$\left[(PS)\!-\!\bigcirc\!\!\!\bigcirc\!\!\!\bigcirc \right]^{-} Na^{+} + (COD)_2PdCl_2 \longrightarrow \left[(PS)\!-\!\bigcirc\!\!\!\bigcirc\!\!\!\bigcirc \right] \cdot Pd \quad (31)$$

$$(PS)\!-\!\bigcirc\!-\!CH_2Cl \xrightarrow{C_6H_5Li} (PS)\!-\!\bigcirc\!-\!CH_2\!-\!\bigcirc \xrightarrow{RLi}$$

(32)

$$(PS)\!-\!\bigcirc\!-\!\underset{Li}{CH}\!-\!\bigcirc \xrightarrow[\text{HMPA-THF}]{(COD)PdCl_2} \xrightarrow{\Delta} \left[(PS)\!-\!\bigcirc\!-\!CH_2\!-\!\bigcirc \right] \cdot Pd$$

(33)

$$(PS)\!-\!\bigcirc\!-\!CH\overset{Si(CH_3)_3}{\underset{\bigcirc}{\diagup}} \xrightarrow{RLi} (PS)\!-\!\bigcirc\!-\!CLi\overset{Si(CH_3)_3}{\underset{\bigcirc}{\diagup}} \xrightarrow[\text{HMPA-THF}]{(COD)PdCl_2}$$

$$\xrightarrow{\Delta} \left[(PS)\!-\!\bigcirc\!-\!CH\overset{Si(CH_3)_3}{\underset{\bigcirc}{\diagup}} \right] \cdot Pd$$

extracted with hot THF for several days to both remove any soluble materials and to thermolyze any σ-bonded carbon–palladium bonds to yield palladium(0) crystallites. The presence of palladium(0) crystallites in the resulting catalysts was confirmed by ESCA spectroscopy and the catalysts were also characterized by electron microscopy. The catalysts derived from equations 30, 32 and 33 all consisted of relatively uniform dispersions of 20 – 30 Å palladium crystallites. Comparison of these catalysts to commercial palladium(0) catalysts such as Pd/C showed them to be very comparable in most reactions. The palladium/polystyrene catalysts gave less isomerization in hydrogenation of terminal alkenes and exhibited higher selectivity in semihydrogenation of alkynes. Solvent effects on hydrogenation rates for alkene hydrogenation in the case of the palladium/polystyrene catalysts were rationalized in terms of both the solvents' polarity and its ability to swell the gel copolymer matrix in which the palladium crystallites were dispersed. Novel catalytic activity observed for these palladium/polystyrene catalysts included their activity as decarbonylation catalysts (equation 34) and their use in allylic substitutions (equation 35). A particularly noteworthy feature of the latter reaction was the synergistic effect of added triphenylphosphine.[49]

$$CH_3(CH_2)_{10}CHO \quad \xrightarrow[\substack{C_6H_5CH_3 \\ \text{reflux, 60 h}}]{\text{PS-b-Pd}} \quad CH_3(CH_2)_9CH_3 \quad \quad (34)$$
$$77\%$$

$$CH_3CO_2CH_2CH=CH_2 \; + \; (C_2H_5)_2NH \xrightarrow[\substack{65\ °C,\ 24\ h}]{\text{PS-b-Pd}} (C_2H_5)_2NCH_2CH=CH_2 \quad (35)$$
$$84\ \%$$

REFERENCES

1. Y. N. Novikov and M. E. Vol'pin, Russ. Chem. Rev., **40**, 733 (1971) and H. B. Kagan, Chemtech, 510 (1976).

2. L. B. Ebert, J. Molecular Catal., **15**, 275 (1982).

3. M. P. Rosynek and J. D. Winder, J. Catal., **56**, 258 (1979).

4. J. M. Lalancette and R. Roussel, Can. J. Chem., **54**, 2110 (1976).

5. H. E. Podal, N. E. Foster and A. R. Giraitis, J. Org. Chem., **23**, 82 (1958).

6. D. Savoia, C. Thrombini, and A. Umani-Ronchi, Tetrahedron Lett., 653 (1977).

7. H. Hart, B. Chen and C. Peng, Tetrahedron Lett., 3121 (1977).

8. D. Savoia, G. Thrombini and A. Umani-Ronchi, J. Org. Chem., **43**, 2907 (1978).

9. D. E. Bergbreiter and J. M. Killough, J. Am. Chem. Soc., **100**, 2126 (1978).

10. Although the reactions of water and alcohols with C_8K have been discussed in terms of the potential Lewis basicity of C_8K in reference 9 and although the postulated basicity of C_8 has been used in deprotonation of weak carbon acids (references 6-8), more recent reports have suggested that potassium-graphite does not behave like a two electron Lewis base, cf. L. B. Ebert, L. Matty, D. R. Mills and J. C. Scanlon, Mater. Res. Bull., **15**, 251 (1980)

and I. B. Rashkov, I. N. Panayoto and V. C. Shishkov, <u>Carbon</u>, **17**, 479 (1980). Thus, the way in which the weakly acidic carbon acids in references 6-8 are deprotonated is not yet completely understood.

11. G. Bram, E. d'Inean and A. Loupy, <u>Nouveau J. Chemie</u>, **6**, 689 (1982).

12. D. Savoia, C. Thrombini and A. Umani-Ronchi, <u>J. Chem. Soc., Perkin I</u>, 123 (1977).

13. S. J. Cristol and G. A. Giat, <u>J. Org. Chem.</u>, **47**, 5186 (1982).

14. P. Beak and T. A. Sullivan, <u>J. Am. Chem. Soc.</u>, **104**, 4450 (1982).

15. D. Savoia, C. Thrombini, A. Umani-Ronchi and G. Verardo, <u>J. Chem. Soc., Chem. Commun.</u>, 540 (1981). D. Savoia, C. Thrombini, A. Umani-Ronchi and G. Verardo, <u>J. Chem. Soc., Chem. Commun.</u> 541 (1981).

16. D. Savoia, E. Tagliavini, C. Thrombini and A. Umani-Ronchi, <u>J. Org. Chem.</u>, **46**, 5340 (1981). D. Savoia, E. Tagliavini, C. Thrombini and A. Umani-Ronchi, <u>J. Org. Chem.</u>, **46**, 5344 (1981).

17. D. Savoia, E. Tagliavini, C. Thrombini and A. Umani-Ronchi, <u>J. Org. Chem.</u>, **47**, 876 (1982).

18. G. P. Boldrini, D. Savoia, E. Tagliavini, C. Thrombini and A. Umani-Ronchi, <u>J. Org. Chem.</u>, **48**, 4108 (1983).

19. M. Morton, "Anionic Polymerization: Principles and Practice," Academic Press, New York, 1983.

20. M. J. Farrall and J. M. J. Frechet, <u>J. Org. Chem.</u>, **41**, 3877 (1976).

21. R. H. Grubbs and S. H. Su, <u>J. Organomet. Chem.</u>, **122**, 151 (1976).

22. J. M. Burlitch and R. C. Winterton, <u>J. Organomet. Chem.</u>, **159**, 299 (1978).

23. Brit. patent 1 172 477 (1969); <u>Chem. Abstr.</u> **72**, 32770d (1970).

24. G. Clouet and J. Brossas, <u>Makromol. Chem.</u>, **180**, 867 (1979).

25. A. Guyot and M. Bartholin, <u>Prog. Poly. Sci.</u>, **8**, 277 (1982).

26. W. Heitz and R. Michels, Makromol. Chem., **148**, 9 (1971).

27. F. Camps, J. Castells, M. J. Fernando and J. Font, Tetra-hedron Lett., 1713 (1971).

28. R. T. Taylor, R. A. Cassell and L. A. Flood, Ind. Eng. Chem. Prod. Res. Dev., **21**, 462 (1982).

29. D. E. Bergbreiter and J. M. Killough, Macromolecules, **13**, 187 (1980).

30. H. Gilman and F. K. Cartledge, J. Organomet. Chem., **2**, 477 (1964).

31. D. E. Bergbreiter and E. Pendergrass, J. Org. Chem., **46**, 219 (1981).

32. B. J. Cohen, M. A. Kraus and A. Patchornik, J. Am. Chem. Soc., **103**, 7620 (1981).

33. D. E. Bergbreiter, J. R. Blanton and B. Chen, J. Org. Chem., **48**, 0000 (1983).

34. D. E. Bergbreiter and J. M. Killough, J. Chem. Soc., Chem. Commun., 319 (1980).

35. G. Greber and G. Egle, Makromol. Chem., **59**, 174 (1963).

36. P. K. Freeman and L. L. Hutchinson, J. Org. Chem., **45**, 1924 (1980).

37. Y. H. Chang and W. T. Ford, J. Org. Chem., **46**, 3756 (1981).

38. Y. H. Chang and W. T. Ford, J. Org. Chem., **46**, 5364 (1981).

39. A. Patchornik and M. A. Kraus, Pure Appl. Chem., **43**, 503 (1975).

40. J. I. Crowley and H. Rapoport, J. Org. Chem., **45**, 3215 (1980).

41. J. M. J. Frechet, W. Amaratunga and J. Halgas, Nouveau J. Chemie, **6**, 609 (1982).

42. C. R. McArthur, J. L. Jiang and C. C. Leznoff, Can. J. Chem., **60**, 2984 (1982) and C. R. McArthur, P. M. Worster, J. L. Jiang and C. C. Leznoff, Can. J. Chem., **60**, 1636 (1982).

43. P. Hodge, B. J. Hunt, E. Khoshdel and J. Waterhouse, Nouveau J. Chemie, **6**, 617 (1982).

44. M. Bernard and W. T. Ford, J. Org. Chem., **48**, 326 (1983).

45. M. Bernard, W. T. Ford and E. C. Nelson, J. Org. Chem., **48**, 3164 (1983).

46. A. Patchornik, Nouveau J. Chemie, **6**, 639 (1982).

47. J. V. Minkiewicz, D. Milstein, J. Lieto, B. C. Gates and R. L. Albright, ACS Sym. Ser., **192**, 9 (1982).

48. D. E. Bergbreiter, B. Chen and T. J. Lynch, J. Org. Chem., **48**, 4179 (1983).

49. D. E. Bergbreiter and B. Chen, J. Chem. Soc., Chem. Commun., 1238 (1983).

SYNTHESIS, CHARACTERIZATION AND APPLICATIONS OF RARE EARTH METAL

ION CHELATING POLYMERS

Y. Okamoto, S.S. Wang, K.J. Zhu, E. Banks, B. Garetz
and E. K. Murphy
Polytechnic Institute of New York
Department of Chemistry and Polymer Research Institute
333 Jay Street, Brooklyn, NY 11201

INTRODUCTION

The fluorescent characteristics of europium chelates are of considerable interest in connection with electronic energy transfer processes and with their use in laser systems.[1,2] The most successful laser systems have demonstrated the use of compounds of the $[(\beta\text{-diketono})_4 Eu]^- C^+$ system as the active species, where C^+ is the cation such as piperidinium or substituted ammonium. Laser activity of the chelate compounds in organic solution was observed initially at temperatures near 100K;[3] however, room temperature operation has also been observed. To permit long term repetitive flashing, externally cooling the solution has been required.[4]

The chelate system is interesting because the pump energy is absorbed by the organic molecule and then is efficiently transferred to the rare earth ion. This differs from conventional systems where the optical pump energy is absorbed by the rare earth ion itself. Because of this energy transfer process in the chelate system, the rare earth ion is pumped more effectively than in the conventional crystal and glass lasers. The europium chelate laser emits at a wavelength of 613 nm in the red portion of the visible spectrum. However, the tetrakis chelates tend to dissociate in organic solution into lower chelate forms; tris- and bischelates, and free diketone. In the case of benzoylacetone and dibenzoylmethane, the tris chelates have been shown to be non-lasing.[5,6,7]. The dissociation of tetrakis chelates into tris chelate in solution has been discussed extensively.[8] The degree of dissociation of a given compound was found to be greatly affected by the nature of the solvent. For example, the

425

piperidium salt of Eu-(dibenzoylmethide)$_4$ was found to be 37%
dissociated into tris or lower forms in 3:1 ethanol-methanol solution at 93°K, while adding dimethylformamide to this system increased the dissociation to 82%.

The dissociation may be minimal in the solid polymer complexes
where the chelate should be, to a certain extent, locked into a
specific configuration.

An attempt to utilize a polymer as a matrix was reported by
Wolff and Pressler.[9] They found that a solid solution of europium-
4,4,4-trifluoro-1-(2-thienyl)-1,3-butenedione (EuTTA) in
poly(methyl methacrylate)(PMMA) had optical maser action. When a
clear plastic fiber containing the complex was excited at 340 nm
using a xenon flash lamp at liquid nitrogen temperature, stimulated
emission was observed around 613 nm.

The stimulated emission of TbTTA in PMMA was also reported
by Huffman[10], the sample was a fiber approximately 0.75 mm in
diameter by 60mm in length. Both ends of the fiber were polished
and one end was silvered. The fluorescence at 545 nm was recorded
with excitation at 335 nm at 77K. A threshold of stimulated
emission was detected at 225 joule input. In addition to PMMA,
polystyrene and epoxy resins were also used as organic polymer
matrices. However, polymeric systems in which rare earth metals
are directly bonded to the polymer have not yet been investigated.
Recently, we have initiated investigations on the synthesis and
characterization of rare earth metal containing polymers.[11-14]
Our primary objectives were (1) to utilize rare earth metal
probes to elucidate the structures of ionomers, (2) to study detailed fluorescence properties of rare earth metal chelating
synthetic polymers and to explore the possible laser action of
the polymers.

In this article, we will discuss our preliminary fluorescence
results, as well as experimental work on the laser properties of
europium and terbium in chelating polymers.

Eu ion - β-diketone ligand containing polymers

Since the fluorescence properties of rare earth metal ions
chelate compounds containing β-diketone moiety have been widely
investigated, we thus first investigated Eu^{3+}-polymer complexes
containing the β-diketone moiety.[13]

Eu^{3+} chelates of dibenzoylmethane (DBM) 1 and β-diketone containing polymers; poly(p-benzoyl acetyl styrene) 2 and poly(aryl-
β-diketone) 3 were prepared and characterized. Polymer (2)was
synthesized from partially (20 mole %) acetylated polystyrene
(Dow Chemical Co. M.W. 30,000). The substitution was found to be

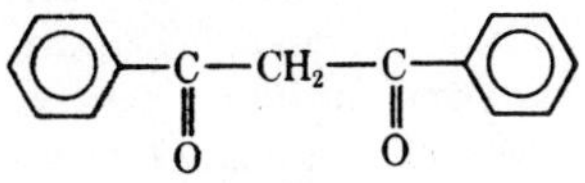

1

dibenzoylmethane

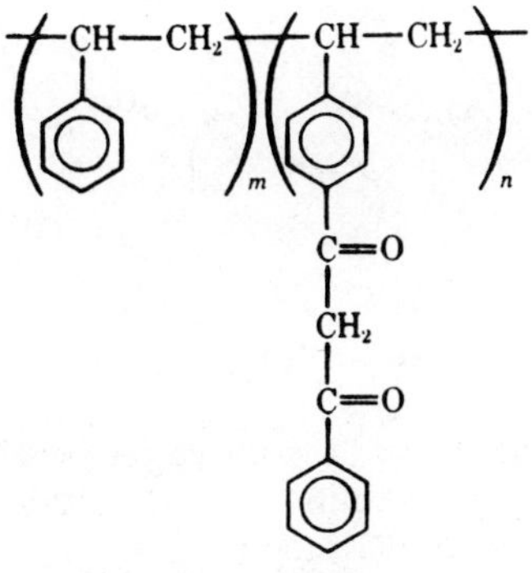

2

poly(p-benzoylacetylstyrene)

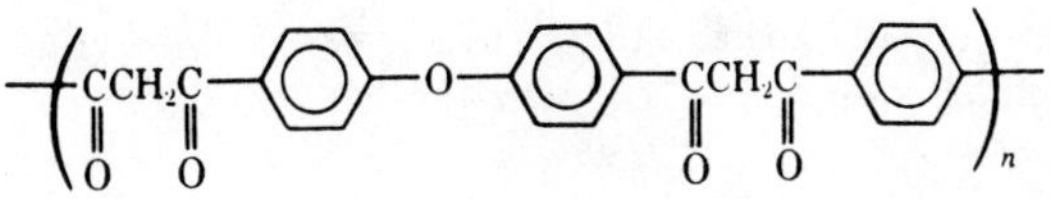

$m > n$

3

poly(aryl β-diketone)

mainly on the para position. The acetylated polymer was further reacted with methyl benzoate to give polymer (2). Polymer (3) was prepared by the condensation reaction of p,p'-diacetyldiphenyl ether and dimethyl terephthalate using a method similar to that reported in the literature.[15]

EuCl$_3$ solution of tetrahydrofuran and methanol(1:1 volume ratio) was added to the tetrahydrofuran solution of polymers 2 or 3(1-2% solution). The pH of the solution was adjusted to slightly basic ($\sim$pH 8) by adding piperidine, and the solution was stirred for 1 hr. After the precipitate was filtered, the solid was washed continuously with methanol and dried under vacuum at 50°C for two days. No chlorine was detected in the Eu^{3+} complexes of the polymers.

The coordination structure of Eu^{3+}-polymer complexes was investigated by measuring the IR spectra. The polymer has a mostly enol structure and a wide absorption band between 1350 and 1700 cm^{-1}

because of the overlap of two effects - intramolecular hydrogen
bonding and conjugated double bond.   Upon the formation of the
chelate with Eu^{3+}, new strong bands appeared at around 1400 and
1530 cm^{-1} corresponding to the asymmetric and symmetrical vibration of two C=O links in the resonance structures. These bands have
been assigned in the copper chelates of acetylacetone[16] and in the
corresponding polymer.[15]

The polymers and their Eu^{3+} complexes were thermally stable
and started to decompose at around 350°C.   Typical TGA data of
these compounds are shown in Figures 1 and 2.   It is of interest
that 3 has unusually high char yield (~45%)(Fig. 1).   This may be
due to formation of crosslinkages via dehydration during heating.
When the Eu-polymer complex was heated in nitrogen or air atmosphere, piperidine was initially eliminated from the complex, resulting in gradual decrease of weight (Fig. 2).

Rare earth ions are incorporated in organic chelates by coordination through a donor atom such as oxygen; when excited with
light absorbed by the ligand, they exhibit narrow-line emission at
approximately the same frequencies as the inorganic crystal system.
This phenomenon is the result of an intramolecular energy transfer
from the electronic states associated with the organic complex to
localized intra-4f shell energy levels of the ions.[17]   The fluorescent properties of europium dibenzoylmethides have been intensely
investigated.[18]   The sharp emission spectrum is detected at 613nm
even at room temperature(Fig.3).   The peak is due to the $^5D_0 \rightarrow {}^7F_2$
transition.

We have prepared several Eu^{3+}-containing coordination polymers
and mixtures in which Eu^{3+}-dibenzoylmethide, $Eu(DBM)_4P^+$ (P^+ is the
piperidinium ion) is uniformly dispersed in a polystyrene matrix.
The preparation of transparent films became difficult for the
polymers containing high metal concentrations.   Thus, in order to
investigate the fluorescent properties of a wide range of metal-
containing polymers, we decided to study the properties of pressed
powders.   The fluorescence intensity measurements on solid samples
are known to be difficult.   Various factors, such as particle size
and reabsorption of the Eu^{3+} emission, may influence the intensity.
Thus, we took three measurements on a sample and took the average
value.   The scatter of these values was at times as much as 15%.

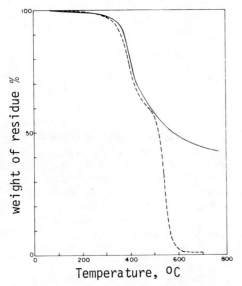

Fig. 1. TGA for poly(aryl β-diketone): (——) under N_2:
(---) under air atmos-
phere; 10°C/min.

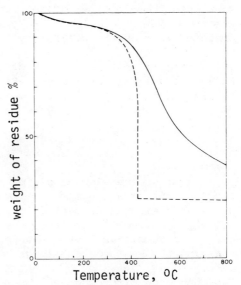

Fig. 2. TGA for poly(aryl β-diketone)-Eu^{3+} complex:
(——) under N_2; (---) under
air atmosphere; 10°C/min.

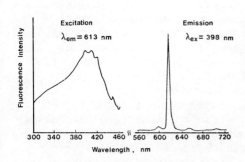

Fig. 3. Fluorescence spectra of
polystyrene-Eu(DBM)$_4$P
(7.0 wt.% Eu) blend,
sensitivity 1,slit
Ex/Em=1.5/1.5.

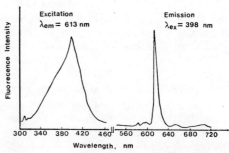

Fig. 4. Fluorescence spectra of
DKPS-Eu (1.5 wt % Eu),
sensitivity 1,slit
Ex/Em=3/2.5.

The typical excitation and emission spectra of the polymers are shown in Figure 4. The sharp emission peak at 613 nm was detected in those Eu-polymer systems at room temperature(Fig.4).

In addition to the lines of the emission spectrum in the Eu(DBM)$_4$ (Fig.3), as shown in Figure 4, a peak at 580 nm was detected for the Eu-polymer complex, corresponding to the forbidden $^5D_0 \rightarrow {}^7F_0$ transition.[19] The appearance of the 0 → 0 transition has been explained in terms of low symmetry around Eu^{3+}.[20] Thus, the spectrum indicated that the Eu-polymer complex may have more asymmetric structure around Eu^{3+} than that of the Eu(DBM)$_4$.

For the composite sample of Eu(DBM)$_4$ in polystyrene, the fluorescence emission intensity was found to increase linearly with increasing Eu content (Fig. 5). However, for the Eu-coordination polymer complexes, the intensity reached a maximum at Eu^{3+} content as small as 1 wt % and remained constant on further increasing the Eu^{3+} content (Figs. 6 and 7).

When the fluorescence intensities of these Eu complexes were compared under the same conditions, viz., 1 wt % Eu, at which Eu-2 and Eu-3 show nearly maximum fluorescence intensity, the order was found to be Eu(DBM)$_4$>Eu-2>Eu-3. The samples' intensities differed by a factor of about 20-30. This was far larger than the experimental error for each sample. The fluorescence emission intensity is known to vary with environmental factors such as the nature of coordination bond, solvent, and temperature.

In order to obtain data on the effect of the coordination number on the fluorescence intensity in the Eu-dibenzoylmethide system, we prepared Eu-tetrakis-, tris-, and bisdibenzoylmethides and found that the fluorescence intensities were in the order Eu(DBM)$_4$>Eu(DBM)$_3$>Eu(DBM)$_2$. The intensity ratios were 40- to 50-fold among these complexes as compared to the same amount of the Eu^{3+} ion.

When Eu ion reacts with excess dibenzoylmethane in basic solution, generally the tetrakis coordination complex is obtained. However, when β-diketone groups are incorporated into the polymer chain, the formation of multiple coordination bonding between Eu and the β-diketone may be restricted by the increasing steric hindrance and the decrease of the freedom of bond rotation.

The fluorescence intensity data discussed above suggest that polymer 2 forms tris-coordination complexes with a small amount of Eu^{3+}, (∿1 wt % Eu) and subsequently forms bis- and mono-coordination complexes with excess Eu^{3+}. Since the fluorescence intensity of the tris-coordination compound Eu(DBM)$_3$ was far greater than those of bis- and mono-coordination compounds, the intensity

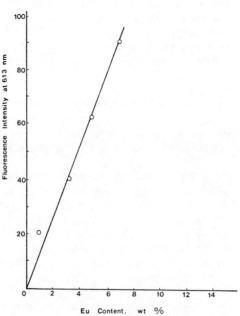

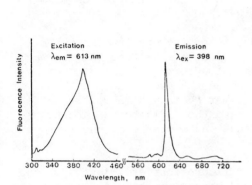

Fig. 4. Fluorescence spectra of DKPS-Eu (1.5 wt. %Eu), sensitivity 1, slit Ex/Em = 3/2.5.

Fig. 5. Relationship between fluorescence intensity at 613 nm and Eu content of polystyrene-Eu(DBM)$_4$P blend, sensitivity 1, slit Ex/Em = 1.5/1.5, λ_{ex} = 398 nm.

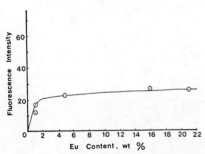

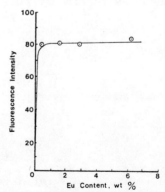

Fig. 6. Relationship between fluorescence intensity and Eu content of poly(aryl β-diketone)-Eu complex; sensitivity 1, slit = 6/6, λ_{ex} = 396 nm.

Fig. 7. Relationship between fluorescence intensity and Eu content of poly(p-benzoylacetyl-styrene)-Eu complex; sensitivity 1, slit Ex/Em = 3/3.5, λ_{ex} = 398 nm.

remained essentially constant after reaching a maximum at low Eu^{3+}
concentration, as shown in Figure 6. In other words, the maximum
intensity obtained is due mostly to the tris-coordination complex
between Eu^{3+} and polymer 2. In the case of polymer 3, the β-diketone groups are incorporated in the linear chain; consequently,
the formation of multiple coordination bonds between Eu ion and
β-diketone moiety is expected to be more restricted as compared
with polymer 2 in which β-diketone is attached on the side chain.
Therefore, polymer 3 may form mainly the bis-coordination complex
with limited amounts of Eu^{3+}, followed by the formation of mono-
coordination complex with excess Eu^{3+}.

Thus, the maximum fluorescence intensity obtained in the Eu-
polymer 3 system may be due to that of Eu-bisdiketone complex, and
the intensity remained nearly constant even with increasing Eu
content (Fig. 7). These results showed that when the β-diketone
group was incorporated into the polymer chain, the formation of
multiple coordination between Eu^{3+} and the β-diketone is restricted by the increasing steric hindrance and decrease of the freedom
of bond rotation.

In order to obtain the maximum fluorescent efficiency for the
Eu^{3+} content, a monomer (Eu^{3+} tetradibenzoyl methide) was synthesized,
in which a polymerizable vinyl group was bonded to one of the phenyl
groups $(Eu\ D'D_3)^-\ P^+$. The synthetic route for the monomer is shown
in Scheme I.

The monomer was dissolved in methylene chloride and MMA and
then copolymerized by AIBN initiator at 60-65°C. The polymer
obtained was repeatedly purified by reprecipitating the methylene
chloride solution of copolymer into hexane.

$$C=C \quad + \quad \underset{CO_2CH_3}{\overset{CH_3}{C=C}} \quad \longrightarrow \quad -C-C-C-\underset{CO_2CH_3}{\overset{CH_3}{C}}-$$

R = tetrakis coordinated compounds.

Various composition ratios of Eu^{3+} to MMA were prepared and
the relationship between the fluorescence intensity at 613 nm and
Eu content in the copolymers is shown in Table 1 and Fig. 8

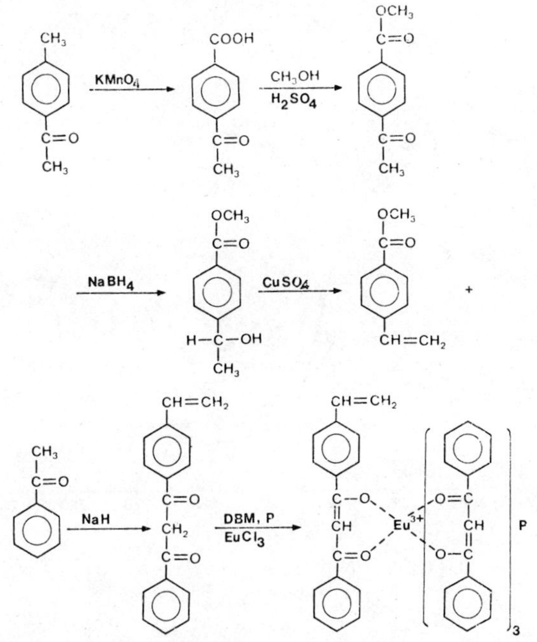

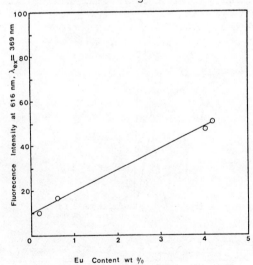

Scheme I

Synthesis of (p-vinylphenyl)-3-phenyl-1,3-propanediones
and europium chelates. (D'EuD$_3$P).

Fig. 8. Fluorescence intensity at 613 nm vs Eu^{+3}
content in D'EuD$_3$P:MMA copolymer powder (1:6).
Slit Ex/Em = 2/2, Sensitivity 3, λ_{ex} = 396 nm.

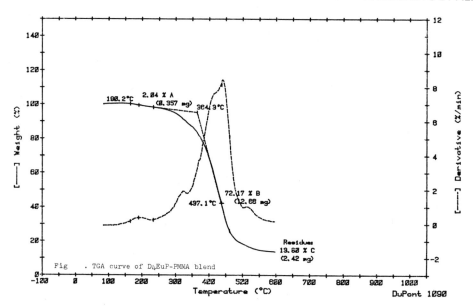

Fig. 9.   TGA for the blend of
D$_4$EuP-PMMA

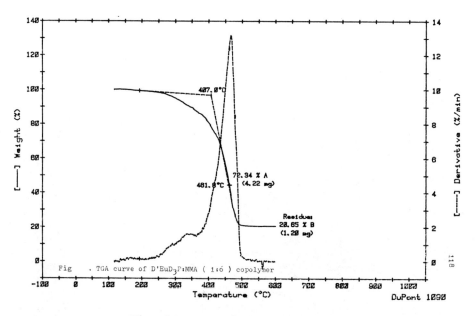

Fig. 10.   TGA for the copolymer
of D'EuD$_3$P and MMA

Table 1.   Fluorescence Intensity at 613 nm vs Eu^{3+} Content in
           $D'EuD_3P:MMA$ Copolymer Powder

Copolymerization ratio of copolymer $(D'EuD_3P:MMA)$	Eu contents in polymer produced	Fluorescence Intensity
1:4	4.18%	51
1:6	4.04%	47
1:40	0.61%	17
1:100	0.19%	10

The fluorescence intensity increased linearly with increasing
Eu content. However, the fluorescence intensity of the mixture of
the monomer   and PMMA was found to be much lower than the corresponding copolymer. This indicated that the vinyl group may
quench the fluorescence.

The thermal stability of the polymers was investigated by TGA
under nitrogen. The results shown in Figures 9 and 10 correspond
to D_4EuP-PMMA blend and $D'EuD_3P:MMA$ copolymer. These have the
same content (4.04%). It is found that the blend decomposes at
$364.3^{\circ}C$, and the copolymer at $407^{\circ}C$, suggesting that the copolymer
is thermally more stable than the blend.

Eu^{3+}-carboxylbenzoyl and carboxylnaphthoyl ligand-containing polymer complexes[14]

Tanner and Thomas prepared Eu^{3+}-2-benzoyl benzoate and
demonstrated that energy transfer occurred from benzoyl benzoate
ligand to Eu^{3+} ion in ethanol-water solution.[21] The fluorescent
properties of Eu^{3+} phthalate and naphthalate were investigated by
Sinha and coworkers who showed that the fluorescence intensity of
Eu^{3+}-naphthalate was about 15-fold greater than that of the
phthalate.[22] This result was attributed to the better energy
donor capability of the naphthyl group than the phenyl group.

Because carboxylic acid forms salt with Eu-ion and the bond
between them is strong, we synthesized partially 2-carboxybenzoyl
(4) and 3-carboxyl-2-naphthoyl-substituted polystyrenes (5)(6).
These polymers were prepared by the Friedel Crafts reaction of
polystyrene with the corresponding dicarboxylic anhydrides. The
carboxyl content in the polymers was in the range of 1.3-2.0 mole %.
The Eu^{3+} salts of (4) and (5) were prepared by the neutralization
of the acids with $Eu(OH)_3$.

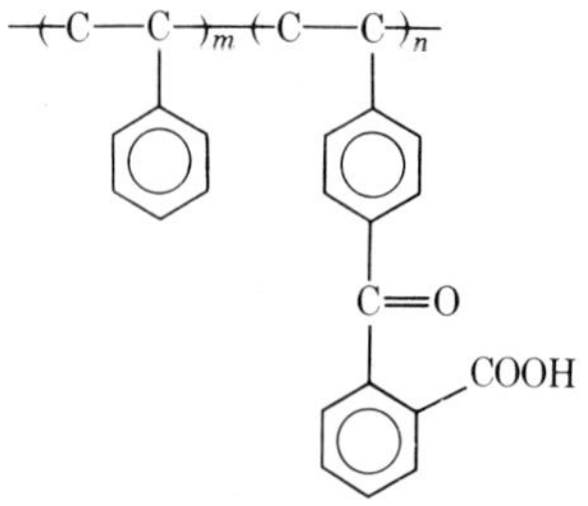

4

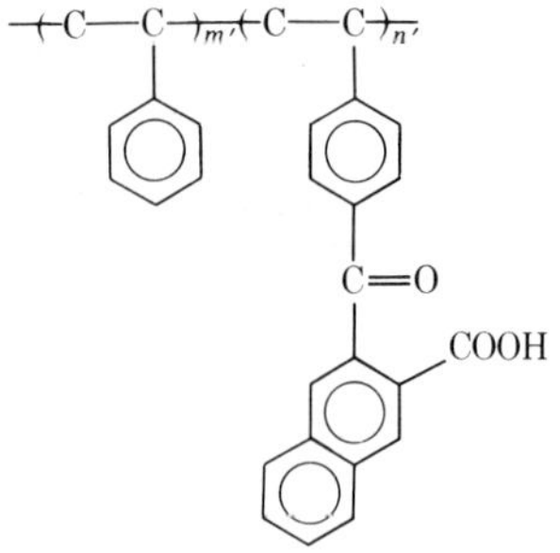

5

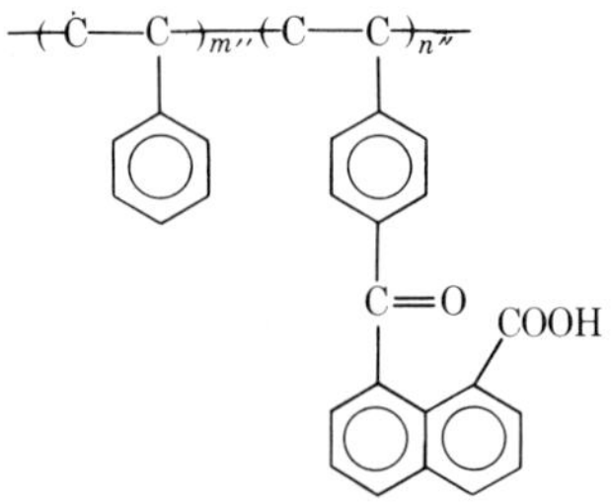

6

Tanner and Thomas recorded the fluorescence spectra of the aqueous solution of o-benzoylbenzoate in the presence of $Eu(ClO_4)_3$ and found that the europium emission peak was at 590 nm, measured with the excitation wavelength set at 320 nm, the maximum in the excitation spectrum.[21] Polymer (4) Eu salt also reaches its excitation maximum at 325 nm and shows a characteristic line emission (616 nm) for Eu^{3+} on excitation at 325 nm (Fig.11). The relationship between fluorescence intensities at 616 nm and Eu content in (4) Eu and (5) Eu salt is shown in Figure 12. For the (4) Eu salt system, however, the increase in intensity with the Eu content levels off at around 0.6 wt% of Eu, as we observed in polymers 2 and 3. The calculated maximum Eu^{3+} content in the Eu^{3+} bonded with three 2-benzoylbenzoate groups in the polymer (4) is 0.94 wt %, whereas for Eu^{3+} associated with only one 2-benzoylbenzoate group it is 2.8 wt %.

When the Eu^{3+} ion reacts with excess 2-benzoylbenzoic acid in a slightly basic solution trisbenzoate is obtained, but when the 2-benzoylbenzoate group is incorporated into the polymeric system such as (4), the formation of multiple links between Eu^{3+} and the benzoate may be restricted by the increasing steric hindrance and a decrease in the freedom of bond rotation.

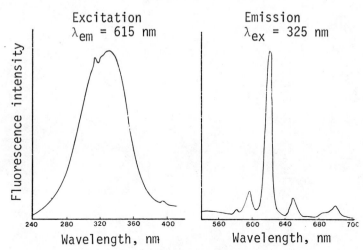

Fig. 11.   Fluorescence spectra of partially
2-carboxylbenzoyl-Eu complex

The fluorescence properties for the (<u>4</u>) Eu^{3+} salt shown in
Figure 12 indicate that polymer (<u>4</u>) initially forms a trisbenzoate
with a small amount of Eu^{3+} and subsequently forms the bis- and
monoEu^{3+} benzoate with excess Eu^{3+}. Because the fluorescence intensity of the tris compound is far greater than that of bis- and
monocompounds, the intensity remained essentially constant after
reaching a maximum at low Eu^{3+} concentration. In other words, the
maximum intensity obtained is due mostly to the trisbenzoate. In
polymer (<u>5</u>) the bulky 3-carboxyl-2-naphthoyl groups are substituted
on the polystyrene; consequently the formation of multiple coordination links between Eu ion and the carboxyl moiety is expected to be
more restricted. Therefore polymer (<u>5</u>) may form mainly monoEu^{3+}
salt. Thus, even though the naphthyl group is a better energy
donor to Eu^{3+}, the fluorescence intensity of the complex is much
smaller than that of the (<u>4</u>)Eu^{3+} system.

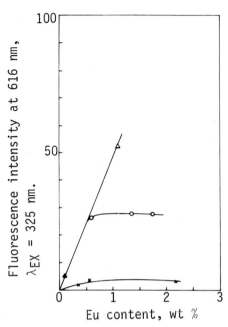

Fig. 12. Relationship between fluorescence intensity of Eu content
of polymer-Eu complexes. (△) polymer 8-Eu salt;
(○) polymer 4-Eu salt; (×) polymer 5-Eu salt.

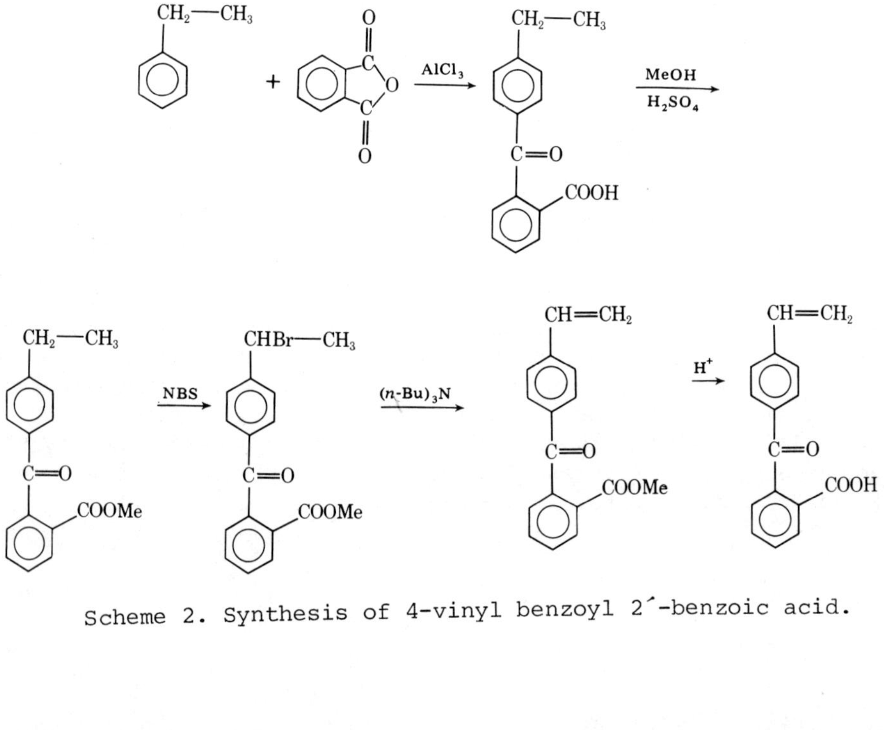

Scheme 2. Synthesis of 4-vinyl benzoyl 2´-benzoic acid.

Preparation of 4-vinylbenzoyl 2'-benzoic acid and europium
chelates

Thus, in order to obtain the maximum fluorescence efficiency
for Eu^{3+} content. 4-Vinyl benzoyl 2' benzoic acid was prepared
using the synthetic route described in Scheme 2.

A Eu-carboxylate monomer (7), in which one of the phenyl
moieties is attached to a polymerizable vinyl group, was prepared
by the reaction of 4-vinyl 2'-benzoyl benzoic acid and 2-benzoyl
benzoic acid in 1:2 molar ratio with $Eu(OH)_3$. After the complex
was purified the monomer was copolymerized with styrene by bulk
polymerization using AIBN as catalyst.

The polymer (8) obtained was dissolved in dichloromethane
and precipitated by adding methanol. This purification procedure
was repeated twice. The fluorescence intensity was found to be
linearly increasing with Eu^{3+} content in the polymer (8)(Fig. 12).

Eu^{3+} and Tb^{3+} bipyridyl complexes containing polymers

Sinha and coworkers synthesized rare earth 2,2'-bipyridyl
complexes and reported their fluorescence properties.(22) In
particular, Eu^{3+} and Tb^{3+} complexes showed a strong characteristic
line emission even at room temperature. Thus, for the further
study of fluorescence properties of Eu^{3+} and Tb^{3+} containing polymeric systems, we prepared various Eu^{3+} and Tb^{3+} bipyridyl complex
containing polymers. 4-Vinyl-4'-methyl 2,2'-bipyridine (VBP) was
synthesized by a similar method to that reported in the literature.(23)(Scheme 3).

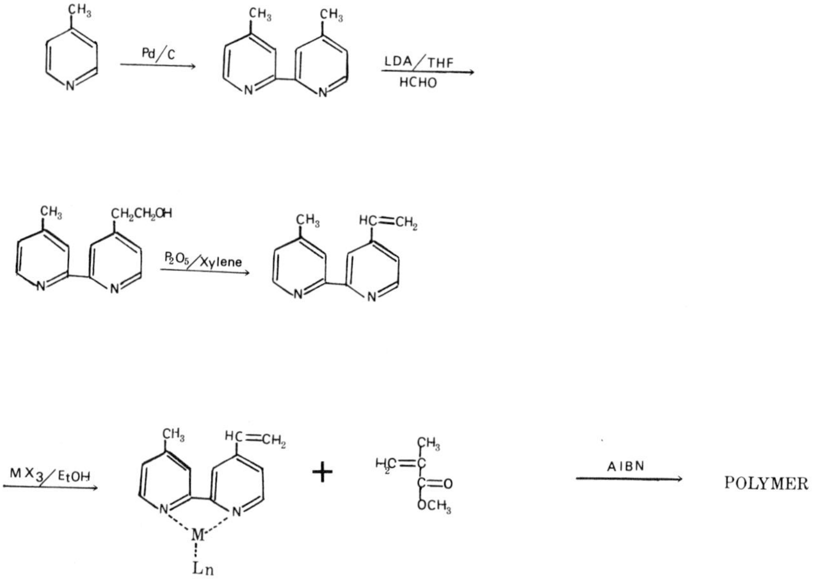

Scheme 3.   Preparation of a vinyl-4'-methyl 2,2'-bipyridine (VBP)
            and the copolymer with MMA.   M is Eu^{3+} or Tb^{3+}

Various Eu^{3+} and Tb^{3+} complexes containing VBP monomer were
prepared. The typical preparation procedure was as follows:

Eu^{3+} (VBP)(TTFA)$_3$ was prepared by dissolving VBP(5.1 mmole) and
TTFA(15.3 mmole) in 40 ml of 95% ethanol. After 1N NaOH was added,
the mixture was stirred and $EuCl_36H_2O$ (5.1 mmole) was added dropwise. Upon evaporation of the solvent, the solid obtained was
washed with water and purified by precipitation from the benzene
solution with hexane (30% yield). The complex obtained was
characterized by measuring IR and NMR. The Eu content was found
to be 14.3% (Calc. for Eu^{3+}(VBP)(TTFA)$_3$ was 15.5%). The copolymerizations of Eu^{3+} or Tb^{3+} complex monomers and MMA were carried
out either in bulk or in methanol solution using AIBN as an initiator. Some complexes were found to be insoluble in methyl
methacrylate and thus we used methanol as the solvent. The typical polymerization conditions are summarized in Table 2. The
polymer obtained was purified by washing with alcohol and dried
under vacuum at 60°C for 24 hrs..

    Typical fluorescence spectra of polymers are shown in Figs.
13 and 14. The excitation spectra were varied with the ligand but
the emission spectra showed Eu^{3+} and Tb^{3+} characteristic bands.
The fluorescence properties of these polymers are summarized in
Table 3. The emission was the result of an intermolecular energy
transfer from excited electronic states of organic ligands to the
localized 4f energy levels of Eu^{3+} and Tb^{3+}.

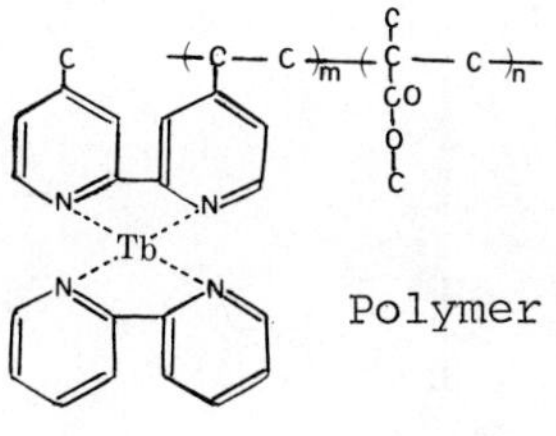

Polymer 9

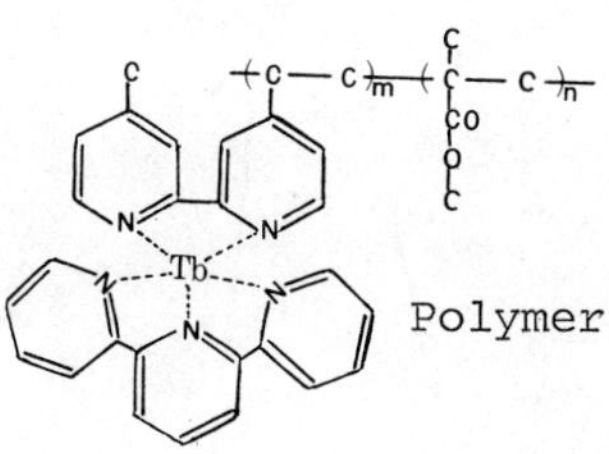

Polymer 10

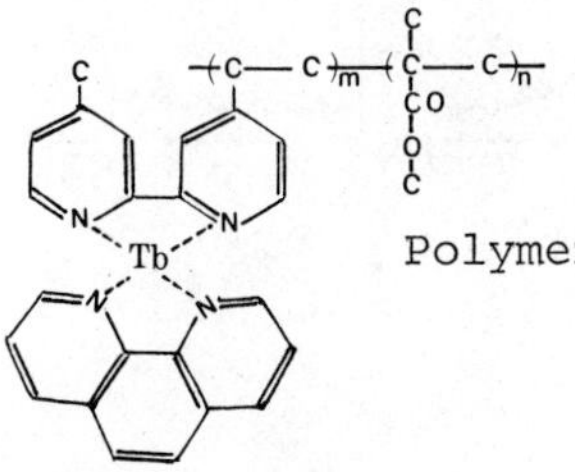

Polymer 11

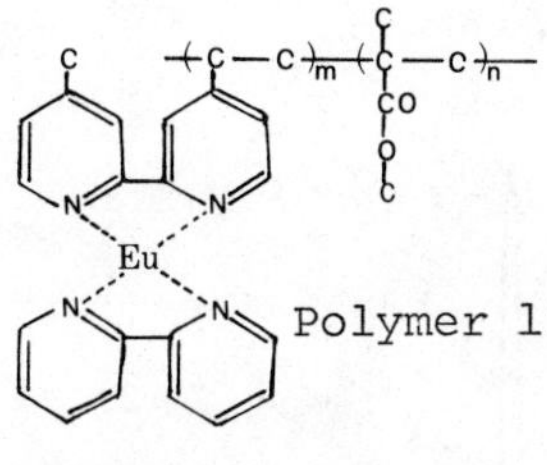

Polymer 12

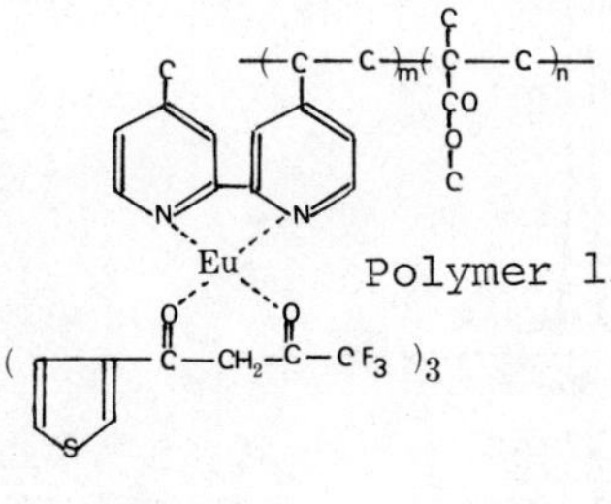

Polymer 13

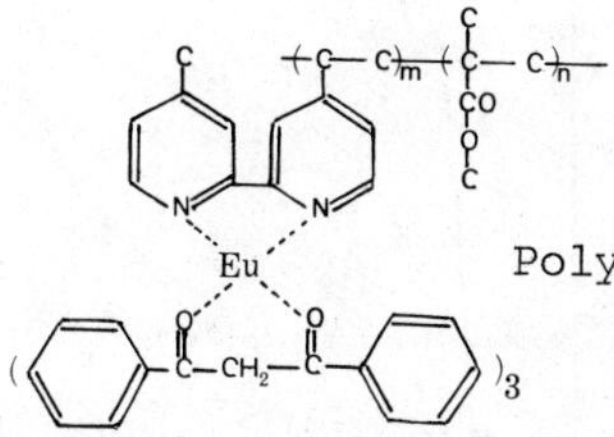

Polymer 14

$m \ll n$

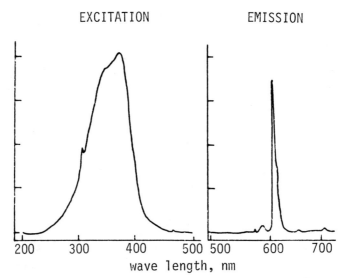

Fig. 13.   Fluorescence Spectra of Polymer 13. λ_{EM} = 612 nm,
λ_{EX} = 369 nm

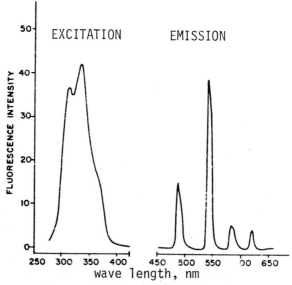

Fig. 14.   Fluorescence Spectra of Polymer 10. $\bar{\lambda}_{EM}$ = 612 nm,
λ_{EX} = 335.

Table 2.   The Polymerization Conditions and the Eu^{3+} and Tb^{3+}
           Contents of the Resulting Polymers*

Polymer (MMA)	Solvent	[complex] :	[MMA] :	[MeOH]	Found Eu^{3+} or Tb^{3+} in polymer (wt.%)
polymer 9	MeOH	0.44 :	100:	250	0.26
"	"	0.68 :	100:	250	0.52
"	"	0.83 :	100:	250	0.90
polymer 10	MeOH	0.39 :	100:	250	0.25
"	"	0.59 :	100:	250	0.50
"	"	0.97 :	100:	250	1.20
polymer 11	MeOH	0.42 :	100:	250	0.30
"	"	0.63 :	100:	250	0.42
"	"	1.05 :	100:	250	0.85
polymer 12	MeOH	0.35 :	100:	250	0.15
"	"	0.56 :	100:	250	0.45
"	"	1.01 :	100:	250	1.21
polymer 13	None	0.23 :	100:	0	0.30
"	"	0.53 :	100:	0	0.82
"	"	1.08 :	100:	0	1.61
polymer 14	None	0.28 :	100:	0	0.41
"	"	0.82 :	100:	0	1.17
"	"	0.95 :	100:	0	1.42

* Polymerization temperature was $55^{O}C$ for 48 hrs and (AIBN)=
$9.2 \times 10^{-4}M$ was used as the catalyst.

Table 3.   Fluorescence properties of Eu^{3+} and Tb^{3+} containing
           polymers

Polymers	Excitation (nm)	Emission (relative intensity)
polymer 9	312	487(45), 497(11), 543(100), 547(91),585(15),589(12),621(12)
polymer 10	335	488(40),492(29),543(100),547(87), 583(13),590(12),621(11)
polymer 11	330	488(49),491(27),543(100),547(78), 587(11),593(9), 625 (8)
polymer 12	312	580(5),590(11),612(100),650(4), 702 (7)
polymer 13	369	580(3),590(4),597(3),612(100) 616(39),623(7),650(2),701(3)
polymer 14	396	580(4),590(4),597(3),611(100) 616(43,623(10),652(2), 703(3)

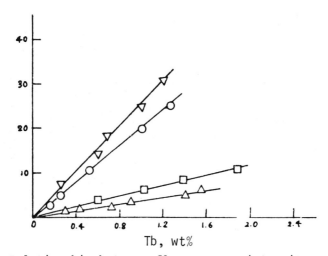

Fig. 15. Relationship between fluorescence intensity
and Tb³⁺ content in polymers.
▽ Polymer 10; ○ Polymer 1; □ Eu(VBP) (BP)Cl₃ in PMMA;
△ Polymer 11

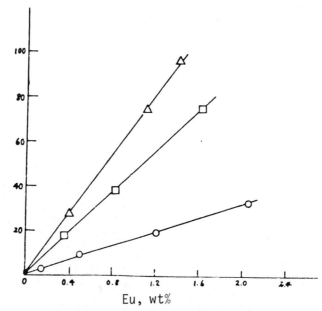

Fig. 16. Relationship between fluorescence intensity
and Eu³⁺ content in polymers
△ Polymer 14; □ Polymer 13; ○ Polymer 12

In order to find the effect of ligands on the fluorescence
intensities of polymers, we have measured the fluorescence properties
containing various concentrations of Eu^{3+} and Tb^{3+} ions (Table 2).
Figs. 15 and 16 show the relationship between the fluorescence intensity and the ion content. It was found that the order was polymer 10> polymer 9> polymer 11 for Tb^{3+} polymer and polymer 14>
polymer 13> polymer 12 for Eu^{3+} polymer. We also found that the
fluorescence intensity of the homogeneously mixed material of monomers and PMMA was much lower than the corresponding polymers. This
also showed that the double bond attached to the monomer quenches
the fluorescence.

Investigation of laser action in Eu^{3+} polymer complexes

In pursuit of possible laser action in polymeric Eu ion complexes, the copolymers of $(D'EuD_3)^-P^+$ and MMA were initially
investigated. The $(D'EuD_3)^-P^+$ was dissolved in a weighed amount
of MMA and then a small quantity of AIBN as a free radical
initiator was added. The mixture was poured into a glass or
stainless tube which had a given diameter and was closed at the
bottom. After the solution was carefully degassed, the tube was
immersed in a thermostat. The temperature of polymerization was
maintained at 60°C for 10—15 hrs. The polymer rod produced was
removed and both ends were polished. The samples were generally
made in the form of translucent rods 4 mm in diameter and 15-20mm
in length.

In order to demonstrate the laser action and the energy
transfer within the europium-ion-polymer system, the measurements
of the characteristic fluorescence spectrum of a europium chelate
excited by a pulse from a xenon flash were done and then the results of the experiments were compared with the solutions of the
rate equations by computer. The apparatus used in the measurements is shown in Fig. 17. The translucent rod of the sample
was substituted into the cavity of a Hadron 104A neodymium laser
head. The pulse from a xenon flash lamp was used as the exciting
source. There were various inputs from 200 J to 800 J in the
xenon flash. The fluorescence intensity in a path at right
angles to the excitation path was filtered, gathered and presented,
as a function of time on an oscilloscope screen. The traces of
the light output versus time were recorded photographically on high
speed film. The sample was cooled by a nitrogen stream at room
temperature.

The samples were excited over a range of input energies in
an attempt to discern the influence of excitation energy intensity
on the fluorescence lifetime. This would include an indication
of possible superfluorescence when the threshold condition was
achieved. This phenomenon should be manifested as a sharp rise

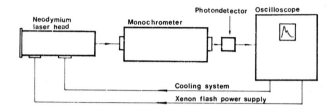

Fig. 17.   Experimental arrangement for oscilloscope traces
of fluorescence intensity.

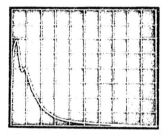

Fig. 18.   Oscilloscope traces of fluorescence of MMA + $DEuD_3P$
copolymer.
Input energy: 520J; Wavelength: 614.2 nm.
Slits: 25 mm. N.D.: 1.5. Termination: 30,000 ohms.
P.M. bias: -350V, 0.2V/div., Sweep speed: o.5ms/div.
Triggered externally through a photodiode.

in fluorescence intensity accompanied by lifetime shortening followed by a return to the characteristic lifetime profile. Figs. 18 and 19 show oscilloscope traces of two samples which reveal such curves. In Fig. 19 there are two curves, the lower at 400 J input energy exhibiting normal fluorescence and the upper curve at 600 J input energy showing lifetime shortening.

In an effort to see if the resultant decay profiles were consistent with actual superfluorescence, the kinetics of absorption and emission were computer-modeled, assuming a four-level system as described in Fig. 19. Kinetic equations similar to those described by Roess (24) were employed, although a spontaneous emission term was also deemed necessary, as in the studies of Sorokin et al. (25).

The coupled set of differential equations were numerically integrated on a digital computer. Parameters in the computer simulation were chosen to match the known characteristics of the rare-earth polymer systems, with the exception of the laser-cavity mirror reflectivities, which, due to computer-time limitations, were made larger than those used in any actual experiments. The computer solutions were obtained for a number of parameter combinations. A computer solution and oscilloscope trace of the fluorescence photons for the copolymer are shown in Fig. 19.

A computer model was also run in which self quenching was simulated. This second order quenching process has been found to take place via an energy transfer between Eu^{3+} ions within a critical distance of 17.3 A$^{\circ}$.   Values of the quenching rate constant over various orders of magnitude were tested. The traces which included quenching terms merely followed the trace of incident photons. There was no lifetime shortening evidence in these cases.

The computer model discussed herein is only a first approximation. The excited complex loses energy from other transitions than the $^5D_0$ - $^7F_2$ transition. A rigorous treatment of the kinetics of the system would necessarily include these terms. However, as a first approximation, the experimental results and the computer results are in good agreement.

Conclusion

Translucent polymers were synthesized by the copolymerization of methyl methacrylate or styrene with Eu^{3+} containing monomers.

In pursuit of possible laser action in the polymers, the fluorescence lifetime was monitored. The samples were excited by a pulse from a xenon flashlamp at room temperature. Since the sample was cooled only by a nitrogen stream, the polymer began to degrade due to the heat produced after several pulses. However, when a threshold excitation energy was reached, a considerable shortening of fluorescence lifetime was observed, followed by a return to the characteristic lifetime profile. Such lifetime shortening is typical of the phenomenon of superfluorescence, which is a precursor to laser action.

Input energy:
  upper: 600 J
  lower: 400 J
Wavelength: 614.2 nm
Slits: 0.25 mm
Termination: 30,000 ohms
P.M. bias: -200 V
  0.1 V/div.
Sweep speed: 0.5 ms/div.
Triggered externally
through a photodiode   **a**

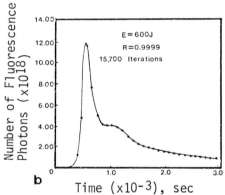

b

Fig. 19. Comparison of an actual oscillogram and a
         computer solution of rate equations.
         (a) Oscilloscope traces of fluorescence
             of MMa + D'EuD$_3$P copolymer.
         (b) A computer solution of number of
             fluorescence photons for the rate
             equations.

REFERENCES

1.   S. I. Weissman, J. Chem. Phys., 10, 214 (1942).

2.   A. Heller and E. Wasserman, J. Chem.Phys., 42, 949(1965);
     S.P.Tanner and D.L. Thomas,J.Am.Chem.Soc., 96, 706 (1974).

3.   G.A. Crosby, R.E.Whan,and R.M. Alire, J.Chem.Phys., 34, 743
     (1961); E.P.Riedel and R.G.Charles, J.Chem.Phys., 45, 1908
     (1966); H.Samelson,A.Lempicki, V.A.Brophy and C.Brecher,
     J.Chem.Phys., 40, 2547 (1964).

4.   E.J. Schmitschek, R.B. Nehrich, J.A. Trias, J.Chim.Phys. 64,
     673 (1967).

5.   M.L. Bhaumik, P.C. Fletcher, S. Higa, S.M. Lee, L.J. Nugent
     C.L.Telk and M. Weinberg, J.Phys.Chem., 68, 1490 (1964).

6.   M. Metlay, J. Chem. Phys., 39, 491 (1963).

7.   E.J. Shimitschek and R.B. Nehrich,Jr., J.Appl.Phys., 35,2786
     (1964).

8.   C. Brecher, H. Samelson and A. Lempicki, J.Chem.Phys., 42,
     1081 (1965).

9.   N.E. Wolff and R.J. Pressley, Appl. Phys. Lett., 2,152(1963).

10. E.H. Huffman, Nature, 200, 158 (1963).

11. E. Banks,Y.Okamoto and Y.Ueba,J.Appl.Polym.Sci.,25,359(1980).

12. Y.Okamoto,Y.Ueba,N.F. Dzhanibekov and E.Banks,Macromolecules,
     14, 17 (1981).

13. Y.Ueba,E.Banks and Y.Okamoto,J.Appl.Polym.Sci.,25,2007(1980).

14. Y.Ueba, K.J. Zhu,E.Banks and Y.Okamoto,J.Polym.Sci.,Chem.Ed.,
     20, 1271 (1982).

15. Y.Nose,M.Hatano and S.Kambara, Makromolek.Chem., 98,136(1966).

16. L.B. Bellamy,J.Chem.Soc.(London),4489(1954); J.Lecomte,
     Discuss. Faraday Soc., 9, 125 (1950).

17. W.F. Sager, N. Filipescu, and F.A. Serafin, J.Phys.Chem. 69,
     1092 (1965).

18. L.R. Melby,N.J. Rose,E.Abramson,and J.C.Caris,J.Am.Chem.Soc., 86, 5117 (1964);H.Bauer, J.Blanc,and D.L.Ross,J.Am.Chem.Soc., 86, 5125(1964); C.Brecher,H.Samelson and A.Lempicki, J.Chem.Phys., 42, 1081 (1965).

19. W.C. Pieuwpoore and G.Blasse, Solid State Commun., 4,227 (1966).

20. M.N. Sundberg,H.V.Lauer and F.K.Fong,J.Chem.Phys., 62, 1853 (1975).

21. S.P. Tanner and D.L.Thomas,J.Am.Chem.Soc.,96,706 (1974).

22. S.P.Sinha,C.K. Jørgensen and R.Pappalardo, Z.Naturforschg., 19a, 434 (1964).

23. P.K. Ghosh, and T.G.Spiro, J.Am.Chem.Soc., 102,5543(1980).

24. D. Roess, J. Appl. Phys., 37, 2004 (1966).

25. P.P. Sorokin, J.R. Lankard, E.C. Hammond, and V.L. Moruzzi IBM Journal, 11, 130-147 (1967).

Acknowledgement

This research was supported by a grant from the National Science Foundation under Polymer Program Grant No. DMR 09764.

INTERMOLECULAR ENERGY TRANSFER AS A MEANS FOR THE CHARACTERIZATION OF POLYMERIC LANTHANIDE COMPLEXES IN SOLUTION

Harry G. Brittain[1]

Department of Chemistry
Seton Hall University
South Orange, NJ 07079

INTRODUCTION

A variety of physical methods exist for the characterization of polymeric materials in the solid state, but the range of techniques suited for the solution phase is more limited. A major consideration associated with inorganic or organometallic compounds in fluid solutions concerns the possible oligomerization of these species. Clear and unequivocal proof for the existence of a monomeric (or polymeric) species is definitely a desirable goal in any study which seeks to correlate solution phase structures with some experimental observable. In addition, any procedure which is capable of indicating the onset of polynuclear association will be of great value in characterization studies.

In considering the complexes formed by members of the lanthanide series (we shall use Ln as a generic label for these metals), determination of the association state of the compounds is of utmost importance. Metal-ligand bonding is essentially ionic in nature (due to a lack of involvement in the bonding by the f-electrons), and hence totally non-directional. The existence of well-defined polyhedra in solution is rare, and lanthanide compounds are normally found to be very labile. Furthermore, the metal ions display a strong affinity for oxygen donors, and in basic aqueous solution extremely strong Ln-OH interactions can result in extensive hydrolysis of the compounds. In many situations, bridging hydroxide groups lead to extensive self-association of the complexes [2].

The development of a spectroscopic technique suited for studies of possible lanthanide compound self-association required

451

the existence of some phenomenon sensitive to intermolecular interactions. One such property is the nonradiative transfer of
electronic energy from one species to another. Clearly, this
transfer would be much more efficient if the donor and acceptor
compounds were physically bound in a polymeric species. Energy
transfer among monomeric compounds would require a collision
during the lifetime of the excited state of the donor, while in
the polynuclear compounds the donor and acceptor would always be
in close proximity. The pioneering work in energy transfer studies
was carried out by Forster [3,4], who developed the electric
dipole-dipole interaction theory which is still used. This theory
was subsequently extended to include higher order multipole and
exchange interactions by Dexter [5]. An excellent review of
energy transfer processes as taking place in concentrated inorganic
systems has been provided by Powell and Blasse [6].

     In many solid state host materials, concentration quenching
of lanthanide ion luminescence takes place at dopant levels above
1%. Since these materials have found widespread use as lamp and
cathode ray tube phosphors, a vast amount of work investigating
the solid state energy transfer phenemona has been performed [6-10].
Intermolecular energy transfer among lanthanide ions in solution
has received considerably less attention, although the process has
been shown to be possible. Peterson and Bridenbaugh observed that
both the Tb(III) emission intensity and lifetime measured for anti-
pyrine complexes diminished on substitution of small amounts of
other lanthanides for Tb(III) [11]. Heller and coworkers found
that energy could be transferred from 4,4'-dimethoxybenzophenone
to Eu(III) only in the presence of Tb(III) [12]. In this work, it
was concluded that the Tb(III) ion functioned as an energy acceptor
with respect to the organic donor, and as a sensitizer of the
Eu(III) ion. Subsequent work clearly demonstrated that electronic
energy could be transferred in a nonradative manner among lanthanide ions in aqueous solution [13,14].

     Subsequent work was able to show that in aqueous solution,
the energy transfer process could be modulated by organic ligands
coordinated to the lanthanide ions [15,16]. The energy transfer
among lanthanide acetylacetonate complexes was also studied, and
it was finally concluded that the most efficient transfer took
place upon formation of mixed-metal dimeric species [17,18]. The
same general method was used to demonstrate the monomeric nature
of the lanthanide derivatives of 2,2,6,6-tetramethyl-3,5-heptane-
dione [19].

     This background material amply shows that measurements of
intermolecular energy transfer are capable of indicating the
state of association experienced by a given lanthanide complex in
solution. A systematic study of the energy transfer efficiencies
as a function of solution conditions (e.g., pH in aqueous solutions)

is capable of providing information regarding the conditions under which formation of polynuclear species would take place. Such information is of crucial importance to the chemist interested in lanthanide ion solution chemistry. The method is, of course, applicable to studies of lanthanide ions bound in solid state polymeric materials. For the purposes of this review, however, we shall consider only the energy transfer phenomena as has been studied in the solution phase.

THEORY

Once a molecule has absorbed a photon of light and has been placed into an excited state:

$$D \; + \; h\nu \; \longrightarrow \; D^* \tag{1}$$

the following processes may occur:

$$D^* \xrightarrow{\; k_e \;} D \; + \; h\nu' \tag{2}$$

$$D^* \xrightarrow{\; k_n \;} D \; + \; \Delta \tag{3}$$

$$D^* \xrightarrow{\; k_g \;} E^* \; + \; \Delta \tag{4}$$

$$D^* \; + \; Q \xrightarrow{\; k_q \;} D \; + \; Q^* \tag{5}$$

$$Q^* \xrightarrow{\; k_e' \;} Q \; + \; h\nu'' \tag{6}$$

$$Q^* \xrightarrow{\; k_n' \;} Q \; + \; \Delta \tag{7}$$

Equation (2) represents the situation where D^* deactivates through the emission of a photon having lower energy than the original excitation energy (the fluorescence process), and thus k_e is the radiative rate constant. Process (3) represents the nonradiative return of D to the ground state, and hence k_n is the nonradiative rate constant. Process (4) represents the internal conversion process in which the excited molecule converts to some other excited state (such as the formation of an excited triplet from an excited singlet). In both processes (3) and (4), the excess energy left over from the excitation step (1) is lost as heat. Process (5) represents the nonradiative transfer of electronic energy from the excited donor (D^*) to the quencher (Q). This process leaves Q in an excited state, returns D^* to its ground state, and is most efficient if the transfer is resonant in nature. The excited quencher subsequently returns to its ground state either by its own fluorescence process (6) or by a nonradiative pathway (7).

We will now consider a luminescent material undergoing irradiation with excitation energy for an illumination period which is long in comparison with the emission lifetime. Under these conditions, a steady state is set up in which the rate of production of excited molecules is exactly balanced by the rate of disappearence of these species. In the absence of any photochemical reactions or intersystem crossings from upper excited states, one may express the rate of production of excited molecules as being equal to the rate of light absorption, I_a:

$$I_a = (k_e + k_n + k_g) [D^*] \tag{8}$$

When the energy transfer process is operative, an additional source of excited state deactivation becomes operative and equation (8) becomes:

$$I_a = (k_e + k_n + k_g + k_q[Q]) [D^*] \tag{9}$$

The luminescence quantum efficiency is proportional to the actual emission intensity, and is given as the number of photons emitted per photon absorbed. This quantity is clearly related to the steady state rate constants, and one obtains that in the absence of any quenching process:

$$\phi^0 = k_e / (k_e + k_n + k_g) \tag{10}$$

Should luminescence quenching also be present:

$$\phi = k_e / (k_e + k_n + k_g + k_q[Q]) \tag{11}$$

Taking the dividend of equations (10) and (11) yields:

$$\phi^0 / \phi = 1 + \frac{k_q [Q]}{(k_e + k_n + k_g)} \tag{12}$$

$$= 1 + K_{sv}^{\phi} [Q] \tag{13}$$

Equation (13) is of the form developed for luminescence quenching by Stern and Volmer [20], and K_{sv}^{ϕ} is termed the Stern-Volmer quenching constant.

In a similar manner, one may develop an expression to describe the decrease in luminescence lifetimes as caused by a suitable quenching agent. In that case, one obtains:

$$\tau^0 / \tau = 1 + K_{sv}^{\tau} [Q] \tag{14}$$

as the Stern-Volmer equation. In equation (14), τ^0 and τ are the emission lifetimes in the absence and presence of quencher whose concentration is [Q]. Equations (13) and (14) are valid as long as the donor and quencher species encounter each other via a collisional process. This mechanism is termed dynamic quenching, and would be anticipated when the donor and quencher are not physically bound. The diffusion of monomeric donors and acceptors which encounter each other solely through collisional processes is seen to fulfil these requirements. Under these conditions, it follows that:

$$K_{sv}^{\phi} = K_{sv}^{\tau} \tag{15}$$

A comparison of energy transfer as studied simultaneously through intensity and lifetime quenching can therefore serve to identify the conditions under which no association of donor and quencher takes place. If the donor is a Tb(III) complex of some type, and the acceptor is a Eu(III) complex of an identical type, then the presence of pure dynamic quenching may be taken as proof of the monomeric nature of the lanthanide complexes.

A very different situation would result when the donor and acceptor species were bound in close proximity as part of a polymeric structure. In that case, the energy transfer process would not require that Q encounter D^* during the excited state lifetime of the donor, as all geometrical considerations would be fulfilled as part of the formation of the polynuclear species. One would anticipate that any DQ pairs which could be formed would be non-luminescent as far as the donor is concerned. The only luminescence which could be observed would have to come from Q itself, or from an uncomplexed D. This quenching mechanism is termed static quenching, and would also result in the observation of reduced emission intensities. It will be shown shortly that such a situation would result in linear Stern-Volmer relations, but that the significance of the quenching constant is very different. Pure static quenching cannot affect the observed luminescence lifetime, as this quantity can only determined for the emissive species. These must correspond only to free D, and it would follow that $\tau^0 = \tau$ since the DQ species is unobservable. It must follow, then, that $K_{SV}\tau = 0$. The presence of pure static quenching can be demonstrated from lifetime experiments, and in the case of lanthanide complexes the existence of static quenching must imply that polynuclear association of the complexes has taken place.

In most situations, the luminescence quenching would take place as a result of simultaneous dynamic and static quenching. The formation of donor/acceptor pairs would follow the usual equilibrium expression:

$$D \quad + \quad Q \rightleftharpoons DQ \tag{16}$$

The association constant corresponding to formation of the DQ pair would be given by:

$$K_c = \frac{[DQ]}{[D]\ [Q]} \tag{17}$$

The intensity of light which would be emitted from a solution in which both static and quenching modes were operative will be given by [21]:

$$I \quad = \quad R \cdot F_D \cdot F_{LUM} \tag{18}$$

where R is a proportionality constant, F_D is the fraction of excitation energy which actually excites the free D, and F_{LUM} is the fraction of free D which is not dynamically deactivated by the quencher Q. To generate equation (18), it has been assumed that D is quenched dynamically through equation (5). In addition, it has been assumed that any DQ pair which is formed is non-emissive but capable of absorbing excitation energy. The free Q is assumed not to absorb any incident radiation, and it is also assumed that the solution is optically dense (hence, all excitation energy will be absorbed by either D or DQ).

F_D is defined as:

$$F_D = \frac{\varepsilon_D [D]}{\varepsilon_D [D] + \varepsilon_{DQ} [DQ]} \tag{19}$$

When dealing with lanthanide complexes, work in our own laboratory has shown that the extinction coefficients of free Tb(III) complexes are essentially the same as for the associated Tb(III)/Eu(III) polynuclear species. In that case:

$$\varepsilon_D = \varepsilon_{DQ} \tag{20}$$

and then the fraction of excitation energy which is capable of exciting the free D is given by:

$$F_D = \frac{[D]}{[D] + [DQ]} \tag{21}$$

The fraction of free D not dynamically quenched by Q is given by:

$$F_{LUM} = \frac{1}{1 + K_{sv}^{\tau} [Q]} \tag{22}$$

Substitution of equations (21) and (22) into (18) yields:

$$I = R \left\{ \frac{[D]}{[D] + [DQ]} \right\} \left\{ \frac{1}{1 + K_{sv}^{\tau} [Q]} \right\} \tag{23}$$

From equation (17) we have that:

$$[DQ] = K_c [D] [Q] \tag{24}$$

and therefore:

$$I = R \left\{ \frac{1}{1 + K_c [Q]} \right\} \left\{ \frac{1}{1 + K_{sv}^{\tau} [Q]} \right\} \tag{25}$$

For [Q] = 0, equation (25) reduces to:

$$I^0 = R \tag{26}$$

A general Stern-Volmer equation covering simultaneous dynamic and static quenching may be obtained by dividing equation (26) by equation (25). One then finds:

$$I^O / I = 1 + (K_{sv}^{\tau} + K_c)[Q]$$
$$+ K_{sv}^{\tau} K_c [Q]^2 \qquad (27)$$

Equation (27) was developed independently by two groups, each attempting to deal with non-linear Stern-Volmer quenching data [22,23]. Additional interpretation of non-linear Stern-Volmer relationships has been provided by Keizer [24]. Simultaneous measurements of emission intensity and lifetime quenching are seen to be invaluable in the determination of the quenching mechanism, and one may easily obtain the association constant of the donor/quencher pair.

On a practical level, the observation of static quenching (where $K_c > 0$) is an indication that the lanthanide complexes have associated into polynuclear species. Since the distance over which energy may be transferred between lanthanide ions (via the Forster mechanism) is quite small [25], the technique permits one to study the conditions under which polymeric species might be formed. In addition, if one makes certain assumptions regarding the structure of the complex species, one may compute the distance between the metal centers from the quenching data [25]. Studies of this type may be performed on either solid-state or solution-phase media.

An interesting implication in equation (27) concerns pure static quenching, where:

$$K_{sv}^{\tau} = 0 \qquad (28)$$

In that instance, equation (27) reduces to:

$$I^O / I = 1 + K_c [Q] \qquad (29)$$

The form of this Stern-Volmer relation is identical to that obtained as equation (14), except that now the Stern-Volmer quenching constant is identified as the formation constant of the donor/quencher complex. Without studies of the possible lifetime quenching, one might assume that a linear Stern-Volmer relation always implies the presence of pure dynamic quenching.

The bimolecular rate constant corresponding to the quenching process may be calculated by:

$$k_q = K_{sv}^{\tau} / \tau^O \qquad (30)$$

In many studies of energy transfer among organic or inorganic donor/quencher systems, the reactions are observed to proceed at the diffusion controlled limit. With Tb(III) donor and Eu(III) acceptor complexes, however, the rate constants measured for dynamic quenching mechanisms invariably proceed at much slower rates.

ENERGY LEVELS OF Tb(III) and Eu(III)

The solution phase energy transfer studies to be described in the next section will all involve Tb(III) compounds as the donor species and Eu(III) compounds as the energy acceptors. When bound in complexes, the energy levels of these two ions are not significantly different than those known for the free ion. This situation arises as a result of the very low degree of covalency present in Ln-ligand bonds, and to the total lack of involvement in bonding on the part of the f-electrons. Crystal field effects in lanthanide spectra are therefore extremely small. An energy level diagram for the lowest f-states of Tb(III) and Eu(III) is provided in Figure 1.

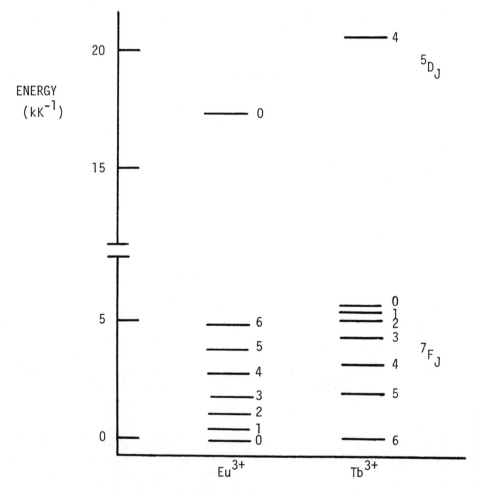

Figure 1. Free ion energy levels for the Eu(III) and Tb(III) ions.

In the solution phase, all Tb(III) luminescence originates from the $^5D_4$ excited state, and all Eu(III) emission originates from the $^5D_0$ state. One may note that the $^5D_0$ level of Eu(III) lies approximately 3000^{-1} below the $^5D_4$ level of Tb(III), and hence one finds that the Eu(III) ion is a good quencher of Tb(III) emission. Luminescence from either ion terminates in the ^7F_J levels of the ground state. For the sake of simplicity, we shall denote the various spectroscopic transitions solely by their J quantum numbers. Each level is characterized by a degeneracy of $(2J + 1)$.

For Tb(III), the most intense transitions are the 4-6 (490 nm), 4-5 (545 nm), 4-4 (585 nm), and 4-3 (620 nm). Of these, the 4-5 band invariably is of the highest intensity, and consequently most workers monitor only the intensity of this band system for a given study. With Eu(III), the 0-1 (595 nm), 0-2 (615 nm), and 0-4 (690 nm) transitions exhibit the largest luminescence intensities. In purely octahedral symmetry only the 0-1 band can be observed, but in lower symmetries the 0-2 band system is found to be the most intense.

In every study which will be discussed in the following section, the Tb/ligand complexes were titrated with identical Eu/ligand complexes at exactly the same pH. The most reliable results were obtained when the Eu/ligand concentration was never allowed to be more than 5% than that of the Tb/ligand concentration. Generally, the solutions were excited at 365 nm, since it had been determined that Tb(III) could be effectively excited at this wavelength and that Eu(III) could not (at least in aqueous phase carboxylate complexes). Another approach is to excite the Tb(III) ion directly within the 6-4 absorption, but this method requires access to an Ar-ion laser and its pumping wavelength of 488 nm. The usual practice was to then follow the decrease in Tb(III) emission at 545 nm as a function of added Eu(III), and to observe whether any Eu(III) luminescence could be observed at 615 nm. Since the Eu(III) quantum yield at the chosen excitation wavelengths was known to be poor (especially at the micromolar concentration levels which were normally used), any observable Eu(III) emission must have resulted from Tb(III)-Eu(III) energy transfer. In most situations, the Tb(III) emission lifetime was also determined. In this manner, the emission intensity and lifetime quenching data could be fitted by the Stern-Volmer relations given in equations (13) and (14). The presence of any static contribution to the overall quenching was indicated when:

$$K_{sv}^{\phi} > K_{sv}^{\tau} \tag{31}$$

When this situation was detected, the data were analyzed by means of equation (27). In favorable instances, one could calculate the association constant corresponding to the mixed Tb/ligand/Eu species.

Given the chemical similarity of common lanthanide complexes, the
formation of significant amounts of polynuclear Tb/Eu species
could be taken to imply that the overall bonding situation in
solution contained an extensive contribution from polymeric species.
On the other hand, when the presence of pure dynamic quenching
could be established, then it was assumed that all the Ln complexes
were totally monomeric in nature. These assumptions now make it
possible to evaluate the experimental quenching results, and provide
the mechanism for the interpretation of the observed trends.

SOLUTION PHASE ENERGY TRANSFER STUDIES ON LANTHANIDE COMPLEXES IN
POLYMERIC SYSTEMS

     Since earlier work [15,16] had indicated that the nonradiative
transfer of electronic energy from Tb(III) complexes to analagous
Eu(III) complexes could be modulated by the presence of coordinated ligands, we felt that a fuller investigation of the phenomenon was required. However, the first study of this type
required the existence of systems whose solution phase behavior
was already known in reasonable detail, since a calibration of
the procedures was needed as a first step. Fortunately, the
required information was available for the family of pyridine
carboxylic acids:

PIC   NIC   DPA

PIC = picolinic acid, NIC = nicotinic acid, and DPA = dipicolinic
acid. It had been established earlier [26,27] that $Ln(DPA)_3^{3-}$
complexes remained monomeric at all pH values, and that these
were known to exhibit well-defined trigonal symmetry in aqueous
solution. While formation constant data was available for Ln/PIC
or Ln/NIC complexes, no hard experimental data existed regarding
the nature of the solution phase complexes. Thus, these ligand
systems appeared to be perfectly suited for an evaluative study
of the energy transfer method [28].

     It was established immediately that multidentate binding of
the ligands by Ln(III) ions was required for the observation of
Tb(III) emission, as no luminescence could be detected at any pH
in the Tb/NIC solutions (at least at 1 mM concentration levels).
Luminescence was only obtained in the Tb/PIC solutions above pH
4, corresponding to ionization of the carboxylate group. The
ionization constants for DPA are significantly larger than that
of PIC, and consequently strong emission was detected in Tb/DPA

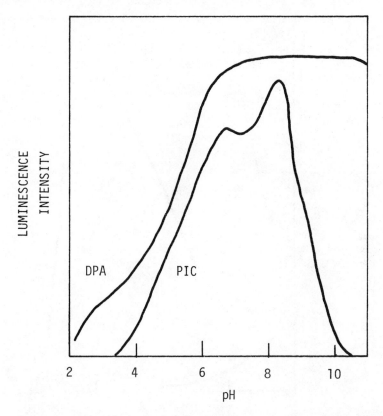

Figure 2. pH dependence of the Tb(III) luminescence obtained from
        Tb/DPA and Tb/PIC solutions, at 1:5 metal/ligand ratios.
        The intensity scale is arbitrary, but the Tb/DPA emission
        is roughly twice as intense as the Tb/PIC emission.

solutions at pH values as low as 3. Examples (taken at 1:5 ratios
of metal/ligand) of the pH dependence of the observed Tb(III)
luminescence in each system are given in Figure 2.
An examination of Figure 2 reveals that the Tb/PIC complexes become
extensively hydrolyzed above pH 9, while the Tb/DPA complexes do
not exhibit such tendencies. The energy transfer results were
found to be in complete accord with these predictions. The Stern-
Volmer quenching constants obtained for the Tb/PIC and Tb/DPA
complexes have been plotted in Figure 3.

     One may note that a very small degree of energy transfer is
noted in the Tb/DPA complexes at low pH values, but even this poor
energy transfer efficiency decreases once the solution becomes
sufficiently basic to fully deprotonate the DPA ligand. These
observations all indicate the presence of purely monomeric complexes
as indicated from the earlier work. The Stern-Volmer quenching

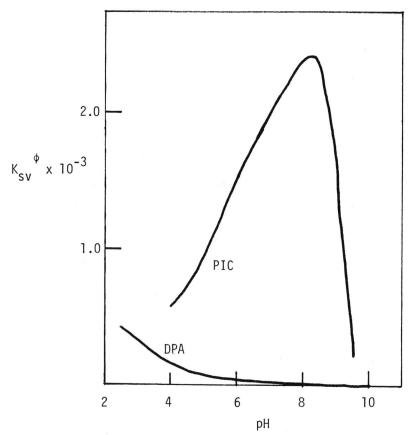

Figure 3. pH dependence of the Stern-Volmer quenching constants
(as obtained from intensity measurements) for 1:5 metal/
ligand ratios of Ln/DPA and Ln/PIC.

constants obtained from lifetime studies were identical to those
obtained from the intensity quenching studies, confirming the
presence of pure dynamic quenching.

With the Tb/PIC complexes, the intensity quenching was found
to be considerably more efficient than that noted for the Tb/DPA
complexes. On the other hand, the Stern-Volmer constants computed
from the lifetime data were relatively constant between pH 5 and
pH 9, and these values were not found to exceed 200 liter/mole,
The presence of a large static component in the quenching is
therefore indicated, and one may thus conclude that extensive
association of the complexes takes place at the low pH values.
It is interesting to note that the hydrolysis reactions taking
place above pH 8.5 destroy the polynuclear species, and eventually
one obtains a precipitate of $Ln(OH)_3$ above pH 11. The low pH
complexes undoubtably contain PIC as a bridging ligand.

After establishing that the energy transfer method could be used to determine the degree of self-association for a given lanthanide complex, the method was then used in an explorative manner for a sequence of studies. The first of these focused on the lanthanide complexes of benzenecarboxylic acids [29], since the complexes of these provide a variety of chelating abilities.

Ln(III) complexes of phthalic (PHT) and hemimellitic (HML) acids were found to remain soluble up to pH 9, the Ln(III) complexes of trimellitic (TML) and pyromellitic (PML) acids precipitated once pH 3 was exceeded, and the Ln(III) complexes of trimesic acid (TMS) were not found to be soluble at any pH value. No data could be obtained on the Ln/TMS system, but energy transfer studies were performed over the full solubility range for the other four ligand systems [29].

The Stern-Volmer quenching constants obtained for the various ligand systems provided interesting information about the nature of the solution phase complexation phenemona. For all Tb(III) complexes, the existence of monomeric species was found to prevail between pH 1.5 and 2.5. Above pH 3, all complexes were found to undergo extensive self-association. The bridging of Ln(III) species was apparently so extensive for Ln/TML and Ln/PML systems that these precipitated above pH 3. The Ln/PHT and Ln/HML compounds remained in solution up to pH 9, but were found to undergo increasing amounts of polynuclear association.

Even though large amounts of static quenching were noted in
the Ln/PHT solutions, the degree of self-association of these
complexes was found to be much less than that of the Ln/HML
complexes. This trend must be related to the extra carboxylate
group of the HML ligand, a functionality which is in a reasonable
position to promote bridging of the compounds. Additional carboxyl
groups are in much better positions to promote intermolecular
bridging in the TML and PML ligands, and it was found that the
degree of self-association increased in a steady manner up to the
point of precipitation. Apparently, in these compounds the formation of insoluble polymeric species is favored, and this bonding
mode dominates the complexation processes above pH 3. The TMS
ligand cannot form any closed chelate rings, and is therefore
forced to bridge metal centers as its carboxyl groups are deprotonated. Consequently, these compounds cannot be rendered soluble
at any pH value.

The next system chosen for study was that of aspartic acid:

$$HOOC - CH_2 - CH - COOH \qquad\qquad ASP$$
$$\underset{\displaystyle NH_2}{|}$$

since previous chiroptical work [30] had shown that the Tb/ASP
complexes exhibited a variety of spectral changes between pH 6 and
pH 10. This behavior is remarkable in that no ligand ionizations
were known to take place within this pH region, and hence it seemed
plausible that polynuclear complex formation was the source of the
variable lineshapes.

Complete energy transfer studies were carried out using both
the Ln/L-ASP and Ln/D,L-ASP complex systems. The very surprising
emission intensity results are illustrated in Figure 4. At pH
values below 5, the Tb(III) $\rightarrow$ Eu(III) energy transfer efficiencies
were found to be identical for the Ln/L-ASP and Ln/DL-ASP systems.
Once the solution pH was raised above 5, however, the energy
transfer was found to be considerably more efficient in the
Ln/DL-ASP system relative to that observed in either the Ln/L-ASP
or Ln/D-ASP systems. Invariably, it was found that the energy
transfer associated with the Ln/L-ASP system was exactly equivalent to that of the Ln/D-ASP system.

Additional evidence for the existence of stereoselectivity
was obtained by following the rise in sensitized Eu(III) emission.
Very little Eu(III) luminescence was noted below pH 5, but once
the polymeric species formed the Eu(III) emission increased
rapidly. Significantly, the Eu(III) emission intensity associated
with the Ln/DL-ASP system was found to be higher than that observed
for an equivalent Ln/L-ASP or Ln/D-ASP system. These observations
are consistent with the presence of greater energy transfer in the
DL-ASP system.

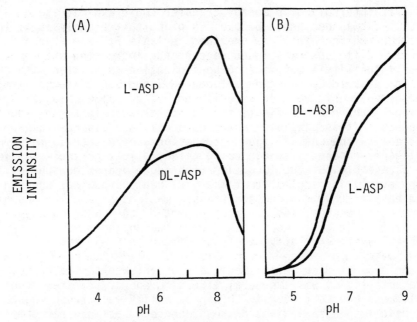

Figure 4. (A) pH dependence of the 4-5 Tb(III) emission intensity in Tb/L-ASP/Eu and Tb/DL-ASP/Eu systems.
(B) pH dependence of the sensitized 0-2 Eu(III) emission intensity in Tb/L-ASP/Eu and Tb/DL-ASP/Eu systems.

Table I. Photophysical Parameters Calculated From Quenching Studies in the Ln/L-ASP and Ln/DL-ASP Systems

pH	Ln/L-ASP $K_{sv}^{\phi} \times 10^{-3}$	Ln/DL-ASP $K_{sv}^{\phi} \times 10^{-3}$	Ln/L-ASP $K_c \times 10^{-2}$	Ln/DL-ASP $K_c \times 10^{-2}$
3.0	0.99	0.99	0.0	0.0
3.5	1.01	1.00	0.0	0.0
4.0	1.11	1.08	0.4	0.4
4.5	1.18	1.17	0.8	0.8
5.0	1.46	1.45	1.2	1.1
5.5	1.62	1.70	1.9	2.3
6.0	1.77	1.80	2.6	2.7
6.5	2.25	2.44	4.8	5.6
7.0	3.79	4.44	11.8	14.7
7.5	5.47	6.87	19.4	25.8
8.0	6.72	8.60	25.1	33.6
8.5	7.50	10.71	28.6	43.2
9.0	7.64	12.33	29.3	50.6

That intermolecular energy transfer could exhibit stereoselectivity was not known before. A quantitative analysis of the
Ln/ASP quenching results is provided in Table I. The stereoselective difference was found to continue increasing with pH up
to the solubility limit. Since no ionization associated with the
ASP ligand would be expected between pH 5 and 9, it seemed likely
that the polymeric species were linked by hydroxide bridges.
Apparently there are definite steric preferences in the polynuclear
compounds. It is possible that the racemic ASP ligands take up
less room on the Ln(III) ion coordination sphere than do resolved
ASP ligands (essentially a packing effect), leaving the metal ion
more accessible for the formation of bridged hydroxide species.
In any event, it is clear that the variable chiroptical spectra
noted earlier [30] reflect the transformations of a multitude of
polymeric compounds. As would be expected, the donor/quencher
association constants were found to be much higher for the
Ln/DL-ASP systems relative to those of the Ln/L-ASP systems.

Several studies were immediately undertaken to learn whether
stereoselectivity was a general effect in energy transfer studies.
The ligands chosen to verify this possibility were those whose
chiroptical spectra indicated the presence of polynuclear complex
species [30,32]. The ligands used in these studies were:

COOH	COOH	COOH
HC — OH	HC — NH_2	HC — NH_2
CH_2	CH_2	HC — CH_3
COOH	OH	OH
MAL	SER	THR

Extremely large stereoselective effects were again noted in the
energy transfer among malic acid (MAL) complexes [33], while less
stereoselectivity was found in the energy transfer among serine
(SER) and threonine (THR) complexes [34]. With the hydroxy amino
acids, the racemic ligands were found to promote the largest
degree of polynuclear association, but the opposite situation was
shown to exist for the MAL compounds. These studies therefore
indicated that while stereoselective energy transfer appeared to
be a general phenemenon when one worked with potentially chiral
ligands, the source of these effects was not immediately apparent.

Much of the work described thus far was carried out at 1:5
metal/ligand ratios. This particular value was chosen since many
previous investigators had worked at such a value, and it seemed
desirable to characterize compounds on which a body of data was
available. To study the effect of the metal/ligand ratio on the
energy transfer efficiencies, a series of studies were carried out
employing histidine (HIS) as the complexing agent [35].

Table II.Stern-Volmer quenching constants for the Ln/HIS system as a function of pH at selected metal/ligand ratios

pH	$K_{sv}^{\phi} \times 10^{-2}$ 1:2.5	$K_{sv}^{\phi} \times 10^{-2}$ 1:5	$K_{sv}^{\phi} \times 10^{-2}$ 1:10
2.5	1.21	1.66	1.15
3.0	1.23	2.95	1.19
3.5	1.30	4.11	1.22
4.0	1.29	5.58	1.84
4.5	1.28	7.88	3.17
5.0	1.69	10.40	3.58
5.5	2.17	13.20	3.90
6.0	2.68	17.58	5.41
6.5	8.99	33.51	5.97
7.0	17.81	50.23	6.98

HIS

The Stern-Volmer quenching constants obtained for the Ln/HIS system were found to exhibit a remarkable dependence on the metal/ligand ratio, as may be seen in Table II.

At metal/ligand ratios of 1:2.5 and 1:5 one anticipates the existence of only Ln(HIS), and the pH dependence of the quenching constants was found to follow the usual pattern (extensive formation of polymeric species at higher pH values). The quenching efficiency was found to be much higher at the 1:5 ratio, where the Ln(HIS) complex is expected to be fully formed. Since the HIS ligand is known to form a variety of hydrogen bonded compounds, it is entirely possible that such a mechanism aids the energy transfer process.

The most interesting result was noted at the 1:10 Ln/HIS ratio. In these solutions, all compounds were found to remain monomeric at all pH values. Since at this metal/ligand ratio it is likely that the most significant Ln(III) species is Ln(HIS)$_2$, one may conclude that the binding of the second HIS ligand destroys the capability of the Ln(HIS)$_2$ compounds to form polynuclear species. This would happen if the HIS ligand occupied more of the Ln(III) coordination sites, leaving less of these accessible for the formation of μ-hydroxide bridges.

The observation that formation of Ln(III) compounds with several multidentate ligands could inhibit the existence of polymeric complexes clearly needed to be investigated further. As another study of how the metal/ligand ratio could affect energy transfer efficiencies, the behavior of Ln(III) complexes with citric acid (CIT) was investigated.

$$HOOC - CH_2 - \overset{\overset{\displaystyle OH}{|}}{\underset{\underset{\displaystyle COOH}{|}}{C}} - CH_2 - COOH \qquad CIT$$

It is apparent from the structure that this particular ligand contains a number of functional groups capable of binding a Ln(III) ion in a variety of possible multidentate manners. Evidence exists in the literature indicating that both Ln(CIT) and Ln(CIT)$_2$ compounds may be formed, although the log K_2 values are approximately 4 orders of magnitude smaller than the log K_1 values.

Energy transfer studies were carried out at several metal/ligand ratios [36], and the Stern-Volmer quenching constants obtained from these are located in Table III.
It is immediately apparent that once the metal/ligand ratio exceeds 1:1, all Ln/CIT compounds remain totally monomeric. From an analysis of the formation constant data, it was determined that formation of the Tb(CIT)$_2$ compound from Tb(CIT) would be 77% complete at the 1:2 metal/ligand ratio. At the 1:3 ratio, the conversion would be essentially 94% complete, and at the 1:5 ratio the conversion would be 98% complete. When one couples this information with the trends of Table III, it becomes apparent that formation of the Ln(CIT)$_2$ compound interferes greatly with the energy transfer. This situation would exist if each CIT ligand bound the Ln(III) ion in a multidentate fashion with no bridging of metal centers.

Table III. pH dependence of the Stern-Volmer Quenching Constants associated with the Ln/CIT complexes, as a function of the metal/ligand ratio

pH	$K_{sv}^{\phi} \times 10^{-2}$ 1:1	$K_{sv}^{\phi} \times 10^{-2}$ 1:2	$K_{sv}^{\phi} \times 10^{-2}$ 1:3	$K_{sv}^{\phi} \times 10^{-2}$ 1:5
3.0	1.67	0.86	2.71	1.50
4.0	5.77	1.52	2.13	1.05
5.0	17.13	0.75	0.83	0.45
6.0	27.48	1.05	0.50	0.33
7.0	16.37	2.61	1.80	0.37
8.0	7.25	2.15	1.91	1.04
9.0	4.18	1.94	1.64	1.28
10.0	2.54	2.15	1.73	1.50
11.0	2.03	1.88	2.02	1.24
12.0	2.39	1.90	1.95	1.15

The results obtained with the 1:1 Ln/CIT ratio stand in sharp contrast to the higher ratios. In this particular system only the Ln(CIT) compound can be formed, and the quenching data reveal the existence of distinct pH regions. The degree of polynuclear association increases up to pH 5, remains fairly constant between pH 5.5 to 6.5, and decreases beyond pH 7. By the time pH 9 is exceeded, the quenching becomes purely dynamic in nature. One would not expect the formation of hydroxide bridges below pH 7, and thus it appears that in acidic solution (at 1:1 metal/ligand ratios) the CIT ligand prefers to bridge metal centers. Above pH 7, the final deprotonation of CIT takes place, and then the ligand is found to wrap around the Ln(III) ion in a multidentate fashion. This last process breaks up the bridged polymeric species and leaves only monomeric compounds.

The studies just described indicate the sensitivity of the energy transfer method to details of metal/ligand bonding. In addition, these works provide a sufficient number of examples which illustrate the generality and utility of the technique as applied to the study of polynuclear complex species in solution. Having reached this point, it may now be assumed that information acquired from energy transfer studies has all the significance implied in the theory section. Once it has been determined that a given complex is indeed monomeric, then it becomes possible to deduce direct correlations between any observed parameter and the plausible solution phase structures. We will now describe one more study for which such correlations were of extreme importance.

The commercial availability of separated lanthanide elements became possible through the use of ion exchange chromatography, and the best methods employed aminopolycarboxylic acids (APC) as complexing agents. As a result, a great deal of work has focused on the study of 1:1 Ln(APC) complexes. It is known that many of the Ln(APC) compounds are not coordinatively saturated, and thus contain labile water molecules. This behavior has been exploited in that the Ln(APC) compounds have found use as aqueous nmr shift reagents [37]. It was noted in this study that once the Ln(APC) compounds formed ternary hydroxo species at high pH, they ceased to function as shift reagents.

A series of energy transfer studies on several Ln(APC) compounds was carried out as a function of pH to learn if any effects due to polynuclear association could be detected [38]. The APC ligands employed were:

$$HOOC-H_2C \diagdown \qquad \diagup CH_2-COOH$$
$$N-CH-CH_2-N$$
$$HOOC-H_2C \diagup \quad \underset{R}{|} \qquad \diagdown CH_2-COOH$$

R = H, EDTA

CH$_3$, PDTA

HOOC — H₂C, N—N, CH₂ — COOH ... CDTA

$$HOOC-H_2C \diagdown N \diagup CH_2-COOH$$

(structures for CDTA, EGTA, DTPA)

CDTA

EGTA

DTPA

The pH dependence of the emission intensities associated with representative Tb(APC) compounds is shown in Figure 5.

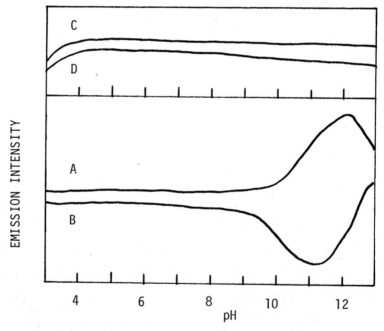

Figure 5. Luminescence intensity of the 4-5 Tb(III) band in (A) Tb(EDTA), (B) Tb(EDTA)/Eu(EDTA), (C) Tb(EGTA), and (D) Tb(EGTA)/Eu(EGTA) as a function of pH. The Tb/Eu mole ratio used was 15:1.

It was found that Tb(EDTA), Tb(PDTA), and Tb(CDTA) exhibited little change in emission intensity up to pH 10, but a significant intensity increase was noted between pH 10 and 12. Above pH 12, hydrolysis reactions begin to break up the complexes and cause a decrease in emission intensity. On the other hand, the Tb(EGTA) and Tb(DTPA) complexes were found to exhibit relatively constant emission intensities over the entire pH range. We have shown [39] that Tb(EDTA), Tb(PDTA), and Tb(CDTA) are not coordinatively saturated, but that Tb(EGTA) and Tb(DTPA) are. One would therefore conclude that Tb(EGTA) and Tb(DTPA) would not form ternary hydroxo species (which might be bridged), but that the other complexes might.

Further probing of the Ln(APC) systems was carried out through energy transfer studies on these compounds [38], and it was found that the same distinction drawn from the emission intensity studies followed through in the energy transfer work. Selected Stern-Volmer quenching constants are found in Table IV.

Table IV . pH dependence of the Stern-Volmer quenching constants associated with the Ln(APC) systems.

pH	EDTA	PDTA	CDTA	EGTA	DTPA
3.0	135	101	83	16	145
4.0	169	118	83	14	130
5.0	143	134	74	18	130
6.0	145	152	65	21	116
7.0	147	152	67	21	101
8.0	157	169	80	19	116
9.0	303	222	175	19	119
10.0	394	588	609	22	118
10.5	638	1204	972	22	103
11.0	1000	1424	1024	30	87
11.5	1340	1412	750	30	78
12.0	1070	858	380	30	75
12.5	706	493	195	28	70
13.0	421	333	185	25	65

It is immediately apparent that the Ln(EGTA) and Ln(DTPA) complexes remain totally monomeric at all pH values between 3 and 13. In contrast, the other Ln(APC) compounds remain monomeric only up to pH 10, and above this pH value become extensively associated. The degree of association drops again at still higher pH values, undoubtably resulting from the effects of the hydrolysis reactions. Association constants corresponding to the formation of the donor/quencher complexes were calculated, and these have been collected in Table V.

It is abundantly clear that the increase in emission intensity

Table V. Formation Constants of the Mixed Tb/Eu/APC Complexes as
         Obtained from the Quenching Data

pH	K_c (EDTA)	K_c (PDTA)	K_c (CDTA)
10.0	244	429	517
10.5	475	1005	667
11.0	816	1214	718
11.5	1137	1201	453
12.0	882	682	296
12.5	539	340	118
13.0	270	190	108

noted above pH 10 is linked to the existence of associated Ln(APC)
compounds, as this effect would probably expel any water molecules
remaining in the Ln(III) inner coordination sphere and thus lower
the degree of solvent quenching. Such a saturation of the coordination sphere would be consistent with the inability of certain
Ln(APC) compounds to function as nmr shift reagents under highly
basic conditions. If the substrate cannot move close enough to the
Ln(III) ion, it cannot experience a paramagnetic shift. Once more,
the energy transfer method has provided information which explains
a variety of other observations and trends.

USE OF FORSTER-TYPE ENERGY TRANSFER AS A MEANS TO DETERMINE DONOR-ACCEPTOR DISTANCES

When the donors and acceptors in an energy transfer process
are physically bound by the same polymer chain, the dynamic quenching
mechanism must cease to be operative. In that case, the quenching
would be purely static in nature and the appropriate equations
developed in the theory section would apply. However, this situation
was exactly that treated by Forster [3,4]. If one assumes that the
dipole-dipole energy transfer mechanism is operative, then in
principle it is possible to use the known radial dependence of the
process to calculate the distance between the donor and the acceptor.

According to Forster, the luminescence quenching resulting
from a dipole-dipole energy transfer process is given by:

$$\frac{I^0}{I} = \frac{1}{1 + (R_0/R)^6} \tag{32}$$

where I and I^0 are defined as the luminescence intensities in the
presence and absence of energy transfer, respectively. R is the
distance between the donor and acceptor, and R_0 is the critical
distance for 50% energy transfer. This last value is obtained

(in units of cm^6) from:

$$R_O^6 = (8.78 \times 10^{-25}) \kappa^2 Q n^{-4} J \qquad (33)$$

κ is the dipole-dipole orientation factor, Q is the donor quantum yield in the absence of energy transfer, and n is the refractive index of the medium between the donor and quencher. J is the spectral overalp integral, and is defined by:

$$J = \int F(\nu) a(\nu) \nu^{-4} d\nu \qquad (34)$$

where $F(\nu)$ is the luminescence intensity of the donor at frequency ν (in cm^{-1}), and $a(\nu)$ is the molar absorptivity (in units of cm^{-1} M^{-1}) of the acceptor.

The original work using energy transfer as a "spectroscopic ruler" was reported at essentially the same time by Horrocks et al. [40], and by Berner et al. [41]. Both of these investigations centered around the protein thermolysin, and both groups used Tb(III) as the donor species and Co(II) as the energy acceptor. The parameters of equations (32)-(34) were evaluated from the known parameters obtained from the existing crystal structure. From the crystal structure, one could calculate a Tb(III)-Co(II) distance of 13.9 A [42,43]. The values of R computed from the two sets of energy transfer measurements (carried out in the solution phase) agreed excellently with this value, thus providing a sound basis for the application of the method and also establishing the mechanism for the energy transfer.

In later works, the group led by Horrocks applied the energy transfer method to the measurement of interlanthanide distances in protein systems capable of binding more than one metal ion. It was shown that the near isotropic nature of the Ln(III) absorption and emission transitions permitted one to take the orientation factor as 2/3 [44]. Detailed studies were carried out on Tb(III), Eu(III), and Ho(III) substituted parvalbumin [45] and thermolysin [46]. Where the interlanthanide distances were known, the computed R values were found to agree quite well. The R_0 values were found to vary from approximately 9 A for the Tb(III)-Ho(III) pair, and down to approximately 5.5 A for the Eu(III)-Ho(III) pair. These works placed the energy transfer method on a sufficiently firm footing that one may now feel confident that distances calculated in less well-defined systems are completely accurate.

SUMMARY AND FUTURE DIRECTIONS

The characterization of lanthanide substituted polymers in solution is a difficult problem, but the energy transfer methods described in the preceding sections are capable of yielding important information. Clarification of the transfer mechanism (dynamic vs. static quenching) may be used as a tool to follow the polymerization processes. Through these studies, one may unequivocally determine the conditions under which the formation of polynuclear species would take place. This information is of vital importance to workers seeking to correlate observed parameters with plausible solution phase structures. At the same time, one could study the mechanisms of polymerization processes. One could easily imagine an experiment where one portion of a copolymer contained the energy donor, and another portion contained the energy acceptor. The reaction pathways associated with the copolymerization process could be characterized through detailed studies of the energy transfer efficiencies.

Once a polymer has been formed, the energy transfer method may be used to evaluate interatomic distances. Normally, such information is only available from x-ray crystallographic studies. With amorphous polymers, this kind of data would be simply unavailable. However, by studying the energy transfer phenemona, one is able to obtain structural information on a suitable organometallic polymer without the need for crystallizing the material. The method has been calibrated on the protein systems, and the results drawn from such studies have been shown to be completely reliable. Obviously this technique is easily extended to include the study of metallo-polymers in the solid state. Okamoto and coworkers [47] have examined the aggregation of ions in polymeric materials through studies of the Tb(III)-Eu(III) and Tb(III)-Co(II) energy transfer processes. For instance, data regarding intermetallic distances were obtained for poly(acrylic acid), and copolymers of styrene-acrylic acid, styrene-maleic acid, and methyl methacrylate-methacrylic acid [48].

The range of applications suitable for probing by the energy transfer method is indeed large. More work needs to be done in the characterization of metallo-protein complexes. The characterization of metal binding sites on membranes and their synthetic analogues (micelles and vesicles) clearly would benefit from such investigations. Studies of the condensation of cations onto polyelectrolytes is yet another area in which extensive work needs to be performed. These areas represent only a partial listing of potential systems for which energy transfer studies would prove invaluable. It should be clear by now that any organometallic polymer system is suited for probing by energy transfer methods, and hopefully workers in this field will incorporate these methods into their arsenal.

ACKNOWLEDGEMENTS

This work has been supported at various times by the Research Corporation, the National Science Foundation, and the Camille and Henry Dreyfus Foundation. Additional thanks are due to Mr. Robert Copeland and Ms. Laura Spaulding for their experimental assistance in several of the investigations.

REFERENCES

1) Teacher-Scholar of the Camille and Henry Dreyfus Foundation, 1980-85.

2) R. Prados, L. G. Stadtherr, H. Donato, Jr., and R. B. Martin, J. Inorg. Nucl. Chem., 36, 689 (1974).

3) T. Forster, Ann. Phys., 2, 55 (1948).

4) T. Forster, Z. Naturforsch., A4, 321 (1949).

5) D. L. Dexter, J. Chem. Phys., 21, 836 (1953).

6) R. C. Powell and G. Blasse, Structure and Bonding, 42, 43 (1980).

7) R. Reisfeld, Structure and Bonding, 13, 53 (1973).

8) R. Reisfeld, Structure and Bonding, 30, 65 (1976).

9) R. Reisfeld and C. K. Jorgensen, "Lasers and Excited States of Rare Earths", Springer-Verlag, New York, 1977, ch. 4.

10) G. Blasse, Philips Res. Repts., 24, 131 (1969).

11) G. E. Peterson and P. M. Bridenbaugh, J. Opt. Sci. Am., 53, 494 (1963).

12) P. K. Gallagher, A. Heller, and E. Wasserman, J. Chem. Phys., 41, 3921 (1964).

13) W. W. Holloway, Jr. and M. Kestigan, J. Chem. Phys., 47, 1826 (1967).

14) F. S. Quiring, J. Chem. Phys., 49, 2448 (1968).

15) C. K. Luk and F. S. Richardson, J. Am. Chem. Soc., 97, 6666 (1975).

16) H. G. Brittain and F. S. Richardson, Inorg. Chem., 15, 1507 (1976).

17) G. D. R. Napier, J. D. Neilson, and T. M. Shepherd, J. Chem. Soc. Far. II, 71, 1487 (1975).

18) J. D. Neilson and T. M. Shepherd, J. Chem. Soc. Far. II, 72, 557 (1976).

19) H. G. Brittain and F. S. Richardson, J. Chem. Soc. Far. II, 73, 545 (1977).

20) O. Stern and M. Volmer, Phys. Z., 20, 183 (1919).

21) J. N. Demas, J. Chem. Ed., 53, 657 (1976).

22) H. Boaz and G. K. Rollefson, J. Am. Chem. Soc., 72, 3435 (1950).

23) E. J. Bowen and W. S. Metcalf, Proc. Royal Soc. London Ser. A, 206, 437 (1951).

24) J. Keizer, J. Am. Chem. Soc., 105, 1494 (1983).

25) W. DeW. Horrocks, Jr., and W. E. Collier, J. Am. Chem. Soc., 103, 2856 (1981).

26) H. Donato, Jr. and R. B. Martin, J. Am. Chem. Soc., 94, 4129 (1972).

27) J. F. Desreux and C. N. Reilley, J. Am. Chem. Soc., 98, 2105 (1976).

28) H. G. Brittain, Inorg. Chem., 17, 2762 (1978).

29) H. G. Brittain, J. Inorg. Nucl. Chem., 41, 561, 567 (1979).

30) H. G. Brittain and F. S. Richardson, Bioinorg. Chem., 7, 233 (1977).

31) H. G. Brittain, Inorg. Chem., 18, 1740 (1979).

32) H. G. Brittain and F. S. Richardson, Inorg. Chem., 15, 1507 (1976).

33) H. G. Brittain, J. Inorg. Nucl. Chem., 41, 721 (1979).

34) H. G. Brittain, J. Inorg. Nucl. Chem., 41, 1775 (1979).

35) H. G. Brittain, J. Lumin., 21, 43 (1979).

36) L. Spaulding and H. G. Brittain, J. Lumin., in press.

37) G. A. Elgavish and J. Reuben, J. Am. Chem. Soc., 98, 4755 (1976).

38) L. Spaulding and H. G. Brittain, Inorg. Chem., 22, 3486 (1983).

39) H. G. Brittain, Inorg. Chim. Acta, 70, 91 (1983).

40) W. DeW. Horrocks, Jr., B. Holmquist, and B. L. Vallee, Proc. Nat. Acad. Sci. USA, 72, 4764 (1975).

41) V. G. Berner, D. W. Darnall, and E. R. Birnbaum, Biochem. Biophys. Res. Comm., 66, 763 (1975).

42) P. M. Colman, J. N. Jansonius, and B. W. Matthews, J. Mol. Biol., 70, 701 (1972).

43) B. W. Matthews, L. H. Weaver, and W. R. Kester, J. Biol. Chem., 249, 8030 (1974).

44) W. DeW. Horrocks, Jr., M.-J. Rhee, A. P. Snyder, and D. R. Sudnick, J. Am. Chem. Soc., 102, 3650 (1980).

45) M.-J. Rhee, D. R. Sudnick, V. K. Arkle, and W. DeW. Horrocks, Jr., Biochem., 20, 3328 (1981).

46) A. P. Snyder, D. R. Sudnick, V. K. Arkle, and W. DeW. Horrocks, Jr., Biochem., 20, 3334 (1981).

47) E. Banks, Y. Ueba, and Y. Okamoto, Ann. NY Acad. Sci., 366, 356 (1981).

48) Y. Okamoto, Y. Ueba, N. F. Dzhanibekov, and E. Banks, Macromolecules, 14, 17 (1981).

MOISTURE-CROSSLINKABLE SILANE-GRAFTED POLYOLEFINS

Dan Munteanu

Chemical Research Institute, Plastic Research
Center, "Solventul" Laboratory
Str.Gàrii 25, R-1900 Timisoara, Romania

INTRODUCTION

Polyolefins, like other thermoplastic polymers, soften
and finally flow at elevated temperatures. The resistance to
thermo-mechanical deformation of polyolefins may be improved
by crosslinking, i.e. by converting the initially thermoplastic material into a thermosetting one. The radical crosslinking of polyolefins is very well studied. Peroxide crosslinking and, to a lesser extent, radiation crosslinking are the
two main methods of crosslinking industrially employed. Peroxide crosslinking, involving the use of organic peroxides
at elevated temperature, is used especially in the cable industry. Radiation crosslinking, involving the polymer treatment with high energy rays, usually accelerated electrons,
appears to be restricted to thin finished articles like films,
light wiring, heat-shrinkable sleevings and foams. Both
methods have production disadvantages in spite of their commercial acceptance.

In the last decade a new crosslinking method has been
developed. The process consists, basically, of two distinctly
separated steps: grafting and crosslinking. Polyolefins are
melt grafted with unsaturated hydrolyzable organosilanes.
After shaping, accomplished on line with grafting or coming
after that, the finished article is crosslinked in the presence of water. The new method avoids some technological disadvantages of the radiation and peroxide crosslinking and im-
proves certain properties of the crosslinked polyolefins.
Therefore, the silane crosslinking technology may be employed
in those areas where radical crosslinked polyolefins are used

already and in areas where they have either gained no entry
or made only a minor inroad.

This paper reviews the literature of polyolefin crosslinking by grafting hydrolyzable unsaturated organosilanes.
As in a previous review (1) an attempt was made to present
the state of the art and future developments of this new
crosslinking method.

CHEMISTRY OF THE SILANE CROSSLINKING METHOD

In the first step of the process, polyolefins (PO) are
grafted with polyfunctional organosilanes (SI) having the
general formula $R'Si(OR)_3$, where R' is a polymerizable group
and OR an easily hydrolyzable functionality. The melt grafting method is employed. This is a relatively simple process
which is accomplished in extruders or internal mixers, at
$140-240°C$, in the presence of a grafting initiator. The grafting of SI onto PO proceeds via a free radical mechanism. The
grafting initiator thermally decomposes to free radicals which
abstract hydrogen atoms from the PO backbone. The SI is then
grafted onto these active sites (figure 1.A).

The silane grafted polyolefins (PO-g-SI) are crosslinkable polymers. In the absence of moisture they are still thermoplastic and can be processed in the same way as normal, uncrosslinked PO. Therefore, crosslinking takes place separately

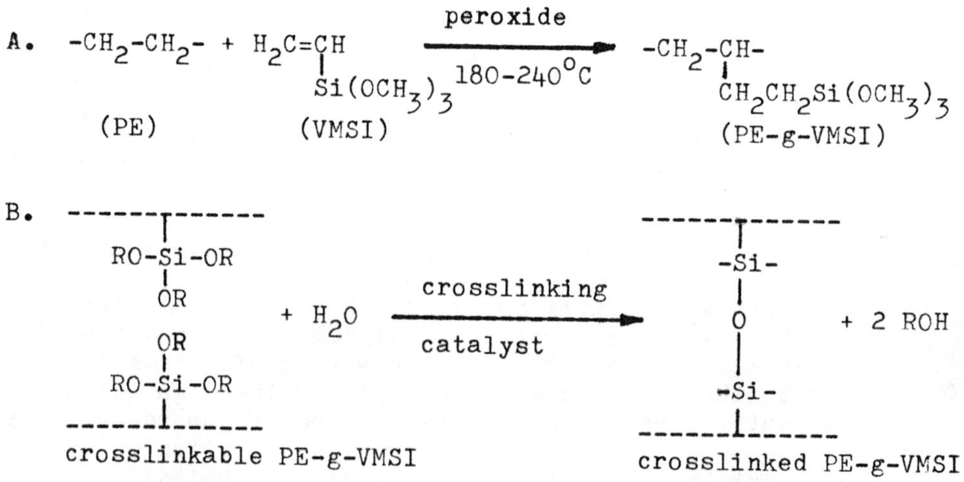

A. $-CH_2-CH_2-$ + $H_2C=CH$ $\xrightarrow[180-240°C]{\text{peroxide}}$ $-CH_2-CH-$
                                $|$ $|$
                          $Si(OCH_3)_3$ $CH_2CH_2Si(OCH_3)_3$

     (PE)   (VMSI)   (PE-g-VMSI)

B.   --------|--------   --------|--------
          $|$ $|$
       $RO-Si-OR$ $-Si-$
          $|$ $|$
          OR + H_2O $\xrightarrow[\text{catalyst}]{\text{crosslinking}}$ O + 2 ROH
                                                             $|$
          OR $-Si-$
          $|$ $|$
       $RO-Si-OR$
       --------|--------   --------|--------

     crosslinkable PE-g-VMSI   crosslinked PE-g-VMSI

Fig. 1. The chemistry of the grafting (A: vinyltrimethoxy-
        silane onto polyethylene) and crosslinking (B) steps.

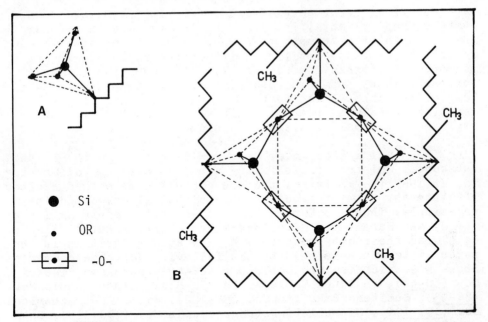

Fig. 2. Possible sterical configuration for polyfunctional
         bridge heads of uncrosslinked (A) and crosslinked
         (B) PO-g-SI (2).

from the grafting process and is accomplished on the shaped
finished article by its exposure to trace amount of water.
Crosslinking proceeds through hydrolysis of alkoxy silane
groups and rapid condensation of the resulting silanols.
Siloxane crosslinks between adjacent PO chains are formed.
The process takes place spontaneously in the presence of
moisture, but is strongly accelerated in the prsence of a
catalyst (figure 1.B).

     The structure of the crosslinked PO-g-SI is very diffe-
rent from that of the PO crosslinked by radical methods. Peroxide and radiation crosslinking involves the formation of a
network in which each crosslink point results from the coupling of two PO chains by C-C bonds. A chain of PO-g-SI is
capable of reacting with two or more similar chains to form
networks with siloxane crosslinks. Although the grafted SI
has three alkoxy groups present at each silicon atom, it is
not probable that all of them react. Based on kinetic and
thermodynamic considerations the most probable hypothesis
seems to be that, in average, only two of the three alkoxy
groups react. Consequently, four polymer chains are bound
in each crosslink point of the network by Si-O-Si bonds.
Voigt (2) postulates a possible structure of such "bunch-like"
crosslinks (figure 2).

The first reported grafting of PO with SI seems to be that claimed by Santelli (3). High density polyethylene was grafted with vinyltrimethoxysilane, in aromatic solvents, for several hours at the boiling point of the solution in the presence of dibenzoyl peroxide. Very low grafting yield was obtained, about 3-5%. Therefore, it is not surprising that PO grafting with SI was developed only when an efficient grafting technique, capable of industrial application, was employed - the melt grafting technique.

The novel method, i.e. melt grafting of PO with SI, and subsequent moisture crosslinking was developed by Dow Corning Co. in 1967 (4-7). Later, the new crosslinking method excited the interest of many researchers, so that it developed increasingly. Thus, in Europe B.I.C.C.-Etablissements Maillefer (8-10) and Kabel- und Metallwerke Guttehofnungshütte (11-20) developed their own processes. A very strong position in the patent literature is held by the following Japanese companies: Showa Electric Wire and Cable Co. (22-49), Furukawa Electric Co. (50-69), Fujikura Cable Works (70-83), Hitachi Cable Co. (84-93), Sumitomo Bakelite Co. (94-102), Mitsui Petrochemicals Co. (103-109), Mitsubishi Petrochemicals Co. (110-115), Seki-sui Chemicals Co. (116-120), Hitachi Chemicals Co. (121-125), Ube Industries (126-129), Dainichi-Nippon Cables Co. (130-132) and Tehnichi Densen K.K. (133-135). Unlike melt grafting, other methods for grafting of SI onto PO, i.e. high energy- and UV-ray- induced grafting and solution grafting, have gained no industrial application (136-139).

POLYOLEFIN GRAFTING WITH SILANES

Although many grafting procedures have been developed, all of them have the same characteristic features. The melt grafting of SI onto PO is usually accomplished in an extruder, but can also be performed in an internal mixer. The mixture of PO, SI and grafting initiator (GI) is passed through the extruder, usually a high shear mixing one, at 180-240°C, the residence time of the reaction mixture being 1-5 min. For the internal mixer grafting, lower temperature (140-180°C) and higher kneading time (5-30 min.) may be used. The antioxidant (AX) and crosslinking catalyst (XLC) may be added before or after the grafting step. Various grafting receptures have been used, with different types and proportions of the individual components: PO, SI, GI and AX.

Polyolefins and Silanes

As backbone for the grafting of SI, homo- and copolymers of the olefins have been used: low-, medium- and high-density polyethylene (LDPE, MDPE and HDPE), polypropylene (PP), ethylene-propylene blockcopolymers (EP), ethylene-propylene (EPM) and ethylene-propylene-unconjugated diene (EPDM) elastomers, ethylene-vinyl acetate (EVA) and other ethylene copolymers.

The unsaturated hydrolyzable organosilanes $R'Si(OR)_3$, employed to graft PO, belong to the polyfunctional silanes used as coupling agents and adhesion promoters. The polymerizable group R' is vinyl or methacryloyloxy, and the hydrolyzable functionality OR is methoxy or ethoxy. The following trialkoxysilanes are used:

(VMSI) vinyltrimethoxysilane $\quad H_2C=CH-Si(OCH_3)_3$

(VESI) vinyltriethoxysilane $\quad H_2C=CH-Si(OC_2H_5)_3$

(MMSI) 3-methacryloyloxypropyl-trimethoxysilane $\quad H_2C=\underset{\underset{CH_3}{|}}{C}-COOCH_2CH_2CH_2Si(OCH_3)_3$

They are colorless liquids with boiling points higher than $120^\circ C$ (table 1).

PO are usually grafted with VMSI. The vast majority of the patents claim the grafting of VMSI onto LDPE. Other PO grafted with VMSI are: HDPE (36,46,48,81,82,106,113,120,124, 125), MDPE (91), ethylene oligomers (103,105), PP (73,122), 1-hexene-propylene copolymers (115), EVA copolymers (35,97, 99,104,119,131), acrylic acid zinc salt-ethylene copolymer (68), EPM and EPDM elastomers (22,65,66,92,101,108). Polyolefin mixtures have been also grafted with VMSI, e.g. 1:1 LDPE:HDPE (57), 9:1 EPM:PP (19), 1:1 EPM:EVA (88) and 1:1 EPDM:PIB (34).

Among the other SI, VESI was grafted onto LDPE (54,118), HDPE (117), EVA copolymers (95), vinyl chloride grafted EVA copolymers (94), EPDM elastomers (76,77), and, MMSI was grafted onto PE (93) and EPM elastomers (43,59). Sometimes PO are

Table 1. Some Physical Properties of Silane Monomers

Silane Type	Molecular Weight	Boiling Point at 760 mm Hg ($^\circ$C)	Specific Gravity
VMSI	148.3	123.0	0.953
VESI	190.3	160.5	0.894
MMSI	248.1	255.0	1.044

grafted with mixtures SI + comonomer, e.g. grafting of EPM
elastomers with MMSI + glycidyl methacrylate (126,129), MMSI
+ calcium or zinc acrylate (127,128), and PE grafting with
MMSI + stearyl acrylate (50).

The proportion of the individual components in the grafting receptures is expressed as parts weight (p.w.) per 100
p.w. PO. Although 0.5-4 p.w. SI is usually added to PO, in
most instances the grafting receptures contain about 2 p.w.
SI. At this level, a high crosslinking degree of PO-g-SI may
be achieved. Sometimes PO are grafted with higher SI levels,
usually up to 10 p.w., but even a proportion of 30 p.w. SI
was used (108).

Practically, no literature is available for the influence
of PO and SI type on the grafting process. Wouters and Woods
(140) studied the effects of several structural and molecular
parameters of the EPM and EPDM elastomers on the gel content
of the crosslinked silane grafted elastomers. The crosslinking
potential of the EPM elastomers with narrow molecular weight
distribution (MWD) decreases with the increasing of the propylene segments content (figure 3,A). The free radicals abstract hydrogen easier from tertiary carbon atoms than from
secondary ones, but the reactivity of the resulted macroradicals is in the reverse order. Therefore, the tendency of
the EPM chains to undergo degradation increases as the ethyl-
ene/propylene ratio decreases, because in propylene segments
the tendency of the relatively stable macroradicals to be
stabilized through a graft or crosslinking mechanism is lower.
The medium molecular weight EPM elastomers with broader MWD
showed no response in their crosslinking potential, to the
fraction of propylene units in the chain (figure 3,B). The
low molecular weight EPM elastomers with very broad MWD have
a lower crosslinking potential (figure 3,C). They contain a
large proportion of oligomers which make very little contribution to the formation of crosslinked networks. The crosslinking potential of the EPDM elastomers is much lower than
that of the EPD elastomers (table 2). This is a most surprising result because EPDM elastomers with such type of unsaturation are easier to be peroxide crosslinked than saturated
EPM elastomers with otherwise similar characteristics. No explanation was given for this behaviour.

Concerning the influence of the SI type on the crosslinking of PO, only   patent information was found. According
to that, grafted EP blockcopolymers (2% ethylene units content)
have higher crosslinking degree if MMSI is used instead of
VMSI (59).

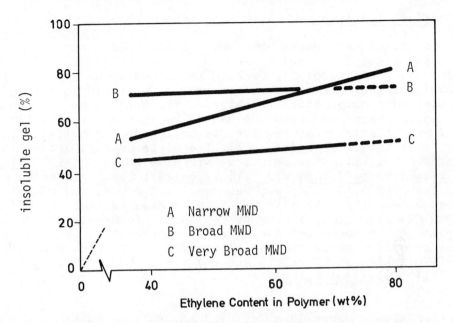

Fig. 3. Effect of ethylene content on the moisture crosslinking of EPM elastomers (140).
Grafting recepture: 100 p.w. EPM elastomer + 3 p.w.
VMSI + 0.192 p.w. dibenzoyl peroxide. Internal mixer
grafting: closed Brabender mixing chamber of 30 cc.,
kneading for 4 min. at 140°C, 0.1 p.w. dibutyltin
dilaurate added after the 3rd min. Crosslinking:
thin sheets immersed in boiling water for 2 hours.

Table 2. Effect of Elastomer Unsaturation on Gel Content of
Moisture-Crosslinked VMSI-Grafted EP Elastomers
(140)

Elastomer Type	EPM	EPDM	EPM	EPDM	EPM	EPDM	EPM	EPDM
MWD		---broad---				---medium---		
Ethylene, %	64	63	44	45	66	66	43	43
Unsaturation, %ENB	0	3.7	0	3.2	0	3.9	0	3.5
Insoluble Gel, %	73	47	73	56	66	48	59	27

All elastomers have medium molecular weight, in Mooney viscosity range 40±3ML(1+8) at 127°C. ENB - ethylidene norbornene
unit. Grafting conditions the same as in figure 3.

Grafting Initiators

The GI decomposes into free radicals which abstract hydrogen atoms from the PO backbone to create the active sites for the grafting of SI. Organic peroxides which may easily abstract hydrogen atoms and thermally decompose in a time interval shorter than the material residence time are employed. Therefore, peroxide has to be selected as a function of grafting temperature, i.e. depending on its half-lifetime at this temperature (table 3). The GI is usually dicumyl peroxide (I), but other peroxides are also employed: (II) (103, 121), (IV) (123,125) and (VI) (126-129). The GI is added at a level of 0.05-0.5 p.w., but usually 0.2 p.w. The use of an excess of SI relative to GI normally assures that SI grafting prevails over crosslinking by C-C bonds, formed by direct coupling of PO macroradicals. For the same reason, use was made of peroxides which are not characteristic for the radical crosslinking of PO: (III) and (V) (11-16,57), (VII) (94) and (VIII) (63,115,140,141).

Table 3.   Peroxides Employed as Grafting Initiators

Peroxide Type	T (°C)
Dialkyl Peroxides ($>$C-O-O-C$<$)	
(I) dicumyl peroxide $C_6H_5(CH_3)_2$C-O-O-C$(CH_3)_2C_6H_5$	180
(II) di-t-butyl peroxide $(H_3C)_3$C-O-O-C$(CH_3)_3$	190
(III) bis(t-butylperoxyisopropyl)benzene	190
$(H_3C)_3$C-O-O-C$(CH_3)_2C_6H_4(CH_3)_2$C-O-O-C$(CH_3)_3$	
Alkyl Peresters ($>$C-O-O-CO-C$<$)	
(IV) t-butyl per-2-ethylhexanoate	130
$(H_3C)_3$C-O-O-OC-CH$(C_2H_5)CH_2CH_2CH_2CH_3$	
(V) t-butyl perisononanoate	160
$(H_3C)_3$C-O-O-OC-CH$_2$CH(CH$_3$)CH$_2$C(CH$_3)_3$	
(VI) t-butyl perbenzoate $(H_3C)_3$C-O-O-OC-C$_6H_5$	170
Diacyl Peroxides (—CO-O-O-OC—)	
(VII) dilauroyl peroxide H$_3$C(CH$_2)_{10}$CO-O-O-OC(CH$_2)_{10}$CH$_3$	120
(VIII) dibenzoyl peroxide H$_5C_6$CO-O-O-OCC$_6H_5$	130

T - temperature for a half-lifetime of 1 min.

Antioxidants

For stabilization during processing and the service life of the finished articles, AX are incorporated into the polymer, before or after the grafting step. Typical AX for PO are usually employed: (IX) 2,6-di-tert-butyl-4-methylphenol (65,67), (X) 4,4'-thiobis(3-methyl-6-tert-butylphenol) (31-33, 41,53,58,113), (XI) 4,4'-thiobis(6-tert-butyl-hydroxybenzyl) (57), (XII) pentaerythritol tetrakis[3-(3,5-di-tert-butyl-4-hydroxyphenyl)propionate] and (XIII) di[3-(3,5-di-tert-butyl-4-hydroxyphenyl)propionate]thioethyleneglycol (56,60-62,69,126-129) (figure 4).

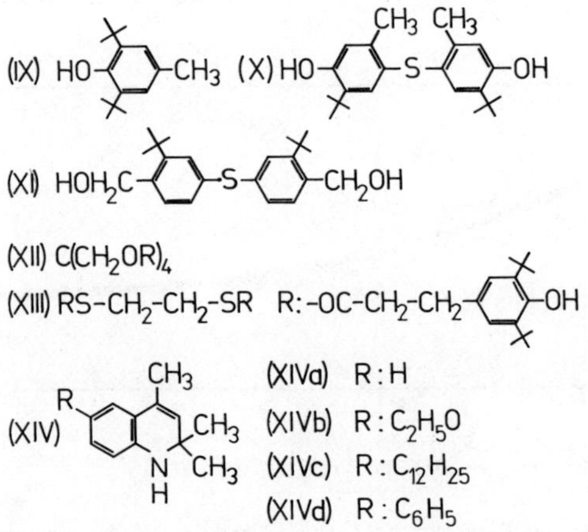

Fig. 4. Antioxidants for the stabilization of PO-g-SI.

The low molecular weight compounds, usually employed, are not permanent AX, their concentration in the polymer decreasing during processing and long-term use of the polymer. The poor compatibility, non-uniform distribution, volatility and extractibility (leaching) of the AX are the main factors responsable for the physical loss of AX from polymers. The permanence of AX in polymers may be improved using high molecular weight AX. Thus, polymerized 2,2,4-trimethyl-1,2-dihydroquinoline (XIVa) was used as "polymeric antioxidant" for PO-g-SI (24,25,61,84-86,9091). A better improvement of the AX permanence may be achieved with antioxidants chemically bound to the polymer, for example by grafting. The same "monomeric antioxidant" (XIVa) and its 6-substituted derivatives (XIV b-d) have been used. They are added before the grafting step, in the mixture PO + SI + GI. In the conditions of the melt grafting, i.e. high temperature and the presence of free radicals, the reactivity of the C=C double bond of the heterocycle assures antioxidant grafting onto PO chains (9-11,13,15,16).

The level of AX is 0.05-1 p.w. and depends on the end use of the PO-g-SI. For many commercial applications involving heat resistance, cable insulation being the typical one, the level of AX has to be high, usually 0.5 p.w. Therefore, it is necessary to incorporate extra AX into the polymer, because most commercial PO grades contain very small additions of AX, usually below 0.1 p.w. If the AX is added before the grafting step, it has an obvious effect on the grafting of SI, because the reaction is initiated by free radicals. The crosslinking degree decreases with AX addition, due to

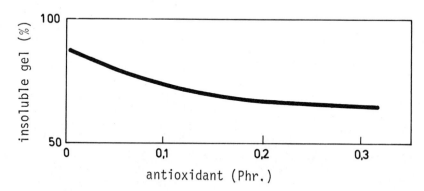

Fig. 5.   Effect of antioxidant addition on crosslinking
          degree (140).
          Antioxidant: octadecyl-3,5-di-tert-butyl-4-hydro-
          cinnamate. Grafting conditions the same as in
          figure 3.

the decreasing of the grafting yield, but this appears to be levelling off at higher additions (figure 5). In practice, therefore, adjustments made to the GI level to compensate for these inhibitive effects can assure high crosslinking degree of the stabilized PO-g-SI.

MOISTURE-CROSSLINKING OF THE SILANE-GRAFTED POLYOLEFINS

The thermoplastic PO-g-SI are crosslinked in the presence of moisture. Water causes hydrolysis of alkoxysilane groups and rapid condensation of the resulting silanols, to form siloxane crosslinks (figure 1,B). Although the process takes place spontaneously in the prsence of moisture, the unaided condensation reaction is relatively slow, so that a small addition of a suitable polymer-soluble Lewis acid is necessary to make the crosslinking rate more attractive in practical terms. Dibutyltin dilaurate (DBTDL) is, practically, the only catalyst employed. The crosslinking process is strongly accelerated in the presence of DBTDL, its concentration being, usually, 0.05-0.15 p.w. The catalyst accelerates both the hydrolysis of alkoxysilane groups to silanols and their condensation to siloxane bonds, but no quantitative data were reported about the kinetics of these two reactions. The crosslinking degree is evaluated by determination of the insoluble gel content, e.g. Soxhlet extraction in boiling xylene for 20 hours.

The crosslinking rate depends on the rate of diffusion of water into the PO. Therefore, the crosslinking temperature, PO crystallinity and the article thickness are the most important factors which determine the time for a full crosslinking, i.e. over 70% gel content. Basically, crosslinking at room temperature is possible. Because the saturation concentration of water in PO is very low (30-50 ppm H_2O in PE, at $20^\circ C$), the quantity of water consumed for hydrolysis has to be compensated by diffusion. High gel contents are obtained only at high temperature, over $80^\circ C$ (figure 6). Full crosslinking at room temperature requires several days, or even weeks, as water or atmospheric moisture permeates into the mass of the finished article (figure 7). Therefore, for industrial purposes, the manufactured articles are crosslinked by immersion in hot water at $90^\circ C$, or by low-pressure steam treatment, for 3-6 hours. The time for full crosslinking increases with the thickness of the manufactured article and the crystallinity degree of the PO (figure 7, table 4).

The moisture crosslinking takes place separately from the grafting process and is accomplished on the shaped finished article. Therefore, many differences appear in regard

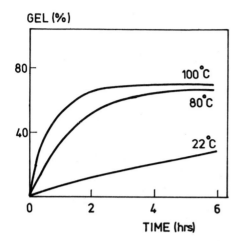

Fig. 6. Crosslinking rate
variation with temp-
erature for LPDE-g-SI
(5).

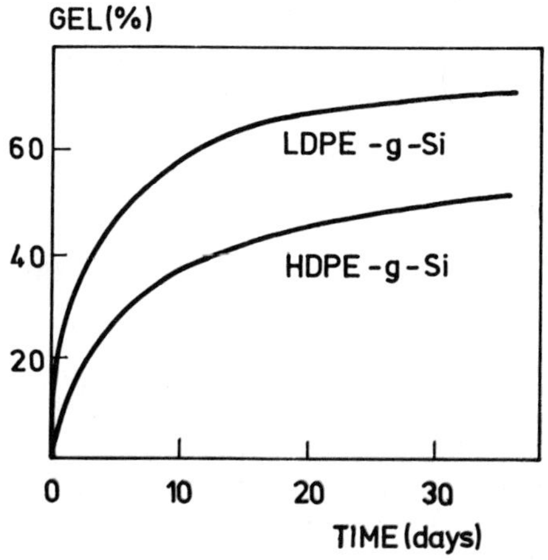

Fig. 7. Crosslinking rate at
room temperature for
LPDE-g-SI and HPDE-g-SI
(1.125 in. thick com-
pression molded plates
immersed in water at
22°C) (4).

to thermal methods of crosslinking. Peroxide crosslinking is
carried out above the crystalline melting point of the PO,
so that its crystallinity decreases due to the constraint
imposed by the crosslinks on subsequent crystallization. PO
with high melting points. like HDPE and PP, are processed sat-
isfactorily only above 160°C. At such temperatures the usual
peroxides for.crosslinking are not sufficiently stable to be
incorporated in the PO melt.

Table 4.   Crosslinking Time of PE-g-SI (5)

Crosslinking Time (in water, at 100°C)	Plate Thickness (in)	
	LDPE-g-SI	HDPE-g-SI
1 sec.	0.0011	0.0006
1 min.	0.0084	0.0049
1 hour	0.0650	0.0380
24 hours	0.3190	0.1860

In comparison with peroxide crosslinking, the moisture crosslinking of PO-g-SI becomes a separate and not a rate-controlling process with regard to the polymer processing, e.g. by extrusion. There is no need to maintain a narrow interval of the processing temperature by the polymer shaping in finished articles. Although crosslinking conditions are determined by the type and crystallinity of the PO, and the thickness of the finished article, in most instances the crosslinking temperature is below the melting point of the polymer. Consequently, the following are the advantages: (i) difficulty in maintaining dimensional accuracy during cross-linking is avoided, (ii) the density of the crosslinked PO is normally the same as the base polymer, and (iii) PO with high densities can be crosslinked. Of course, because the rate of moisture crosslinking is slower than peroxide cross-linking, the crosslinknig time for thick articles is long and this is a limitation of the process.

PROPERTIES OF CROSSLINKED SILANE-GRAFTED POLYOLEFINS

The crosslinked PO-g-SI contain Si-O-Si crosslinks between adjacent PO chains. Therefore, these networks will be expected to behave differently from the networks with C-C crosslinks formed by peroxide and radiation crosslinking methods. The specific structure of the network (figure 2) gives better thermo-mechanical properties than peroxide cross-linked PO. The dependence of hot elongation, deformation under mechanized load under isothermal conditions, and swelling value on the gel content of the crosslinked PE emphasizes the better behaviour of the crosslinked PE-g-SI (figure 8). In comparison with peroxide crosslinked PE, for samples with the same gel value, all these properties have lower values, indicating better resistance to deformation and solvents. Consequently, the same value of a certain property, for example hot elongation, is obtained with a lower gel content. The residual strength above the crystalline melting point is sufficient to maintain geometrical shape at temperatures that melt objects made from uncrosslinked polymer (5).

Other properties, including electrical properties, are as good as, if no better than, peroxide crosslinked PO. Resistance to heat aging and UV radiation seems to be better. Some of these properties are presented in tables 5 and 6. Other specific properties, like environmental stress cracking and adhesion will be discussed in connection with the applications of the PO-g-SI.

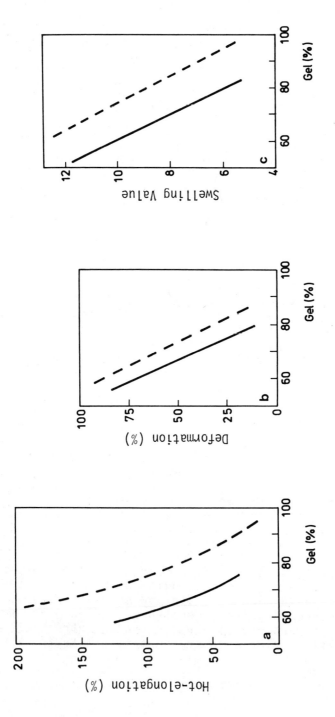

Fig. 8. Plots of (A) hot elongation at 150°C, (B) deformation under mechanical load
at 150°C, and (C) swelling value, versus insoluble gel content, for peroxide
crosslinked PE (------) and crosslinked PE-g-SI (——) (2).

Table 5. Typical Properties of Sioplas E Polyethylenes (4)

Property	Unit	E231	E651
Density	g/ml	0.932	0.967
Melt flow index	g/10 min.	1	1
Tensile yield			
at room temperature	MN/m^2	14.3	30.1
at 60°C	MN/m^2	6.7	20.8
Falling-weight impact strength	J	no break	9.7
Shore hardness	D scale	54	71
Vicat softening point	°C	94.5	128.5
Water absorbtion at 100°C	%	0.14	0.05
Permittivity at 10^3-10^6 Hz	-	2.32	2.36
Loss tangent			
at 10^3 Hz	-	0.0005	0.0005
at 10^6 Hz	-	0.0002	0.0002

Sioplas E: trade mark of Dow Corning Corp. for LDPE-g-SI (E231) and HDPE-g-SI (E651).

Table 6. Aging Properties of Peroxide and Silane Cross-linked Polyethylenes (5)

Aging temperature (°C)	Aging time (weeks)	Peroxide XLPE T(%)	Peroxide XLPE E(%)	Sioplas XLPE T(%)	Sioplas XLPE E(%)
135	1	102	100	90	93
135	2	81	93	100	85
135	4	85	90	86	85
135	8	75	80	99	90
135	12	57	51	92	86
175	1	38	11	90	95
175	2	33	9	100	100
175	3	discontinued		101	90
175	4			87	86
175	5			87	83

T(%) and E(%) - Percentage retention of tensile strength (TS) and elongation at break (EB). Peroxide XLPE: TS=15.5 MN/m^2 and EB=550%. Sioplas XLPE: TS=15.0 MN/m^2 and EB=300%.

TECHNOLOGICAL PROCESSES

In all the applications of the silane crosslinking technology the crosslinking is normally done in a separate step,
so that differences appear only by the grafting and shaping
steps. Production methods have been developed, initially for
wire and cable insulation. There are three basic technological processes: one- and two-step extrusion processes and internal mixer processes, each of them with several variants.

Two-Step Extrusion Processes

Dow Corning was the first company which developed a technological process for the grafting of LDPE and HDPE with SI,
and subsequent moisture crosslinking of the grafted PE. The
basic two-step extrusion process is known commercially as
the Sioplas E technology (4,5). Sioplas E polyethylene is
designed as a two-component system to eliminate rapid crosslinking of the grafted component by trace amounts of moisture
in the presence of the catalyst (figure 9).

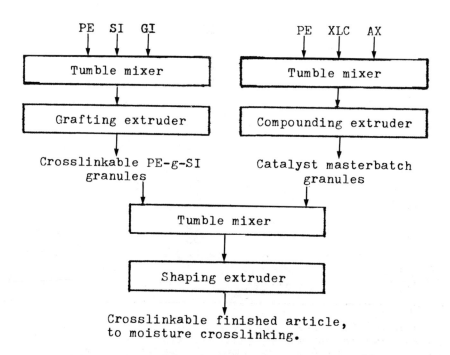

Fig. 9.   Flow diagram for a basic two-step
          extrusion process.

In the first step, PE-g-SI and a catalyst masterbatch are obtained. The grafted PE is prepared by passing the mixture PE + SI + GI, obtained by tumble mixing, through the grafting extruder, under conditions that assure SI grafting onto PE. Because tumble mixing is not a very efficient process, high shear compounding extruders or Ko-kneaders are necessary. Except the PE grades for cables, no AX is incorporated into the mixture. To maintain low atmospheric concentrations of SI (<10 ppm), a very good extract ventilation at the hopper end of the extruder is necessary. The palletized crosslinkable grafted polymer has to be carefully stored away from moisture to avoid crosslinking before its further shaping. Under conditions of careful dry storage, shelf-life of three months is possible with no deterioration in processability. The second component is obtained by normal mixing and compounding, to dispense the XLC and AX through the same PO, i.e. PE. The masterbatch is also collected as dry pellets.

In the second step, the two components, i.e. grafted PE and masterbatch, are tumble mixed, then compounded in an extruder with a die which will shape the compound into the finished article, e.g. cable insulation or pipe. The types of shaping extruders are those normally used for PE, i.e. extruders fitted with screws suitable for processing PE: L:D of at least 20:1 and preferably 25:1, compression ratio of about 3:1. Temperature profile and extrusion rate are the same as for the base polymer.

Modifications of this basic two-step extrusion process are also successfully used on a commercial scale:
- discontinuous or continuous metering and mixing of ingredients into the extruder hopper, provided with a stirrer,
- direct injection of a   solution, containing the GI and, sometimes, the AX, either in the base of the feeding hopper or in the extruder barrel, into the melted polymer.
Such modifications increase automation and also decrease atmospheric contamination with SI.

Kabel-Metal Co. (11-20) developed an essential modification of the basic two-step process. The first step involves dispensing of all additives into the solid PO, as a distinct and separate operation, carried out in a high speed mixer. The PO pellets or powder are stirred at high speed (500-3000 pm) so that their temperature increases. After reaching a temperature of 80°C, a SI solution containing GI + AX + XLC is added. The mixture is stirred 3-20 min. at 80-100°C, i.e. at a temperature below the softening point of the PO, so that all ingredients penetrate into the solid polymer, which retains its geometrical shape. The second step involves grafting and shaping as one operation within the same extruder.

The dry PO granules or powder, containing all the ingredients, are passed through the extruder under conditions suitable for the reaction to take place. The melt of PO-g-SI is then extruded directly to obtain the finished article, ready for crosslinking. Because additives are incorporated into the PO, there is no need for a high shear mixing extruder. Single screw extruders with L:D=25:1 are employed. Void-free articles are obtained by degassing the PO-g-SI melt, to eliminate volatile compounds: ungrafted SI, included air and decomposition products of the PO. From the grafting extruder (D=150mm and L:D=25:1) the melt of PO-g-SI is passed directly into the shaping extruder (D=150mm and L:D=15:1), to obtain the finished article, e.g. extrusion onto electrical wire and cable. The shaping extruder is provided with a vacuum pump, at the connection point (L:D=3:1) with the grafting extruder. This system allows the production of cable insulations for use at voltages higher than 10 KV (13,15,17).

One-Step Extrusion Processes

The B.I.C.C. and Maillefer companies (8-10)developed equipment that is capable of carrying out the grafting and extrusion onto the conductor in the same extruder. This one-step extrusion process is known commercially as the Monosil process. The extruder is loaded with PO and all other ingredients: SI, GI, AX and XLC. The reaction mixture may be obtained with the same methods as those employed by the two-step extrusion processes. In the first part of the extruder the grafting reaction takes place. In the second one, kneading and extrusion of the PO-g-SI melt are carried out. Therefore, a high-shear mixing extruder is usually employed.

Table 7. Physical Properties of Crosslinked Polyethylenes (8)

Physical Property	Unit	XL.PE-g-SI (Monosil)	XL.PE-g-SI (Sioplas)	Peroxide XLPE
Tensile Yield	MN/m^2	17-19	11-16	16-18
Elongation at Break	%	360	255-350	300
Crosslinking Degree	%	73-76	65-72	75-77
Hot Elongation at 200°C under Load of 0.2 MN/m^2	%	55	60-100	70-160
Remanent Deformation after Hot Elongation	%	0	2.5	0-5

In comparison with the two-step processes the following are the advantages:
- there is no danger of moisture crosslinking, because the grafted PO is not stored before shaping,
- the investment costs are only about 20%, as great
- the space requirement for the production unit is only about 40% as great.
Table 7 presents some properties of crosslinked PE-g-SI, obtained by one-step (Monosil) and two-step (Sioplas) processes, in comparison with peroxide crosslinked PE.

Internal Mixer Processes

Internal mixers can be also employed to graft PO with SI. The SI should have a boiling point above the processing temperature to avoid its evaporation. PO pellets are tumble blended with a SI solution of GI, before mixer loading. The loading of the mixer with PO and a PO masterbatch containing SI + GI is also possible. The XLC and AX are added pure or in the form of masterbatch, usually at the end of the mixing cycle. In comparison with the extrusion processes, the internal mixer grafting is especially suitable for:
- direct grafting of bale elastomers,
- grafting of high SI levels, e.g. 5-30 p.w.,
- compounding of PO-g-SI with other ingredients (polymers, resins, waxes, fillers, flame retardants) in the same mixing cycle.
For the internal mixer grafting, lower temperature (140-180°C) and higher reaction time (5-30 min.) may be used than for the extruder grafting (50,103,104,131).

Moisture Crosslinking of the Finished Articles

The crosslinking of the shaped finished articles in the presence of moisture is accomplished separately from the two preceding steps, i.e. grafting and shaping. The shaped article usually contain the XLC (DBTDL) incorporated into the crosslinkable PO-g-SI. The crosslinking is conveniently carried out in a batchwise process, where the finished article is exposed to either hot water or, usually, to low-pressure steam (about 1 bar). Therefore, moisture crosslinking is much more energy efficient than conventional crosslinkable methods, e.g. peroxide crosslinking requires much more thermal energy which is supplied by high pressure steam (16-18 bar). For cable covering, because of the low rate of moisture crosslinking, it is not possible to carry out the crosslinking in line with normal insulation thickness and line speed. Although crosslinking begins, theoretically, in the cooling bath of the extrusion line, the whole drum of insulated cable has to be exposed to water.

Separating the crosslinking step means that extrusion
lines can be run at maximum output with no restriction in
terms of in-line crosslinking. In comparison with peroxide
crosslinking, where residence time in the crosslinking tube
is usually the rate-controlling step, this can mean an increase in output of up to several times. For other processing
techniques too, separating the crosslinking step allows a
greater degree of freedom. For example, extrusion blow molding and injection molding are processing techniques that
have received little successful attention from peroxide crosslinked PO, because they require a thermal crosslinking cycle
in the mold. The PO-g-SI can be processed as are the base PO
and then crosslinked outside the mold. Thus, the productivity
of the processing machine is not retarded by the need to
crosslink the material.

Some modifications of the basic crosslinking processes
are described in the patent literature. Cable insulations
with higher smoothness are obtained by immersion in a xylene
solution of DBTDL, at 50-60°C (25,58). Water may be injected
in the barrel of the shaping extruder, by the processing of
a two-component system: PE-g-VMSI + PE masterbatch of DBTDL
(23). The crosslinking rate of PE-g-VMSI may be accelerated
by compounding with a compatible hydrophylic polymer, e.g.
0.5 p.w. Surlyn A (ethylene ionomer) (112). Plates extruded
from LDPE-g-VMSI are crosslinked at temperature over 100°C,
by immersion in a mixture of 9:1 ethyleneglycol:water (102).
A compound of PE-g-VMSI and 0.1% DBTDL "microencapsulated"
in a styrene-divinylbenzene copolymer show no crosslinking
in one month storage at room temperature (55).

Advantages and Limitations of the Silane Crosslinking Technology

As stated before, in comparison with conventional crosslinking methods, the silane crosslinking method has many advantages with regard to both the properties of crosslinked
polymers and the crosslinking technology. Most of these advantages were presented in the former chapters and are the
consequence of separating the crosslinking step from the
shaping one. PO with a wide range of melt flow indices and
densities can be employed, because crosslinking can take
place below the polymer melting point. With minor modifications, suitable thermoplastic extruders can be adapted for
grafting. The grafting process is relatively easy to accom-
plish; it is reproducible and requires low manpower invest-
ments, since it is usually a highly automated process. The
process is very flexible because the grafting may be accomplished separately from the other operations, or may be made
in the same extruder which produces the finished articles.

The productivity of the processing machine and the processing conditions for the crosslinkable PO-g-SI are, basically, the same as for the parent PO.

The silane crosslinking technology has, of course, certain limitations. The main disadvantage is the extreme sensitivity to water, after XLChas been added into the crosslinkable PO-g-SI. Therefore, the catalysed PO-g-SI must be immediately shaped into finished articles, so that the XLC must be added, preferably, during the last step before the shaped articles are moisture crosslinked, e.g. incorporated in the shaping extruder. All the ingredients to be incorporated must be dry, in order to avoid crosslinking during processing. This requirement can cause problems, especially when fillers are added for reinforcement or dilution. Although PO-g-SI without XLC can be stored for several months in reasonably dry conditions, scraps from the finished compounds are not normally re-processable.

ACTUAL AND POTENTIAL APPLICATIONS

The silane crosslinking technology is applied in those areas where peroxide and radiation crosslinked PO are already used, but may be also applied in areas where these conventional crosslinking procedures have either gained no entry, or made only a minor inroad. Besides the very good properties of the crosslinked PO-g-SI, the technological advantages of the new crosslinking technology are decisive in expanding application fields. Developed, initially, for cable covering with PE-g-SI, the new crosslinking technology was later applied to other PO and processing techniques.

Cable insulations

Low-medium voltage wire and cable insulations with SI grafted LDPE and HDPE is the actual main area of commercial application for the moisture-crosslinking technology. The Sioplas E two-step extrusion process of Dow Corning Co. has been offered for license since 1972 and has already been accepted in Europe and Japan (5). Later, the Monosil one-step extrusion process was developed. Other companies tailor individual versions of the product for their own markets. Some information can be found in the patent literature. The main purpose of the researchers is to improve both the grafting and crosslinking technology, and the properties of insulation.

The melt flow index of the backbone PO decreases by grafting due to the sterical hinderance imposed by the grafted silane groups. Therefore, some efforts have been made to improve the processability of PO-g-SI. Thus, higher insulation

speed is possible without melt fracture, and insulation with
better gloss and smoothness. For such purposes
the grafting recepture contains lubricants, specific antioxidants, polymerization inhibitors or other additives (31,
34,56,60,61,69,79,90,118,133,134). The same goal is achieved
by controlling the MWD of LDPE (72) or the particle size
distribution of HDPE powder (124).

    Insulations with better properties and lower cost are
obtained by compounding PO-g-SI with certain fillers, added
before or after the grafting step, at a level of 30-100 p.w.
Carbon black, clay, talc powder and hydrated alumina are employed (27,34,53,76,77,97-99). Other improvements may be
achieved by grafting PO mixtures or by compounding PO-g-SI
with certain polymers. Thus, the stress cracking resistance
of a cable insulation (in oil, at 50°C) is better for a VMSI
grafted 1:1 HDPE:LDPE mixture than for HDPE-g-VMSI (57). The
heat distorsion of a cable covering made from a mixture of
20:1 LDPE-g-VMSI:PP is lower than that of a similar covering
made in absence of PP (62). Fire-resistant, self-extinqui-
shing wire insulations with non-drip properties are obtained
by compounding PE with PO-g-SI (EVA-g-VMSI, EPDM-g-VMSI),
inorganic fillers (clay, hydrated alumina), inorganic and
organic flame retardants (Sb_2O_3, chlorinated paraffin, deca-
bromodiphenylether) (97-99,161).

    In most instances cable insulations are made from VMSI
grafted LDPE and HDPE. Other PO than PE have been grafted
with VMSI, e.g. EPM elastomers (65,92), EPDM terpolymers
(34,101), EVA copolymers (97-99,131) and acrylic acid zinc
salt-ethylene copolymer (68). Other SI than VMSI are rarely
employed, e.g. VESI grafting onto PE (54) and EPDM elastomers (76,77), and MMSI grafting onto PE (93) and EP blockcopolymers (59). The silane crosslinking technology may be applied to obtain many types of wire and cable insulations,
e.g. low-medium voltage cable insulations, auto ignition
cables, nuclear cable insulations and mining cable insulation or submersible insulated wires.

Films

    Crosslinkable films and tapes are used especially in
electrical and packaging applications (26,35,40,42,48,49,70,
111,120,121,123,125,131). In most instances PE-g-SI is employed. Because the content of fisheyes represents a problem
of the film area, efforts have been made to reduce it by op-
timization of the grafting receptures and conditions (108,
111,125). Crosslinkable PE-g-VMSI tapes are wound onto electrical cables to give, after crosslinking, a heat-resistant
insulation (26,42). Films of EVA-g-VMSI may adhere onto cables

at lower temperature than PE-g-VMSI films (35). Heat-shrinkable tubiform films adhere well on metallic and non-metallic
surfaces, to protect them against the environment (40,70).
Non-woven fabrics made from crosslinked HDPE-g-VMSI are heat-
and oil-resistant, and have a very good breakdown resistance
and dimensional stability. Therefore, they are useful as e-
lectric insulating paper substitutes, e.g. for oil-filled
cables in condensers and transformers (46,48,54).

Plates and Sheets

Crosslinked plates with 65-75% gel content are employed
especially for electrical applications. They are obtained by
extrusion and press molding. The following PO-g-SI are used:
LDPE-g-VMSI (25,33,102), PE-g-MMSI (50), PP-g-VMSI (122),
EPM-g-VMSI (22) and (EVA-g-VC)-g-VESI (94). Because of the
very good electrical properties, for example high treeing
resistance and breakdown voltage (35-40 KV/mm) (22,50), such
crosslinked sheets are employed as electric insulator sheets.

Laminates

In comparison with ungrafted PO, the PO-g-SI have very
good adhesion properties. Therefore, various types of laminates are obtained by press molding and extrusion of PO-g-SI
onto the following supports: paper, polymeric and metallic
films and sheets. After cooling, the laminate is immersed in
hot water or treated with low-pressure steam, for the crosslinking of the PO-g-SI layer. The high bonding strength in
these laminates assures their use even at elevated temperatures or in the presence of organic solvents.

Different types of laminated paper are obtained by la-
minating electrical condenser- and kraft-paper, with PO-g-SI
(10,36,37,39,43-45,73,81,82). PE-g-SI is usually employed,
but other PO-g-SI may be also used, e.g.PP-g-VMSI (73), or
compounds PP + HDPE-g-VMSI (81,82). Such laminates have very
useful properties, i.e. high peel strength, heat-resistance,
resistance to water and oil, good electrical insulating properties. Therefore, they are usually employed as insulating
papers.

The PO-g-SI are used as intermediate layer in metal/
metal and metal/ungrafted PO laminates (88,126-129). In alu-
minium/PE or PP, and aluminium/stainless steel laminates
(plates, foils) the adhesive layer consists of the following
PO-g-SI: EP blockcopolymers grafted with a mixture of MMSI
and glycidylmethacrylate, calcium or zinc acrylate (126-129),
VMSI grafted 1:1 EPM:EVA mixture (88). Steel pipes are protected by extrusion of HDPE-g-VMSI as the outside layer of
the laminate (113).

Polymer films and sheets with good surface hardness and
gloss are obtained by their coating with PO-g-SI, e.g. sheets
of 1-hexadecene:4-methyl-1-pentene:1-octadecene copolymer
(4.5:91:4.5) have been immersed in a 1% toluene solution of
EP-g-VMSI (108), and polyester films have been extrusion
coated with PE-g-VMSI (114).

Pipes

The very good thermo-mechanical properties, including
improved pressure-bearing resistance, and the high environmental stress-cracking resistance make crosslinked PO-g-SI
very attractive for pipes in contact with cold and hot water,
detergents, chemicals and chemical waste. Pipes subjected to
the conventional hoop stress tests, to predict suitable
design pressure ratings, show that crosslinking eliminates
the conventional "knee" in the plot for the base polymer.
This permits making a linear extrapolation to the standard
10 or 50 years time value used for design calculation (4).
Rigid and flexible pipes are crosslinked by immersion in hot
water or low-pressure steam. The time for crosslinking may
be reduced by passing hot water or steam through the pipes
(47). Pipes with improved surface uniformity are obtained
from LDPE-g-VMSI containing 0.05% stearic acid (109). The
PO-g-SI is extruded in the presence of a blowing agent to
obtain foamed pipes (116).

Foams

The PO-g-SI can be expanded with conventional blowing
agents to produce foams that are crosslinked upon exposure
to moisture. Thus, crosslinked foams with 75% gel content
and 50 Kg/m^3 density have been obtained from EP-g-VMSI (66),
EVA-g-VMSI (119), PP-g-VMSI and PP-g-MMSI (141). The PO-g-SI
may be also employed as PO foam improver. For example, by
adding 25 p.w. PE-g-VMSI to 100 p.w. PE, stable foams with
uniform cells are obtained (107).

Polymer Modifiers

Employed as modifiers, PO-g-SI are added to certain polymers like PO and PVC to improve some of their properties, especially the thermo-mechanical ones (24,81,82,95,96). Thus,
pipes made from mixtures PVC + EVA-g-VESI (95) and PE +
PE-g-VESI (96) do not sag when are heated for one hour at
200°C and at 150°C, respectively. Heat-resistant hot melt
adhesives are obtained by adding 40% EVA-g-VMSI in the resin
composition (104).

CONCLUDING REMARKS

With some few exceptions the literature of PO cross-linking by grafting with SI consists of patents. This situation is most likely a consequence of the technological character of the method. No detailed data have been reported on the dependence of the grafting process on the reaction conditions and grafting formulations (type and level of PO, SI, GI and AX). In fact, the result of the grafting is evaluated only as the gel content of the crosslinked PO-g-SI, the ratio reacted/unreacted SI being not reported.

Here is another question not yet reported in the published data: is SI grafted on the PO backbone as individual units or as grafted chains ? The schematic of the process chemistry, presented in some papers (2,4,5), describes it as a grafting of individual SI units, i.e. vinyl addition. This may be right for low SI levels, i.e. 2 p.w., usually employed in most applications. However, the formation of grafted chains may not be excluded, especially for higher SI levels, i.e. 10 p.w. or even more, employed in some special applications. Under the condition of melt grafting, i.e. high temperature and the presence of free radicals, the unsaturated SI may behave like other vinyl monomers that are grafted onto PO backbone. The radical homopolymerization of these unsaturated SI is also known. Therefore, studies for a better understanding of the grafting process should be necessary and, certainly, they will be very useful.

Despite such unclarified theoretical aspects, the moisture-crosslinking of PO-g-SI became an industrial method employed on commercial scale. In comparison with peroxide and radiation crosslinking, the new crosslinking technology has its own particular balance of advantages and limitations. However, the high productivity and the low cost of the cross-linking are the most important economic advantages. The very good properties of the crosslinked polymers and the technological advantages of the new crosslinking method will assure its further development. Actual production units are more based on the extrusion of PO-g-SI for longitudinally extended articles, such as cable insulations, pipes, profiles, films, sheets and the like. It is expected from other processing techniques, e.g. injection- and blow-molding, to be also employed on a large scale. Although in most applications PE-g-SI have been used, other PO-g-SI are increasingly employed, for example the silane grafted EP elastomers in the rubber industry. Thus, new applications should be found. Some of such developments are in progress and others are being expected in the near future.

ACKNOWLEDGMENT

The author is greatly indebted to Prof. John E. Sheats
for his continuous interest and kindness, for his support
and encouragement.

REFERENCES

1. D. Munteanu, Mater. Plast.(Bucharest), 19, 75(1982);
Chem. Abstr., 97, 163,875(1982).
2. H. U. Voigt, Kautschuk Gummi Kunstst., 29, 17(1976).
3. T. Santelli, U.S. Pat. 3,075,948(1963); Chem.Abstr.,
58, 10,326(1963).
4. H. G. Scott and J. F. Humpries, Mod. Plast., 50,
82(1973).
5. B. Thomas and M. Bowrey, Wire J., 10, 88(1977).
6. B. Thomas, S.G.F. Publ., 52, 10(1978).
7. B. Thomas and M. Bowrey, Jt. Conf. Proc.-Wire Assoc.
Int., Nonferrous Electr. Div., 1977, p. 13; Chem. Abstr., 91,
158,635(1979).
8. M. E. Kertscher, Rev. Gen. Caoutch. Plast., 55, 67
(1978).
9. Neth. Appl. 14,222(1976); Chem. Abstr., 86, 30,589
(1977).
10. D. R. Edwards, Brit. Pat. 1,536,562(1978); Chem.
Abstr., 91, 22,039(1979).
11. D. Kenper, D. Sack, W. Schmidt and H. U. Voigt, Ger.
Offen. 2,439,513(1976); Chem. Abstr., 84, 165,599(1976).
12. H. U. Voigt and D. Sack, Ger. Offen. 2,439,514(1976);
Chem. Abstr., 84, 165,600(1976).
13. H. U. Voigt, M. Völker and H. P. Stehman, Ger. Offen.
2,458,776(1976); Chem. Abstr., 84, 224,132(1976).
14. H. M. Schmidtchen, Ger. Offen. 2,528,691(1977); Chem.
Abstr., 87, 136,847(1977).
15. H. P. Stehman, D. Kenper and H. U. Voigt, Ger. Offen.
2,529,260(1977); Chem. Abstr., 86, 91,236(1977).
16. F. Glander and H. U. Voigt, U.S. Pat. 4,058,583(1977);
Chem. Abstr., 88, 153,573(1978).
17. M. Voelker, Ger. Offen. 2,636,709(1978); Chem. Abstr.,
88, 122,325(1978).
18. H. U. Voigt, M. Voelker and H. P. Stehman; U.S. Pat.
4,117,063(1978); Chem. Abstr., 90, 39,844(1979).
19. R. Winter and H. U. Voigt, Ger. Offen. 2,736,003(1979)
Chem. Abstr., 90, 170,004(1979).
20. H. U. Voigt, U.S.S.R. Pat. 686,627(1979), Chem. Abstr.
91, 176,266(1979).
21. J. Hendrix, D. Kurth, H. Poepel and H. Schnapka, Ger.
Offen. 2,924,623(1981); Chem. Abstr., 94, 104,437(1981).

22.   H. Nishizawa, S. Kon and J. Fukahari, Jap. Pat.
144,733(1975); Chem. Abstr., 84, 75,438(1976).
23.   M. Fumyu, M. Morita and H. Shimanuki, Jap. Pat.
146,645(1975); Chem. Abstr., 84, 91,206(1976).
24.   H. Nishizawa, M. Morita and I. Nishikawa, Jap. Pat.
8,355(1976); Chem. Abstr., 84, 136,989(1976).
25.   M. Hanay, F. Aida and H. Hashimoto, Jap. Pat.
33,143(1976); Chem. Abstr., 85, 64,035(1976).
26.   H. Hashimoto and S. Suyama, Jap. Pat. 80,371(1977);
Chem. Abstr., 88, 39,052(1978).
27.   H. Hashimoto, H. Shimanuki and T. Hanai, Jap. Pat.
103,462(1977); Chem. Abstr., 88, 62,943(1978).
28.   H. Hashimoto, H. Shimanuki and T. Hanai, Jap. Pat.
103,463(1977); Chem. Abstr., 88, 23,969(1978).
29.   H. Nishizawa, M. Kato, C. Takasugi and K. Kurada,
Jap. Pat. 28,654(1978); Chem. Abstr, 89, 111,195(1978).
30.   H. Nishizawa and M. Kato, Jap. Pat. 105,562(1978);
Chem. Abstr., 90, 24,424(1979).
31.   H. Hashimoto, H. Shimanuki and M. Kato, Jap. Pat.
120,770(1978); Chem. Abstr., 90, 55,963(1979).
32.   H. Hashimoto, H. Shimanuki and M. Kato, Jap. Pat.
120,771(1978); Chem. Abstr., 90, 56,505(1979).
33.   H. Nishizawa, M. Kato, H. Shimanuki and Y. Nakamura,
Jap. Pat. 120,796(1978); Chem. Abstr., 90, 39,734(1979).
34.   H. Nishizawa, M. Nishikai and S. Kon, Jap. Pat.
134,044(1978); Chem. Abstr., 90, 105,235(1979).
35.   M. Ota, K. Kuroda, M. Kato and H. Shimanuki, Jap.
Pat. 147,745(1978); Chem. Abstr., 90, 169,670(1979).
36.   K. Kojima and A. Kinoshita, Jap. Pat. 63,400(1979);
Chem. Abstr., 91, 142,357(1979).
37.   M. Ota and Y. Shiray, Jap. Pat. 68,406(1979); Chem.
Abstr., 91, 125,214(1979).
38.   K. Takeuchi, O. Shimizu, S. Kinoshita and Y. Kato,
Jap. Pat. 76,676(1979); Chem. Abstr., 91, 158,698(1979).
39.   O. Shimizu, S. Kinoshita, K. Takeuchi and Y. Kato,
Jap. Pat. 78,783(1979); Chem. Abstr., 91, 159,296(1979).
40.   O. Shimizu, K. Takeuchi, S. Kinoshita and Y. Kato,
Jap. Pat. 95,682(1979); Chem. Abstr., 92, 7,580(1980).
41.   H. Hashimoto and M. Kato, Jap. Pat. 93,049(1979);
Chem. Abstr., 91, 212,322(1979).
42.   O. Shimizu, S. Kinoshita, K. Takeuchi and Y. Kato,
Jap. Pat. 100,473 (1979); Chem. Abstr., 92, 23,841(1980).
43.   T. Kojima, S. Kinoshita and K. Takeuchi, Ger.
Offen. 2,834,082(1979); Chem. Abstr., 90, 139,300(1979).
44.   T. Kojima, S. Kinoshita and K. Takeuchi, Jap. Pat.
40,840(1980); Chem. Abstr., 93, 28,105(1980).
45.   Jap. Pat. 132,246(1980); Chem. Abstr., 94, 66,904
(1981).
46.   Jap. Pat. 148,268(1980); Chem. Abstr., 94, 86,641
(1981).

47.   Jap. Pat. 43,331(1981); Chem. Abstr., 95, 63,268 (1981).

48.   Jap. Pat. 5,956(1982); Chem. Abstr., 96, 124,493 (1982).

49.   Jap. Pat. 57,409(1982); Chem. Abstr., 97, 57,409 (1982).

50.   K. Ohtani and E. Oda, Jap. Pat. 65,154(1976); Chem. Abstr., 85, 109,687(1976).

51.   H. Hirukawa, Jap. Pat. 154,872(1977); Chem. Abstr., 89, 112,076(1978).

52.   S. Irie, Jap. Pat. 12,944(1978); Chem. Abstr., 88, 192,361(1978).

53.   T. Takei, Jap. Pat. 21,248(1978); Chem. Abstr., 89, 60,518(1978).

54.   Y. Ito, Jap. Pat. 39,360(1978); Chem. Abstr., 89, 75,881(1978).

55.   S. Irie and K. Uesugi, Jap. Pat. 92,857(1978); Chem. Abstr., 90, 7,103(1979).

56.   K. Otani, S. Nagai, E. Saito and H. Hirukawa, Jap. Pat. 144,995(1978); Chem. Abstr., 90, 169,564(1979).

57.   K. Otani, E. Saito and H. Hirukawa, Jap. Pat. 146,747(1978); Chem. Abstr., 90, 153,055(1979).

58.   H. Hirukawa, K. Otani and E. Saito, Jap. Pat. 11,154(1979); Chem. Abstr., 90, 205,221(1979).

59.   A. Nojiri, T. Sawazaki, T. Koreeda and H. Shimokawa, Jap. Pat. 36,356(1979); Chem. Abstr., 91, 58,166(1979).

60.   H. Hirukawa, K. Otani and E. Saito, Jap. Pat. 36,357(1979); Chem. Abstr., 91, 40,416(1979).

61.   H. Hirukawa, K. Otani, H. Nishiyama and Y. Sasaki, Jap. Pat. 114,554(1979); Chem. Abstr., 92, 59,732(1980).

62.   H. Eikawa, K. Otani, H. Nishiyama and Y. Sasaki, Jap. Pat. 142,255(1979); Chem. Abstr., 92, 129,978(1980).

63.   A. Nojiri, T. Sawasaki and T. Koreeda, Eur. Pat. Appl. 4,034(1979); Chem. Abstr., 92, 23,606(1980).

64.   H. Kiumura, K. Uesugi, A. Furuda and K. Shibata, Jap. Pat. 40,701(1980); Chem. Abstr., 93, 73,234(1980).

65.   Jap. Pat. 71,723(1980); Chem. Abstr., 93, 151,346 (1980).

66.   Jap. Pat. 75,432(1980); Chem. Abstr., 93, 187,421 (1980).

67.   Jap. Pat. 82,628(1980); Chem. Abstr., 93, 187,440 (1980).

68.   Jap. Pat. 143,234(1981); Chem. Abstr., 96, 53,453 (1982).

69.   Jap. Pat. 149,453(1981); Chem. Abstr., 96, 86,491 (1982).

70.   I. Fujimoto and S. Isshiki, Jap. Pat. 104,673(1978); Chem. Abstr., 90, 7,290(1979).

71.   I. Fujimoto, S. Isshiki, Y. Kurita and Y. Sato, Jap. Pat. 18,857(1979); Chem. Abstr., 90, 205,404(1979).

72. I. Fujimoto, S. Isshiki, H. Sunazuka and Y. Sato, Jap. Pat. 81,357(1979); Chem. Abstr., 91, 194,322(1979).

73. Jap. Pat. 164,154(1980); Chem. Abstr., 94, 209,926 (1981).

74. Jap. Pat. 5,854(1981); Chem. Abstr., 94, 209,757 (1981).

75. Jap. Pat. 103,818(1981); Chem. Abstr., 95, 221,004 (1981).

76. Jap. Pat. 163,142(1981); Chem. Abstr., 97, 39,828 (1982).

77. Jap. Pat. 163,143(1981); Chem. Abstr., 97, 39,829 (1982).

78. Jap. Pat. 167,731(1981); Chem. Abstr., 96, 144,104 (1982).

79. Jap. Pat. 8,203(1982); Chem. Abstr., 96, 182,228 (1982).

80. Jap. Pat. 28,107(1982); Chem. Abstr., 96, 182,585 (1982).

81. Jap. Pat. 65,613(1982); Chem. Abstr., 97, 129,375 (1982).

82. Jap. Pat. 65,614(1982); Chem. Abstr., 97, 146,453 (1982).

83. Jap. Pat. 87,438(1982); Chem. Abstr., 97, 199,068 (1982).

84. S. Kawawada, M. Azuma, M. Sato, K. Hanawa, H. Kurimoto, S. Ogata and S. Kashiwazaki, Jap. Pat. 65,342(1978); Chem. Abstr., 89, 147,643(1978).

85. S. Kawawada, M. Azuma, M. Sato, K. Hanawa, H. Kurimoto, S. Ogata and S. Kashiwazaki, Jap. Pat. 65,343(1978); Chem. Abstr., 89, 147,644(1978).

86. S. Kawawad , M. Azuma, M. Sato, K. Hanawa, H. Kurimoto, S. Ogata and S. Kashiwazaki, Jap. Pat. 65,344(1978); Chem. Abstr., 89, 147,645(1978).

87. T. Mizugami, T. Suzuki, T. Yamagiwa, H. Takano and S. Kato, Jap. Pat. 37,167(1979); Chem. Abstr., 91, 22,028 (1979).

88. Y. Matsuga, Jap. Pat. 52,142(1979); Chem. Abstr., 91, 75,361(1979).

89. M. Sato, R. Hanawa, Y. Ando, H. Kurimoto and S. Ogata, Jap. Pat. 77,655(1979); Chem. Abstr., 91, 176,380(1979).

90. Jap. Pat. 77,656(1979); Chem. Abstr., 91, 158,753 (1979).

91. Jap. Pat. 22,085(1981); Chem. Abstr., 95, 134,037 (1981).

92. Jap. Pat. 126,204(1981); Chem. Abstr., 96, 37,042 (1982).

93. Jap. Pat. 126.213(1981); Chem. Abstr., 96, 53,421 (1982).

94. S. Kokau and M. Fukuoka; Jap. Pat. 88,050(1978), Chem. Abstr., 89, 216,345(1978).

95. M. Godo, T. Ogawa, M. Fukuoka and H. Morita, Jap.
Pat. 54,162(1979); Chem. Abstr., 91, 92,515(1979).
96. M. Kamikado, T. Ogawa and M. Fukuoka, Jap. Pat.
64,545(1979); Chem. Abstr., 91, 176,181(1979).
97. M. Kando, T. Ogawa, M. Fukuoka and H. Morita,
Jap. Pat. 129,399(1979); Chem. Abstr., 92, 111,878(1980).
98. M. Shinnon, T. Ogawa and H. Morita, Jap. Pat.
1,021(1980); Chem. Abstr., 93, 9,210(1980).
99. K. Nakagawa and T. Ogawa, Jap. Pat. 3,429(1980);
Chem. Abstr., 92, 182,122(1980).
100. Jap. Pat. 80,203(1980); Chem. Abstr., 93, 187,447
(1980).
101. Jap. Pat. 104,336(1980); Chem. Abstr., 92, 4,615
(1981).
102. Jap. Pat. 122,817(1981); Chem. Abstr., 96, 7,280
(1982).
103. H. Inagaki, C. Igarashi, T. Tomoshige and M. Ynasa,
Jap. Pat. 145,785(1979); Chem. Abstr., 92, 111,746(1980).
104. H. Ikeda, S. Toda and T. Kobayashi, Jap. Pat.
40,721(1980); Chem. Abstr., 93, 73,233(1980).
105. Jap. Pat. 73,716(1980); Chem. Abstr., 93, 169,314
(1980).
106. Jap. Pat. 75,415(1980); Chem. Abstr., 93, 169,136
(1980).
107. Jap. Pat. 109,229(1981); Chem. Abstr., 95, 205,122
(1981).
108. Jap. Pat. 164,858(1981); Chem. Abstr., 96, 182,897
(1980).
109. Jap. Pat. 23,650(1981); Chem. Abstr., 97, 39,832
(1982).
110. K. Iwatani, S. Yoshida and M. Ito, Jap. Pat.
63,492(1978); Chem. Abstr., 89, 130,390(1978).
111. M. Gocho, Jap. Pat. 105,588(1978); Chem. Abstr.,
90, 24,172(1979).
112. Jap. Pat. 8,446(1981); Chem. Abstr., 95, 8,268
(1981).
113. Jap. Pat. 11,248(1981); Chem. Abstr., 94, 193,208
(1981).
114. Jap. Pat. 151,561(1981); Chem. Abstr., 96, 86,695
(1982).
115. Jap. Pat. 12,051(1982); Chem. Abstr., 96, 200,716
(1982).
116. S. Nakata, K. Shinkai and N. Chiba, Jap. Pat.
75,664(1975); Chem. Abstr., 83, 180,649(1975).
117. Y. Morikama, T. Yano and Y. Takayama, Jap. Pat.
45,378(1978); Chem. Abstr., 89, 76,040(1978).
118. H. Harayama, K. Shinkai, H. Takahashi and M. Mizu-
sako, Eur. Pat. Appl. 2,910(1979); Chem. Abstr., 91, 194,104
(1979).
119. Jap. Pat. 57,868(1981); Chem. Abstr., 95, 117,240
(1981).

120. Jap. Pat. 166,231(1981); Chem. Abstr., 96, 123,991 (1981).

121. S. Ohta and M. Asaoka, Jap. Pat. 52,447(1976); Chem. Abstr., 85, 193,570(1976).

122. S. Ohta and M. Yamada, Jap. Pat. 40,555(1977); Chem. Abstr., 87, 24,229(1977).

123. M. Yamada and S. Ohta, Jap. Pat. 49,256(1977); Chem. Abstr., 87, 69,199(1977).

124. M. Yamada and S. Ohta, Jap. Pat. 132,095(1978); Chem. Abstr., 91, 21,790(1979).

125. S. Ohta and M. Yamada, Jap. Pat. 56,655(1979); Chem. Abstr., 91, 75,434(1979).

126. J. Kinugasa, K. Kirimoto, O. Fukui and T. Miyamoto, Jap. Pat. 65,185(1976); Chem. Abstr., 85, 125,301(1976).

127. J. Kinugasa, K. Kirimoto, O. Fukui and T. Miyamoto, Jap. Pat. 69,586(1976); Chem. Abstr., 85, 125,321(1976).

128. J. Kinugasa, K. Kirimoto, O. Fukui and T. Miyamoto, Jap. Pat. 84,881(1976); Chem. Abstr., 85, 144,316(1976).

129. J. Kinugasa, K. Kirimoto, O. Fukui and T. Miyamoto, Jap. Pat. 84,882(1976); Chem. Abstr., 85, 144,317(1976).

130. T. Fujita, Jap. Pat. 33,938(1977); Chem. Abstr., 86, 191,032(1977).

131. T. Fujita, Jap. Pat. 33,967(1977); Chem. Abstr., 87, 24,384(1977).

132. T. Ojima and M. Takejima, Jap. Pat. 69,245(1978); Chem. Abstr., 89, 164,436(1978).

133. M. Maruyama, T. Ishioroshi, T. Hiyama, M. Shimizu, S. Yokoyama and T. Shinozaki, Jap. Pat. 50,056(1979); Chem. Abstr., 91, 92,499(1979).

134. Jap. Pat. 86,544(1979); Chem. Abstr., 92, 7,411 (1980).

135. Esso Research Co., Neth. Appl. 14,931(1967); Chem. Abstr., 67, 65,239(1967).

136. F. Engelhardt, U.S. Pat. 3,485,660(1970); Chem. Abstr., 72, 44,491(1970).

137. V. V. Belova, V. S. Tikhomirov, A. I. Popova and V. I. Srenkov, Plast. Massy, 1975, 22.

138. I.C.I. Australia Ltd., Brit. Pat. 1,542,543(1979); Chem. Abstr., 91, 108,490(1979).

139. J. R. Cowan, G. J. Field and W. G. Barton (to I.C.I. Australia Ltd.), S. African Pat. 5,788(1979); Chem. Abstr., 91, 158,357(1979).

140. F. Woods and G. Wouters, Rubbercon 81, Int. Rubber Conf., 1981, C.6.1.; Chem. Abstr., 95, 116,814(1981).

141. A. Nojiri, T. Sawazaki, T. Konishi, S. Kudo and S. Onobiri, Furukawa Denko Jiho, 1981, 81; Chem. Abstr., 96, 53,137(1982).

142. L. B. Klimanova, T. N. Khvatova, Yu. I. Firsov and E. I. Evdokimov, Kompozitsion Materialy na Osnove Termoplastov, 1980, 109; Chem. Abstr., 96, 69,558(1982).